Lecture Notes in Mathematics

Edited by A. Dold and B. Eckmann

851

Stochastic Integrals

Proceedings of the LMS Durham Symposium,
July 7 – 17, 1980

Edited by D. Williams

Springer-Verlag
Berlin Heidelberg New York 1981

Editor

David Williams
Department of Pure Mathematics, University College of Swansea
Singleton Park, Swansea SA2 8PP, Wales, United Kingdom

AMS Subject Classifications (1980): 33-XX, 35-XX, 53-XX, 60-XX, 81-XX

ISBN 3-540-10690-1 Springer-Verlag Berlin Heidelberg New York
ISBN 0-387-10690-1 Springer-Verlag New York Heidelberg Berlin

Printed in Germany

Printing and binding: Beltz Offsetdruck, Hemsbach/Bergstr.
2141/3140-543210

PREFACE

1°. There are many people and organisations to thank, including:

The London Mathematical Society; and especially Heini Halberstam, John Williamson, and Tom Willmore, for 'pre-natal' care on behalf of LMS;

The Science Research Council, for generous financial support; and especially John Kingman, SRC assessor for this symposium, for his interest and sound advice;

The Durham Mathematics Department; and especially Tom Willmore, Peter Green, and of course, Ed. Corrigan;

Grey College, my old home, for its usual warm hospitality;

Paul-André Meyer, for several valuable suggestions;

My wife Sheila, and her father, the late great Edward Harrison, for a lot of work and for unlimited patience;

Robert Elliott, my co-organiser, for effective troubleshooting in some moments of minor crisis, and for much helpfulness throughout two years;

And Chris Rogers and Margaret Brook, whose very hard work somehow defeated my determined efforts to surpass Haydn, Wiener, and Itô, in achieving a representation of chaos.

But, above all, thanks are due to all participants: for a marvellous time; for fine mathematics; and, no less importantly, for fun and friendship.

2°. The three introductory articles - by Elliott, Rogers, and myself - are intended to help make some of the later material accessible to a wider audience.

At the symposium, there was much interest in the Malliavin calculus. My introductory effort is intended to provide some background material for this topic and for related topics.

David Williams

PARTICIPANTS (WITH ADDRESSES)

Convention:

CAMBRIDGE = Department of Pure Mathematics and Mathematical Statistics; University of Cambridge; 16 Mill Lane; CAMBRIDGE CB2 1SB; England.

HULL = The University of Hull; 22 Newland Park; Cottingham Road; HULL HU6 2DW; England.

PARIS VI = Laboratoire de calcul des probabilités; Université de Paris VI; 4, place Jussieu, Tour 56; 75230, PARIS Cédex 05; France.

STRASBOURG = Département de Mathématique; Universite Louis Pasteur de Strasbourg; 7, rue René Descartes; 67084 STRASBOURG, France.

SWANSEA = University College of Swansea; Singleton Park; SWANSEA SA2 8PP; Wales. United Kingdom.

\- -

L. ACCARDI; Istituto Matematico Federico Enriques; Universita di Milano; Via L. Cicognara; 20129 MILANO, Italy.

S. ALBEVERIO; Institut für Mathematik; Gebäude NA; Universitätsstr.150; Postfach 2148; 463 BOCHUM; W. Germany.

D.J. ALDOUS; Department of Statistics; University of California; BERKELEY; California 94720, U.S.A..

A.N. AL-HUSSAINI; Department of Mathematics; The University of Alberta; EDMONTON T6G 2G1; Canada.

J. AZEMA; PARIS VI.

A.J. BADDELEY; CAMBRIDGE.

A. BARBOUR; CAMBRIDGE.

M.T. BARLOW; LIVERPOOL - now at CAMBRIDGE.

T. BARTH; Department of Pure Mathematics; HULL.

J.A. BATHER; School of Mathematical and Physical Sciences; The University of Sussex; Falmer; BRIGHTON BN19QH; England.

P. BAXENDALE; Department of Mathematics; King's College; University of Aberdeen; High Street, ABERDEEN AB9 2UB; Scotland, United Kingdom.

D. BELL; Department of Pure Mathematics; HULL.

K. BICHTELER; Department of Mathematics; University of Texas at Austin; AUSTIN; Texas 78712; U.S.A.

N.H. BINGHAM; Department of Mathematics; Westfield College; Kidderpore Avenue; LONDON NW3 7ST; England.

J.M. BISMUT; Département de Mathématique; Université de Paris-Sud; Bât.425; ORSAY 91405; Paris, France.

T.C. BROWN; School of Mathematics; University of Bath; Claverton Down; BATH BA2 7AY; England.

T.K. CARNE; CAMBRIDGE.

Mireille CHALEYAT-MAUREL; PARIS VI.

L. CHEVALIER; Laboratoire de Mathématiques Pures; Institut Fourier, Université de Grenoble; B.P. 116-38402 Saint Martin d'Hères; GRENOBLE, France.

J.M.C. CLARK; Department of Computing and Control; Imperial College; 180 Queen's Gate; LONDON SW7 2BZ; England.

R.W.R. DARLING; Mathematics Institute; University of WARWICK; COVENTRY CV4 7AL; England.

A.M. DAVIE; Department of Mathematics; University of Edinburgh; James Clarke Maxwell Building; The King's Buildings; Mayfield Road; EDINBURGH EH9 3JZ; Scotland, U.K.

M.H.A. DAVIS; Department of Computing and Control; Imperial College, 180 Queen's Gate; LONDON SW7 2BZ; England.

C. DELLACHERIE; Départment de Mathématique; Université de ROUEN; B.P. No. 67; 76130 MONT-SAINT-AIGNAN; Rouen; France.

R.A. DONEY; The Manchester-Sheffield School of Probability and Statistics; Statistical Laboratory; Department of Mathematics; The University, MANCHESTER M13 9PL; England.

H. DOSS; PARIS VI.

E.B. DYNKIN; Department of Mathematics; CORNELL University; White Hall, Ithaca; NEW YORK 14853; U.S.A.

D.A. EDWARDS; Mathematical Institute; University of Oxford; 24-29 St. Giles; OXFORD OX1 3LB; England.

R.J. ELLIOTT; Department of Pure Mathematics; HULL.

K.D. ELWORTHY; Mathematical Institute; University of WARWICK; COVENTRY CV4 7AL; England.

P. EMBRECHTS; Departement Wiskunde KUL; Celestijnenlaan 200-B; B-3030 HEVERLEE, Belgium.

M. EMERY; STRASBOURG.

H. FÖLLMER; Mathematik; ETH-Zentrum; CH-8092 ZÜRICH; Switzerland.

M. FUKUSHIMA; College of General Education; Osaka University; 1-1 Machikanayama-cho; Toyonaka-shi; OSAKA 560; Japan.

D.J.H. GARLING; CAMBRIDGE.

G.R. GRIMMETT; School of Mathematics; University of Bristol; University Walk; BRISTOL BS8 1TW.

B. HAJEK; Coordinated Science Laboratory; College of Engineering; University of Illinois at Urbana-Champaign; URBANA, Illinois 61801; U.S.A.

J.M. HAMMERSLEY; Institute of Economics and Statistics; University of Oxford; St. Cross Building; Manor Road; OXFORD OX1 3UL; England.

J. HAWKES; Department of Statistics; SWANSEA.

R. HOLLEY; Department of Mathematics; University of Colorado; Boulder; COLORADO 80309; U.S.A.

M. JACOBSEN; Institute of Mathematical Statistics; University of Copenhagen; 5 Universitetsparken; DK-2100, COPENHAGEN Ø; Denmark.

J. JACOD: Laboratoire de Probabilités (C.N.R.S.); Université de Rennes; Avenue du Général Leclerc; Rennes Beaulieu, 35042 RENNES Cédex; France.

T. JEULIN; PARIS VI.

K. JANSSEN; DÜSSELDORF.

KARKYACHARIAN; Université de NANCY; France .

D.G. KENDALL; CAMBRIDGE.

W.S. KENDALL; Department of Mathematical Statistics, HULL.

H. KESTEN; Department of Mathematics; White Hall; CORNELL University; Ithaca; NEW YORK 14853; U.S.A.

P.E. KOPP; Department of Pure Mathematics; HULL.

P. KOTELENEZ; Forschungsschwerpunkt Dynamische Systeme; Universität Bremen; Bibliothekstrasse; Postfach 330440; 2800 BREMEN 33; West Germany.

H. KUNITA; Department of Applied Science; Faculty of Engineering; KYUSHU University; Hakozaki; FUKUOKA 812; Japan.

A. KUSSMAUL; Mathematisches Institut der Universität Tübingen- Auf der Morgenstelle 10; 7400 TÜBINGEN 1; West Germany.

E. LENGLART; Département de Mathématique; Université de ROUEN; B.P. no.67; 76130 MONT-SAINT-AIGNAN; Rouen; France.

J.T. LEWIS; School of Theoretical Physics; Dublin Institute for Advanced Studies; 10, Burlington Road; DUBLIN 4; Eire.

T. LYONS; Mathematical Institute; University of Oxford; 24-29 St. Giles; OXFORD OX1 3LB; England.

P. MALLIAVIN; Département de Mathématique; Université de Paris VI; 4 place Jussieu, Tour 56; 75230 PARIS Cédex 05; France.

P. McGILL; Department of Mathematics; University of Ulster; COLERAINE; Northern Ireland; U.K.

P.A. MEYER; STRASBOURG.

S. MOHAMMED; School of Mathematical Sciences, University of KHARTOUM, Sudan.

J. NEVEU; PARIS VI.

F. PAPANGELOU; The Manchester-Sheffield School of Probability and Statistics; Statistical Laboratory; Department of Mathematics; The University; MANCHESTER M13 9PL; England.

J. PELLAUMAIL; I.N.S.A.; 20 Avenue des Buttes de Coësmes; B.P. 14A; 35031 RENNES; France.

M. PINSKY; Department of Mathematics; College of Arts and Sciences; NORTHWESTERN University; Evanston; ILLINIOS 60201; U.S.A.

G.C. PRICE; Department of Pure Mathematics; SWANSEA.

P. PROTTER; Department of Mathematics and Statistics; PURDUE University; Lafayette; IND 47907; U.S.A.

B. RIPLEY; Department of Mathematics; Huxley Building; Imperial College; Huxley Building; Imperial College; 180 Queen's Gate; LONDON SW7 2BZ; England.

L.C.G. ROGERS; SWANSEA - now at Department of Statistics; University of WARWICK; COVENTRY CV4 7AL; England.

M.J. SHARPE; Department of Mathematics; University of California, SAN DIEGO; P.O.Box 109; LA JOLLA, California 92093; U.S.A.

R.F. STREATER; Department of Mathematics; Bedford College; University of London; Regents Park; LONDON NW1 4NS; England.

C. STRICKER; STRASBOURG.

D.W. STROOCK, Department of Mathematics; University of Colorado; Boulder; COLORADO 80309; U.S.A.

J.C. TAYLOR; Department of Mathematics; Burnside Hall; 805 Sherbrooke Street West; MONTREAL PQ; Canada H3A 2K6.

L.C. THOMAS; Department of Decision Theory; University of Manchester; MANCHESTER M13 9PL; England.

G. VINCENT-SMITH; Mathematical Institute; University of Oxford; 24-29 St Giles; OXFORD OX1 3LB; England.

J.B. WALSH; Department of Mathematics; University of BRITISH COLUMBIA; 2075 Westbrook Hall; VANCOUVER, B.C. V6T 1W5; Canada.

S. WATANABE; Department of Mathematics; Faculty of Science; Kyoto University; KYOTO; Japan.

J. WATKINS; Free University of BERLIN.

D. WILLIAMS; Department of Pure Mathematics; SWANSEA.

T.J. WILLMORE; Department of Mathematics, University of Durham, Science Laboratories; South Road, DURHAM DH1 3LE.

E. WONG; Department of Electrical Engineering and Computer Sciences; University of California, BERKELEY, California 94720; U.S.A.

M. YOR; PARIS VI.

M. ZAKAI; Department of Electrical Engineering; TECHNION-Israel Institute of Technology; TECHNION CITY; HAIFA 32000; Israel.

CONTENTS

CONTENTS

Introductory articles

Part: based on main talks and lectures

Papers based on [illegible] group talks

"TO BEGIN AT THE BEGINNING: ..."

by

David Williams

Some readers may be helped by this more-or-less self-contained introduction to some important concepts: continuous semimartingales and the associated stochastic integrals; the Stroock-Varadhan theorem and its consequences for martingale representation; the Girsanov theorem; ...; and, as a main theme, the modern theory of the Kolmogorov forward (or Fokker-Planck) equation, involving hypoellipticity and all that.

Comments on notation and terminology. The symbol '$\equiv$' signifies 'is defined to be equal to'. By a smooth function we shall always mean an infinitely differentiable (C^∞) function. We use C_K^∞ to denote the space of smooth functions of compact support.

The summation convention is used throughout the paper, so that, for example, in equation (1.3), it is understood that the first term on the right-hand side is summed over the (repeated) indices i and j, while the last term is summed over i. Note especially: $\sigma_q^i\sigma_q^j$ will mean $\sum_q \sigma_q^i\sigma_q^j$. ‡

A brief Appendix at the end of this paper collects some information about Schwartz distributions and hypoelliptic operators.

Part I. Fifty years of the forward equation.

I think it best to start by trying to motivate things via this account of various approaches to diffusion theory even though it means speaking of certain concepts before recalling their definitions.

§1. KOLMOGOROV (1931).

Roughly speaking, a diffusion process X on $\mathbb{R}^n$ is a path-continuous process $X = (X^1, X^2, \ldots, X^n)$ such that for $t \geq 0$ and $h > 0$,

‡ The 'one up, one down' convention does not work well for transposes!

$$E[X^i_{t+h} - X^i_t \mid X_s : s \le t] = b^i(X_t)h + o(h),$$

(1.1)

$$E[\{X^i_{t+h} - X^i_t - b^i(X_t)h\}\{X^j_{t+h} - X^j_t - b^j(X_t)h\}] = a^{ij}(X_t)h + o(h),$$

for some functions b^i $(1 \le i \le n)$ on $\mathbb{R}^n$ called 'drift coefficients', and some functions a^{ij} $(1 \le i, j \le n)$ on $\mathbb{R}^n$ called 'diffusion coefficients'. Note that for each x in $\mathbb{R}^n$, the matrix $a(x)$ is positive semi-definite. [Let me mention one technical difficulty: as I have stated (1.1), the integrals determining the expectations could blow up; so we need to truncate. I skip this now because it is subsumed and superceded via the later use of local martingales.]

Various heuristic arguments (turned into precise proofs and theorems below) suggest that, under suitable conditions, X must be a Markov process possessing a transition density function $p_t(x,y)$:

(1.2) $$P[X_{u+t} \in dy \mid X_s : s \le u;\ X_u = x] = p_t(x,y)dy,$$

where p satisfies the Kolmogorov backward and forward equations now to be described. Let $\mathcal{G}$ be the operator defined as follows:

(1.3) $$(\mathcal{G}f)(x) \equiv \tfrac{1}{2}a^{ij}(x)\partial_i\partial_j f(x) + b^i(x)\partial_i f(x), \qquad \partial_i \equiv \partial/\partial x^i.$$

If, for example, the functions a^{ij} and b^i are smooth, then for f and h in C^∞_K, we have

$$\int_{\mathbb{R}^n} h(x)(\mathcal{G}f)(x)dx = \int_{\mathbb{R}^n} f(x)(\mathcal{G}^*h)(x)dx,$$

where $\mathcal{G}^*$ is the adjoint operator with

(1.4) $$\mathcal{G}^*h(y) = \tfrac{1}{2}\partial_i\partial_j[a^{ij}(y)h(y)] - \partial_i[b^i(y)h(y)], \qquad \partial_i \equiv \partial/\partial y^i.$$

Then the Kolmogorov backward and forward equations take the form:

(B) $$\frac{\partial}{\partial t}p_t(x,y) = \mathcal{G}_x p_t(x,y) \quad \text{for fixed } y,$$

(F) $$\frac{\partial}{\partial t}p_t(x,y) = \mathcal{G}^*_y p_t(x,y) \quad \text{for fixed } x.$$

The subscripts on $\mathcal{G}_x$ and $\mathcal{G}^*_y$ are meant to indicate the variables in which these operators act; but it is neater to speak of the forward equation by saying that $p = p.(x,\cdot)$ satisfies

$$\text{(F)} \qquad \frac{\partial p}{\partial t} = \mathcal{G}^* p.$$

(As usual, $p.(x,\cdot)$ is the function $(t,y) \mapsto p_t(x,y)$.)

The early work of Kolmogorov, Feller, and others used partial-differential-equation (PDE) theory to establish (under suitable conditions) the existence of a Markov transition density function p satisfying (B) and (F); the existence of X, as a process 'proper' carried by some $(\Omega,\mathcal{F},P)$, could then be deduced from the Kolmogorov-Daniell theorem supplemented by Kolmogorov's criterion for path continuity.

§2. STROOCK-VARADHAN (1969). We jump on to the Stroock-Varadhan approach because it exactly captures the spirit of (1.1). The point is that (1.1) may be formulated precisely as follows:

$$(2.1) \quad \begin{cases} M^i_t \equiv X^i_t - X^i_0 - \int_0^t b^i(X_s)ds & \text{defines a local martingale } M^i, \\ M^i_t M^j_t - \int_0^t a^{ij}(X_s)ds & \text{is a local martingale,} \end{cases}$$

The (generalised) Itô formula implies that the conditions (2.1) are exactly equivalent to the following statement:

$$\text{(SV)} \qquad \boxed{\forall f \in C^\infty_K, \quad C^f_t \equiv f(X_t) - f(X_0) - \int_0^t (\mathcal{G}f)(X_s)ds \quad \text{defines a martingale } C^f.}$$

Stroock and Varadhan make (SV) the defining condition for a diffusion process. One advantage is clear: if $\mathcal{G}$ is a second-order elliptic operator on a manifold, then (SV) makes perfect sense as a condition on a process X with values in the manifold.

Let us be more specific about the Stroock-Varadhan approach. We stick to the manifold $\mathbb{R}^n$. We now insist that X be a canonical process with values in $\mathbb{R}^n$. Thus, let W be the set of continuous maps w from $[0,\infty)$ to $\mathbb{R}^n$; and set $X_t(w) \equiv w(t)$. Define $A_t \equiv \sigma\{X_s : s \le t\}$, the smallest σ-algebra of subsets of W which makes A_t measurable all maps $w \mapsto X_s(w)$ with $s \le t$; and set $A = \sigma\{X_s : s < \infty\}$.

Fix $x \in \mathbb{R}^n$. Let $\mathcal{G}$ (equivalently: a and b) be given. We say that a probability measure P^x on (W, A) solves the martingale problem for $\mathcal{G}$ starting from x if $P^x[X_0 = x] = 1$ and (SV) holds in that for each f in C_K^∞, C^f is a martingale relative to the set-up $(W, A, \{A_t\}, P^x)$. (See §7 for definition of "martingale relative to a 'set-up'".)

(2.2) THEOREM (Stroock and Varadhan). Suppose that each $a^{ij}(\cdot)$ is continuous, and that for each y, the matrix $a(y)$ is strictly positive definite. Suppose that each b^i is measurable. Assume that for some constant K, and all i,j,

(2.2A) $$a^{ij}(y) \le K(1 + |y|^2), \qquad b^i(y) \le K(1 + |y|^2).$$

Then, for each x in $\mathbb{R}^n$, there is precisely one solution P^x of the martingale problem for $\mathcal{G}$ starting from x. Moreover, each P^x is Markovian: for $0 < t_1 < t_2 < \dots < t_n$,

$$P^{y_0}[X_{t_i} \in dy_i : 1 \le i \le k] = \prod_{i=1}^{k} P^{y_{i-1}}[X_{t_i - t_{i-1}} \in dy_i].$$

Further, the formula:

$$P_t f(x) \equiv \int P_t(x, dy) f(y), \quad \text{where } P_t(x, dy) \equiv P^x[X_t \in dy],$$

defines the unique Feller semigroup $\{P_t : t \ge 0\}$ with generator extending $\mathcal{G}$ on $C_K^\infty(\mathbb{R}^n)$.

Clarification of the last sentence. Note that we now write $P_t f(x)$ instead of the clearer, but too cumbersome, $(P_t f)(x)$. The statement that the Markov transition function $\{P_t\}$ has the Feller property means that:

$$P_t : C_b(\mathbb{R}^n) \to C_b(\mathbb{R}^n)$$

where $C_b(\mathbb{R}^n)$ is the banach space of bounded continuous functions on $\mathbb{R}^n$. The statement that the generator of $\{P_t\}$ extends $\mathscr{G}$ on C_K^∞ means that:

$$\text{(2.3)} \qquad \forall f \in C_K^\infty, \quad P_t f(x) - f(x) = \int_0^t P_s \mathscr{G} f(x) ds.$$

Equation (2.3) obviously represents a weak form of the forward equation.

- -

The ellipticity assumption that a is strictly positive-definite everywhere actually implies the stronger conclusion that $\{P_t : t \geq 0\}$ is strong Feller in that each P_t maps bounded (measurable) functions into continuous functions.

- -

The deep part of Theorem 2.2 is the uniqueness result that there is at most one P^x, and this is true if assumption (2.2A) is dropped. For the 'existence' part of the theorem, some such condition as (2.2A) is necessary to preclude explosion. More refined conditions precluding explosion are given in the Stroock-Varadhan book [23].

Proof of the Stroock-Varadhan theorem is very difficult. In Williams [25], I tried to give a clue to it.

Let me mention an important connection between uniqueness of solution of the martingale problem and the very important matter of martingale representation via stochastic integrals. As a very special case of general results of Jacod, and Jacod and Yor (for which see Jacod [8 , 9], we have the following result. (Note. We shall not need to use the theorem until after we have recalled the definition of previsible process, etc.)

(2.4) THEOREM (Jacod). Suppose that there is one and only one solution P^x of the martingale problem for $\mathcal{G}$ starting from x. Then, if N is any martingale relative to $(W, \mathcal{A}, \{\mathcal{A}_t\}, P^x)$, we can represent N as follows:

$$N_t - N_0 = \int_0^t C_i d_I M^i$$

where each C_i $(i = 1,2,\ldots,n)$ is a previsible process.

We are using d_I to signify the use of the Itô (-Kunita-Watanabe) integral as opposed to the Stratonovich integral which (when it exists) is signified via the use of d_S.

§3. ITÔ (1946). In this section, it is assumed that the reader is familiar with the classical theory of Itô integrals relative to Brownian motion (which is, of course, a special case of the theory presented in §8).

In the Stroock-Varadhan theory, the key assumption is the ellipticity assumption that $a(\cdot)$ is everywhere strictly positive definite; and then (for S-V theory) $a(\cdot)$ need only be assumed continuous and $b(\cdot)$ measurable. The Itô theory requires that both $a(\cdot)$ and $b(\cdot)$ satisfy some slightly stronger 'smoothness' assumption than continuity; but then (for Itô theory) the ellipticity assumption becomes irrelevant; and there are important cases in which $a(x)$ is singular for some (or even all) x. A famous Girsanov example (McKean [13], Stroock and Varadhan [23]) shows that we can no longer expect to have uniqueness of solution of the martingale problem if we sacrifice ellipticity and require only continuity of $a(\cdot)$ and $b(\cdot)$.

Suppose that $X = (X^i : 1 \leq i \leq n)$ is a process on $\mathbb{R}^n$ which solves the stochastic differential equation:

(I) $$dX^i = \sigma^i_q(X_t) d_I B^q + b^i(X_t) dt, \quad X_0 = x,$$

where:

$(B^q : 1 \leq q \leq r)$ is an r-dimensional Brownian motion starting at 0;

the σ^i_q $(1 \leq i \leq n,\ 1 \leq q \leq r)$ and b^i $(1 \leq i \leq n)$ are functions on $\mathbb{R}^n$;

x is a point in $\mathbb{R}^n$.

Set $a^{ij}(y) \equiv \sigma^i_q(y)\sigma^j_q(y)$, so that $a = \sigma\sigma^*$, where σ^* denotes the transpose of σ; and define $\mathcal{G}$ by (1.3).

As we shall check later, Itô's formula shows that for $f \in C^\infty_K(\mathbb{R}^n)$,

$$(3.1) \qquad C^f_t \equiv f(X_t) - f(X_0) - \int_0^t \mathcal{G}f(X_u)du = \int_0^t U_q f(X_u)d_I B^q,$$

where U_q is the vector field (first-order differential operator):

$$(3.2) \qquad U_q f(y) \equiv \sigma^i_q(y)\partial_i f(y), \qquad \partial_i \equiv \partial/\partial y^i.$$

Hence, for $f \in C^\infty_K(\mathbb{R}^n)$, the process C^f, as a stochastic integral relative to Brownian motion, is a local martingale - and indeed a martingale because it is bounded on each finite interval $[0,t_0]$. In other words, X solves the martingale problem for $\mathcal{G}$ starting from x (or, if you prefer, the law P^x of X does).

Let us now recall the fundamental existence-and-uniqueness theorem of Itô's theory. The theorem is true if only Lipschitz (or even only suitable Hölder) conditions are imposed on σ^i_q and b^i, but it is that case in which these functions are smooth which chiefly concerns us.

(3.3) THEOREM (ITÔ). Suppose that the functions σ^i_q and b^i are smooth, and that (say) Assumption (2.2.A) holds (with $a = \sigma\sigma^*$) - to preclude explosions. Then equation (I) has a pathwise-unique solution X which may be constructed from B by successive approximation in the spirit of Picard. The solution X is Markovian with generator extending $\mathcal{G}$ on $C^\infty_K(\mathbb{R}^n)$. ‡

The statement that X is <u>pathwise-unique</u> means that if Y is another solution of (I), then

$$P[X_t = Y_t, \forall t] = 1.$$

This is a different form of uniqueness from the <u>uniqueness in law</u> studied in Stroock-Varadhan theory. An examination of the relationship between the two concepts of uniqueness, and of the related matter of the difference between

‡ The transition semigroup of X is Feller, but, without the ellipticity assumption, we cannot say that it is strong Feller.

strong and weak solutions of stochastic differential equations (SDEs), is made in §11 below. We shall need to be a little more precise about the concept of pathwise uniqueness, and shall then be able to assert the Yamada-Watanabe result that

(3.4) **Pathwise uniqueness implies uniqueness in law.**

Thus, provided we have smooth coefficients σ and b in (I), we can establish both pathwise uniqueness of the solution to (I) and the uniqueness of solution to the corresponding martingale problem (in which $a = \sigma\sigma^*$), all this without any assumption that $a(y)$ is non-singular for each y. Indeed, if $r < n$, $a(y)$ is singular for every y!

In what has just been said, it was assumed that we are presented with equation (I) as a model for some system. But, as explained in §1, the 'data' with which we are presented often takes the form of the diffusion and drift coefficients: a and b. If $a(y)$ is non-singular for each y, then we can take $r = n$, and define $\sigma(y)$ to be the unique positive-definite square root of $a(y)$; and then it is easily shown that σ is smooth if a is smooth. In this way, we can prove by Itô theory that if the hypotheses of Theorem 2.2 are satisfied and if, further, a and b are smooth, then the law P^x of the pathwise-unique solution X of (I) (where $\sigma = a^{\frac{1}{2}}$) is the unique solution of the martingale problem for $\mathcal{G}$ starting from x.

§4. The relevance of hypoellipticity.

Suppose now that:

X is a solution of equation (I) where the coefficients σ and b are SMOOTH.

As always, $a \equiv \sigma\sigma^*$, and $\mathcal{G}$ is defined as at (1.3).

Itô's formula allows us to generalise (3.1) as follows:

if $f \in C_K^\infty((0,\infty) \times \mathbb{R}^n)$, then

$$D_t^f \equiv f(t,X_t) - f(0,X_0) - \int_0^t \left(\frac{\partial}{\partial s} + \mathcal{G}\right) f(s,X_s)\,ds = \int_0^t U_q f(s,X_s)\,d_I B^q,$$

where U_q is at (3.2). Hence, D^f is a (uniformly bounded) martingale. If t_0 is so large that support $(f) \subset (0,t_0] \times \mathbb{R}^n$, then

$$E[f(t_0,X_{t_0}) - f(0,X_0)] = E[0 - 0] = 0.$$

Hence

(4.1) $$\forall f \in C_K^\infty((0,\infty) \times \mathbb{R}^n), \qquad \alpha(\frac{\partial f}{\partial t} + \mathcal{G}f) = 0,$$

where, for $h \in C_K^\infty((0,\infty) \times \mathbb{R}^n)$,

(4.2) $$\alpha(h) \equiv E\int_0^\infty h(s,X_s)ds = \int_{s=0}^\infty \int_{y\in\mathbb{R}^n} h(s,y)P[X_s \in dy]ds.$$

Now α is induced by (or <u>is</u>!) a measure on $(0,\infty) \times \mathbb{R}^n$ which is finite on compacts, and hence is a Schwartz distribution. (Recall that I have very nobly recalled a few basic facts about distributions in the Appendix to this paper.) In the language of distribution theory, (4.1) takes the form:

(4.3) $$\left(-\frac{\partial}{\partial t} + \mathcal{G}^*\right)\alpha = 0.$$

If the operator $(-\partial/\partial t + \mathcal{G}^*)$ is <u>hypoelliptic</u> (in the sense explained in the Appendix), we can conclude that α is in fact a smooth function p on $(0,\infty) \times \mathbb{R}^n$, so that

(4.4) $$\alpha(h) = \int_{s=0}^\infty \int_{y\in\mathbb{R}^n} p_s(y)h(s,y)dsdy.$$

On comparing (4.2) and (4.4), we conclude that

if $(-\partial/\partial t + \mathcal{G}^*)$ is <u>hypoelliptic, then</u>

$$P[X_t \in dy] = p_t(y)dy,$$

<u>where</u> p <u>is a smooth function on</u> $(0,\infty) \times \mathbb{R}^n$ <u>satisfying the forward equation</u>:

(F) $$\frac{\partial p}{\partial t} = \mathcal{G}^* p;$$

<u>and moreover</u>, p <u>is the fundamental solution with pole at</u> x <u>of (F)</u>.

The statement that p is the fundamental solution of (F) with pole at x means:

(i) p is smooth on $(0,\infty) \times \mathbb{R}^n$ and satisfies (F);

(ii) p is nonnegative and

$$\lim_{\varepsilon \downarrow 0} \int_G p_\varepsilon(y)dy = 1 \qquad \text{for every neighbourhood } G \text{ of } x;$$

(iii) if p^* also has properties (i) and (ii), then $p \le p^*$.

Readers familiar with McKean's marvellous book [13] will realise that McKean bases his treatment on Weyl's Lemma: <u>if</u> $a(y)$ <u>is non-singular for each</u> y, <u>then</u> $(-\partial/\partial t + \mathcal{G}^*)$ <u>is hypoelliptic</u>.

<u>§5.</u> <u>HÖRMANDER (1967).</u> In a profound paper, Hörmander ([5]) gave what is for most practical purposes the complete solution of the problem of deciding when $(-\partial/\partial t + \mathcal{G}^*)$ is hypoelliptic.

We continue to assume that X solves the equation

$$\text{(I)} \qquad \boxed{dX^i = \sigma^i_q(X_t)d_I B^q + b^i(X_t)dt, \qquad X_0 = x,}$$

where the functions σ^i_q and b^i are smooth. Recall the notations:

$$U_q = \sigma^i_q \partial_i, \quad \partial_i \equiv \partial/\partial y^i; \quad a = \sigma\sigma^* ;$$

$$\mathcal{G} = \tfrac{1}{2}a^{ij}\partial_i\partial_j + b^i\partial_i.$$

It is immediately checked that

$$\text{(5.1)} \qquad \boxed{\mathcal{G} = \tfrac{1}{2}\sum_q U_q^2 + V,}$$

where

$$\text{(5.2)} \qquad V \equiv \beta^i\partial_i, \quad \text{where } \beta^i \equiv b^i - \tfrac{1}{2}U_q(\sigma^i_q).$$

Equation (5.1) is closely linked to the Stratonovich form of (I).[‡] It is easily checked that

---

‡ See §21

(5.3)
$$\boxed{\mathcal{g}^* = \frac{1}{2}\sum_q U_q^2 - \widetilde{V} + c,}$$

where

(5.4)
$$\widetilde{V} = V - h_q U_q, \qquad \text{where } h_q \equiv \partial_i(\sigma_q^i),$$

and

$$c(y) = \frac{1}{2}\partial_i\partial_i a^{ij}(y) - \partial_i b^i(y).$$

The vector fields on $\mathbb{R}^n$ are the first-order differential operators on $\mathbb{R}^n$ with smooth coefficients. If a vector field W has the form:

$$W = e_W^i \partial_i,$$

then we can think of W as assigning to each point y of $\mathbb{R}^n$ the 'classical vector'

$$e_W(y) = (e_W^1(y),\ldots,e_W^n(y))$$

or, in differential-geometry language, the 'tangent vector':

$$W(y) = e_W^i(y)\partial_i \quad \text{with } \partial_i \text{ evaluated at } y,$$

which is just the derivative at y along $e_W(y)$. The tangent vectors at y form a vector space $T_y(\mathbb{R}^n)$ over $\mathbb{R}$ of dimension n (which from the classical point of view is just $\mathbb{R}^n$). If Λ is a collection of vector fields on $\mathbb{R}^n$, we shall mean by the statement:

Λ <u>is full at each point</u>

that:

<u>for each</u> y, <u>the vectors</u> $\{W(y): W \in \Lambda\}$ <u>span the space</u> $T_y(\mathbb{R}^n)$.

If W_1 and W_2 are vector fields on $\mathbb{R}^n$, then the <u>Lie bracket</u>:

$$[W_1,W_2] \equiv W_1W_2 - W_2W_1$$

defines a new vector field $[W_1,W_2]$ because the second-order terms on the right-hand side cancel out.

Now introduce

(5.5) $$\tilde{\Lambda}_+ \equiv \mathrm{Lie}(U_1,U_2,\ldots,U_r,\tilde{V})$$

the Lie algebra generated by $U_1,U_2,\ldots,U_r,\tilde{V}$; that is, the smallest vector space of vector fields which contains $U_1,U_2,\ldots,U_r,V$, and which is closed under the Lie-bracket operation.

(5.6) THEOREM(Hörmander). If $\tilde{\Lambda}_+$ is full at each point, then $\mathcal{G}^*$ is hypoelliptic.

[Note. Set $\Lambda_+ \equiv \mathrm{Lie}(U_1,U_2,\ldots,U_r,V)$. Then, by using (5.4), you can check that

(5.7) Λ_+ is full at each point $\Longleftrightarrow$ $\tilde{\Lambda}_+$ is full at each point.]

We shall examine the probabilistic significance of Theorem 5.6 below; but recall that for the forward (Fokker-Planck) equation, we need hypoellipticity not of $\mathcal{G}^*$ but of $(-\partial/\partial t + \mathcal{G}^*)$. Now

$$(-\partial/\partial t + \mathcal{G}^*) = \frac{1}{2}\sum_q U_q^2 - (\tilde{V} + \partial/\partial t) + c,$$

so that we can apply Theorem 5.6 to $(-\partial/\partial t + \mathcal{G}^*)$ by 'going up one dimension' (to the space-time process (X_t,t)). Now it is easy to verify that

$\mathrm{Lie}(U_1,U_2,\ldots,U_r,\tilde{V} + \partial/\partial t)$ is full at each point of $(0,\infty) \times \mathbb{R}^n$

if and only if $\tilde{\Lambda}_-$ is full at each point of $\mathbb{R}^n$ where

$$\tilde{\Lambda}_- \equiv \mathrm{Lie}(U_1,U_2,\ldots,U_r,[U_1,\tilde{V}],\ldots,[U_r,\tilde{V}]).$$

The point is that $\tilde{V} + \partial/\partial t$ is 'used up' on the time-coordinate, and cannot 'help out' for the space-coordinates. (Think of determinants!)

(5.8) COROLLARY TO HÖRMANDER's THEOREM. If $\tilde{\Lambda}_-$ is full at each point of $\mathbb{R}^n$, then $(-\partial/\partial t + \mathcal{G}^*)$ is hypoelliptic.

So, the business of the forward equation is settled.

We shall turn to the probabilistic significance of Theorem 5.6 in a moment. But first we look at two examples. Professor Dynkin insists that these must be the simplest possible!

(5.9) **Example.** Take $r = 1$, $n = 1$, and let (I) take the form:

$$dX = dt, \quad X_0 = 0.$$

Then $\mathcal{G}^*f(y) = -\partial f/\partial y$, $U = 0$, $\tilde{V} = -\partial/\partial y$. Obviously, $\tilde{\Lambda}_- = \{0\}$; and indeed, since $X_t = t$, X_t does not have a smooth density. But $\tilde{\Lambda}_+$ is obviously full at each point, so (as you can no doubt prove without Hörmander's help!) $\mathcal{G}^*$ is hypoelliptic.

(5.10) **Example.** Take $r = 1$, $n = 2$, and let (I) take the form:

$$\begin{cases} dX^1 = dB^1, & X_0^1 = 0, \\ dX^2 = X^1 dt, & X_0^2 = 0. \end{cases}$$

Then $\mathcal{G}^* = \frac{1}{2}\partial_1^2 - y_1\partial_2$, $U_1 = \partial_1$, $\tilde{V} = y_1\partial_2$. Since $U_1 = \partial_1$ and $[U_1, \tilde{V}] = \partial_2$, Λ_- is full at each point and so equation (F) holds. Of course,

$$X_t^1 = B_t^1, \quad X_t^2 = \int_0^t B_s^1 ds,$$

so that (X_t^1, X_t^2) has the Gaussian density on $\mathbb{R}^2$ of zero mean and covariance matrix

$$\begin{pmatrix} t & t^2/2 \\ t^2/2 & t^3/3 \end{pmatrix}.$$

Hence, you can write down $p_t(y)$ explicitly.

(5.11) **Exercise.** Check that the Corollary to Hörmander's Theorem does the right thing for the illuminating Example 2.10 in Stroock's paper in this volume. □

What probabilistic consequences can we derive if we know only that $\tilde{\Lambda}_+$ is full at each point? Let T be an exponentially distributed variable (of rate $\lambda > 0$) which is independent of X. If λR_λ is the probability distribution of X_T considered as a Schwartz distribution, then, by very familiar arguments,

$$(5.12) \qquad (\lambda - \mathcal{G}^*)R_\lambda = \delta_x.$$

Now, if $\tilde{\Lambda}_+$ is full at each point, so that $(\lambda - \mathcal{G}^*)$ is hypoelliptic, then the probability distribution λR_λ of X_T has smooth density except at the point $x = X_0$.

§6. **MALLIAVIN (1976).** An exciting possibility is suggested by the preceding sentence: if one could prove directly by probabilistic methods that, if $\tilde{\Lambda}_+$ is full at each point, then X_T has smooth density on $\mathbb{R}^n \setminus \{x\}$, then one would have a probabilistic proof of Hörmander's Theorem and a probabilistic understanding of the forward equation. Malliavin's remarkable paper [12] took a giant step in this direction by setting up a calculus of variations for Brownian motion which provides an entirely natural way of showing that, under suitable conditions, X_t has a smooth density for t fixed. (The distribution of X_t is $P \circ X_t^{-1}$, where P is the (Wiener-measure) law of the Brownian motion B. So, since P is 'smooth', then, provided we can assert that X_t is a 'smooth' map from $(\Omega, \mathcal{F}, P)$ to $\mathbb{R}^n$, then we are home.)

Stroock (in [22] and in his paper in this volume) gives a great deal of insight into the Malliavin calculus, and also indicates that it has applications which lie outside the scope of PDE theory and which are likely to prove important in physics. Bismut's paper in this volume develops the Malliavin calculus through to the full Hörmander theorem (even with some improvements). I understand that there is an excellent account of the Malliavin calculus in a book [7] by Ikeda and Watanabe; and that that book will appear before this one.

I shall say just a few words about the Malliavin calculus, trying to link Malliavin's approach, based on the infinite-dimensional Ornstein-Uhlenbeck process, with Stroock's more axiomatic approach in this volume, and with the different variational technique used by Bismut. But it is no part of my task to poach from other contributors to this volume.

Note. At Durham, and in correspondence since, Professor Stroock has rightly stressed that our justified pride in the Malliavin calculus must be accompanied by a proper respect for the contributions of analysts. Not only did one L. Hörmander discover his theorem; but his theorem has been proved (for example in Kohn [10]) by techniques much less formidable than the Malliavin calculus, and has undergone truly significant extension (see [20] and other work by Oleinik and Radkevič).

PART II. Continuous semimartingales and the associated stochastic integrals.

References: Meyer [16], Dellacherie and Meyer [3], Métivier and Pellaumail [15]. For an excellent survey, see Dellacherie [2].

§7. Basics. We work with a set-up $(\Omega,\mathcal{F},\{\mathcal{F}_t\},P)$ satisfying the usual conditions. This means that: $(\Omega,\mathcal{F},P)$ is a complete probability triple; $\{\mathcal{F}_t : t \geq 0\}$ is a filtration (increasing family of sub-σ-algebras of $\mathcal{F}$); $\mathcal{F}_0$ contains all P-null sets in $\mathcal{F}$; $\{\mathcal{F}_t\}$ is right-continuous in that $\mathcal{F}_{t+} = \mathcal{F}_t$, where $\mathcal{F}_{t+} \equiv \bigcap_{u>t}\mathcal{F}_u$. (We shall sometimes write $\mathcal{F}(t)$ for $\mathcal{F}_t$.) As usual, $E[\cdot]$ denotes expectation relative to P; and for a sub-σ-algebra $\mathcal{H}$ of $\mathcal{F}$, $E[\cdot|\mathcal{H}]$ denotes conditional P-expectation given $\mathcal{H}$.

For the moment, we deal only with real-valued random variables and processes. A process X is, of course, a family $\{X_t : t \geq 0\}$ of random variables ($\mathcal{F}$ measurable maps from Ω to $\mathbb{R}$). The map $t \mapsto X_t(\omega)$ is the sample path of X corresponding to ω in Ω. We sometimes write $X(t)$ instead of X_t. Two processes X and Y are called modifications of each other if $P[X_t = Y_t] = 1$, $\forall t$; and are called indistinguishable if the stronger condition: $P[X_t = Y_t, \forall t] = 1$ holds. A process is called right-continuous [respectively, left-continuous, etc.] if its sample paths are right-continuous [left-continuous, etc.] A process X is called cadlag (or, by some, 'Skorokhod', or even 'corlol') if X is right-continuous and if the left limits

$$X_{t-} \equiv \lim_{s<t, s\to t} X_s$$

exist for $t > 0$. Of course, 'cadlag' stands for 'continu à droite et pourvu de limites à gauche'. (We sometimes give ourselves licence to 'ignore null sets' and, for example, to call a process cadlag if almost all its paths are cadlag; but be warned that in parts of the modern theory, one must be cautious in doing such things. Anyway, we shall continue to do what we have done in Part I, and suppress 'almost surely' phrases which are obviously implicit.)

A process X is called adapted (to $\{\mathcal{F}_t\}$) if $\forall t$, X_t is $\mathcal{F}_t$ measurable.

A martingale M (relative to the set-up $(\Omega,\mathcal{F},\{\mathcal{F}_t\},P)$) is a process

$\{M_t : t \geq 0\}$ such that

<u>7.1 (i)</u> M is adapted (to $\{\mathcal{F}_t\}$);

<u>7.1 (ii)</u> $E|M_t| < \infty$, $\forall t$;

<u>7.1 (iii)</u> whenever $s \leq t$, $E[M_t | \mathcal{F}_s] = M_s$ (with probability 1).

For <u>supermartingale</u> [respectively, <u>submartingale</u>] replace '=' in 7.1(iii) by '$\leq$' ['$\geq$']. <u>Every martingale</u> (and every supermartingale X for which the map $t \mapsto E[X_t]$ is right-continuous) has a <u>cadlag modification</u>.

A <u>stopping time</u> T is a map $T: \Omega \to [0,\infty]$ such that

$$\{\omega : T(\omega) \leq t\} \in \mathcal{F}_t, \forall t.$$

If M is a continuous adapted process with $M_0 = 0$, then we say that M is a continuous local martingale null at 0, and write $M \in \mathcal{M}^{0,c}_{loc}$ if for some sequence $\{T_n\}$ of stopping times with $T_n \uparrow \infty$, each stopped process $\{M(t \wedge T_n) : t \leq 0\}$ is a martingale. Of course, $u \wedge v \equiv \min(u,v)$. (For this 'continuous' case with $M_0 = 0$, either the sequence $\{T_n\}$ with $T_n \equiv \inf\{t : |M_t| = n\}$ will do the trick, or nothing will.) For a general continuous adapted process M, we write $M \in \mathcal{M}^c_{loc}$ and call M a <u>continuous local martingale</u> if $M - M_0 \in \mathcal{M}^{0,c}_{loc}$.

It must be appreciated that many of the results which will now be stated depend heavily on the fact that we are dealing with <u>continuous</u> local martingales. See the papers by Rogers and Elliott for how to deal with jumps.

Let $M \in \mathcal{M}^c_{loc}$. It is a consequence of the <u>Meyer decomposition theorem</u> that there is a unique continuous adapted process $\langle M,M \rangle$ with non-decreasing paths such that $M^2 - \langle M,M \rangle \in \mathcal{M}^{0,c}_{loc}$. Note that $\langle M,M \rangle_0 \equiv M_0^2$. The uniqueness assertion refers of course to 'uniqneness modulo indistinguishability' and is equivalent to the result that a <u>continuous local martingale with paths of finite variation is constant</u>. The process $\langle M,M \rangle$ is equal to the dyadic-quadratic-variation process of M. Thus, for any sequence Γ of integers tending to ∞ 'sufficiently quickly', it is true with probability 1 that

$$(7.2) \qquad \langle M,M \rangle_t = \lim_{n \in \Gamma} \sum_{i \geq 1} [M(t_n^i) - M(t_n^{i-1})]^2 + M_0^2,$$

where

$$(7.3) \qquad t_n^i \equiv (i2^{-n})\wedge t\,,$$

the limit in (7.2) existing uniformly on compact t-intervals. (Note. It is easy to see that if M has paths of finite variation, then (7.2) implies that $\langle M,M\rangle_t = M_0^2$, so $M_t = M_0$, $\forall t$.) It is an immediate consequence of (7.2) that

$$(7.4) \qquad \lim_{n\in\Gamma} \sum_{i\geq 1} 2M(t_n^{i-1})[M(t_n^i) - M(t_n^{i-1})] = M_t^2 - \langle M,M\rangle_t,$$

and it is of course trivial that

$$(7.5) \qquad \lim_{n} \sum_{i\geq 1} [M(t_n^{i-1}) + M(t_n^i)][M(t_n^i) - M(t_n^{i-1})] = M_t^2 - M_0^2\,.$$

Formulae (7.4) and (7.5) are the keys to the transformation rules (Itô formulae) of the Itô and Stratonovich calculi.

Let $M,N \in \mathcal{M}_{loc}^c$. Define $\langle M,N\rangle$ by polarization:

$$(7.6) \qquad \langle M,N\rangle \equiv \tfrac{1}{4}\langle M+N,M+N\rangle - \tfrac{1}{4}\langle M-N,M-N\rangle.$$

Then $\langle M,N\rangle$ <u>is the unique continuous adapted process with paths of finite variation such that</u> $MN - \langle M,N\rangle \in \mathcal{M}_{loc}^{0,c}$. For sequences Γ of sufficient growth

$$(7.7) \qquad \begin{aligned} &M_tN_t - \langle M,N\rangle_t \\ &= \lim_{n\in\Gamma}\Big\{\sum_{i\geq 1} M(t_n^{i-1})[N(t_n^i) - N(t_n^{i-1})] + \sum_{i\geq 1} N(t_n^{i-1})[M(t_n^i) - M(t_n^{i-1})]\Big\}.\end{aligned}$$

This consequence of (7.4) is the key to <u>integration by parts</u> in the Itô calculus.

A <u>continuous semimartingale</u> X is a continuous adapted process which may be decomposed as follows:

$$(7.8) \qquad X = X_0 + M + A,$$

where $M \in \mathcal{M}_{loc}^{0,c}$ and A is a process (necessarily continuous adapted and null at 0) with paths of finite variation. If, once again, we invoke the result that a continuous local martingale with paths of finite variation in constant, we see that <u>the decomposition (7.8) of a continuous semimartingale</u> X <u>is unique</u> (modulo indistinguishability).

§8. The ITÔ integral. It is now known in the maximum generality possible when it makes sense to talk of an Itô integral $\int C d_I X$. The process C must be previsible (in the sense explained below), X must be a semimartingale, and certain 'boundedness' conditions must hold. (See the paper by Rogers.) We are interested here only in the case when X is continuous; but we will need to allow for discontinuities in C.

It will not surprise you to know that if X is our continuous semimartingale at (7.8), then we shall define $\int C d_I X$ to be the continuous semimartingale with canonical decomposition:

$$\int_{[0,t]} C d_I X = C_0 X_0 + \int_{(0,t]} C d_I M + \int_{(0,t]} C_s dA_s. \tag{8.1}$$

Now if C is previsible, $s \mapsto C_s(\omega)$ is measurable for every ω (or maybe for for almost every ω, depending on convention), and

$$\left(\int_{(0,t]} C_s dA_s\right)(\omega) \equiv \int_{(0,t]} C_s(\omega) dA_s(\omega)$$

exists as a Lebesgue-Stieltjes integral; and so, we know all about it. (We must however beware of making too cavalier statements. See the beginning of §1 of Meyer [16]!) All we need to recall therefore are the definition of previsible process and the definition of $\int C d_I M$ when C is previsible, M is a local martingale, and appropriate boundedness conditions hold.

Definition of 'previsible process'. Think of a process X as a map $(t,\omega) \mapsto X(t,\omega)$ from $[0,\infty) \times \Omega$ to $\mathbb{R}$. Let $\mathcal{P}$ be the smallest σ-algebra on $[0,\infty) \times \Omega$ such that every left-continuous adapted process is $\mathcal{P}$ measurable. Now call a process C previsible if C is $\mathcal{P}$ measurable. (I skip discussion of whether a process indistinguishable from a previsible process should, or should not, be counted as being previsible. It will not really matter to us in this paper.) A previsible process C is called elementary if for some n, C is constant on every interval of the form $(\overline{i-1}\,2^{-n}, i2^{-n}]$ for $i \geq 1$. Thus

$$C(t) = \xi_n^{i-1} \text{ on } (\overline{i-1}\,2^{-n}, i2^{-n}], \quad i \geq 1, \tag{8.2a}$$

where, since C is adapted and $\{\mathcal{F}_t\}$ is right-continuous,

(8.2b) $$\xi_n^{i-1} \text{ is } \mathcal{F}(\overline{i-1}\,2^{-n}) \text{ measurable.}$$

Now suppose for a moment that C is a bounded elementary previsible process. (Saying that C is 'bounded' means that for some K, $|C(t,\omega)| \le K, \forall t, \forall\omega$.) Suppose also that M is a bounded continuous local martingale (and hence in particular a true martingale). Then if we make the obvious definition:

(8.3) $$\int_{[0,t]} C\, d_I M = C_0 M_0 + \sum_{i\ge 1} \xi_n^{i-1}[M(t_n^i) - M(t_n^{i-1})],$$

we find easily from such calculations as:

$$E\{\xi_n^{i-1}[M(t_n^i) - M(t_n^{i-1})] \,|\, \mathcal{F}(\overline{i-1}\,2^{-n})\}$$

$$= \xi_n^{i-1}\, E\{[M(t_n^i) - M(t_n^{i-1})] \,|\, \mathcal{F}(\overline{i-1}\,2^{-n})\} = 0,$$

that $\int C\, d_I M$ is a martingale, and for any bounded continuous martingale N,

$$\left\langle \int C\, d_I M, N \right\rangle_t = \int_{[0,t]} C\, d\langle M,N\rangle_s, \quad \forall t.$$

(8.4) THEOREM, DEFINITION (Kunita, Watanabe). Let M be a continuous local martingale, and let C be a previsible process such that with probability 1,

(8.5) $$\int_{[0,t]} C_s^2\, d\langle M,M\rangle_s < \infty \quad \forall t.$$

Then there exists one and (up to indistinguishability) only one continuous local martingale $\int C\, d_I M$ such that for every continuous local martingale N,

(8.6) $$\left\langle \int C\, d_I M, N \right\rangle_t = \int_{[0,t]} C_s\, d\langle M,N\rangle_s, \quad \forall t.$$

(Since $\langle M,N\rangle$ is of finite variation, the integral on the right-hand side of (8.6) is a Stieltjes integral.) Then $\int C\, d_I M$ <u>is called the Itô</u> (-Kunita-Watanabe) <u>stochastic integral of</u> C <u>with respect</u> to M.

Note that

(8.7) $$\left\langle \int C\, d_I M, \int C\, d_I M \right\rangle_t = \int_{[0,t]} C_s^2\, d\langle M,M\rangle_s.$$

This shows that condition (8.5) cannot be relaxed.

The existence part of theorem 8.4 is proved by extending the definition from bounded elementary processes via a Hilbert space isometry (in effect, the P-integrated form of (8.7)) and then using localisation.

Note. The papers by Rogers and by Elliott introduce the general theory of stochastic integrals (relative to semimartingales with jumps).

§9. ITÔ'S FORMULA. Let M be a continuous local martingale ($M \in M^c_{loc}$). Let C be a continuous adapted process, so that, in particular, C is previsible and (8.5) automatically holds. Then, for any sequence Γ of integers tending to ∞ sufficiently quickly, we have with probability 1,

$$(9.1) \qquad \int_{[0,t]} C d_I M = C_0 M_0 + \lim \sum_{i \geq 1} C(t_n^{i-1})[M(t_n^i) - M(t_n^{i-1})],$$

the limit existing uniformly over compact intervals. Note the characteristic feature of the Itô integral that "the M increments point into the future of the C value", or rather that "the C-value lies in the past of the M increment".

From (7.7) we therefore find that for $M, N \in M^c_{loc}$,

$$(9.2) \qquad M_t N_t - \langle M,N \rangle_t = \int_{(0,t]} \{M d_I N + N d_I M\}.$$

Note the range of integration: $(0,t]$, not $[0,t]$. Now it is easily seen that dyadic quadratic variation is unaffected by the addition of a continuous process of finite variation. Hence, for our continuous semimartingale $X = X_0 + M + A$, we put

$$\langle X,X \rangle \equiv X_0^2 + \langle M,M \rangle, \quad \langle M,A \rangle \equiv 0, \quad \langle A,A \rangle \equiv 0.$$

Then we have the integration-by-parts formula:

for two continuous semimartingales X and Y,

$$(9.3) \qquad X_t Y_t - \langle X,Y \rangle_t = \int_{(0,t]} \{X d_I Y + Y d_I X\}.$$

(Of course, $\langle X,Y \rangle = \frac{1}{4}\langle X+Y, X+Y \rangle - \frac{1}{4}\langle X-Y, X-Y \rangle$.)

In particular,

$$X_t^2 - \langle X,X\rangle_t = 2\int_{(0,t]} X d_I X ,$$

or in differentiated form:

$$d(X^2) = 2XdX + d\langle X,X\rangle .$$

This is the particular case when $f(x) = x^2$ of (9.5).

(9.3) ITÔ'S FORMULA. If X is a continuous semimartingale, and if f is a smooth (or, more generally, C^2 function on $\mathbb{R}$, then

$$f(X_t) - f(X_0) = \int_{(0,t]} f'(X_s) d_I X + \frac{1}{2}\int_{(0,t]} f''(X_s) d\langle X,X\rangle_s .$$

In differentiated form, Itô's formula reads:

(9.5) $$df(X) = f'(X)d_I X + \frac{1}{2}f''(X)d\langle X,X\rangle .$$

More generally, if $X = (X^1, X^2, \ldots, X^n)$ is an $\mathbb{R}^n$-valued semimartingale (each X^i is a semimartingale), then, for f $C^2(\mathbb{R}^n)$,

(9.6) $$df(X) = \partial_i f(X) d_I X^i + \frac{1}{2}\partial_i \partial_j f(X) d\langle X^i, X^j\rangle .$$

In particular, $f(X)$ is a semimartingale with canonical decomposition:

$$f(X_t) = f(X_0) + \int_{(0,t]} \partial_i f(X) d_I X^i + \left\{\int_{(0,t]} \partial_i f(X) dA^i + \frac{1}{2}\int_{(0,t]} \partial_i \partial_j f(X) d\langle X^i, X^j\rangle\right\},$$

where $X^i = X_0^i + M^i + A^i$ is the canonical decomposition of X^i.

Now, the integration-by-parts formula (9.3) is just Itô's formula (9.6) applied with $f(x,y) = xy$. However, it is not difficult to deduce (9.6) from (9.3). The reason is that we can use (9.3) to prove (9.6) for polynomial functions f by induction; and then we can on a compact cube in $\mathbb{R}^n$ approximate f uniformly by polynomials p_k in such a way that the $\partial_i f$ and $\partial_i \partial_j f$ are uniformly approximated by $\partial_i p_k$ and $\partial_i \partial_j p_k$.

§10. LÉVY's theorem. We now prove the following fundamental characterization of Brownian motion due to Lévy.

(10.1) THEOREM (Lévy). Let $B = (B^1, B^2, \ldots, B^n)$ be a process on $\mathbb{R}^n$ such that each B^i is a continuous local martingale null at 0 and that $\langle B^i, B^j \rangle_t = t\delta^{ij}$. Then B <u>is an</u> $(\{\mathcal{F}_t\}, P)$ <u>Brownian motion</u>: (B is $\{\mathcal{F}_t\}$ adapted and) for $t < u$, the variable $B_u - B_t$ is independent of $\mathcal{F}_t$ and has the multivariate Gaussian distribution of mean 0 and covariance matrix $\{(u-t)\delta^{ij}\}$.

<u>Note</u>. Since $\mathcal{F}_t$ may contain more information than $\sigma\{B_s : s \leq t\}$, the conclusion is stronger than the assertion that the law of B is Wiener measure.

<u>Proof</u> (Kunita-Watanabe). Let $\theta \in \mathbb{R}^n$, and let $\theta.B$ denote the scalar product $\theta_k B^k$. Then (with i now denoting $\sqrt{(-1)}$), put

$$Z_t^\theta \equiv \exp(i\theta.B + \tfrac{1}{2}|\theta|^2 t).$$

By Itô's formula,

$$dZ_t^\theta = (iZ_t^\theta)\theta.d_I B,$$

so that Z_t^θ is a complex-valued continuous local martingale. But Z_t^θ is bounded on finite intervals, so that Z_t^θ is a <u>martingale</u>. Hence, for $u > t$, $E[Z_u^\theta | \mathcal{F}_t] = Z_t^\theta$. On rearranging, we obtain

$$E[\exp\{i\theta.(B_u - B_t)\} | \mathcal{F}_t] = \exp\{-\tfrac{1}{2}|\theta|^2(u-t)\};$$

and the result follows. □

§11. Martingale problems and weak solutions of SDEs.

The purpose of this section is to clarify and extend remarks made in Part I about the relationship between Itô and Stroock-Varadhan approaches to diffusion theory, and in particular to explain the concept of a weak solution of an SDE.

First, let me make a technical point. The theory of stochastic integration is almost always developed under the assumption of the usual conditions on $(\Omega, \mathcal{F}, \{\mathcal{F}_t\}, P)$. However, the martingale problem is about measures on $(W, \mathcal{A}, \{\mathcal{A}_t\})$, and the usual conditions will not hold. However, this difficulty is illusory in our context.

Suppose that Ω is a set, $\mathcal{F}^o$ is a σ-algebra on Ω, and $\{\mathcal{F}_t^o : t \geq 0\}$ is a filtration on $(\Omega,\mathcal{F}^o)$ (that is, an increasing family of sub-σ-algebras of $\mathcal{F}^o$). Suppose that Y is a given $\{\mathcal{F}^o_t\}$ adapted process with <u>all</u> its paths right-continuous. Let P be any measure on $(\Omega,\mathcal{F}^o)$ such that Y is an $(\{\mathcal{F}^o_t\},P)$ martingale, so that (with E as P-expectation) $E|Y_t| < \infty$, $\forall t$, and for $s < t$, $E[Y_t|\mathcal{F}^o_s] = Y_s$ (with probability 1). Then, for $s < u < t$,

$$E[Y_t|\mathcal{F}^o_u] = Y_u \qquad \text{(with probability 1).}$$

Let $u \downarrow s$ through a sequence. Use the martingale-convergence theorem on the left-hand side, and use the right-continuity of Y on the right-hand side to see that

$$E[Y_t|\mathcal{F}^o_{s+}] = Y_s \qquad \text{(with probability 1).}$$

Now let $(\Omega,\mathcal{F},P)$ be the completion of $(\Omega,\mathcal{F}^o,P)$ in the usual measure-theoretic sense. Let $\mathcal{F}_t$ be the smallest σ-algebra extending $\mathcal{F}^o_{t+}$ and including all P-null sets in $\mathcal{F}$. Then $(\Omega,\mathcal{F},\{\mathcal{F}_t\},P)$, called the <u>usual</u> P-<u>augmentation</u> of $(\Omega,\mathcal{F}^o,\{\mathcal{F}^o_t\},P)$, satisfies the usual conditions. Our argument shows that Y is an $(\{\mathcal{F}_t\},P)$ martingale.

I included the above 'technical' discussion early so as not to interrupt the following discussion. The purpose of the above remarks will become clear when we consider the move 'from Stroock-Varadhan in the direction of Itô' below.

- -

<u>From "Itô" to "Stroock-Varadhan"</u>. Let $(\Omega,\mathcal{F},\{\mathcal{F}_t\},P)$ be a set-up satisfying the usual conditions. Let $B = \{B^q : 1 \leq 1 \leq r\}$ be an r-dimensional Brownian motion starting at 0. Suppose that an $\mathbb{R}^n$-valued process X satisfies equation:

$$\text{(I)} \qquad dX^i = \sigma^i_q(X_t)d_I B^q + b^i(X_t)dt, \qquad X_0 = x,$$

where for the moment, we need only assume that σ and b are Borel measurable. (We are <u>assuming</u> that a solution X exists.) Then each X^i is a semimartingale with canonical decomposition:

$$X^i_t = x^i + \int_{(0,t]} \sigma^i_q(X_t)d_I B^q + \int_{(0,t]} b^i(X_t)dt$$

so that the martingale part of M^i of X^i is given by:

$$M_t^i = \int_{(0,t]} \sigma_q^i(X_t) d_I B^q = X_t^i - x^i - \int_{(0,t]} b^i(X_t) dt$$

From the Kunita-Watanabe formula (8.6) and the fact that $\langle B^q, B^r \rangle_t = \delta^{qr} t$, we find that

$$\langle M^i, M^j \rangle = \left\langle \int \sigma_q^i(X_s) d_I B^q, \int \sigma_r^j(X_s) d_I B^r \right\rangle$$

$$= \int \sigma_q^i(X_s) \sigma_q^j(X_s) ds = \int a^{ij}(X_s) ds.$$

Itô's formula now establishes (3.1), and, as explained there, X satisfies property:

(SV) $\forall f \in C_K^\infty$, $C_t^f \equiv f(X_t) - f(X_0) - \int_0^t \mathcal{G} f(X_s) ds$ defines a martingale C^f.

The law of X is of course the measure $P \circ X^{-1}$ on (W, A^0) where X is regarded as the map $\omega \mapsto X(\cdot, \omega)$ of Ω to W. (Recall that W is the set of continuous maps w from $[0,\infty)$ to $\mathbb{R}$, that A^0 is the smallest σ-algebra such that each map $w \mapsto w(t)$ is A^0 measurable). Since X satisfies (SV), the law of X solves the martingale problem for $\mathcal{G}$ starting from x.

As stated in Theorem 3.3, when σ and b are smooth and some condition is imposed to preclude explosion, there is one and only one solution X constructible from B by successive approximation. In particular, X is $\{\mathcal{B}_t\}$ adapted, where (we may as well have the usual conditions!) the set-up $(\Omega, \mathcal{B}, \{\mathcal{B}_t\}, P)$ is the usual P-augmention of $(\Omega, \mathcal{B}^0, \{\mathcal{B}_t^0\}, P)$, where $\mathcal{B}_t^0 \equiv \sigma\{B_s : s \le 0\}$. We say that X is a strong solution.

When we speak of a strong solution of (I), we understand that B is regarded as a given process, and that X must be "solved in terms of B" in the sense that X is $\{\mathcal{B}_t\}$ adapted.

From "Stroock-Varadhan" in the direction of "Itô". Take the canonical set-up: $W \equiv C([0,\infty); \mathbb{R}^n)$, $X_t(w) \equiv w(t)$, $A_t \equiv \sigma\{X_s : s \le t\}$, $A \equiv \sigma\{X_s : s \ge 0\}$. Let $\mathcal{G}$ be our operator:

$$\mathcal{G} = \tfrac{1}{2} a^{ij}(x) \partial_i \partial_j + b^i(x) \partial_i,$$

where the a^{ij} and b^i are Borel measurable and where each $a(x)$ is a strictly positive-definite matrix. Let $\sigma(x)$ be the unique positive-definite square root of $a(x)$.

Suppose that P is some solution of the martingale problem for $\mathcal{G}$ starting from x. Then, by that boring argument at the beginning of this section, $\forall f \in C_K^\infty$, C^f is a martingale relative to the P-augmentation $(W,\bar{A},\{\bar{A}_t\},P)$ of $(W,A,\{A_t\},P)$, where, as usual,

$$C_t^f \equiv f(X_t) - f(X_0) - \int_0^t \mathcal{G}f(X_s)ds.$$

So, with the usual conditions restored, we can apply stochastic integral theory without any worries. It is clear that

(11.1) $\forall f \in C^\infty$, C^f is a local martingale ("relative to $(\{A_t\},P)$" is now understood).

In particular, for each i,

$$(11.2) \qquad \boxed{M_t^i \equiv X_t^i - x^i - \int_0^t b^i(X_s)ds}$$

is a local martingale. (Take $f(x) = x^i$).

Now comes a calculation which is echoed several times in this volume. For $f \in C^\infty$, we exhibit $F_t \equiv f(X_t)$ as the semimartingale:

$$F_t \equiv f(X_t) = f(x) + C_t^f + \int_0^t \mathcal{G}f(X_s)ds.$$

Similarly, for $g \in C^\infty$, we have:

$$G_t \equiv g(X_t) = g(x) + C_t^g + \int_0^t \mathcal{G}g(X_s)ds.$$

On applying the same argument to fg, we obtain:

$$F_tG_t = f(X_t)g(X_t) = f(x)g(x) + C_t^{fg} + \int_0^t \mathcal{G}(fg)oX_s ds.$$

But Itô tells us that

$$d(FG) = FdG + GdF + d\langle F,G\rangle.$$

Hence,

$$dC_t^{fg} + \mathcal{G}(fg)oX_t dt = \{F_t dC_t^g + G_t dC_t^f\} + (f\mathcal{G}g + g\mathcal{G}f)oX_t dt + d\langle F,G\rangle_t .$$

Hence, since $\langle F,G\rangle = F_0G_0 + \langle C^f, C^g\rangle$, we deduce from the uniqueness of canonical decompositions that

$$(11.3)\qquad \boxed{\begin{array}{c} d\langle C^f, C^g\rangle = \langle f,g\rangle(X_t)dt, \\ \underline{\text{where}}\ \langle f,g\rangle\ \underline{\text{is the function on}}\ \mathbb{R}^n\ \underline{\text{defined by}} \\ \langle f,g\rangle \equiv \mathcal{G}(fg) - f\mathcal{G}g - g\mathcal{G}f \end{array}}$$

[Note that if $\mathcal{G} = \frac{1}{2}\Delta$, then $\langle f,g\rangle = \operatorname{grad} f \cdot \operatorname{grad} g$, which checks with the Kunita-Watanabe formula because, in that case,

$$f(X_t) - f(X_0) = \int_{(0,t]} \operatorname{grad} f(X_s) \cdot dB_s + \frac{1}{2}\int_{(0,t]} \Delta f(X_s)ds] .$$

On applying (11.3) to the case when $f(x) = x^i$ and $g(x) = x^j$, we obtain

$$(11.4)\qquad d\langle M^i, M^j\rangle = a^{ij}(X_t)dt$$

where M^i is at (11.2). Now set

$$(11.5)\qquad \boxed{B_t^q \equiv \int_{(0,t]} \alpha_i^q(X_s)dM^i} \qquad (1 \le q \le n),$$

where $\{\alpha_i^q(y)\}$ is the inverse of the matrix $\{\sigma_q^i(y)\}$. By the Kunita-Watanabe formula,

$$d\langle B^q, B^r\rangle = \alpha_i^q \alpha_j^r d\langle M^i, M^j\rangle = \delta^{qr}dt,$$

since $\sigma = \sigma^*$ and $\sigma^2 = a$. Hence B <u>is an</u> $(\{A_t\}, P)$ <u>Brownian motion</u>. As you can easily believe, and as is immediate from the Kunita-Watanabe characterisation, (11.5) implies that

$$(11.6)\qquad M_t^i = \int_{(0,t]} \sigma_q^i(X_s)dB^q .$$

Hence, from (11.2) and (11.6), we obtain:

$$(I)\qquad dX^i = \sigma_q^i(X_t)dB^q + b^i(X_t)dt , \qquad X_0 = x.$$

<u>We constructed</u> B <u>from</u> X via (11.4) and (11.2). Now, if σ and b are smooth, then, by Itô's theorem, X is $\{\mathcal{B}_t\}$ adapted, where $(W,\mathcal{B},\{\mathcal{B}_t\},P)$ is the usual P-augmentation of $(W,\mathcal{B}^0,\{\mathcal{B}_t^0\},P)$ where $\mathcal{B}_t^0 \equiv \sigma\{B_s : s \le t\}$, $\mathcal{B}^0 \equiv \sigma\{B_s : s \le 0\}$. However, <u>if</u>, <u>for example</u>, a <u>and</u> b <u>are merely assumed continuous</u>, <u>it is not known whether</u> X <u>is</u> $\{\mathcal{B}_t\}$ <u>adapted</u>. Now if X is not $\{\mathcal{B}_t\}$ adapted, then it is clearly

impossible to construct X from B.

By a weak solution of equation (I), we mean: a pair (X,B) of processes on some set-up $(\Omega,\mathcal{F},\{\mathcal{F}_t\},P)$ such that B is an $(\{\mathcal{F}_t\},P)$ Brownian motion, X is a continuous $\{\mathcal{F}_t\}$ adapted process, and equation (I) is satisfied. We have just proved that if $a(x)$ is non-singular for each x, then, whenever there is a solution of the martingale problem for $\mathcal{G}$ starting from x, then equation (I) has a weak solution. Our move "from Itô to Stroock-Varadhan" establishes the converse fact that if (I) has a weak solution (X,B), then the law of X is a solution of the martingale problem for $\mathcal{G}$ starting from x.

The notorious Tsirel'son example of a stochastic differential equation which has a weak solution which is not a strong solution, is described in §13.

- -

Pathwise uniqueness is said to hold for (I) if whenever there exists a set-up $(\Omega,\mathcal{F},\{\mathcal{F}_t\},P)$ carrying a triple $(X,\tilde{X},B)$ of continuous $\{\mathcal{F}_t\}$ adapted processes such that B is an $(\{\mathcal{F}_t\},P)$ Brownian motion and both (X,B) and $(\tilde{X},B)$ solve (I), then

$$P[\tilde{X}_t = X_t, \forall t] = 1.$$

Yamada and Watanabe ([26]) proved that pathwise uniqueness implies uniqueness of solution of the martingale problem. Stroock and Varadhan ([23]) showed that pathwise uniqueness of (I) implies that any weak solution (X,B) of (I) is a strong solution: X is $\{\mathcal{B}_t\}$ adapted where $\{\mathcal{B}_t\}$ is the P-augmented filtration generated by B.

- -

For a profound study of these topics in a much wider context, see the paper in this volume by Jacod and Mémin.

§12. GIRSANOV's theorem, martingale representation.

Once again, let

$$W \equiv C([0,\infty);\mathbb{R}^n); \quad X_t(w) \equiv w(t); \quad A_t \equiv \sigma\{X_s : s \le t\}, \quad A \equiv \sigma\{X_s : s \ge 0\}.$$

Let P_W denote Wiener measure of (W,A), so that P_W is the unique measure on (W,A) such that $(\forall i)$ X^i is an $(\{A_t\},P)$ martingale and $(\forall i,\forall j)$ $X^iX^j - t\delta^{ij}$

is an $(\{A_t\},P)$ martingale.

Let us now agree to write $(W,A,\{A_t\},P_W)$ for the P-augmentation of the set-up just defined. Since the new set-up satisfies the usual conditions, every martingale has a cadlag modification (§7). By the principle of Jacod and Yor mentioned in § , we have the following theorem.

(12.1) THEOREM. Let M be a cadlag local martingale relative to $(W,A,\{A_t\},P_W)$. Then there exists a previsible process C with values in $\mathbb{R}^n$ such that

$$M_t - M_0 = \int_{(0,t]} C_i dX^i .$$

In particular, M is (almost surely) continuous. The well-known 0-1 law implies that M_0 is (almost surely) constant on W.

It was a striking proof given by Dellacherie of the fact that Theorem 12.1 follows naturally from Lévy's theorem which motivated the work of Jacod and Yor. Dellacherie and Meyer ([3]) have a nice account of these ideas. Meyer [16] has a direct proof of Theorem 12.1 echoed by Rogers in this volume.

- -

We now use Theorem 12.1 to characterise all laws on (W,A) which are equivalent to P_W. Let Q be a probability measure on (W,A) which is equivalent to (that is: has the same null sets as) P_W. Let M_∞ be the Radon-Nikodym derivative dQ/dP on A. Then, almost by definition,

$$M_t \equiv E[M_\infty | \mathcal{F}_t] = dQ/dP_W \text{ on } A_t .$$

Since Q is equivalent to P_W, $M_t > 0$ $(\forall t)$ with probability 1. Now, by Theorem 12.1,

$$M_t - M_0 = \int C_i dX^i$$

for some previsible C with values in $\mathbb{R}^n$. By the Kunita- Watanabe formula, $d\langle M,M\rangle = \|C\|^2 dt$. Put $V_t \equiv \log M_t$. Then

$$dV_t = M_t^{-1} dM - \tfrac{1}{2} M_t^{-2} \|C\|^2 dt = u_i dX^i - \tfrac{1}{2}\|u(t)\|^2 dt,$$

where $u_i(t) = M_t^{-1} C_i(t)$. Now put

$$B_t^i \equiv X_t^i - \int_{(0,t]} u_i(s)ds$$

Then $d\langle M,B^i\rangle = c_i(t)dt$, and

$$\begin{aligned} d(M_tB_t^i) &= M_t dB^i + B_t^i dM + d\langle M,B^i\rangle \\ &= M_t(dX^i - u_i(t)dt) + B_t^i dM + c_i(t)dt \\ &= M_t dX^i + B_t^i dM, \end{aligned}$$

so that $M_tB_t^i$ is an $(\{A_t\},P_W)$ local martingale. Thus, (almost by definition) B_t^i is an $(\{A_t\},Q)$ local martingale. Now because of (7.7), the value of $\langle B^i,B^j\rangle$ for the Q set-up is the same as its value in the P_W set-up, and since B^i differs from X^i only by a finite-variation part, we have $\langle B^i,B^j\rangle_t = \langle X^i,X^j\rangle_t = t\delta^{ij}$. We have proved the following result.

(12.2) THEOREM (Girsanov, Cameron-Martin, ...). Let Q be a law on (W,A) equivalent to Wiener measure P_W. Then there exists a previsible process u and an $(\{A_t\},Q)$ Brownian motion B such that

$$dX_t^i = dB_t^i + u_i(t)dt, \tag{12.3}$$

$$\frac{dQ}{dP_W} = \exp\left[\int_0^t u_i dX^i - \frac{1}{2}\int_0^t \|u\|^2 ds\right] \quad \text{on } (W,A_t). \tag{12.4}$$

Conversely, if u is such that the right-hand side of (12.4) has P_W expectation 1 for all t (as will happen if for example, u is bounded), then we can define Q via (12.4) and thus produce the weak solution (X,B) of (12.3) on the set-up $(W,A,\{A_t\},Q)$.

Notes.

1. The value of $u_i(t)$ will generally depend on the behaviour of X throughout the interval $[0,t]$, but (12.3) is even so a perfectly good stochastic differential equation. The solution X will normally not be Markovian under Q.

2. It is not obvious that the P_W expectation of the right-hand side of (12.4) is equal to 1 when u is bounded. I must skip the proof due to shortage of space. See §6.2.2 of Liptser and Shiryayev [11].

- -

We shall see in §15 that we can use Theorem 12.2 to put Theorem 12.1 in a 'practicable' explicit form for an important class of martingales M.

§13. The TSIREL'SON example; and an open problem. Let $n = 1$, so that $W \equiv C([0,\infty);\mathbb{R})$, etc.. Let $t_0, t_{-1}, t_{-2}, \ldots$ be numbers such that

$$t_0 > t_{-1} > t_{-2} > \ldots, \qquad \lim t_{-n} = 0.$$

With $\{\cdot\}$ denoting "fractional part of", define for $w \in W$:

$$u_t(w) \equiv \left\{\frac{w(t_i) - w(t_{i-1})}{t_i - t_{-1}}\right\} \text{ for } t \in (t_{i-1}, t_i],$$

$$\equiv 0 \text{ for } t = 0 \text{ and for } t > t_0.$$

Then u is a left-continuous $\{A_t\}$ adapted process, and hence is previsible. Moreover, u is bounded. Hence, since also $u = 0$ on $[t_0,\infty)$, we may define a law Q on (W,A) by the formula:

$$\frac{dQ}{dP} = \exp\left[\int_0^\infty u\,dB - \frac{1}{2}\int_0^\infty u^2\,dt\right].$$

Then the equation:

$$B_t \equiv X_t - \int_0^t u(s)\,ds$$

defines an $(\{A_t\},Q)$ Brownian motion B. Thus the pair (X,B) on the set-up $(W,A,\{A_t\},Q)$ is a weak solution of the equation:

(TSIR) $$\boxed{dX_t = dB_t + u(X_{[0,t]})dt, \quad X_0 = B_0 = 0.}$$

The significance of the notation $u(X_{[0,t]})$ should be clear.

Tsirel'son proved that (TSIR) has no strong solution. Yor made this much more dramatic and mind-boggling by showing that if (X,B) is any weak solution of (TSIR), then for each $t > 0$, the variable $u(X_{[0,t]})$ is independent of the process B. So much then for any hope of constructing B from X !! (For Yor's proof and for many interesting things, see Stroock and Yor [24]).

- -

Open problem. Let X be the canonical Brownian motion of this section, and let L be its local time at 0. Does $X + L$ generate the same (P_W augmented) filtration as X? This has been driving many of us crazy. Please let me know if you solve it. (I believe that the problem originates from Berkeley.)

Last-minute addendum. I understand that Philip Protter and Burgess Davis have now made substantial progress with this problem.

PART III. Prelude to the Malliavin calculus.

§14. Part III is not an account of the Malliavin calculus. Rather, it consists of a few words of encouragement for anyone who might think that the subject looks technical and who is hesitant to read what the experts have to say.

At the symposium, Dynkin's advice "Discuss the simplest case" was often heard to thunder (in the most friendly way) around the main lecture-room. But in the case of Stroock's lectures, Dynkin was pre-empted. For, in Example 2.10 of Stroock's paper, we find the illuminating "ridiculous exercise" of showing how the Malliavin-Stroock calculus can establish the following result: if B is a Brownian motion on $\mathbb{R}$ starting at 0, then B_1 has smooth density! In §15, I shall explain how this result may be proved by the Malliavin-Bismut calculus.

Let me remind you that each of the various versions of the Malliavin calculus rests on a calculus of variations technique: we perturb the Brownian path $\{B(t)\}$, and look at the derivative of (possibly conditional) expectations of Brownian functionals along the perturbation. Malliavin ([12]), and Stroock in [22], perturb $\{B(t)\}$ by introducing a second time-parameter τ so that $\{B_\tau(t)\}$ is a Gaussian sheet with each $B_\tau(\cdot)$ a Brownian motion. Bismut perturbs B by adding in a drift term and exploiting the Girsanov theorem. In his paper in this volume, Stroock axiomatises the Malliavin variational procedure. I shall say just a few words about each approach.

§15. Calculus of variations via Girsanov's theorem (BISMUT). Let X on $(W,\{\mathcal{A}_t\},P_W)$ be canonical Brownian motion on $\mathbb{R}$ starting at 0. Let u be a bounded previsible process, and, for $\varepsilon \in \mathbb{R}$, define Q^ε on $(W,\mathcal{A})$ via:

$$\frac{dQ^\varepsilon}{dP_W} = \eta_t^\varepsilon \equiv \exp[\varepsilon\int_0^t u\,d_I X - \tfrac{1}{2}\varepsilon^2\int_0^t u^2 ds] \quad \text{on} \quad (W,\mathcal{A}_t).$$

Put $B_t^\varepsilon \equiv X_t - \varepsilon\int_0^t u\,ds$. Then B^ε is a Brownian motion under Q^ε. Now let $h \in C_K^\infty$. Then, with the functional notation $Q^\varepsilon[\cdot]$ for Q-expectation and with E for P_W expectation:

(15.1) $$E[h(X_1)] = Q^\varepsilon[h(B_1^\varepsilon)] = E[\eta_1^\varepsilon h(X_1 - \varepsilon\int_0^1 uds)].$$

Differentiate with respect to ε and put $\varepsilon = 0$ to get:

(15.2) $$0 = E[h(X_1)\int_0^1 ud_I X] - E[h'(X_1)\int_0^1 uds].$$

In particular, on taking $u_t = 1, \forall t$, we find that

$$E[h'(X_1)] = E[X_1 h(X_1)],$$

so that, by induction, for $\alpha = 1,2,3,\ldots,$

$$E[h^{(\alpha)}(X_1)] = E[\Pi_\alpha(X_1)h(X_1)],$$

where Π_α is the α-th Hermite polynomial, and

(15.3) $$|E[h^{(\alpha)}(X_1)]| \le \|h\|_\infty E[|\Pi_\alpha(X_1)|] = \text{constant}_\alpha \|h\|_\infty,$$

where $\|h\|_\infty \equiv \sup|h(y)|$ as usual. Hence, the map $h \mapsto E[h^{(\alpha)}(X_1)]$ on C_K^∞ extends to a bounded linear map on the Banach space C_0 of continuous functions on $\mathbb{R}$ which vanish at ∞. By the Riesz theorem,

$$\int h^{(\alpha)}(y)P_W[X_1 \in dy] = E[h^{(\alpha)}(X_1)] = \int h(y)\nu_\alpha(dy)$$

for some signed measure ν_α. Integrate by parts α times on the right-hand side to show that X_1 has a density which is an α-fold integral and hence of class $C^{\alpha-2}$. But α is arbitrary, and our ridiculous exercise of proving that X_1 has smooth density is complete.

The very beginning of Stroock's paper explains a neat Fourier way of proceeding from (15.3) (for $h \in C_b^\infty$).

§16 (Continuation of §15). Martingale representation made explicit. I digress for a moment - though it is not really a digression - to mention the simplest case of Bismut's treatment of martingale representation. (And, after all, there is more to probability theory than proving hypoellipticity!)

It is convenient to pretend in this section that X, u and B^ε are defined only on the parameter-set $[0,1]$.

We can apply the method of §15 not just to Brownian functionals of the form $h(X_1)$ but also to certain functionals $g(X)$ which depend on the whole path $\{X_t : t \in [0,1]\}$. Let g be a strongly differentiable function on $C[0,1]$: thus, for fixed y in $C[0,1]$, there is a bounded linear functional dg^y from $C[0,1]$ to $\mathbb{R}$ such that

$$g(y+z) - g(y) = dg^y(z) + o(\|z\|).$$

By the Riesz theorem,

$$(16.1) \qquad dg^y(z) = \int_0^1 z(t)\mu_g^y(dt)$$

for some signed measure μ_g^y on $[0,1]$ of finite total variation $\|\mu_g^y\| = \|dg^y\|$. We assume that g is uniformly Lipschitz in that $\sup_y \|dg^y\| < \infty$. Then

$$E[g(X)] = Q^\varepsilon[g(B^\varepsilon)] = E[\eta_1^\varepsilon g(X_\cdot - \varepsilon\int_0^\cdot u ds)].$$

Differentiate with respect to ε, and put $\varepsilon = 0$ to get:

$$0 = E[g(X)\int_0^1 u d_I X] - E[dg^X(\int_0^\cdot u ds)],$$

so that

$$(16.2) \qquad E[g(X)\int_0^1 u d_I X] = E[\int_0^1 \mu_g^X(dt)\int_0^t u ds] = E[\int_0^1 u_s \mu_g^X(s,1] ds].$$

Now we know from Theorem 12.1 that

$$M_t \equiv E[g(X) | A_t] = E[g(X)] + \int_0^t C d_I X$$

for some previsible process C, and that (Kunita-Watanabe)

$$(16.3) \qquad E[g(X)\int_0^1 u d_I X] = E[\int_0^1 u_s C_s ds].$$

It is a standard fact (Dellacherie-Meyer [3]) that there exists a previsible process C^* called the previsible projection of the process $s \to \mu_g^X(s,1]$ such that (in particular) for each s, the equation

$$C_s^* = E[\mu_g^X(s,1] | A_{s^-}]$$

holds with probability 1. On comparing (16.2) with (16.3), and remembering that, of course,

$$E\{u_s\mu_g^X(s,1]\} = E\{E[u_s\mu_g^X(s,1]|\mathcal{A}_s\} = E[u_s C_s^*],$$

we see that (almost surely) $C_t = C_t^*$ for almost all t. In other words, we can take $C = C^*$ in (16.3). Thus we have proved the important explicit martingale representation due to Clark:

$$(16.4) \qquad E[g(X)|\mathcal{A}_t] = E[g(X)] + \int_0^t E[\mu_g^X(s,1]|\mathcal{A}_s]d_I X$$

where μ_g^X is as at (16.1), and we take a previsible version of $s \to E[\mu_g^X(s,1]|\mathcal{A}_s]$.

Now, of course, there is an obvious 'formula' for finding C in the representation:

$$M_t = M_0 + \int_0^t C d_I X,$$

namely: $C_t = d\langle M,X\rangle_t/dt$. In practice, this is not as usefulas (16.4).

- -

Bismut's paper contains a proof of Haussman's deep martingale representation for

$$E[g(X)|\mathcal{F}_t]$$

where X is the (pathwise-unique strong) solution of

$$(I) \qquad dX^i = \sigma_q^i(X_t)d_I B^q + b^i(X_t)dt, \quad X_0 = x,$$

where B is an r-dimensional Brownian motion on some set-up $(\Omega,\{\mathcal{F}_t\},P)$, and σ and b are smooth. This splendid application of the 'Girsanov' variational calculus invokes other important ideas one of which I mention in §22.

Forgetting for the moment about hypoellipticity, we switch attention to ...

§17. Variation along a Gaussian sheet (MALLIAVIN, STROOCK; I understand that IKEDA, SHIGEKAWA and WATANABE have also done fundamental work on this topic).

In trying to understand Malliavin [12] and Stroock [22], several people, including Walsh and myself, have independently found it helpful to interpret things in terms of the familiar Brownian sheet (the definition of which is recalled below). At the present time, the Brownian-sheet approach, while it provides a very clear intuitive picture of the beginnings of the Malliavin calculus, runs into technical difficulties once it gets underway. Indeed, these difficulties are what motivated Stroock to adopt the axiomatic approach, and Bismut the 'Girsanov' approach. But in case it helps someone, I describe briefly the Brownian sheet picture. I do not rule out the possibility that, perhaps through the use of two-parameter techniques and new martingale inequalities, the Brownian-sheet picture may one day yield simpler proofs of hypoellipticity results. What is certainly true is that it raises interesting problems in other directions.

- -

The Brownian sheet. Let Q be the Gaussian white-noise measure on the quadrant $[0,\infty)^2$. Thus, for a Borel subset Λ of $[0,\infty)^2$, $Q(\Lambda)$ has the normal distribution $N(0,|\Lambda|)$ of mean 0 and variance $|\Lambda|$, the area of Λ; and, for disjoint Borel subsets $\Lambda_1,\Lambda_2,\ldots$ of $[0,\infty)^2$, the variables $Q(\Lambda_1),Q(\Lambda_2),\ldots$ are independent. Define :

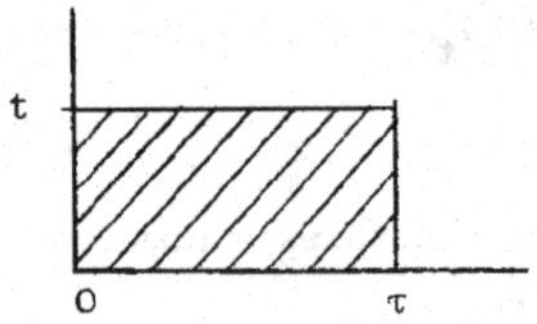

$$S^0_\tau(t) \equiv Q([0,\tau]\times[0,t])$$

(In our pictures, Greek time goes → and Roman time goes ↑.) By Kolmogorov's criterion, we can choose a ('jointly') continuous modification $\{S_\tau(t):(\tau,t)\in[0,\infty)^2\}$ of $\{S^0_\tau(t):(\tau,t)\in[0,\infty)^2$. Then $S=\{S_\tau(t)\}$ is Chentsov's Brownian sheet.

Malliavin's Markov process V on $W \equiv C([0,\infty);\mathbb{R})$. We now set :

$$V_\tau(t) \equiv e^{-\frac{1}{2}\tau}S_{e^\tau}(t), \qquad (\tau\in\mathbb{R},\ t\geq 0).$$

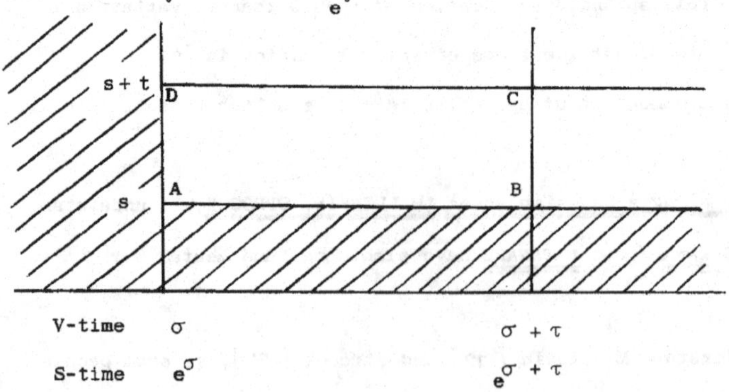

Figure 1

With a look at Figure 1, note that

$$V_{\sigma+\tau}(s+t) - V_{\sigma+\tau}(s) - e^{-\frac{1}{2}\tau}[V_\sigma(s+t) - V_\sigma(s)]$$

$$= e^{-\frac{1}{2}(\sigma+\tau)}\{S_{\sigma+\tau}(s+t) - S_{\sigma+\tau}(s) - S_\sigma(s+t) + S_\sigma(s)\}$$

$$= e^{-\frac{1}{2}(\sigma+\tau)}Q(\text{rectangle ABCD in S-time}).$$

Using the independence properties of Q and the fact that V is Gaussian, we see that

(17.1) for fixed σ in $\mathbb{R}$ and $s \geq 0$, the process

$$\{V_{\sigma+\tau}(s+t) - V_{\sigma+\tau}(s) - e^{-\frac{1}{2}\tau}[V_\sigma(s+t) - V_\sigma(s)]: \tau \geq 0,\ t \geq 0\}$$

is independent of the V-values in the shaded region in Figure 1, and is identical in law to the process

$$\{e^{-\frac{1}{2}\tau}[S_{e^\tau}(t) - S_1(t)]: \tau \geq 0,\ t \geq 0\}.$$

Provided we do not worry about technicalities, everything follows from (17.1).

Write $W \equiv C([0,\infty);\mathbb{R})$, and with $w(t)$ thought of as the map $w \to w(t)$, set $\mathcal{A}_t \equiv \sigma\{w(s): s \leq t\}$, $\mathcal{A} \equiv \sigma\{w(s): s \geq 0\}$. Write V_τ for the element $t \to V_\tau(t)$ of W. Thus we can regard $\{V_\tau: \tau \in \mathbb{R}\}$ as a W-valued process. Moreover, if we put on W the usual topology of uniform convergence on compacts, then $\{V_\tau: \tau \in \mathbb{R}\}$ is a continuous W-valued process.

It is (now) immediate from (17.1) that

(17.2) $\{V_\tau: \tau \in \mathbb{R}\}$ is a continuous time-homogeneous Markov process on W and that the transition semigroup $\{T_\tau: \tau \geq 0\}$ of $\{V_\tau\}$ is given by the fact that for bounded $\mathcal{A}$ measurable f on W,

$$T_\tau f(w) = E[f(V_{\sigma+\tau})|V_\sigma = w;\ V_\rho : \rho \leq \sigma]$$

$$= E[f\{e^{-\frac{1}{2}\tau}[w(\cdot) + S_{e^\tau}(\cdot) - S_1(\cdot)]\}]$$

$$= E[f\{e^{-\frac{1}{2}\tau}w(\cdot) + (1 - e^{-\tau})^{\frac{1}{2}}B(\cdot)\}]$$

where B is a Brownian motion starting at 0.

The process $V = \{V_\tau\}$ is the fundamental process which Malliavin [12] and Stroock [22] construct via different methods. Malliavin calls it an "Ornstein-Uhlenbeck process" for reasons which will become apparent here and in Stroock's paper in this volume. For the standpoint of McKean's paper [14] the correct way to think of V is as Brownian motion on the 'sphere' W.

The operator $\mathcal{L}$. For each fixed τ, the distribution of V_τ is just Wiener measure P_W. Moreover, the process $\{V_\tau\}$ has exactly the same law as $\{V_{-\tau}\}$. In other words,

(17.3) the process V is time-reversible with stationary invariant measure P_W. Hence the semigroup $\{T_\tau : \tau \geq 0\}$ has a unique self-adjoint extension on $L^2(W,\mathcal{A},P_W)$.

We now define:

(17.4) $\mathcal{L}$ is the infinitesimal generator of the semigroup T_τ on $L^2(W,\mathcal{A},P_W)$.

Then $\mathcal{L}$ is self-adjoint (and closed) with dense domain. The critical importance of the self-adjointness of $\mathcal{L}$ for 'integration by parts' is very clearly explained by Stroock. (Of course, it is not obvious that $\mathcal{L}$ as defined in (17.4) - that is, as in Stroock [22] - is the same as that defined in Stroock's paper in this volume, but we shall shortly discover why it is.)

Note. The final expression in (17.2) is reminiscent of Neveu's marvellous proof of Nelson's hypercontractivity theorem (Neveu [19]). Indeed, $\{T_\tau\}$ has the hypercontractive properties discovered by Nelson [18].

Martingales in 'Greek' time. This part of the discussion should investigate problems concerning the nature of the domain of $\mathcal{L}$. Now Stroock has thought all these through thoroughly, both in [22], where for certain purposes he looks at extensions of $\mathcal{L}$, and here, as his very careful axiomatisation reveals. In order to indicate the probabilistic counterparts of some of Stroock's axioms, I employ the usual cheap Williams device for the avoidance of real mathematics: the use of the word 'nice'.

For a nice element f of $\mathcal{D}(\mathcal{L})$, the equation:

(17.5) $$C_\tau^f \equiv f(V_\tau) - f(V_0) - \int_0^\tau \mathcal{L}f(V_\sigma)d\sigma \qquad (\tau \geq 0)$$

defines a continuous martingale C^f. The relevant filtration here is of course the augmentation $\{\mathcal{F}_\tau:\tau \geq 0\}$ relative to the underlying probability measure of the filtration $\{\mathcal{F}_\tau^0:\tau \geq 0\}$, where

$$\mathcal{F}_\tau^0 \equiv \sigma\{V_\sigma:\sigma \leq \tau\} = \sigma\{V_\sigma(t):\sigma \leq \tau;\ t \geq 0\}.$$

<u>Note</u>. For <u>every</u> f in $\mathcal{D}(\mathcal{L})$, (17.5) defines a martingale C^f which has a <u>continuous</u> modification. There are several ways of proving this. See Stroock [22] for one of them. □

For nice f and g in $\mathcal{D}(\mathcal{L})$, let $\{\langle C^f,C^g\rangle_\tau\}$ be the unique continuous $\{\mathcal{F}_\tau\}$ adapted process such that

$$\{C_\tau^f C_\tau^g - \langle C^f,C^g\rangle_\tau : \tau \geq 0\}$$

is an $\{\mathcal{F}_\tau\}$ martingale null at 0. The argument leading to (11.3) shows that for nice f and g in $\mathcal{D}(\mathcal{L})$ for which fg is also in $\mathcal{D}(\mathcal{L})$,

(17.6) $$d\langle C^f,C^g\rangle_\tau = \langle f,g\rangle(V_\tau)d\tau ,$$

where $\langle f,g\rangle$ is the function on W defined by the equation:

(17.7) $$\langle f,g\rangle \equiv \mathcal{L}(fg) - f(\mathcal{L}g) - g(\mathcal{L}f).$$

For nice f in $\mathcal{D}(\mathcal{L})$ and a nice function φ in $\mathbb{R}$, Itô's formula produces from (17.5) and (17.6) the following canonical decomposition (written in differential form) of the semimartingale $\varphi of(V_\tau) = \varphi(f(V_\tau))$:

$$d[\varphi of(V_\tau)] = \varphi' of(V_\tau)dC^f + [\varphi'(f)\mathcal{L}f + \tfrac{1}{2}\varphi''(f)\langle f,f\rangle] o V_\tau d\tau.$$

But we will also have (assuming that $\varphi of \in \mathcal{D}(\mathcal{L})$):

$$d[\varphi of(V_\tau)] = dC^{\varphi of} + \mathcal{L}(\varphi of) o V_\tau d\tau.$$

Hence, 'modulo null sets of some kind',

(17.8) $$\mathcal{L}(\varphi of) = \varphi'(f)\mathcal{L}f + \tfrac{1}{2}\varphi''(f)\langle f,f\rangle,$$

(17.9) $$\langle \varphi of,f\rangle = \varphi'(f)\langle f,f\rangle.$$

<u>Action of $\mathcal{L}$ on stochastic integrals.</u> Let Φ and Ψ be (say) bounded functions on W such that

$$\Phi \in A_s \equiv \sigma\{w(r): r \le s\}, \qquad \Psi \in \sigma\{w(s+t) - w(s): t > 0\}.$$

(The notation $\Phi \in A_s$ means: Φ is A_s measurable.) Then the property (17.1) implies that for $\sigma \in \mathbb{R}$, $\tau > 0$,

(17.10) $\Phi(V_{\sigma+\tau})$ <u>and</u> $\Psi(V_{\sigma+\tau})$ <u>are conditionally independent given</u> $\{V_\rho : \rho \le \sigma\}$.

Hence

$$T_\tau(\Phi\Psi) = T_\tau(\Phi)T_\tau(\Psi). \tag{17.11}$$

In particular, if $\alpha(s)$ is bounded A_s measurable on W, and $h > 0$, then

$$T_\tau[\alpha(s)\{w(s+h) - w(s)\}] = e^{-\frac{1}{2}\tau}T_\tau[\alpha(s)]\{w(s+h) - w(s)\},$$

because it is obvious from (17.2) that $T_\tau[w(t)] = e^{-\frac{1}{2}\tau}w(t)$. It is easy to believe that for a nice $\{A_t\}$ previsible process $\{\alpha(t)\}$, we shall have

$$T_\tau\int_0^t \alpha(s)d_I w = e^{-\frac{1}{2}\tau}\int_0^t T_\tau[\alpha(s)]d_I w,$$

so that on differentiating with respect to τ and putting $\tau = 0$:

$$\boxed{\mathcal{L}\int_0^t \alpha(s)d_I w = \int_0^t \{\mathcal{L}[\alpha(s)] - \tfrac{1}{2}\alpha(s)\}\, d_I w .} \tag{17.12}$$

It is convenient to remember (17.12), a centrally important result of Malliavin and Stroock, as stating that

$$[\mathcal{S}, \mathcal{L}] = \tfrac{1}{2}\mathcal{S} \tag{17.13}$$

where $\mathcal{S}$ is the 'operator of stochastic integration'. The connections with the number operator on Fock space (see Stroock's paper) are immediately suggested by (17.13).

<u>§18.</u> Suppose that a 1-dimensional process X solves:

$$dX = \sigma(X_t)d_I B + b(X_t)dt, \qquad X_0 = x, \tag{18.1}$$

where B is a 1-dimensional Brownian motion, and σ and b are smooth functions on $\mathbb{R}$ with σ never zero. By the argument in §15, we can prove that, for fixed t, the distribution of X_t has a density relative to Lebesgue measure provided we can establish an estimate of the type: for $\varphi \in C_K^\infty$,

(18.2) $$E[|\varphi'(X)|] \leq \text{constant}. \|\varphi\|_\infty.$$

One trivial pointer to this type of result in Bismut's theory was given in §15, and we find another by noting that in the notation of (16.1),

$$\mu^y_{\varphi \circ g} = \varphi'(g(y))\mu^y_g.$$

In Stroock's approach, it is equation (17.9) which is the key to obtaining results such as (18.2). Of course, the techniques of the Malliavin calculus become applicable to equation (18.1) only because $\mathcal{L}$ acts on stochastic integrals in the simple direct way described in equation (17.12)... .

...But I promised to do my best not to poach. I have explained in a rough way some of the probabilistic motivation for Stroock's axioms. And surely, it is clear that careful axiomatisation allows for effective treatment both of the many technical details which I completely ignored in §17, and of the theory proper when the ideas of §17 are applied. Stroock's paper is very clearly written, and requires no background knowledge other than that of classical Itô theory.

My earlier remark that Bismut's paper contains a proof of the full Hörmander theorem, together with other comments I have made, constitute advertisement enough.

Note added in proof. By the use of some very deep techniques, Professor Meyer has just obtained splendid results which greatly clarify some of the foundations of the Malliavin calculus. There is every possibility that this work will also provide some most remarkable - indeed rather wonderful - new insights into the H^1 and BMO spaces for Brownian motion.

Part IV.

§19. A striking feature of modern probability theory, well illustrated at this symposium (and not least by Bismut's paper), is its rapidly increasing involvement with differential geometry. Anyone who wishes to work with diffusions (even on $\mathbb{R}^n$) will have to become acquainted with basic differential geometry. As someone who knows almost nothing about differential geometry, I can express the deep gratitude which very many will feel, that Meyer (in a paper here and in an extended treatment [17]) and Pinsky (in his paper here) have presented important developments in a way which uses probability theory to make differential geometry much more accessible. (I feel an even deeper sense of gratitude that these authors thereby spare me any need to attempt an expository task for this 'Introduction' which would have been impossible for me.) In particular, both Meyer and Pinsky (in different ways) provide very clear understanding of the concept of a Brownian motion on a Riemannian manifold M, and hence of the Laplacian (Laplace-Beltrami operator) on M.

PINSKY defines on M a process which is the exact analogue of Lord Rayleigh's random flight, and, via the usual scaling (properly formulated) obtains Brownian motion as a weak limit. His method provides a simple analytic way of thinking about the Laplace operator. Let us take M to be $\mathbb{R}^n$ for the moment. Let $x \in \mathbb{R}^n$, and let ξ be a unit ('tangent') vector at x. Let $\gamma_{x,\xi}: \mathbb{R} \to \mathbb{R}^n$ be the unique geodesic in $\mathbb{R}^n$ with $\gamma_{x,\xi}(0) = x$, $\gamma'_{x,\xi}(0) = \xi$. Of course, since we are working with $\mathbb{R}^n$, $\gamma_{x,\xi}(t) = x+t\xi$. Let Δ_ξ be "the second derivative at x in the direction ξ":

$$\Delta_\xi f(x) \equiv \frac{d^2}{dt^2} f(\gamma_{x,\xi}(t))\Big|_{t=0} .$$

Now let $\Delta f(x)$ be the average value of $\Delta_\xi f(x)$ as ξ ranges over the unit sphere in the tangent space at x. You can immediately check that $\Delta = n^{-1}\Sigma\partial_i^2$, so that modulo a constant factor which is a matter of convention, Δ is the Laplacian. (There are obvious interrelations with theorems of Gauss and Green.)

Now, on a general finite-dimensional Riemannian manifold M, we have the concepts of unit tangent vector and geodesic, and there is a natural 'invariant' probability measure on the unit sphere in the tangent space $T_x(M)$. Hence we can define Δ for M in the way described above.

Pinsky extends the random-flight construction to obtain the Brownian motion on the orthonormal frame bundle, $O(M)$, of M. Let me mention one reason for interest in Brownian motion on $O(M)$. By utilising this concept, one can describe any non-singular diffusion (with smooth coefficients) on any manifold via a single global stochastic differential equation - see §23. This is very satisfying; but let me add that it is not fully clear to me to what extent the messy business of 'patching' can really be avoided - in the working, as opposed to in the final answer. (Of course, there is no patching in Pinsky's approach).

§20. (MEYER). Let M be a (final-dimensional, smooth) manifold. By a semimartingale X with values in M is meant the obvious thing: a process X such that $f(X)$ is a semimartingale for every $f \in C^{\infty}(M)$.

Meyer's paper explains that that 'difficult' concept of a connection on a manifold M is nothing more (and nothing less) than what is needed to define the concept of a local martingale with values in M. For a manifold with connection, Meyer defines the Itô integral of a 1-form along the path of a semimartingale. In analogy with classical theory, the Itô integral of a 1-form along the path of a local martingale is a local martingale.

Meyer explains that the Stratonovich integral (of a 1-form along the path of a semimartingale) may be defined for any M - it does not require a connection - and is a much more natural geometric object. (For classical 'Stratonovich' theory, see below.)

In Meyer's paper here and in his long paper [17], you will find several ideas which will profoundly influence probability theory, all presented with a clarity which would make any further comments from me superfluous.

§21. The STRATONOVICH integral (classical version). For continuous SEMIMARTINGALES X and Y, we define the Stratonovich integral:

$$\int_0^t Y d_S X \equiv \int_0^t Y d_I X + \tfrac{1}{2}(\langle Y,X\rangle_t - \langle Y,X\rangle_0).$$

If f is a C^∞ function (or, more generally, a C^3 function) and X is a continuous semimartingale, then

$$f'(X_t) = f'(X_0) + \int_{(0,t]} f''(X_u) d_I X_u + \tfrac{1}{2}\int_{(0,t]} f'''(X_u) d\langle X,X\rangle_u,$$

and, by the Kunita-Watanabe characterisation,

$$\langle f'(X),X\rangle_t - \langle f'(X),X\rangle_0 = \int_{(0,t]} f''(X_u) d\langle X,X\rangle_u.$$

Hence, Itô's formula for $f(X)$ takes the Stratonovich form:

$$f(X_t) = f(X_0) + \int_{(0,t]} f'(X_u) d_S X.$$

In other words, the Stratonovich calculus obeys the same rules as the 'ordinary' Newton-Leibniz calculus. See Meyer's work for more profound thoughts on this.

You can easily check that our Itô stochastic differential equation:

(I) $$dX^i = \sigma^i_q(X_t) d_I B^q + b^i(X_t) dt$$

(where σ and b are smooth) takes the Stratonovich form:

(S) $$dX^i = \sigma^i_q(X_t) d_S B^q + \beta^i(X_t) dt$$

or, equivalently,

(21.1) $$df(X) = U_q f(X_t) d_S B^q + V f(X_t) dt.$$

We have used the notations of §5:

$$U_q \equiv \sigma^i_q \partial_i, \quad \beta^i \equiv b^i - \tfrac{1}{2} U_q(\sigma^i_q), \quad V = \beta^i \partial_i.$$

You can also easily check that (21.1) translates into the Itô form:

(21.2) $$df(X) = U_q f(X_t) d_I B^q + \mathcal{L} f(X_t) dt,$$

where $\mathcal{L}$ is the generator of X in Hörmander's form:

$$\mathcal{L} = \frac{1}{2}\sum_q U_q^2 + V.$$

From what we have just done, you can sense a fundamental point of Meyer's paper: that the Stratonovich calculus best conveys the geometry, but that in order to interpret drift, and generally to understand the rôle of martingales (for example, in connection with the martingale problem for $\mathcal{L}$), we need an Itô calculus.

§22. The diffeomorphism theorem. Let me mention a result, the diffeomorphism theorem, which plays a key part in the modern theory of diffusions. The first work in this direction was done by GIHMAN and SKOROKHOD, and important contributions have come from Warwick (BAXENDALE, EELLS, ELWORTHY), Paris (BISMUT, MALLIAVIN), and Japan (KUNITA). The precise history is not known to me. Amongst much else of great interest in Kunita's paper in this volume, you will find a nice account of the diffeomorphism theorem expressed clearly in non-technical language. Bismut's paper makes heavy use of the theorem.

I state the 1-dimensional case. The extension of the statement to the n-dimensional case is obvious.

THEOREM. Let B be a Brownian motion on $\mathbb{R}$. Let σ and b be smooth functions on $\mathbb{R}$ with $\sigma(x) > 0$ $(\forall x)$ and such that a diffusion associated with σ and b does not explode. Then we can construct a (two-parameter) process $\{X_t^x : x \in \mathbb{R}, t \geq 0\}$ such that

(i) for each x, the process $\{X_t^x : t \geq 0\}$ solves the Itô equation

(22.1) $$dX^x = \sigma(X_t^x)d_I B + b(X_t^x)dt, \quad X_0^x = x ;$$

(ii) with probability 1, the map $(t,x) \to X_t^x(\omega)$ is continuous;

(iii) with probability 1: for every t, the map $x \to X_t^x(\omega)$ is a diffeomorphism of $\mathbb{R}$ onto $\mathbb{R}$;

(iv) the process $\{Y_t^x : t \geq 0\}$ with $Y_t^x(\omega) \equiv \frac{\partial}{\partial x}X_t^x(\omega)$ satisfies the equation obtained by formally differentiating (22.1) relative to x:

(22.2) $$dY^x = \sigma'(X_t^x)Y_t^x d_I B + b'(X_t^x)Y_t^x dt, \quad Y_0^x = 1.$$

You should check that the Stratonovich form of (22.2) is obtained by formally differentiating the Stratonovich form of (22.1).

23. Lifting of diffusions (ITÔ, DYNKIN, McKEAN, EELLS and ELWORTHY, MALLIAVIN, ...).

In this section we need a little of the terminology of Riemannian geometry. Pinsky's paper contains a clear explanation of 'orthonormal frame bundle', etc; and Meyer's paper also explains clearly some of the concepts needed.

Let M be a smooth n-dimensional manifold. Let $\mathcal{G}$ be a second-order elliptic operator on M which on a local-coordinate neighbourhood may be expressed in the form:

(23.1) $$\mathcal{G} = \frac{1}{2}a^{ij}(x)\partial_i\partial_j + b^i(x)\partial_i ,$$

where each $a(x)$ is symmetric and strictly positive-definite, and where a and b are smooth. (It is clear that $\mathcal{G}$ will take the same form in another coordinate system, so that this description is intrinsic.)

The traditional way of studying the diffusion process on M associated with $\mathcal{G}$ was to set up the Itô equation:

$$dX^i = \sigma^i_q(X_t)d_I B^q + b^i(X_t)dt$$

on a local-coordinate neighbourhood, where $\sigma(x) = a(x)^{\frac{1}{2}}$; and then to patch together the various 'local' diffusions. This approach is non-intrinsic, and is plagued by many technical difficulties.

It is very satisfying that, thanks to the work on 'lifting' by the people acknowledged at the heading of this section, one can describe the diffusion X globally on M via a single stochastic differential equation. But, as I mentioned earlier, the exact extent to which localisation can really be avoided in the 'working' of the subject, is not yet clear to me. The lifting technique is essential, however, in many applications.

The operator $\mathcal{G}$ induces in a natural way a Riemannian metric on M. Indeed, in the local-coordinate system corresponding to (23.1), put $(g_{ij}(x)) = (a^{ij}(x))^{-1}$. Then g is a positive-definite covariant tensor which we can - and do - take as Riemannian metric tensor. (Here we see one of Meyer's themes: the 'second-order' part of $\mathcal{G}$ at (23.1) behaves properly under coordinate changes. The 'first-order' part does not: it is not contravariant, and so does not define a vector field.)

On the manifold M now made Riemannian, $\mathcal{G}$ takes the form:

$$\mathcal{G} = \frac{1}{2}\Delta + e \tag{23.2}$$

where Δ is the Laplace-Beltrami operator on M and e is a vector field on M. Formula (23.2) is clear because (in the usual normalisation)

$$\frac{1}{2}\Delta = \frac{1}{2}g^{ij}(x)(\partial_i\partial_j - \Gamma^k_{ij}(x)\partial_k)$$

where $(g^{ij}(x)) = (g_{ij}(x))^{-1} = (a^{ij}(x))$. (Note that $\mathcal{G} - \frac{1}{2}\Delta$ has no 'pure second-order' part and so <u>does</u> transform in contravariant fashion. Thus e is a bona-fide vector field).

The <u>orthonormal frame bundle</u>, O(M), of M is the set of (n+1)-tuples

$$r = (x,[e_1,e_2,\dots,e_n]), \tag{23.3}$$

where $x \in M$ and $[e_1,e_2,\dots,e_n]$ is an orthonormal basis for the tangent space $T_x(M)$ at x. As usual, we write $\pi r = x$. (Do note that (23.3) is a statement of the kind of object which r is, and does not involve any coordinization.) The bundle O(M) has a natural smooth manifold structure. Note that the dimension of O(M) is $n + \dim O(n) = \frac{1}{2}(n^2+n)$.

Now let $r \in O(M)$, and let $x(t)$ be a smooth curve on M with $x(0) = x = \pi r$. By letting $e_1,e_2,\dots,e_n$ move by parallel displacement (see the Meyer and Pinsky papers) along the curve $x(\cdot)$, we obtain the <u>horizontal lift</u>

$$r(t) = (x(t),[e_1(t),e_2(t),\dots,e_n(t)])$$

of $x(\cdot)$ to O(M). (Recall that parallel displacement preserves inner products.) If v is the tangent to $x(\cdot)$ when $t = 0$, let L_v be the tangent to $r(\cdot)$ when $t = 0$. Then, for each $r \in \pi^{-1}x$, the map $v \to L_v$ is well-defined from $T_x(M)$ to $T_r(O(M))$, and is called the <u>horizontal lift</u>.

Let $(B^1,B^2,\dots,B^n)$ be a Brownian motion on $\mathbb{R}^n$. For $r \in O(M)$, the Stratonovich differential equation:

$$(23.4) \qquad df(R) = L_{e_q} f(R_t) d_S B^q + L_e f(R_t) dt, \quad R_0 = r,$$

thought of as holding simultaneously for all f in $C^{\infty}(O(M))$, defines a unique diffusion (R_t) on $O(M)$. Equation (23.4) is exactly analogous to equation (21.1), and as the argument at (21.2) shows, the infinitesimal generator of (R_t) takes the Hörmander form:

$$\frac{1}{2}\sum_q L_{e_q}^2 + L_e = \frac{1}{2}\Delta_{O(M)} + L_e,$$

where $\Delta_{O(M)}$ is Bochner's horizontal Laplacian on $O(M)$.

Now, $X = \pi R$ is the desired diffusion on M with generator $\mathcal{G}$, and R is the horizontal lift of X. (Parallel displacement along the path of X is described by the usual Levi-Civita equations - with the Stratonovich interpretation.)

APPENDIX : DISTRIBUTIONS AND HYPOELLIPTIC OPERATORS

References: Hörmander [6], Rudin [21].

A1. Let G be a non-empty open subset of $\mathbb{R}^n$. (The case when $G = \mathbb{R}^n$ is our main concern, but we do need the greater generality.)

A multi-index α is an n-tuple $\alpha = (\alpha_1, \alpha_2, \ldots, \alpha_n)$ of non-negative integers. For such an α, we put:

$$|\alpha| \equiv \alpha_1 + \alpha_2 + \ldots + \alpha_n,$$

$$D^\alpha \equiv \partial_1^{\alpha_1} \partial_2^{\alpha_2} \ldots \partial_n^{\alpha_n}, \quad \partial_i \equiv \partial/\partial x^i.$$

The space $\mathscr{D}(G)$ of test functions on G. A test function on G is a (C^∞) smooth complex-valued function φ defined on G such that the support of φ is a compact subset of G. The space $\mathscr{D}(G)$ is the space of test functions on G, equipped with a certain topology τ discussed below. (In our notation of the remainder of the paper, $\mathscr{D}(G)$ is just 'complex' $C_K^\infty(G)$. But we use $\mathscr{D}(G)$ to emphasise the topologisation; and we shall need K to denote particular compact sets.)

Sequential convergence in $\mathscr{D}(G)$. A sequence (φ_i) of elements of $\mathscr{D}(G)$ converges to 0 for the topology τ if

(i) there is some compact subset K of G which contains the support of every φ_i,

(ii) for every multi-index α,

$$\sup_{x \in G} |D^\alpha \varphi_i(x)| \to 0 \qquad (i \to \infty).$$

(Of course, we could have written $\sup_{x \in K}$ instead.)

The space $\mathscr{D}'(G)$ of distributions on G. A distribution Λ on G is a linear map $\Lambda : \mathscr{D}(G) \to \mathbb{C}$ which is continuous in the sense that

if $\varphi_i \to 0$ (for the topology τ), then $\Lambda\varphi_i \to 0$.

Remark. The topology τ imposed on $\mathscr{D}(G)$ by functional analysts is non-metrizable. Even so, the 'sequential' definition of continuity of a linear functional Λ just given is strictly equivalent to the 'official' definition. For our purposes therefore, the precise way in which $\mathfrak{D}(G)$ is topologised is not of particular interest. Rudin [21] has a nice account of the functional analysis.

Examples of distributions.

(i) For $x \in G$, the delta function δ_x, with $\delta_x(\varphi) \equiv \varphi(x)$, is obviously a distribution.

(ii) If f is a smooth function on G (not necessarily of compact support) then the formula:

$$\Lambda_f(\varphi) \equiv \int_G \varphi(x) f(x) dx \qquad (\varphi \in \mathfrak{D}(G))$$

obviously defines a distribution. We normally say that "Λ_f is the smooth function f".

(iii) If μ is a measure with $\mu(K) < \infty$ for each compact subset K of G, then

$$\Lambda_\mu(\varphi) \equiv \int_G \varphi(x) \mu(dx) \qquad (\varphi \in \mathfrak{D}(G))$$

obviously defines a distribution.

Localisation. If H is an open subset of G, then any element of $\mathfrak{D}(H)$ extends to an element of $\mathfrak{D}(G)$ which is zero on $G \backslash H$. Hence $\mathfrak{D}(H)$ forms a subspace of $\mathfrak{D}(G)$. If Λ is a distribution on G, restricting Λ to $\mathfrak{D}(H)$ produces a distribution on H: "Λ on H". In particular, we can now interpret the statement: "on H, Λ is a smooth function".

A2. We use the conventions:

$$h, f \in C^\infty(G); \quad u, \Lambda \in \mathfrak{D}'(G); \quad \varphi \in \mathfrak{D}(G).$$

We shall not employ the summation convention for multi-indices α.

Since

$$\int_G \varphi(x)[h(x)f(x)]dx = \int_G [h(x)\varphi(x)]f(x)dx,$$

we have $\Lambda_{hf}(\varphi) = \Lambda_f(h\varphi)$. But, in order that distributions should be a sensible generalisation of functions, we want $h\Lambda_f = \Lambda_{hf}$; so we put $(h\Lambda_f)(\varphi) = \Lambda_f(h\varphi)$. Generally, we define

$$(h\Lambda)(\varphi) \equiv \Lambda(h\varphi).$$

It is a little tricky to prove that $h\Lambda$ is a distribution. See Rudin's book.

Since

$$\int\varphi(x)(D^\alpha f)(x)dx = (-1)^{|\alpha|}\int(D^\alpha\varphi)(x)f(x)dx,$$

we define

$$(D^\alpha\Lambda)(\varphi) = (-1)^{|\alpha|}\Lambda(D^\alpha\varphi).$$

It <u>is</u> obvious that $D^\alpha\Lambda$ is a distribution.

Let A be a differential operator:

$$A = \sum_\alpha a_\alpha(x)D^\alpha$$

where each a_α is smooth, and a_α is the zero function for all but finitely many α. Then

$$\int(A\varphi)(x)f(x)dx = \int\varphi(x)(A^*f)(x)dx.$$

where

$$A^*f = \sum_\alpha(-1)^{|\alpha|}D^\alpha[a_\alpha(x)f] = \Sigma b_\alpha(x)D^\alpha f$$

for some coefficients $b_\alpha(x)$. Exactly similarly,

$$u(A\varphi) = (A^*u)(\varphi),$$

where

$$A^*u = \Sigma b_\alpha(x)D^\alpha u.$$

<u>A3.</u> <u>Hypoelliptic operators</u>. Let A be a differential operator on G with smooth coefficients. Then

A is called <u>HYPOELLIPTIC</u> if, whenever u is a distribution on G, then u <u>is a smooth function on any open set on which</u> Au <u>is a smooth function</u>.

Here is a trivial example (of obvious significance in the theory of Brownian motion. Consider

$$\left(\lambda - \frac{1}{2}\frac{d^2}{dx^2}\right)u = \delta_0 \text{ on } \mathbb{R}, \text{ where } \lambda > 0.$$

The operator $A = (\lambda - \frac{1}{2}D^2)$ is hypoelliptic by Hörmander's theorem (or by immediate verification). Since δ_0 is a smooth function (zero!) on $\mathbb{R}\setminus\{0\}$, u must be smooth on $\mathbb{R}\setminus\{0\}$, and must not be smooth at 0. Of course, u is the function:

$$u(x) = \gamma^{-1}e^{-\gamma|x|} + c_1e^{\gamma x} + c_2e^{-\gamma x}$$

where $\gamma = (2\lambda)^{\frac{1}{2}}$.

IN CONCLUSION

This article has attempted to introduce one of the themes which aroused much interest at the symposium: the use of 'differential geometry' techniques. These techniques are here to stay. The marvellous paper by Takahashi and Watanabe in this volume shows that deep concrete results - of striking intuitive content and of importance in physics - can be obtained by their use. The paper [4] by De Witt-Morette, Elworthy, Nelson, and Sammelson, is another illustration of the way things are going. Of course, as W.S. Kendall's paper in this volume shows, the traffic between probability theory and differential geometry is by no means all one way.

So, I must learn differential geometry. And if you are in the same predicament, you too can look forward to reading that magnificient book: Abraham and Marsden [1].

- -

Of course, there are very interesting and very important papers in this volume on other themes. To have contrived to mention them all in this article would have been artificial.

- -

Acknowledgements. I must again thank Margaret Brook: this time for her superb typing of this article (and of so much other symposium material).

For their helpful comments on this article, many thanks to John Lewis and, especially, Daniel Stroock. They have seen only parts of the article; and all the responsibility for its 'manifold sins and weaknesses' is mine.

REFERENCES

[1] R. ABRAHAM and J.E. MARSDEN, Foundations of Mechanics, second edition (Benjamin/Cummings), 1978.

[2] C. DELLACHERIE, Un survoi de la théorie de l'intégrale stochastique, Stoch. Proc. Appl. 10, 115-144 (1980).

[3] C. DELLACHERIE and P.-A. MEYER, Probabilités et Potentiel (Hermann): Vol.I, 1975; Vol.II, 1980.

[4] C. DE WITT-MORETTE, K.D. ELWORTHY, B.L. NELSON, and G.S. SAMMELSON, A stochastic scheme for constructing solutions of the Schrödinger equations, Ann. Inst. H. Poincaré, Section A, Vol.XXXII, 327-341, 1980.

[5] L. HÖRMANDER, Hypoelliptic second-order differential equations, Acta Math. 119, 147-171, 1967.

[6] L. HÖRMANDER, Linear Partial Differential Operators (Springer), 1963.

[7] N. IKEDA and S. WATANABE, Stochastic Differential Equations and Diffusion Processes, (Kodansha, Wiley), 1980.

[8] J. JACOD, A general representation theorm for martingales, in Probability (ed. J.L. Doob), Proc. Symp. Pure Math. XXXI, (American Mathematical Society), 1977.

[9] J. JACOD, Calcul Stochastique et Problèmes de Martingales, Springer Lecture Notes in Math. 714, 1979.

[10] J.J. KOHN, Pseudo-differential operators and hypoellipticity, in Partial Differential Equations, Proc. Symp. Pure Math. XXIII (American Mathematical Society), 1973.

[11] R.S. LIPTSER and A.N. SHIRYAYEV, Statistics of Random Processes, I : General Theory (Springer), 1977.

[12] P. MALLIAVIN, Stochastic calculus of variation and hypoelliptic operators, Proc. Intern. Symp. SDE (ed. K. Itô).

[13] H.P. McKEAN, Stochastic Integrals (Academic Press), 1969.

[14] H.P. McKEAN, Geometry of differential space, Ann. Prob. 1, 197-206, 1973.

[15] M. MÉTIVIER and J. PELLAUMAIL, Stochastic Integration (Academic Press) 1980.

[16] P.-A. MEYER, Un cours sur les intégrales stochastiques, Séminaire de Probabilités X, Springer Lecture Notes in Math. 511, 245-400, 1976.

[17] P.-A. MEYER, Geometrie stochastique sans larmes, to appear in Séminaire de Probabilités XV, Springer Lecture Notes in Math.

[18] E. NELSON, The free Markov field, J. Funct. Anal. 12, 211-227 (1973).

[19] J. NEVEU, Sur l'espérance conditionelle par rapport à un mouvement brownien, Ann. Inst. H. Poincaré, Section B, Vol.XII No.2, 105-109, 1976.

[20] O.A. OLEINIK and E.V. RADKEVIČ, Second order equations with nonnegative characteristic form (English translation), Plenum Press, 1973.

[21] W. RUDIN, Functional Analysis (McGraw-Hill), 1973.

[22] D.W. STROOCK, The Malliavin calculus and its application to parbolic differential equations,

[23] D.W. STROOCK and S.R.S. VARADHAN, Multidimensional Diffusion Processes (Springer), 1979.

[24] D.W. STROOCK and M. YOR, On extremal solutions to martingale problems (to appear).

[25] D. WILLIAMS, Review of [23] in Bull. Amer. Math. Soc. (New Series) 2, 493-503, 1980.

[26] T. YAMADA and S. WATANABE, On the uniqueness of solutions to stochastic differential equations: I, II, J. Math. Kyoto Univ. 11, 155-167 and 553-563, 1971.

STOCHASTIC INTEGRALS: BASIC THEORY

by

L.C.G. Rogers

University College of Swansea †

1. Introduction, notation and definitions

The aim of this paper is to provide a brief summary of the construction and fundamental properties of stochastic integrals, leading up to Itô's formula, the most useful single result in the whole theory. To illustrate the scope of these techniques, we apply them to Lévy's characterisation of Brownian motion and the Brownian martingale representation result. With very few exceptions, proofs will be omitted. For these, the reader is referred to the works of Kunita-Watanabe, Meyer, Dellacherie and Meyer, Métivier-Pellaumail,.. .

1.1 A filtered probability space $(\Omega, \mathcal{F}, (\mathcal{F}_t)_{t\geq 0}, P)$ satisfies the usual conditions if

(i) the σ-fields $\mathcal{F}_t$ are increasing; $\mathcal{F}_s \subseteq \mathcal{F}_t$ for $s\leq t$

(ii) the σ-fields $\mathcal{F}_t$ are right continuous; $\mathcal{F}_t = \mathcal{F}_{t+} \equiv \bigcap_{s>t} \mathcal{F}_s$

(iii) each $\mathcal{F}_t$ contains all P-null sets of $\mathcal{F}$.

[A P-null set of $\mathcal{F}$ is a subset $B \subseteq \Omega$ for which $\exists A\in\mathcal{F}$, such that $A \supseteq B$ and $P(A)=0$].

1.2 A process $X \equiv (X_t)_{t\geq 0}$ is said to be

- càdlàg (or RCLL, Skorokhod, corlol, zipfo,...) if all its paths are right continuous with left limits; we write $(X_-)_t \equiv X_{t-}$, $\Delta X \equiv X-X_-$.

- adapted if X_t is $\mathcal{F}_t$-measurable for every $t\geq 0$

† Now at Department of Statistics, University of Warwick.

- increasing if càdlàg adapted, $X_{0-} \equiv 0 \leq X_s \leq X_t$ for $0 \leq s \leq t$.

- of finite variation if $\exists$ increasing X^+, X^- such that $X = X^+ - X^-$

- of integrable variation if of finite variation and $X_\infty^+, X_\infty^- \in L^1$.

- bounded in L^p if $\|X\|_p \equiv \sup_t \|X_t\|_p < \infty$ $(p \geq 1)$

- a submartingale if $X_t \in L^1(\Omega, \mathcal{F}_t, P)$ and $X_t \leq E(X_{t+u} | \mathcal{F}_t)$ a.s. for all $t, u \geq 0$.

- a martingale if X, $-X$ are both submartingales.

An optional (= stopping) time is a r.v. $T:\Omega \to [0,\infty]$ such that $\{T \leq t\} \in \mathcal{F}_t$ for all t. We shall write X^T for the process X stopped at T;

$$(X^T)_t \equiv X_{T \wedge t}.$$

We shall write $[T,S)$ ($\equiv [\![T,S [\![$) for the process

$$I_{\{T(\omega) \leq t < S(\omega)\}},$$

and will use $[T]$ for $[T,T]$.

1.3 Henceforth, we shall assume the usual conditions; this assures us, amongst other things, that every martingale has a càdlàg modification, so we shall also assume that all martingales are càdlàg. We then have the martingale convergence theorem; if $(M_t)_{t \geq 0}$ is a martingale, then

for p>1: $\|M\|_p < \infty \iff M_t \xrightarrow[L^p]{a.s.} M_\infty \iff M_t = E(M_\infty | \mathcal{F}_t),\ M_\infty \in L^p$

for p=1: $\|M\|_1 < \infty \implies M_t \xrightarrow{a.s.} M_\infty$. (The convergence is in L^1 iff M is uniformly integrable).

One further result we shall use is Doob's inequalities; if M is a martingale, $M^* \equiv \sup_t |M_t|$, then for $p>1$,

$$\|M^*\|_p \leq \frac{p}{p-1} \|M\|_p .$$

<u>1.4</u> For our last two preliminary definitions, it is helpful to think of a process X as a mapping from $\Omega\times[0,\infty)$ to $\mathbb{R}$. The real line obviously carries its Borel σ-field, but what σ-field should we put on $\Omega\times[0,\infty)$? There are two natural (and important) possibilities;

The <u>optional</u> σ-field is defined to be

$$\mathcal{O} \equiv \sigma(\{\text{càdlàg adapted processes}\})$$

$$= \sigma(\{[T,\infty);\ T \text{ an optional time}\}),$$

and the <u>previsible</u> (predictable) σ-field is defined to be

$$\mathcal{P} \equiv \sigma(\{\text{left continuous adapted processes}\})$$

$$= \sigma(\{(T,\infty);\ T \text{ an optional time}\} \cup \{I_A[0]\ ;\ A\in\mathcal{F}_0\}).$$

It is obvious from the second characterisations of $\mathcal{O}$ and $\mathcal{P}$ that $\mathcal{P} \subseteq \mathcal{O}$. We call a process X <u>optional</u> (resp. <u>previsible</u>) if it is measurable w.r. to $\mathcal{O}$ (resp. $\mathcal{P}$).

2. <u>Stochastic integrals with respect to L^2 - bounded martingales</u>

In this section, we shall define the stochastic integral of a process C satisfying certain measurability and integrability properties with respect to an L^2 - bounded martingale M. One approach is to try to define a stochastic integral for particularly simple processes C, and to extend the definition. So if the process C is of the form

$$C = \sum_{j\geq 0} Y_j(j2^{-n},(j+1)2^{-n}], \tag{2.1}$$

where $|Y_j|\leq 1$ and Y_j is $\mathcal{F}$-measurable for each j, then only an idiot (or a genius!) would propose a definition of $\int_0^t C_s\, dM_s$ which was other than

(2.2) $$\int_0^t C_s \, dM_s = \sum_{j\geq 0} Y_j \{M(t_n^{j+1}) - M(t_n^j)\},$$

where $t_n^j \equiv j2^{-n} \wedge t$. The idea is now to approximate an arbitrary (bounded, measurable) integrand C by integrands of the form (2.1), and hope that the stochastic integrals (2.2) converge in some sense. But let us first consider an example. Suppose $\{B_t; t\geq 0\}$ is Brownian motion, $B_0=0$ and write

$$\Delta_{jn} \equiv B\big((j+1)2^{-n}\big) - B(j2^{-n}).$$

We define the integrands $C^{(n)}$ by

$$C^{(n)} \equiv \sum_{j\geq 0} Y_{nj} (j2^{-n}, (j+1)2^{-n}],$$

where

$$Y_{nj} = n^{-1} I_{\{\Delta_{jn}>0\}} \, .$$

Thus the integrands $C^{(n)}$ converge uniformly to zero as $n\to\infty$. On the other hand, using (2.2), we learn that

$$\int_0^1 C_s^{(n)} \, dB_s = n^{-1} \sum_{j=0}^{2^n-1} \Delta_{jn}^+ ,$$

and since $\Delta_{1n}^+, \Delta_{2n}^+, \ldots$ are i.i.d. for each n with mean $(2^{n+1}\pi)^{-\frac{1}{2}}$ and variance $2^{-n-1}(1-1/\pi)$ (which follows from the fact that Δ_{jn} are i.i.d. $N(0,2^{-n})$ random variables), simple estimation combined with the first Borel-Cantelli Lemma shows that

$$\int_0^1 C_s^{(n)} \, dB_s \longrightarrow \infty \qquad \text{a.s.!}$$

The reader will see how to construct integrands C of the form (2.1) converging uniformly to zero for which the corresponding stochastic integrals converge a.s. to $-\infty$. In the light of this, only an idiot (or a genius) would continue to trust his intuition! Yet there is a way forward; just as in the construction of Lebesgue measure we must relax our demand for a measure defined on <u>all</u> subsets of $[0,1]$, so here, if we are content to integrate a restricted class of processes, C, we can get something meaningful. The problem arises because we have allowed our integrand to "anticipate" the martingale. Call a process C of the form (2.1) <u>elementary</u> if Y_j is $\mathcal{F}_{j2^{-n}}$ - measurable for all j. If C is elementary, the stochastic integral (2.2) is easily shown to be a martingale and, in the case where M is Brownian motion,

$$E[(\int_0^t C_s \, dM_s)^2] = E\,[\int_0^t C_s^2 \, ds].$$

Thus the mapping of $L^2(\Omega \times \mathbb{R}^+, \mathcal{P}, P\times Leb)$ to $L^2(\Omega, \mathcal{F}, P)$ which takes an elementary integrand to its stochastic integral is an <u>isometry</u>, and it was by extending this isometry that Itô constructed stochastic integrals. Notice that each elementary integrand is left continuous and adapted, and therefore <u>previsible</u>. Accordingly, this method will only construct stochastic integrals for previsible integrands.

The fact that the stochastic integral (2.2) is a martingale for elementary C is the key to the modern (Kunita-Watanabe) development of stochastic integrals, where the stochastic integral is characterised directly. It is this approach which we shall now follow, but, as we shall see, the two techniques agree, confirming our belief that we have defined something "natural"!

<u>2.1</u> Later, we shall need the following definition to extend our stochastic calculus. A càdlàg adapted process $(M_t)_{t\geq 0}$ is a <u>local martingale</u> if $\exists$ optional times $T_n \uparrow \infty$ such that for each n

$$M^{T_n} \text{ is a uniformly integrable martingale.}$$

We need the definition immediately to state the following key result.

Theorem (Doob-Meyer decomposition).

Let Z be a (càdlàg) submartingale. Then there exists a unique previsible increasing process A and local martingale M with $M_0 = A_0 = 0$ a.s. and

$$Z_t = Z_0 + M_t + A_t .$$

Remarks (i) Submartingales "rise on average": the Doob-Meyer decomposition shows that this tendency to rise can be exactly "balanced" in a previsible way.

(ii) The theorem is usually stated for a submartingale of class (D), when both M and A are uniformly integrable.

(iii) We shall henceforth assume all submartingales are càdlàg. There exist some which are not (obviously!), but they are not interesting.

2.2 Now let $\mathcal{M} \equiv \{\text{martingales } M;\ \|M\|_2 < \infty\}$, $\mathcal{M}_0 \equiv \{M \in \mathcal{M}; M_0 = 0\}$. $\mathcal{M}$ is a Hilbert space when equipped with the inner product $(M,N) \mapsto E\, M_\infty N_\infty$. For $M \in \mathcal{M}$, M^2 is a uniformly integrable submartingale so, by the Doob-Meyer decomposition, there exists a unique previsible increasing process $\langle M,M\rangle$, called the angle brackets process of M such that

(i) $\langle M,M\rangle_0 = M_0^2$

(ii) $M_t^2 - \langle M,M\rangle_t$ is a uniformly integrable martingale.

For $M,N \in \mathcal{M}$, we define $\langle M,N\rangle$ by polarisation:

$$\langle M,N\rangle \equiv \tfrac{1}{2}\,\{\langle M+N,M+N\rangle - \langle M,M\rangle - \langle N,N\rangle\} .$$

<u>2.3</u> Define $\mathcal{M}^c \equiv \{M\in\mathcal{M};\ M \text{ is continuous}\}$. It follows from Doob's inequalities that $\mathcal{M}^c$ is a closed subspace of the Hilbert space $\mathcal{M}$ so we may define the subspace $\mathcal{M}^d$ of <u>discontinuous</u> martingales (also called purely discontinuous) by

$$\mathcal{M}^d = (\mathcal{M}^c)^\perp ,$$

whence each $M\in\mathcal{M}$ can be uniquely expressed as

$$M = M^c + M^d$$

for some $M^c\in\mathcal{M}^c$, $M^d\in\mathcal{M}^d$.

If T is any optional time, and $M\in\mathcal{M}^c$, then M^T is still continuous, so $M^T\in\mathcal{M}^c$. Hence the same property holds for $\mathcal{M}^d$; $M\in\mathcal{M}^d \Longrightarrow M^T\in\mathcal{M}^d$. It follows easily that if $M\in\mathcal{M}^c$, $N\in\mathcal{M}^d$, then MN is a martingale. This is a special case of the following important theorem, which is proved by approximating N by martingales of integrable variation for which the result is elementary (but not trivial).

<u>Theorem.</u> <u>If</u> $M\in\mathcal{M}$ <u>and</u> $N\in\mathcal{M}^d$, <u>then</u>

$$M_tN_t - \sum_{s\leq t} \Delta M_s \Delta N_s \ \underline{\text{is a martingale}}.$$

We now come to a crucial definition.

<u>2.4</u> <u>Definition.</u> <u>For</u> $M\in\mathcal{M}$, <u>the square brackets process</u> $[M,M]$ <u>is defined to be</u>

$$[M,M]_t = \langle M^c,M^c\rangle_t + \sum_{s\leq t} \Delta M_s^2 . \tag{2.3}$$

<u>Remarks</u> (i) For $M = M^c+M^d\in\mathcal{M}$,

$$M^2 - [M,M] = (M^c)^2 - \langle M^c,M^c\rangle + 2M^cM^d + (M^d)^2 - \Sigma\Delta M_s^2$$

which is a martingale, by the previous theorem. The angle brackets process $\langle M,M\rangle$ for an arbitrary square integrable martingale is the dual predictable projection of the increasing process $[M,M]$, though we shall not use this.

(ii) $[M,M]$ is an integrable increasing process.

(iii) We define $[M,N]$ by polarisation as before.

Having made these defintions, we are almost ready to define stochastic integrals, but first comes a result whose significance will only be clear once we have seen the definition of stochastic integrals.

Kunita-Watanabe inequalities. For $N,M\in\mathcal{M}$, measurable H,K,

(2.4) $$\int_{[0,\infty)} |H_sK_s|\,|d[M,N]_s| \leq \left(\int_{[0,\infty)} H_s^2\,d[M,M]_s\right)^{\frac{1}{2}} \left(\int_{[0,\infty)} K_s^2\,d[N,N]_s\right)^{\frac{1}{2}}$$

which implies

(2.5) $$E\int_{[0,\infty)} |H_sK_s|\,|d[M,N]|_s \leq \left(E\int_{[0,\infty)} H_s^2\,d[M,M]_s\right)^{\frac{1}{2}} \left(E\int_{[0,\infty)} K_s^2\,d[N,N]_s\right)^{\frac{1}{2}}.$$

For measurable processes H, define $\|H\|_{L^2(M)} \equiv \left(E\int_{[0,\infty)} H_s^2\,d[M,M]_s\right)^{\frac{1}{2}}$, and let

$$L^2(M) \equiv \{\text{previsible processes } H \text{ with } \|H\|_{L^2(M)} < \infty\}.$$

2.5 Definition of stochastic integral.

For $M\in\mathcal{M}$, $H\in L^2(M)$, the stochastic integral of H w.r.t. M is the unique $H.M\in\mathcal{M}$ for which for all $t\geq 0$,

(2.6) $$[H\cdot M,N]_t = H\cdot[M,N]_t \equiv \int_{[0,t]} H_s\,d[M,N]_s \quad \forall N\in\mathcal{M}.$$

Equivalently,

(2.7) $$E(H\cdot M)_\infty N_\infty = E\int_{[0,\infty)} H_s \, d[M,N]_s \qquad \forall N\in\mathcal{M}.$$

Remarks and nice properties. (i) The existence and uniqueness of $H\cdot M$ follows from the Kunita-Watanabe inequality (2.5); indeed,

$$N \longmapsto E\int_{[0,\infty)} H_s \, d[M,N]_s$$

is a continuous linear functional on the Hilbert space $\mathcal{M}$.

(ii) If M is of integrable variation,

$$(H\cdot M)_t = \int_{[0,t]} H_s \, dM_s \quad \text{– a pathwise Stieltjes integral.}$$

(iii) If the integrand H is elementary (2.1), then the stochastic integral is given by the formula (2.2) which we want it to be.

(iv) $(H\cdot M)^c = H\cdot M^c$

(v) $\Delta(H\cdot M) = H\Delta M$

(vi) If $H = [0,T]$, then $H\cdot M = M^T$.

(vii) $H \longmapsto (H\cdot M)_\infty$ is a linear isometry of $L^2(M)$ into $L^2(\Omega,\mathcal{F},P)$.

It was through this last property that Itô originally constructed stochastic integrals.

Though we now know how to construct stochastic integrals with respect to elements of $\mathcal{M}$, the calculus is as yet too clumsy to handle calculation efficiently. For one thing, checking the integrability condition $||H||_{L^2(M)} < \infty$ is irksome in practice, and for another, there are many processes with respect to which we want to form stochastic integrals which are not in $\mathcal{M}$. The next task, therefore, is to extend the stochastic integral.

3. Extension of the stochastic integral.

The first extension is to local martingales. This will not only have the advantage of giving us many more processes with respect to which we can integrate, but it will also allow us to relax almost completely the integrability conditions on the integrand; we shall be able to take any locally bounded previsible process as integrand (a process H is locally bounded if $\exists$ optional $T_n \uparrow \infty$ such that $H\,(0,T_n]$ is bounded).

The result which makes it all work is the following.

3.1 Theorem. Let M be a local martingale. Then $\exists$ optional $T_n \uparrow \infty$, $U^n \in \mathcal{M}_0$, V^n integrable variation martingales for which $V_0^n = 0$ such that

$$M^{T_n} = M_0 + U^n + V^n .$$

Remark. We know how to form stochastic integrals with respect to square-integrable martingales, and with respect to integrable variation martingales (the Stieltjes integral) so this result tells us how to define stochastic integrals, at least locally. There are a lot of things to check; in particular, the decomposition of M^{T_n} is not unique, but we can say

(3.1) $\exists$ unique continuous local martingale M^c such that

$$M_0^c = 0,\ (M^c)^{T_n} = (M^{T_n})^c \text{ for all } n;$$

(3.2) $\exists$ unique previsible increasing process $\langle M^c, M^c\rangle$ such that for all n,

$$\langle M^c, M^c\rangle^{T_n} = \langle (M^{T_n})^c, (M^{T_n})^c\rangle \quad .$$

As in the case of square-integrable martingales, then, we define the increasing optional process

$$[M,M]_t \equiv \langle M^c, M^c\rangle_t + \sum_{0 \leq s \leq t} \Delta M_s^2 \quad .$$

3.2 Definition. Let M be a local martingale, H a locally bounded previsible process. The stochastic integral of H with respect to M is the unique local martingale H·M such that

(3.3) $\quad [H \cdot M, N]_t = H \cdot [M,N]_t \qquad$ for all bounded martingales N.

Nice properties

The nice properties (ii) - (vi) which held for stochastic integrals with respect to square-integrable martingales hold also for stochastic integrals with respect to local martingales. Finally, we extend the definition of stochastic integrals to semimartingales; a semimartingale is a càdlàg adapted process X which can be written as

(3.4) $$X_t = X_0 + M_t + A_t,$$

where M_t is a local martingale and A_t is a process of finite variation, $M_0 = A_0 = 0$.

Notice (i) the decomposition of X is not unique;

(ii) however, the continuous part $X^c \equiv M^c$ is the same for all decompositions, so that the definition

$$[X,X] \equiv \langle X^c, X^c\rangle + \Sigma\, \Delta X_s^2$$

is meaningful;

(iii) if X has continuous paths, then there is a unique decomposition of the form (3.4) of X in which the local martingale M and the finite variation process A are continuous.

Finally, we define a stochastic integral in the greatest generality which we shall require.

3.4 Definition. Let X be a semimartingale, and let H be a locally bounded previsible process. The stochastic integral of H with respect to X is the process

(3.5) $$H\cdot X \equiv H_0X_0 + H\cdot M + H\cdot A.$$

Remarks (i) This definition does not depend on the particular decomposition of X used.

(ii) $(H\cdot X)^c = H\cdot X^c$, $\Delta(H\cdot X) = H\Delta X$, and if H is elementary,

(3.6) $$H = \sum_{j\geq 0} Y_j\, (j2^{-n},(j+1)2^{-n}],$$

then

(3.7) $$(H\cdot X)_t = \sum_{j\geq 0} Y_j\{X(t_n^{j+1}) - X(t_n^j)\} \quad .$$

From this, it is easy to show that if $H^{(n)}$ are elementary processes for which

$$(3.8)\begin{cases} (H^{(n)})_{n\geq 0} \text{ is a Cauchy sequence in } L^{\infty}(\Omega\times\mathbb{R}^{+}), \\ \text{then for each } t\geq 0 \\ (H^{(n)}\cdot X_t)_{n\geq 0} \text{ is Cauchy in probability.} \end{cases}$$

(iii) Not only are stochastic integrals with respect to semimartingales the most general we shall require, they are the most general we can have! More precisely, the following result has been proved by Bichteler and, independently, by Dellacherie - Mokobodski - Métevier - Pellaumail.

Theorem. Let X be a càdlàg adapted process with property (3.8). Then X is a semimartingale.

So if we want to extend the definition of stochastic integral further, we must accept something bizarre, which violates either the "natural" stochastic integral (3.7) of elementary processes, or the weakest imaginable continuity condition (3.8). Luckily, for all practical purposes (as we shall soon see, there are plenty!), the definition 3.4 performs admirably.

(iv) $H\cdot X$ is, of course, a semimartingale.

4. Itô's formula (change of variables formula).

Let $f\colon \mathbb{R}^n \to \mathbb{R}$ be C^2, and suppose $X = (X^1, X^2, \dots, X^n)$ is a vector semimartingale (all its components are semimartingales). Then

$$f(X_t) - f(X_0) = \sum_i \int_{(0,t]} D_i f(X_{s-})\,dX^i_s + \tfrac{1}{2}\sum_i\sum_j \int_0^t D_i D_j f(X_{s-})\,d\langle X^{i^c}, X^{j^c}\rangle_s$$

$$(4.1)\qquad + \sum_{0<s\leq t} \{f(X_s) - f(X_{s-}) - \sum_i D_i f(X_{s-})\Delta X^i_s\}$$

This is the famous Itô's formula.

Remarks (i) The first term on the right of (4.1) is a well defined stochastic integral; the integrand is left continuous adapted, therefore previsible, and is obviously locally bounded.

(ii) Since the second term on the right is continuous, we see the ugly third term on the right for what it is - a correction term to ensure that the jumps match on each side of (4.1). The special case where n = 1, and X is continuous is worthy of note;

$$f(X_t) - f(X_0) = \int_0^t f'(X_s)dX_s + \tfrac{1}{2}\int_0^t f''(X_s)\, d\langle X^c,X^c\rangle_s.$$

(iii) Another special case worthy of note is that where n = 2, $f(x,y) = xy$. We get the integration by parts formula

(4.2)
$$X_t Y_t = \int_{(0,t]} X_{s-}\, dY_s + \int_{(0,t]} Y_{s-}\, dX_s + [X,Y]_t \quad .$$

(iv) It is possible to prove the integration by parts formula directly, and, in fact, this provides a proof of Itô's formula (4.1). Indeed, if

$$\mathcal{A} = \{f\in C^2;\ \text{Itô's formula holds}\},$$

then the integration by parts formula implies, by induction, that all polynomials are in $\mathcal{A}$; but every C^2 function f can be uniformly approximated on compact sets by polynomials whose derivatives uniformly approximate the derivatives of f, and a standard localisation argument implies $\mathcal{A} = C^2$. So Itô's formula is really no more complicated than the integration by parts formula (4.2) which, if one of the semimartingales has no martingale continuous part, is formally exactly what one would get for (deterministic) functions of finite variation!

5. Lévy's characterisation of Brownian motion and martingale representation

As an illustration of the power of Itô's formula, we prove the following well-known result.

Theorem. Let $(\Omega, \mathcal{F}, (\mathcal{F}_t)_{t\geq 0}, P)$ satisfy the usual conditions, and suppose that $B \equiv (B_t)_{t\geq 0}$ is a continuous adapted process such that

(a) B_t is a martingale

(b) $B_t^2 - t$ is a martingale

Then B is Brownian motion.

If $(\mathcal{G}_t)_{t\geq 0}$ is the usual augmentation of the filtration generated by B, and if $X_\infty \in L^2(\mathcal{G}_\infty)$, then $\exists\, H \in L^2(B) \equiv \{(\mathcal{G}_t)\text{-previsible processes } H \text{ s.t. } E\int_0^\infty H_s^2 ds < \infty\}$ s.t.

$$E(X_\infty | \mathcal{G}_t) \equiv X_t = E(X_\infty | \mathcal{G}_0) + \int_0^t H_s dB_s \ . \tag{5.1}$$

Proof

Fix $\theta \in \mathbb{R}$, and apply Itô's formula to the function $f(x,y) = \exp(i\theta x + \frac{1}{2}\theta^2 y)$ and the semimartingales B_t, t; if $N_t^\theta \equiv f(B_t, t)$ we have

$$dN_t^\theta = i\theta N_t^\theta\, dB_t + \tfrac{1}{2}\theta^2 N_t^\theta\, dt - \tfrac{1}{2}\theta^2 N_t^\theta\, d\langle B,B\rangle_t$$

$$= i\theta N_t^\theta\, dB_t$$

Thus $\exp(i\theta B_t + \frac{1}{2}\theta^2 t)$ is a local martingale; since it is bounded on $[0,t]$ for each t, it is even a martingale. This proves the first assertion.

Turning to the second, $Z \equiv \{\int_0^\infty H_s\, dB_s\ ;\ H \in L^2(B)\}$ is a closed subspace of $L^2(\mathcal{G}_\infty)$, since it is the isometric image of $L^2(B)$. Suppose $X_\infty \in Z^\perp$; without loss of generality, $X_0 = 0$ since $L^2(\mathcal{G}_0) \subseteq Z^\perp$. Now Z is closed under stopping so BX

is a martingale, and $\langle B,X\rangle = 0$, which implies that for each $\theta\in\mathbb{R}$,

$$X_t\, N_t^\theta \text{ is a martingale.}$$

Reworking the martingale property, we have

$$E\left[X_{t+s}\exp\left(i\theta(B_{t+s}-B_t)\right)\,\big|\mathcal{G}_t\right] = X_s\, e^{-\frac{1}{2}\theta^2 s} \qquad (t,s\geq 0). \tag{5.2}$$

Use (5.2) repeatedly for $0<t_1<t_2<\ldots<t_n$;

$$E\left[X_\infty \exp\left\{i\theta_1 B_{t_1}+\ldots+i\theta_n(B_{t_n}-B_{t_{n-1}})\right)\right] = E\, X_0 \exp\left(-\tfrac{1}{2}\theta_1^2 t_1-\ldots-\tfrac{1}{2}\theta_n^2(t_n-t_{n-1})\right)$$

(5.3)

$$= 0$$

This is enough; if $X_\infty = Y^+ - Y^-$, with $Y^+, Y^-\geq 0$, and if we define $d\mu^+=Y^+dP$, $d\mu^- = Y^-dP$, equation (5.3) implies that μ^+ and μ^- agree on the field generated by $\{B_t;t\geq 0\}$, hence they agree on $\mathcal{G}_\infty$, and $X_\infty = 0$.

This theorem has an enormous range of consequences and applications; we record just two obvious corollaries.

<u>5.1</u> Every continuous martingale is a time change of Brownian motion.

<u>5.2</u> Every $(\mathcal{G}_t)$-martingale is <u>continuous</u>; so $(\mathcal{G}_t)$-optional times are $(\mathcal{G}_t)$-previsible.

This, then, is a very brief description of the basics of stochastic integral theory; now read on....!

STOCHASTIC INTEGRATION AND DISCONTINUOUS MARTINGALES

Robert J. Elliott

0. INTRODUCTION

This article describes the various kinds of stochastic integration, that is integration with respect to processes of finite variation, continuous martingales and random measures. Integration with respect to a process of finite variation is just Lebesgue-Stieltjes integration on each sample path. Integration with respect to a continuous martingale can be defined by Hilbert space arguments. Indeed, a continuous local martingale is a (stopped) Brownian motion after a time change. These ideas are discussed in section 2 below.

New ideas arise when we consider stochastic integration with respect to a discontinuous local martingale. Jumps occur at stopping times,and the two basic kinds of stopping time are the totally inaccessible and predictable stopping times. The projection theorems are stated in section 3, and the dual projections defined. Using the dual predictable projection elementary jump martingales are described. The stochastic integral with respect to a discontinuous local martingale can then be characterized, and this integral is the same as the Stieltjes integral if the discontinuous local martingale is of finite variation.

Finally, by discussing the single jump process, integration with respect to a random measure is illustrated in section 4.

CONVENTIONS:

Predictable ≡ previsible

corlol ≡ cadlag

(continuous on the right, limits on the left.)

1. SEMIMARTINGALES.

(Ω, F, P) is a probability space and $(F_t, t \geq 0)$ is a filtration of F satisfying the USUAL CONDITIONS of Chris Rogers' paper. That paper indicates why the most useful processes to take as integrators are the SEMIMARTINGALES.

DEFINITION 1.1.

A process $X = \{X_t, t \geq 0\}$ is a semimartingale if it is adapted, (X_t is F_t measurable), and it can be written in the form

$$X = X_0 + M + V .$$

Here $M = M_t$, $t \geq 0$, is a local martingale with $M_0 = 0$; V_t is an adapted process with paths of finite variation, and $V_0 = 0$. V is not necessarily predictable.

This decomposition is, in general, not unique. However, the local martingale M does have a unique decomposition

$$M = M^c + M^d,$$

where M^c is a continuous local martingale and M^d is a discontinuous local martingale.

DEFINITION 1.2.

A local martingale L is said to be a discontinuous local martingale if LN^c is a local martingale for every continuous local martingale N^c. (Discontinuous local martingales are called totally discontinuous local martingales in the Strasbourg notes.)

As we shall see, there are discontinuous martingales with paths of finite variation. This is why the semimartingale decomposition is not unique. However, the only continuous martingales of finite variation are constants, so if we further decompose our semimartingale as

$$X = X_0 + M^c + M^d + V$$

then at least M^c is unique in this decomposition.

Recall that the set of processes for which the stochastic integral is initially defined is the set of bounded, predictable processes bP. For $H \in bP$ the stochastic integral $H.X$ is the process

$$H_0 X_0 + H.M + H.V$$
$$= H_0 X_0 + H.M^c + H.M^d + H.V.$$

Let us investigate this definition further. The integral

$$H.V = \int_0^t H_s dV_s$$

is the process obtained by taking the (Lebesgue) - Stieltjes integral along each sample path of V which, fortunately and by definition, is of finite variation. Explicitly

$$(H.V)_t(\omega) = \int_0^t H_s(\omega) dV_s(\omega) \quad .$$

We shall sketch a proof below of the fact that, if M is both a martingale and a process of bounded variation, then the stochastic integral $H.M$ is the same as the Stieltjes integral $\int_0^t H_s\, dM_s$. Therefore, suppose

$$X = X_0 + M^c + N^d + W$$

is another decomposition of the semimartingale X . (The continuous martingale part M^c is unique).

Then $\quad M^d + V = N^d + W ,$

so $\quad M^d - N^d = W - V$

and is both a martingale and a process of bounded variation. By the above remark

$$H.\ (M^d - N^d) = H.\ (W - V)$$

so
$$\begin{aligned} H.X &= H_0X_0 + H.M^c + H.M^d + H.V \\ &= H_0X_0 + H.M^c + H.N^d + H.W \ . \end{aligned}$$

Therefore, the stochastic integral $H.X$ is well defined, and is independent of the decomposition of X .

2. THE INTEGRAL $H.M^c$

Recall that the process $\langle M^c,M^c\rangle$ is the unique predictable increasing process, null at $t = 0$, such that

$$(M^c)^2 - \langle M^c,M^c\rangle$$

is a local martingale.

For any process H one can consider the random variable

$$H^2.\langle M^c,M^c\rangle_\infty = \int_0^\infty H_s^2\, d\langle M^c,M^c\rangle$$

where the integral is interpreted as a Stieltjes integral for each ω .

DEFINITION 2.1.

$L^2(M^c)$ is the set of predictable process $\{H\}$ such that

$$\|H\|_{L^2(M^c)} < \infty ,$$

where

$$\|H\|_{L^2(M^c)} = \|H^2.\ \langle M^c,M^c\rangle\|_2 \ .$$

Recall that M^2 is the space of corlol martingales M such that

$$\|M\|^2 = \sup_t E(|M_t|^2) < \infty ,$$

and that M^2 is identified with the space $L^2(\Omega, F_\infty, P)$ by the map

$$M \to M_\infty = \lim M_t \in L^2 ,$$

with inverse $M_\infty \to M$, where

$$M_t = E[M_\infty | F_t].$$

Each $H \in L^2(M^c)$ gives a linear functional ϕ_H on the space M^2 by putting

$$\phi_H(N) = E[H.\langle M^c, N\rangle]$$
$$= E[\int_0^\infty H_s d\langle M^c, N\rangle_s] ,$$

for $N \in M^2$. By a generalization of the Schwarz (or Hölder) inequality due to Kunita and Watanabe

$$|\phi_H(N)| \leq \|H\|_{L^2(M^2)} \|N\| .$$

Therefore, ϕ_H is a continuous linear functional on L^2, and so is given by the inner product with respect to some element K_∞ of L^2. The corlol version of the corresponding martingale K , given by $K_t = E[K_\infty | F_t]$, is then defined to be the stochastic integral $H.M^c$. For an expanded version of the above see page 40 of [4] .

A second way to interpret the stochastic integral with respect to the continuous local martingale M^c is to observe that, after a time change, M^c becomes a (possibly stopped) Brownian motion. The integral $H.M^c$ then becomes just the original integral of Ito.

DEFINITION 2.2.

A TIME CHANGE on the filtered space (Ω, F, P), $(F_t, t \geq 0)$ is a corlol increasing process $\tau = (\tau_t)_{t \geq 0}$, such that τ_t is a stopping time for each $t \geq 0$.

For $t \in [0,\infty[$ define

$$C_t = \inf \{s : \tau_s > t\} .$$

Then $$\tau_t = \inf \{s : C_s > t\} .$$

Put

$$C_\infty = \lim_{t\to\infty} C_t, \quad \tau_\infty = \lim_{t\to\infty} \tau_t .$$

If τ is a time change the corresponding filtration becomes

$$(\tau F)_t = F_{\tau_t}$$

and if X is a process, then the process after the time change becomes

$$(\tau X)_t = X_{\tau_t}$$

Now suppose M^c is a continuous martingale such that

$\langle M^c, M^c\rangle = \infty$. Define $C_t = \langle M^c, M^c\rangle_t$ and $\tau_t = \inf\{s : C_s > t\}$. Then it is not difficult to check, (using Doob's optional sampling theorem), that M^c is a local martingale with respect to the filtration $(\tau F)_t$. Furthermore,

$$\langle \tau M^c, \tau M^c\rangle = \tau(\langle M^c, M^c\rangle)$$
$$= \tau(C)$$

However, τ and C are inverse maps, so $\tau(C)_t = t$, and by Lévy's characterization τM^c is a Brownian motion with respect to the filtration $(\tau F)_t$.

If $C_\infty = \langle M^c, M^c\rangle_\infty$ is not ∞ $a.s$ then τM^c is a Brownian motion stopped at time C_∞.

3. THE INTEGRAL $H.M^d$

The constructions of the above section allow us to define, by localization, the stochastic integrals $H.M^c$ for $H \in L^1_{\text{loc}}(M^c)$ and any continuous local martingale M^c .

Recall that a stopping time T is predictable iff there is a sequence $\{T_n\}$ of stopping times such that $T_n \to T$ and $T_n < T$ on $\{T > 0\}$.(The sequence $\{T_n\}$ ANNOUNCES T).

DEFINITION 3.1.

A stopping time T is TOTALLY INACCESSIBLE if for every predictable stopping time S

$$P(\{\omega : T(\omega) = S(\omega) < \infty\}) = 0$$

EXAMPLES:

A constant stopping time is predictable.

A stopping time with a continuous distribution function is totally inaccessible , (with respect to the filtration $F_t = \sigma\{I_{s\geq T}, s\leq t\}$.)

DEFINITION 3.2.

A stopping time T is ACCESSIBLE if there is a sequence $\{T_n\}$ of predictable stopping times such that

$$P(\bigcup_n \{\omega : T_n(\omega) = T(\omega) < \infty\}) = 1 \quad .$$

That is, T is made up of pieces of predictable stopping times.

Recall that $\mathcal{O}$(resp. $\mathcal{P}$), the optional (resp. predictable) σ-field on $[0,\infty) \times \Omega$, is generated by all sets of the form $[\![T,\infty [\![$ where T is an arbitrary (resp. predictable) stopping time.

THEOREM 3.3.

If X is a bounded measurable process there is a bounded optional processes $\Pi_o X$, (resp. a bounded predictable process $\Pi_p X$), such that

$$E[X_T I_{\{T<\infty\}}] = E[(\Pi_o X)_T I_{\{T<\infty\}}]$$

for every stopping time T (resp. $E[X_T I_{\{T<\infty\}}] = E[(\Pi_p X)_T I_{\{T<\infty\}}]$ for every predictable stopping time T).

$\Pi_o X$ (resp. $\Pi_p X$) is called the optional (resp. predictable) projection of X.

PROOF.

See Chapter 5 of [5] .

For bounded measurable processes X and suitable corlol increasing processes A we can consider the Stieltjes integrals $(X.A)_t = \int_0^t X_s \, dA_s$, and the inner product

$$\langle X | A \rangle = E[(X.A)_\infty] .$$

DEFINITION 3.4.

The dual optional (resp. predictable) projection of the increasing process A is the unique optional (resp. predictable) increasing process $\Pi_o^* A$ (resp. $\Pi_p^* A$) such that

$$\langle \Pi_o X | A \rangle = \langle X | \Pi_o^* A \rangle$$

(resp. $\langle \Pi_p X | A \rangle = \langle X | \Pi_p^* A \rangle$)

The dual projections Π_o^* and Π_p^* can be extended to corlol processes of bounded variation

THEOREM 3.5.

If A is an adapted corlol process of bounded variation then $A - \Pi_p^* A$ is a martingale.

Recall that if T is a stopping time

$$F_T \equiv \{A \in F_\infty : A \cap \{\omega : T(\omega) \leq t\} \in F_t, \forall t\} .$$

If T is predictable and announced by the sequence T_n

$$F_{T-} \equiv \vee_n F_{T_n} .$$

We can now construct some basic discontinuous martingales.

THEOREM 3.6.

Suppose T is a totally inaccessible stopping time and $\Phi \in L^2(F_T)$. Consider the process

$$A_t = \Phi I_{t \geq T}$$

Then $\Pi_p^* A_t$ is continuous,

and $M_t = A - \Pi_p^* A$

$$M_t^2 - \Delta M_T^2 I_{t \geq T}$$

are martingales.

THEOREM 3.7.

Suppose T is a predictable stopping time and $\Phi \in L^2(F_T)$ is such that

$$E[\Phi | F_{T-}] = 0 .$$

Consider the process

$$A_t = I_{t \geq T}$$

Then $\Pi_p^* A_t = 0$, and $M_t = A_t$, $M_t^2 - M_T^2 I_{t \geq T}$ $(=0)$ are martingales.

The above results, proved in section 2 of [5] , are used in the proof of the following theorem:

THEOREM 3.8.

Suppose X is an optional process. Then there is a local martingale M such that

$$X_t = (\Delta M)_t = M_t - M_{t-}$$

if and only if there is a sequence of stopping times $\{T_n\}$, $T_n \to \infty$, such that

$$E(\sum_{t \leq T_n} X_t^2)^{\frac{1}{2}} < \infty \text{ for each } n ,$$

and $\quad \Pi_p X = X_0 I_{[\![0]\!]}$.

Under the above conditions there is a unique discontinuous local martingale M such that $X_t = M_t - M_{t-}$.

SKETCH PROOF

(For details see page 44 of [4]).

The finiteness of the above sum implies that for each ω the set of t where $X_t(\omega) \neq 0$ is countable. In fact there are two sequences $\{S_n\},\{T_n\}$ of stopping times, where each S_n is totally inaccessible and each T_n is predictable, such that

$$\{X \neq 0\} \subset \bigcup_n [\![S_n]\!] \cup \bigcup_n [\![T_n]\!] .$$

Writing $\quad A^n = X_{S_n} I_{t \geq S_n}$

$$N^n = A^n - \Pi_p^* A^n$$

and $\quad \tilde{N}^n = X_{T_n} I_{t \geq T_n}$,

both N^n and $\tilde{N}^n$ are martingales from Theorems 3.6 and 3.7. Taking care with convergence one shows that the sum of the martingals N^n and $\tilde{N}^n$ is a martingale M with the required properties. Uniqueness follows because a discontinuous martingale is determined by its jumps.

DEFINITION 3.9.

Suppose M^d is a discontinuous local martingale and H is a predictable process such that $H_t(M_t^d - M_{t-}^d)$ satisfies the same condition as X in Theorem 3.8. Then the stochastic integral $H.M^d$ is the unique local martingale, given by Theorem 3.8, such that

$$(H.M^d)_t - (H.M^d)_{t-} = H_t(M_t^d - M_{t-}^d) \quad .$$

REMARK 3.10

If the local martingale M^d is also of bounded variation then the integral $H.M^d$ coincides with the Stieltjes integral by the uniqueness property.

4. THE SINGLE JUMP PROCESS

Some of the concepts relating to the discontinuous martingales discussed in section 3 can be illustrated by the basic example of a process which has a single random jump at a random time. An example of a random measure is also given.

Consider the process $\{X_t\}$ which has values in a Blackwell space (X, S), and which remains at its initial point $z_o \in X$ until a random time T when it jumps to some position z. The underlying probability space can be taken to be

$$\Omega = [0,\infty] \times X$$

with the σ-field $B * S$. (B denotes the Borel field on $[0,\infty]$). A sample path of the process is

$$X_t(\omega) = \begin{cases} z_o & \text{if } t < T(\omega) \\ z(\omega) & \text{if } t \geq T(\omega). \end{cases}$$

Suppose a probability measure μ is given on Ω and that

$$\mu([0,\infty] \times \{z_o\}) = 0 \text{ and } \mu(\{0\} \times X) = 0 ,$$

so that the probabilities of a zero jump and a jump at time zero are zero.

Write F_t for the completed σ-field generated by x_t up to time t, so F_t is generated by $B([0,t]) * S$. Note that $]t,\infty] \times X$ is an atom in F_t. For $A \in S$ write

$$F_t^A = \mu(]t,\infty] \times A)$$

$$F_t = F_t^X$$

and $$c = \inf \{t : F_t = 0\} .$$

F_t is right continuous and monotonic decreasing, so there are only countably many points of discontinuity $\{u\}$ where

$$\Delta F_u = F_u - F_{u-} \neq 0 .$$

If F_t is continuous then T is totally inaccessible.

Any constant time is a predictable stopping time, so each time u, where $\Delta F_u \neq 0$, is predictable. The set of ω where $T(\omega) = u$ and $\Delta F_u \neq 0$, therefore, corresponds to the accessible part of T. Indeed the following result is easily established (see [1]).

LEMMA 4.1.

Suppose τ is an (F_t) stopping time. Then there is a $t_o \in [0,\infty[$ such that $\tau \wedge T = t_o \wedge T$.

The Stieltjes measure on $([0,\infty], B)$ given by F_t^A is absolutely continuous with respect to that given by F_t so there is a Radon-Nikodym derivative $\lambda(A,s)$ such that

$$F_t^A - F_o^A = \int_{]0,t]} \lambda(A,s)\, dF_s \quad .$$

DEFINITION 4.2.

The pair (λ,Λ) is the Levy system for the jump process, where

$$\Lambda(t) = -\int_{]0,t]} dF_s/F_{s-} \quad .$$

$\lambda(A,t)$ is the probability that $z \in A$ given that the jump occurs at time t .

If $\tilde{\Lambda}(t) = \Lambda(t \wedge T)$ then, roughly, $\tilde{d}\Lambda(t)$ is the probability that the jump occurs at time t , given that it has not happened so far.

For $A \in S$ consider the increasing processes

$$p(t,A) = I_{t \geq T}\, I_{z \in A}$$

$$\tilde{p}(t,A) = -\int_{[0,t\wedge T]} \lambda(A,s)dF_s/F_{s-}$$

$$= \int_{]0,t]} \lambda(A,s)\tilde{d}\Lambda(s) \quad .$$

Clearly $t \wedge T$ is left continuous and

$$-\int_{]0,t]} \lambda(A,s)\,\frac{dF_s}{F_{s-}} \text{ is a Borel function,}$$

so $\tilde{p}(t,A)$ is predictable. Indeed, $\tilde{p}(t,A)$ is the dual predictable projection of $p(t,A)$ and elementary integration verifies that

$$q(t,A) = p(t,A) - \tilde{p}(t,A)$$

is an F_t martingale.

The jump of $q(t,A)$ at a discontinuity u of F_t is

$$\Delta q(u,A) = I_{T=u}\, I_{z\in A} + \lambda(A,u)\,\frac{\Delta F_u}{F_{u-}}\, I_{T\geq u} \quad .$$

However,

$$E[I_{T=u} I_{z\in A} | F_{u-}] = E[I_{z\in A} | T=u]P(T=u | F_{u-})$$

$$= -\lambda(A,u)\frac{\Delta F_u}{F_{u-}} I_{T\geq u}$$

so $\quad E[\Delta q(u,A) | F_{u-}] = 0$.

From Theorem 3.7. $q^{\Delta u} = \Delta q(u,A)\ I_{t \geq u}$ is a square integrable martingale orthogonal to every square integrable martingale continuous at u . Furthermore, the predictable quadratic variation of $q^{\Delta u}$ is, from Theorem 3.7

$$\langle q^{\Delta u}, q^{\Delta u}\rangle_t = E[\Delta q(u,A)^2 | F_{u-}] I_{t \geq u}$$

$$= -\lambda(A,u)\ \frac{\Delta F_u}{F_{u-}}\ (1 + \lambda(A,u)\ \frac{\Delta F_u}{F_{u-}}\ )\ I_{T \geq u}\ I_{t \geq u}$$

THEOREM 4.3.

The predictable quadratic variation $\langle q,q\rangle(t,A)$ of $q(t,A)$ is $\tilde{p}(t,A) - r(t,A)$, where $r(t,A) = \sum_{0<u\leq t\wedge T} \lambda(A,u)^2 \frac{\Delta F_u^2}{F_{u-}^2}$.

PROOF

Decompose F_t into the sum of its continuous part F_t^c and its sum of jumps $F_t^d = \sum_{0<u\leq t} \Delta F_u$.

Then $p(t,A)$ can be similarly decomposed as

$$\tilde{p}^c(t,A) = - \int_{]0,t\wedge T]} \lambda(A,s)\ \frac{dF_s^c}{F_{s-}}$$

and $\quad \tilde{p}^d(t,A) = - \sum_{0<u\leq t\wedge T} \lambda(A,u)\ \frac{\Delta F_u}{F_{u-}}$.

If $\quad p^d(t,A) = \sum_{0<u\leq t\wedge T} I_{T=u} I_{z\in A}$

and $\quad p^c(t,A) = p(t,A) - p^d(t,A)$

then we can write

$$q(t,A) = q^c(t,A) + q^d(t,A)$$

where $\quad q^c(t,A) = p^c(t,A) - \tilde{p}^c(t,A)$

and $\quad q^d(t,A) = p^d(t,A) - \tilde{p}^d(t,A)$

$$= \sum_u q^{\Delta u}\ .$$

Now $q^{\Delta u}$ is orthogonal to $q^{\Delta u'}$ if $u \neq u'$ (meaning their product is a martingale), and $q^d(t,A)$ is othogonal to $q^c(t,A)$. The quadratic variation of a sum of orthogonal martingales is the sum of the quadratic variations, so

$$\langle q,q\rangle(t,A) = \langle q^c,q^c\rangle(t,A) + \langle q^d,q^d\rangle(t,A)$$

and
$$\langle q^d,q^d\rangle(t,A) = \sum_u \langle q^{\Delta u},q^{\Delta u}\rangle(t,A)$$
$$= \tilde{p}^d(t,A) - r(t,A).$$

Because F_t^c is continuous

$$\langle q^c,q^c\rangle(t,A) = \tilde{p}^c(t,A).$$

Therefore,

$$\langle q,q\rangle(t,A) = \tilde{p}^c(t,A) + \tilde{p}^d(t,A) - r(t,A)$$
$$= \tilde{p}(t,A) - r(t,A)$$

REMARK 4.4.

The above form of predictable quadratic variation, with second order terms arising from the accessible part, (where $\Delta F_u \neq 0$), of the jump time T , is very typical of what happens in more complicated situations.

Consider the set $L^1(\Omega,\ \mathcal{B}\times S,\ \mu)$. Because $p(t,A)$ and $\tilde{p}(t,A)$ are countably additive in A we can define for $g \in L^1(\mu)$ the Stieltjes integrals

$$\int_\Omega g(s,x)p(ds,dx) = g(T,z)$$
$$\int_\Omega g(s,x)\tilde{p}(ds,dx) = \int_{]0,T]} \int_X g(s,x)\lambda(s,dx)d\Lambda(s) \quad .$$

Define

$$\int_\Omega g dq = \int_\Omega g dp - \int_\Omega g d\tilde{p} \quad .$$

Elementary integration shows that for $g \in L^1(\mu)$ $M_t^g = \int_\Omega I_{s\leq t}\, g dp$ is an F_t martingale. Conversely, the martingale representation result of [1] states the following:

THEOREM 4.5.

Suppose $h \in L^1(\mu)$ is such that $E h = 0$. Then the uniformly integrable martingale

$$M_t = E[h(T,z)|\mathsf{F}_t]$$

can be represented as

$$M_t^g = \int_\Omega I_{s\le t}\, g(s,x)q(ds,dx) \qquad a.s.$$

where

$$g(s,x) = h(s,x) + I_{s<c}\, F_s^{-1} \int_{]0,s]\times X} h(\tau,z)d\mu(\tau,z) \quad .$$

That is, every uniformly integrable F_t martingale can be represented as a stochastic integral with respect to q .

Arguing as in Theorem 4.3 we have

THEOREM 4.6.

For $g \in L^1(\mu)$ the predictable quadratic variation process of the martingale M_t^g is

$$\langle M^g, M^g\rangle_t = \int_\Omega I_{s\le t}\, g^2 d\tilde{p} - \sum_{0<u\le t\wedge T} (\int_X g(u,x)\lambda(u,dx))^2 \frac{\Delta F_u^2}{F_{u-}^2} \quad .$$

REMARK 4.7.

$p(ds,dx)$, and its dual predictable projection $\tilde{p}(ds,dx)$, are examples of random measures. For a full discussion of random measures, and integration with resp-respect to them, see the work of Jacod [4] .The single jump process is discussed in detail in [3].

REFERENCES

1. M.H.A. DAVIS,The representation of martingales of a jump process. S.I.A.M. Jour. of Control 14 (1976), 623-638.
2. C. DELLACHERIE, Capacités et processus stochastiques, Springer-Verlag.Berlin, Heidelberg, New York 1972.
3. R.J. ELLIOTT, Innovation projections of a jump process and local martingales. Proc. Camb. Phil. Soc. 81(1977), 77-90.
4. J. JACOD,Calcul stochastique et problèmes de martingales. Lecture notes in mathematics 714 Springer-Verlag. Berlin, Heidelberg, New York 1979.
5. P.A. MEYER, Un cours sur les intégrales stochastiques. Lecture notes in mathematics 511, Springer-Verlag. Berlin, Heidelberg, New York. 1976.

Martingales, the Malliavin calculus and Hörmander's Theorem

by

Jean-Michel Bismut

Université Paris-Sud

Département de Mathématiques

91405 Orsay

Abstract: The purpose of this paper is to give a new approach to the Malliavin calculus of variations using martingale techniques. The first part of the paper is devoted to the proof of an integration by parts formula, which is closely related to a martingale representation result of Haussmann. The second part of the paper is devoted to the proof of the existence of densities for the semi-group of a diffusion under slightly more general conditions than Malliavin, and the existence of densities for the resolvent operators under the general conditions of Hörmander.

In a remarkable series of papers [10]-[11], Malliavin has developped probabilistic techniques to prove hypoellipticity for second order differential operators written in Hörmander's form

(0.1) $L = X_o + \frac{1}{2} \sum_1^m X_i^2$

where $X_o \ldots X_m$ are smooth vector fields on R^d. These techniques have been exploited by Stroock in a paper in which various applications are also given [12].

The reasoning in [10]-[11] proceeds in the following way. First, it is possible to associate to L the diffusion on R^d given by the stochastic differential equation

(0.2) $dx = X_o(x)\, dt + X_i(x).dw^i$

$x(0) = x$

where $w = (w^1 \ldots w^m)$ is a given Brownian motion. The measure Q on $C(R^+;R^d)$ associated to equation (0.2) is then the image measure of the brownian measure P on $C(R^+;R^m)$ by the mapping

$\omega = (w_t) \to x_.(\omega)$

At time t, the fundamental solution $P_t(x,.)$ associated to L is the probability law of x_t. To prove that $P_t(x,.)$ is given by a smooth density, it suffices to prove that the differentials of $P_t(x,.)$ in the sense of distributions are bounded measures. The key idea in [10]-[11] is to use the fact that P is a gaussian measure to prove an integration by parts formula on $(C(R^+;R^m),P)$ and obtain the smoothness of $P_t(x,.)$. This uses essentially a generalized Ornstein-Uhlenbeck operator which is an unbounded self-adjoint operator on $L_2(P)$.

In this paper, the integration by parts formula is established by a simpler argument. We essentially use two facts:

a) x given by (0.2) is a very regular function of w, in the sense that if $u \in L_1(R^+;R^m)$, and if x^u is the solution of the stochastic differential equation

$$(0.3) \quad dx^u = X_0(x^u)dt + X_i(x^u).(dw^i + u^i dt)$$
$$x^u(0) = x$$

it is possible to define a **regular** version of x^u, such that for a.e. $w, u \to x^u_.(w)$ is smooth, and this using the theory of stochastic flows developped by Malliavin [10], Baxendale [1] ,Elworthy [8] and ourselves [2] , [3] ,[4] .

b) If u is a bounded adapted process defined on $C(R^+;R^m)$ with values in R^m, the probability law of the process $w_t + \int_0^t u ds$ is equivalent to P on any σ-field $B(w_s; s \leq t)$, and the density is explicitly given by the Cameron-Martin-Maruyama-Girsanov formula [13] .

This permits us to establish directly a formula of integration by parts. This formula is in fact equivalent to the martingale representation result of Haussmann [9] , [19] for general diffusions which extends the result of Clark [6] for the brownian motion. The infinite dimensional aspect of the Malliavin calculus is then taken care of by the Cameron-Martin-Maruyama-Girsanov formula, while all the differential analysis is in fact strictly finite-dimensional.

Using these techniques, we derive the existence of a density for the resolvent operators of (0.2) under the general conditions of Hörmander [15] . This requires the lifting techniques of Rothschild and Stein [17] (see also Hörmander and Melin [16]) and time change. In the special case where the distribution generated by $X_1 \ldots X_m$, $[X_i, X_j]_{0 \leq i,j \leq m} \ldots$ generates a foliation of R^d

by closed submanifolds, the diffusion is shown to be factored in the "product" of a deterministic motion along X_0 and transversal to the leaves of the foliation, and of a diffusion with a regular semi-group in the leaves. These results are related to the results of Ichihara and Kunita [20] who classified hypoelliptic diffusions using Hörmander's theorem [15].

The paper is divided in five sections. In section 1, some results concerning stochastic flows are recalled. In section 2, an integration by parts formula is established, which leads to the Malliavin formula of integration by parts in section 3. In section 4, we derive the existence of a density of transition for the process x given by (0.2) when L verifies restricted Hörmander's conditions, as in Malliavin [10] [11]. The integration by parts is given in a completely explicit form, which exhibits Lie brackets of vector fields constructed via the flow associated to the stochastic differential equation. The main result is proved in section 5. The paper is expository, and the proofs are only briefly indicated. The complete proofs will appear in [22].

1. Stochastic flows

Ω denotes the space $C(R^+;R^m)$ of continuous functions defined on R^+ with values in R^m. The trajectory of a point ω is $w_t = (w_t^1 \ldots w_t^m)$. F_t is the σ-field $B(w_s; s \leq t)$. P is the brownian measure on Ω with the condition $P(w_0=0)=1$.

$X_0 \ldots X_m$ is a family of m+1 vector fields on R^d, which are C^∞, bounded with bounded differentials of all orders.

Consider the stochastic differential equation

$$(1.1) \qquad dx = X_0(x)\, dt + X_i(x) . dw^i$$
$$x(0) = x$$

where dw^i is the Stratonovitch differential of w^i, or the equivalent equation

$$(1.2)\quad dx = (X_o + \frac{1}{2}\frac{\partial X}{\partial x}iX_i)(x)dt + X_i(x)\,.\delta w^i$$
$$x(0) = x$$

where δw^i is the Ito differential of w^i.

Using a result of Kolmogorov, it is possible to prove elementarily that a mapping $(\omega, t, x) \to \varphi_t(\omega, x)$ exists such that

a) $\varphi_t(\omega, x)$ is measurable in ω and continuous in x.

b) For any x, $t \to \varphi_t(\omega, x)$ is the essentially unique solution of (1.1).

Moreover, the following is also proved in [2] (see also Malliavin [10], Elworthy [8], Baxendale [1]):

__Theorem 1.1__ : A.s. , for every $t \geq 0$, $\varphi_t(\omega, x)$ is a C^∞ diffeomorphism of R^d onto R^d, and the differentials $\frac{\partial^m \varphi}{\partial x^m}{}_t(\omega, x)$ are continuous on $R^+ X\, R^d$. For any $x \in R^d$, $Z_t = \frac{\partial \varphi}{\partial x}{}_t(\omega, x)$ and $Z_t' = [\frac{\partial \varphi}{\partial x}{}_t(\omega, x)]^{-1}$ are the unique solutions of the stochastic differential equations

$$(1.3)\quad dZ = \frac{\partial X_o}{\partial x}(x_t)Z\,dt + \frac{\partial X_i}{\partial x}(x_t)\,Z.dw^i$$
$$Z(0) = I$$
$$dZ' = -Z'\frac{\partial X_o}{\partial x}(x_t)dt - Z'\frac{\partial X_i}{\partial x}.dw^i$$
$$Z'(0) = I$$

If L is an adapted locally integrable process with values in R^d, if $z_o \in R^d$, and if $z_t = z_o + \int_o^t L ds$, $\varphi_t(\omega, z_t)$ is a continuous semi-martingale which may be written

$$(1.4)\quad \varphi_t(\omega, z_t) = z_o + \int_o^t X_o(\varphi_s(\omega, z_s))ds + \int_o^t X_i(\varphi_s(\omega, z_s)).dw^i + \int_o^t \frac{\partial \varphi}{\partial x}{}_s(\omega, z_s).dz_s$$

We now have the following technical result:

<u>Theorem 1.2</u> : If the vector fields $X_o \ldots X_m$ have compact support, there exists a negligible set N in Ω, such that for $\omega \notin N, x \in R^d$, if $t \to v_t$ is a Borel function defined on $[0,T]$ with values in R^m, which is in $L_1([0,T]; dt)$, then the differential equation

$$(1.5)\quad dy = (\varphi^{*}{}_t^{-1} X_i)(y_t)v^i dt$$
$$y(0)=x$$

has a unique solution $y_t^v(\omega)$; the mapping

$$(1.6)\quad v \in L_1([0,T]; dt) \to y^v_{\cdot}(\omega) \in C([0,T]; R^d)$$

is infinitely differentiable, and its differential at v=0 is given by the linear mapping

$$(1.7)\quad v \in L_1([0,T]; dt) \to \int_0^t (\varphi^{*}{}_s^{-1} X_i)(x) v^i ds$$

<u>Proof</u>: Since $X_o \ldots X_m$ have compact support, it is not hard to prove that for a.e. ω, the r.h.s. of (1.5) is uniformly bounded by $C|v|$, so that the differential equation (1.5) has a unique solution. (1.7) is then an easy exercise in differential analysis.

<u>Corollary</u> : Under the assumptions of Theorem 1.2, if u is a bounded adapted process defined on $\Omega \times R^+$ with values in R^m, $\varphi_t(\omega, y_t^{u(\omega)})$ is the essentially unique solution of the stochastic differential equation

$$(1.8)\quad dx = (X_o(x) + X_i(x)u^i)dt + X_i(x).dw^i$$
$$x(0) = x$$

<u>Proof</u>: This follows easily from Theorems 1.1 and 1.2.

<u>2. Integration by parts and martingale representation</u>

Consider the unnormalized gaussian law $e^{-|x|^2/2} dx$ on R^k. If g is a C^∞ bounded function on R^k with values in R, which has bounded differentials, we have for any $a \in R^k$

$$(2.1)\quad \int \langle f'(x), a\rangle e^{-|x|^2/2} dx = \int f(x) \langle a,x\rangle e^{-|x|^2/2} dx$$

which is obtained by integration by parts, which essentially uses the invariance of the Lebesgue measure dx on R^k. Since nothing like the Lebesgue measure exists in infinite dimensions, we now derive (2.1) by a subtler sort of argument. For $a \in R^k$, if R^k is endowed with the Gauss measure, the distribution of $x+a$ is equivalent to the Gaussian distribution, and its density with respect to this distribution is given by $\exp(\langle a,x\rangle - \frac{1}{2}|a|^2)$ so that

$$(2.2) \qquad \int f(x+a)e^{-|x|^2/2}dx = \int f(x)e^{\langle a,x\rangle - \frac{1}{2}|a|^2} e^{-|x|^2/2}dx$$

so that (2.1) follows by differentiation.

We will now use the same argument on the probability space $(C(R^+;R^m),P)$, using in particular the quasi-invariance of P. We will in fact use more than that, namely that if u is a bounded adapted process with values in R^m, then the distribution of $w_t + \int_0^t u\,ds$ is equivalent to P on any σ-field $B(x_s; s \leq T)$. This will permit us to avoid using the Ornstein-Uhlenbeck operator as in Malliavin[10], [11].

The result which follows is in fact a martingale representation result of Haussmann [9], [19], extending Clark [6], and uses a technique closely related to [19].

__Theorem 2.1:__ If $u = (u^1 \ldots u^m)$ is a bounded adapted process defined on $\Omega \times R^+$ with values in R^m, then if g is a bounded differentiable Lipchitz function defined on $C([0,T];R^d)$ with values in R, if for $y \in C([0,T];R^d)$, $d\mu^y(t)$ is the bounded measure on $[0,T]$ which is the differential of g at y, then the following equality holds

$$(2.3) \qquad E(g(\varphi_{\cdot}(\omega,x)) \int_0^T u^i \cdot \delta w^i) = E(\int_0^T u^i ds \langle X_i(\varphi_s(\omega,x)),$$

$$\int_s^T \varphi^*_s(\omega,x)\varphi^{*\,-1}_v(\omega,x)\, d\mu^{\varphi_{\cdot}(\omega,x)}(v)\rangle)$$

Proof: In formula (2.3), note that in the left hand-side, $g(\varphi_{\cdot}(\omega, x))$ represents the value of g on the trajectory $\varphi_{\cdot}(\omega, x)$ in $C([0,T];R^d)$. The same can be said on the r.h.s. of (2.3). It suffices to prove (2.3) when u is a step function such that on any interval $[k/2^n,(k+1)/2^n[$, u is constant and $F_{k/2^n}$-measurable(F_t is the σ-field $B(w_s; s \leq t)$). We may even assume that for $t \in [k/2^n,(k+1)/2^n[$,

$$(2.4) \qquad u_t = u^k(w_{t_1} \dots w_{t_{l_k}}) \qquad 0 \leq t_1 \leq t_2 \dots \leq t_{l_k} \leq k/2^n$$

where u^k is a smooth function with bounded differentials.

For $l \in R$, consider the functional equation

$$(2.5) \qquad r_t = i_t - \int_0^t l u_s(r) ds \qquad r, i \in C([0,T];R^m)$$

Note that (2.5) is not a stochastic differential equation, and may be solved by induction for each i in the unknown r. The unique solution of (2.5) is written $r^l(i)$, and is clearly a smooth function of l.

Let then n_t^l be the process

$$(2.6) \qquad n_t^l = w_t + \int_0^t l u_s(w) ds$$

From what has been previously said, we see that

$$(2.7) \qquad w_{\cdot} = r^l(n_{\cdot}^l)$$

so that

$$(2.8) \qquad u_s(w) = u_s(r_{\cdot}^l(n_{\cdot}^l))$$

Now by the Cameron-Martin-Maruyama-Girsanov formula [13], we know that the distribution of $n_{\cdot}^l$ on $C(R^+;R^m)$, which is written Q^l is equivalent to the brownian measure P on F_T, and we have

$$(2.9) \qquad \frac{dQ^l}{dP} F_T = \exp\{\int_0^T \langle l u_s(r_{\cdot}^l(w)), dw\rangle - \frac{1}{2}\int_0^T |l u_s(r_{\cdot}^l(w))|^2 ds\}$$

Let Z_T^{lu} be the density (2.9).

Consider now the stochastic differential equation

$$(2.10) \qquad dx = X_0(x)dt + X_i(x)(dw^i + u^i dt)$$
$$x(0) = x$$

Assume for the moment that $X_0 \dots X_m$ have compact support. By Theorem 1.2, we know that if z^{lu} is the solution of the differential equation

(2.11) $dz^{lu} = (\varphi^{*}{}_{t}^{-1} X_i)(z_t^{lu}) lu^i(w) dt$

$z^{lu}(0) = x$

then the unique solution x^{lu} of (2.10) is given by

(2.12) $x_t^{lu} = \varphi_t(\omega, z_t^{lu})$

Now using (2.9),(2.10) and (2.12), we get

(2.13) $E^P(g(\varphi_{\cdot}(\omega, z_{\cdot}^{lu}(\omega)))) = E^P(g(\varphi_{\cdot}(\omega, x)) Z_T^{lu}(\omega))$

Note now that (2.13) is clearly the generalization of (2.2). To obtain (2.3) it suffices to differentiate both sides in the variable l at l=0. The only difficulty is to show that the differentiation is possible under E^P. (1.7) is of course used to obtain the differential of the l.h.s. of (2.13). The extension to the case where $X_0 \dots X_m$ do not necessarily have compact support is easy.

A consequence of Theorem 2.1 is the original result of Haussmann [9] ,[19] :

<u>Theorem 2.2</u>: Under the assumptions of Theorem 2.1, let M_t be the continuous martingale

(2.14) $M_t = E^{F_t} g(\varphi_{\cdot}(\omega, x))$

Then if H_t is the optional projection of

(2.15) $U_t = \int_t^T \varphi_t^{*}(\omega, x) \varphi^{*}{}_v^{-1}(\omega, x) d_\mu \varphi_{\cdot}(\omega, x)(v)$

M_t has the representation

(2.16) $M_t = \int_0^t \langle H_s, X_i(\varphi_s(\omega, x))\rangle \delta w^i + E(g(\varphi_{\cdot}(\omega, x)))$

<u>Proof</u>: From a fundamental result of Ito [13] , we know that any square integrable martingale for the brownian motion may be written as the sum of a constant and of stochastic integrals with respect to w. (2.16) follows from (2.3).

3. The Malliavin formula of integration by parts

We now apply Theorem 2.1 to obtain the Malliavin formula of integration by parts [10] [12].

f is a C^∞ function defined on R^d with values in R, whose all differentials are uniformly bounded. h is a continuous function defined on $C([0,T];R^d)$ with values in the cotangent plane $T_x^*(R^d)$, which is bounded, differentiable and uniformly Lipschitz. If $y \in C([0,T];R^d)$, the differential $dh(y)$ may be identified to a finite measure $dn^y(t)$ on $[0,T]$ with values in $R^d \otimes R^d$

(3.1) $$z \in C([0,T];R^d) \to dh(y)(z) = \int_0^t dn^y(t)(z_t)$$

From the point of view of differential geometry, $dn^y(t)$ may be identified to a generalized linear mapping from $T_{y_t}(R^d)$ in $T_x^*(R^d)$. In particular, if $h \in T_x(R^d)$, we define the action of $\varphi^{*}{}_t^{-1} dn^y(t)$ on h by

(3.2) $$\int_0^t (\varphi^{*}{}_t^{-1} dn^y(t))(h) = \int_0^t dn^y(t)\varphi^{*}{}_t(h) \in T_x^*(R^d)$$

We then have the key result

<u>Theorem 3.1</u> : The following equality holds

(3.3) $$E(f(\varphi_T(\omega,x)) < h(\varphi_.(\omega,x)), \int_0^T (\varphi^{*}{}_s^{-1}X_i)(x)\,\delta w^i >) =$$

$$= E < h(\varphi_.(\omega,x)), \int_0^T (\varphi^{*}{}_s^{-1}X_i)(x) > < \varphi^{*}{}_s^{-1}X_i(x)ds, (\varphi^{*}{}_T^{-1})df(x) > + E(f(\varphi_T(\omega,x)) \int_0^T < (\varphi^{*}{}_s^{-1}(\omega,x)X_i(x),$$

$$\int_s^T (\varphi^{*}{}_v^{-1}(\omega,x)dn^{\varphi_.(\omega,x)}(v))(\varphi^{*}{}_s^{-1}X_i)(x) > ds)$$

<u>Proof</u>: Let $e_1,\dots,e_d$ be a basis in $T_x(R^d)$, $e^1,\dots e^d$ the dual basis in $T_x^*(R^d)$. We may write

(3.4) $$h(.) = h_k(.)e^k$$

$$\varphi^{*}{}_s^{-1}X_i(x) = (\varphi^{*}{}_s^{-1}X_i)^k e_k$$

Use now equality (2.3) with

$$(3.5) \quad g(y) = h_k(y)f(y_T) \qquad u^i = (\varphi^*{}_s{}^{-1}X_i)^k(x)$$

Summing in k, we obtain (3.3) .

Remark: Note that (3.3) is coordinate invariant. Moreover, using (1.3), it must be noted that by enlarging the state space, we may as well take functions of the trajectory of $\frac{\partial\varphi}{\partial x}{}_t(\omega, x)$ in (3.3) This will be essential in the next sections.

4. Properties of the semi-group under restricted Hörmander's conditions

We make now the following assumption

H_1 : For any $x \in R^d$, the vector space in $T_x(R^d)$ spanned by $X_1, \ldots X_m$, $[X_i, X_j]_{0 \le i,j \le m}$, $[X_i, [X_j, X_k]]_{0 \le i,j,k \le m} \ldots$ is equal to $T_x(R^d)$.

Assumption H_1 is exactly assumption (E) in Ichihara and Kunita [20] .

Proposition 4.1: For any x, a.s., for any T> 0

$$(4.1) \quad \int_0^T (\varphi^*{}_s{}^{-1}X_i)(x) > < \varphi^*{}_s{}^{-1}X_i)(x) \quad ds = C(\omega)$$

is a positive definite form on $T_x(R^d)$.

Proof: This result is an extension of Malliavin [10] . Let U_s be the vector space in $T_x(R^d)$ spanned by $\{\varphi^*{}_s{}^{-1}X_i(x)\}_{1 \le i \le m}$ and V_s be the vector space

$$(4.2) \quad V_s = \bigcup_{t \le s} (U_t)$$

We now define V_s^+ by

$$(4.3) \quad V_s^+ = \bigcap_{t > s} V_t$$

By the zero-one law, we know that V_0^+ is a.s. a fixed space not depending on ω. Assume that $V_0^+ \neq T_x(R^d)$. If $\tilde{T}$ is the stopping time

(4.4) $\tilde{T} = \inf\{t > 0; V_t^+ \neq V_0^+\}$

a.s. $\tilde{T}$ is > 0. Let f be a non zero element in $T_x^*(R^d)$ orthogonal to V_0^+. On $[0,\tilde{T}]$, we have

(4.5) $\langle f, (\varphi^{*}{}_s^{-1} X_i)(x)\rangle = 0$

Now we see easily that

(4.6) $$(\varphi^{*}{}_t^{-1} X_i)(x) = X_i(x) + \int_0^t (\varphi^{*}{}_s^{-1}[X_0, X_i])(x)ds + \\ + \int_0^t (\varphi_s^{*-1}[X_j, X_i])(x) . dw^j$$

or equivalently

(4.7) $$(\varphi^{*}{}_t^{-1} X_i)(x) = X_i(x) + \int_0^t \varphi^{*}{}_s^{-1}([X_0, X_i] + \tfrac{1}{2}[X_j,[X_j, X_i]]) \\ (x)ds + \int_0^t (\varphi^{*}{}_s^{-1}[X_j, X_i])\,\delta w^j$$

Now from (4.5) and (4.7), by canceling the martingale term in $\langle f, (\varphi^{*}{}_t^{-1} X_i)(x)\rangle$, we see that

(4.8) $\langle f, (\varphi^{*}{}_s^{-1}[X_j, X_i])(x)\rangle = 0 \quad s \le \tilde{T},\ 1 \le i, j \le m$

Now using (4.8) and formula (4.7) used on $[X_j, X_i]$, we see that in particular

(4.9) $\langle f, (\varphi^{*}{}_s^{-1}[X_j,[X_j, X_i]])(x)\rangle = 0 \quad s \le \tilde{T}$

so that, by canceling the bounded variation term in the semi-martingale decomposition of $\langle f, (\varphi^{*}{}_s^{-1} X_i)(x)\rangle$, we obtain from (4.7)

(4.10) $\langle f, (\varphi^{*}{}_s^{-1}[X_0, X_i])(x)\rangle = 0 \quad s \le \tilde{T},\ 1 \le i \le m$

By iteration, we see that if $X_{[I]}$ is any of the brackets defined in H_1, we have

(4.11) $\langle f, (\varphi^{*}{}_s^{-1} X_{[I]})(x)\rangle = 0 \quad s \le \tilde{T}$

By taking s=0 in (4.11) and using H_1, we see that f = 0. This is a contradiction, so that $V_0^+ = T_x(R^d)$. The result is proved.

Let b be a C^∞ function defined on $R^d \times R^d$ with values in R such that

a) $0 \leq b \leq 1$

b) b is equall to one on the set of invertible elements A in $R^d \otimes R^d$ such that $\|A^{-1}\| \leq C$, and to zero on the set of invertible elements A in $R^d \otimes R^d$ such that $\|A^{-1}\| \geq 2C$, and on the set of non invertible elements.

We then have the following

<u>Theorem 4.2</u>: If f is a bounded C^∞ function defined on R^d with values in R which has bounded differentials of any order, if Y is a C^∞ vector field on R^d which has bounded differentials of any order, the following equality holds

$$(4.12)\quad E(b(C(\omega))(Yf)(\varphi_T(\omega,x))) = E(b(C(\omega))f(\varphi_T(\omega,x))$$

$$\langle \varphi^{*}{}_T^{-1}Y,\ C^{-1}(\omega)\int_0^T \varphi^{*}{}_s^{-1}X_i\,\delta w^i\rangle) - E(b(C(\omega))f(\varphi_T(\omega,$$

$$x))\int_0^T \langle C^{-1}(\omega)[\varphi^{*}{}_s^{-1}X_i,\ \varphi^{*}{}_T^{-1}Y],\ \varphi^{*}{}_s^{-1}X_i\rangle)\,ds\ +$$

$$+\ E(b(C(\omega))f(\varphi_T(\omega,x))\int_0^T \langle C^{-1}(\omega)\varphi^{*}{}_s^{-1}X_i,\varphi^{*}{}_T^{-1}Y\rangle\,ds$$

$$\int_0^s \langle C^{-1}(\omega)[\varphi^{*}{}_v^{-1}X_j,\ \varphi^{*}{}_s^{-1}X_i],\ \varphi^{*}{}_v^{-1}X_j\rangle\,dv)\ +$$

$$+\ E(b(C(\omega))f(\varphi_T(\omega,x))\int_0^T ds\langle C^{-1}(\omega)\varphi^{*}{}_s^{-1}X_i,$$

$$\int_0^s \langle[\varphi^{*}{}_v^{-1}X_j,\varphi^{*}{}_s^{-1}X_i],\ C^{-1}(\omega)\varphi^{*}{}_T^{-1}Y\rangle\varphi^{*}{}_v^{-1}X_j\rangle\ dv)$$

$$-\ E(f(\varphi_T(\omega,x))\langle\frac{\partial b}{\partial x}(C(\omega)),\ \int_0^T ds\,\varphi^{*}{}_s^{-1}X_i \langle \int_0^s [\varphi^{*}{}_v^{-1}X_j,$$

$$\varphi^{*}{}_s^{-1}X_i]\ (\langle C^{-1}(\omega)\varphi^{*}{}_T^{-1}Y,\varphi^{*}{}_v^{-1}X_j\rangle)\ .\ dv + \int_0^T ds(\int_0^s$$

$$\langle C^{-1}(\omega)\varphi^{*}{}_T^{-1}Y,\varphi^{*}{}_v^{-1}X_j\rangle[\varphi^{*}{}_v^{-1}X_j,\varphi^{*}{}_s^{-1}X_i]\,dv\langle\varphi^{*}{}_s^{-1}X_i,\cdot\rangle))$$

<u>Proof</u>: (4.12) is a consequence of Theorem 3.1 and of the Remark which follows this Theorem. In fact, as noted in this Remark, it is feasible to take h in (3.3) to be

$$(4.13)\quad h = b(C(\omega))\ C^{-1}(\omega)\varphi^{*}{}_T^{-1}Y$$

To compute the r.h.s. of (3.3), assume for the moment that $X_0 \dots X_m$ have compact support. For v Borel and bounded defined on $[0,T]$ with values in R^m, consider the stochastic differential equation

$$(4.14) \qquad dx = X_0(x)\,dt + X_i(x)(dw^i + v^i dt)$$
$$x(0) = x$$

and the associated differential equation (1.5)

$$(4.15) \qquad dy = (\varphi^{*}{}_t^{-1} X_i)(y_t) v^i dt$$
$$y(0) = x$$

so that in (4.14), x_t is given by $\varphi_t(\omega, y_t^v)$, where y^v is the unique solution of (4.15). Let $\varphi_t^v(\omega, x)$ and $\psi_t^v(\omega, x)$ be the flows on R^d associated to (4.14) and (4.15), so that $\varphi_t^v(\omega, x) = \varphi_t(\omega, \psi_t^v(\omega, x))$. Everything boils down to compute the differential at $s = 0$ of the mapping

$$(4.16) \qquad s \to (\varphi_t^{sv*-1} T)(x)$$

when T is a C^∞ vector field on R^d. Clearly

$$(4.17) \qquad (\varphi_t^{sv*-1} T)(x) = (\psi_t^{sv*-1}(\varphi^{*}{}_t^{-1} T))(x)$$

so that classically the differential at $s=0$ of (4.16) is given by the Lie bracket

$$(4.18) \qquad [\int_0^t (\varphi^{*}{}_s^{-1} X_i)(x) v^i ds,\ (\varphi^{*}{}_t^{-1} T)(x)]$$

Formula (4.12) is then an easy consequence of (4.18), still using the techniques of Theorems 2.1 and 3.1.

As noted by Malliavin, (4.12) is enough to see that the distribution of $\varphi_T(\omega, x)$ has a density. In fact a result in harmonic analysis shows that the distribution of $\varphi_T(\omega, x)$ under the measure $b(C(\ \ ))dP$ has a density, so that using Proposition 3.1 the stated result follows easily(see [10]). But more can be said, namely that if $\| C(\omega)^{-1} \|$ is in all the L_p spaces($1 \le p \le +\infty$), (4.14) is valid with $b = 1$, so that if Z is another C^∞ vector field on R^d having the same properties than Y, formula (4.12) may be applied to the function Zf with $b=1$.

In the r.h.s. of (4.12) the function Zf will appear. Now the same procedure may be used so as to change the r.h.s. of (4.12) in such a way that only f appears . As noted by Malliavin in [11] , by iterating this procedure, the distribution of $\varphi_T(\omega, x)$ may then be shown to have a C^∞ density.

Moreover, following the technique of Stroock[12] , this technique may be extended to manifolds, so that the existence of a density for the distribution of $\varphi_T(\omega, x)$ still holds in this case.

5. Hypoellipticity under general Hörmander's conditions

We now apply the previous techniques to study the hypoellipticity of a second order differential **differential** operator under general Hörmander's conditions [15] ,i.e. when, at each $x \in R^d$, the vector space generated by $X_0(x), X_1(x), \ldots X_m(x)$ and all their brackets at x is equal to $T_x(R^d)$. The difference with section 4 is that X_0 comes in explicitly as a generating member of the family. Special attention is given to the distribution T' generated by $X_1, \ldots X_m$ and all the brackets of $X_0 \ldots X_m$, especially when it fibrates.

a) Sweeping the tangent space

In the whole section, we make the general assumption of Hörmander [15] :

H_2: At each $x \in R^d$, the vector space spanned by $X_0 \ldots X_m$ and all their brackets is equal to $T_x(R^d)$.

The distribution T' has been previously defined.

We now give a few results concerning the invertibility of $C(\omega)$

which has been defined in (4.1).

Theorem 5.1: For any $x \in R^d$, if $H_T(\omega,x)$ is the subspace in $T_x(R^d)$ which is the image of $T_x^*(R^d)$ by the linear mapping $A_T(\omega,x)$

(5.1) $p \in T_x^*(R^d) \to \int_0^T \langle p, (\varphi_s^{*-1} X_i)(x)\rangle (\varphi_s^{*-1} X_i)(x)\, ds$

then a.s. , for any $T > 0$, the vector space in $T_x(R^d)$ spanned by $\bigcup_{s \le T} \varphi_s^{*-1}(\omega,x) T'_{\varphi_s(\omega,x)}$ is equal to $H_T(\omega,x)$.

Proof: The transpose of $A_T(\omega,x)$ is equal to itself. Moreover $A_T(\omega,x)$ defines a symetric positive form on $T_x(R^d)$. It is clear that Ker $A_T(\omega,x)$ decreases with T, and is a left-continuous function of T. By orthogonality, we see that $H_T(\omega,x)$ increases with T. The result stated in the Theorem is equivalent to proving that a.s., for every $T > 0$

(5.2) $\text{Ker } A_T(\omega,x) = \bigcap_{s \le T} ((\varphi_s^{*-1}(\omega,x) T'_{\varphi_s(\omega,x)})^{\perp})$

If $X_{[I]}$ is any of the brackets generating T', $(\varphi_s^{*-1} X_{[I]})(x)$ is a continuous process, so that it is equivalent to prove that for every T, (5.2) holds a.s.. For p to be in Ker $A_T(\omega,x)$, it is necessary and sufficient that

(5.3) $\langle p, (\varphi_s^{*-1} X_i)(x)\rangle = 0$ for every $s \le T$

The r.h.s. of (5.2) is then trivially contained in the l.h.s.. We now prove the reverse inclusion. By a measurable selection theorem, it is possible to find a family of random variables $f_1 \ldots f_d$ with values in $T_x^*(R^d)$ so that for a.e. ω, $f_1(\omega), \ldots f_d(\omega)$ generate Ker $A_T(\omega,x)$. Using (4.7) and (5.3), it is not hard to prove that

(5.4) $\langle f_k(\omega), \varphi_s^{*-1}[X_j, X_i](x)\rangle = 0 \quad 0 \le s \le T, 1 \le i,j \le m$ a.s.

(5.4) is in fact not as easy as (4.8), since in (4.8) f is constant, while here $f_k(\omega)$ is random (for the detailed proof, see [22]). From (5.4), it is possible to obtain the equivalent of (4.9),(4.10) and (4.11), so that a.s., $f_k(\omega)$ is orthogonal to $(\varphi_s^{-1} X_{[I]})(x)$ for $s \le T$, when $X_{[I]}$ is any bracket in T'.

The Theorem is then proved.

Corollary: For any $x \in R^d$, a.s. , for any s,T, $s \le T$, then

(5.5) $\quad {\varphi^*_s}^{-1} X_o(x) - X_o(x) \in H_T(\omega, x)$

Proof: Use formula (4.6) with i=0 and the previous Theorem to show that if $f_k(\omega)$ is chosen as previously, a.s. $f_k(\omega)$ is orthogonal to ${\varphi^*_s}^{-1} X_o(x) - X_o(x)$.

Define $T_{(\omega, x)}$ by

(5.6) $\quad T_{(\omega, x)} = \inf\{ t \ge 0;\ A_t(\omega, x) \text{ is invertible}\}$

For every x, $T_{(\omega, x)}$ is a F_t^+-stopping time. By the zero-one law, either $T_{(\omega,x)} = 0$ a.s. either $T_{(\omega,x)} > 0$ a.s.. Let F be the set

(5.7) $\quad F = \{ x \in R^d;\ T_{(\omega,x)} = 0 \text{ a.s.}\}$

Theorem 5.2: F is a G_δ set in R^d which is finely closed for the fine topology induced by the strong Markov process (1.1). For every $x \in R^d$ such that $T'_x \ne T_x(R^d)$, the following identities hold a.s.

(5.8) $\quad T_{(\omega,x)} = \inf\{t \ge 0\ ;\ \varphi_t(\omega, x) \in F\} = \inf\{ t \ge 0;$

$\quad {\varphi^*_t}^{-1} X_i(x) \notin T'_x$ for one of the $i, 1 \le i \le m\} = \inf\{ t \ge 0;$

$\quad {\varphi^*_t}^{-1}(\omega, x) T'_{\varphi_t(\omega,x)} \ne T'_x \}$

Proof: From Theorem 5.1, we know that a.s. , for every $t \ge 0$, $H_t(\omega, x)$ contains T'_x. Since for $t \le T_{(\omega,x)}$, $A_t(\omega,x)$ is not invertible, and since the codimension of T'_x is 1 (this because of H_2 , and the fact that $T_x(R^d) \ne T'_x$), for $t \le T_{(\omega,x)}$, it is clear that

(5.9) $\quad H_t(\omega, x) = T'_x$

Using Theorem 5.1, we see that ${\varphi^*_t}^{-1}(\omega, x) T'_{\varphi_t(\omega,x)}$ is included in T'_x. Since the dimension of $T'_{\varphi_t(\omega,x)}$ is at least d-1, this is possible only if equality holds. If ${\varphi^*_t}^{-1}(\omega, x) T'_{\varphi_t(\omega,x)} \ne$

T'_x , since the codimension of T'_x is 1, the union of these two spaces spans $T_x(R^d)$. The last equality in (5.8) is proved. The second is then trivial, since by (5.3), $H_t(\omega, x)$ is spanned by $\varphi^{*\,-1}_s X_i(x)(s \le t, 1 \le i \le m)$. The first equality in (5.8) is a consequence of the strong multiplicative property of the flow $\varphi_t(\omega, .)$ [2], [3], [22] . F is trivially finely closed for the process (1.1). For the proof that F is a G_δ set, we refer to [22] .

The invertibility of $A_t(\omega, x)$ is clearly related to the behavior of the distribution of $\varphi_t(\omega, x)$. For the moment F seems to depend on $X_0 \ldots X_m$, and not only on the generator $L = X_0 + \frac{1}{2} X_i^2$, i.e. on the Markov process (1.1). Let A be the open set

(5.9) $A = \{ x \in R^d; T'_x = T_x(R^d) \}$

It is easily seen that A only depend on L. We now have

Theorem 5.3: F is equal to the fine closure A^f of A. For every $x \in R^d$, we have

(5.10) $$T(\omega, x) = \inf \{ t \ge 0;\ X_0(\varphi_t(\omega, x)) \in T'_{\varphi_t(\omega, x)} \} =$$

$$\inf \{ t \ge 0; X_0(x) \in \varphi^{*\,-1}_t(\omega, x) T'_{\varphi_t(\omega, x)} \}$$

Proof: This result not being used in what follows, we refer for a proof to [22] .

Remark: The space R^d has been divided in two regions:

a) If $x \in A^f$, using the techniques of section 4, it is easy to see that for any $t > 0$, $P_t(x, .)$ has a density.

b) If $x \notin A^f$, a control of the differentials of P_t may only be obtained in the directions of T'.

b) The semi-group P_t: the degenerate case

In this subsection, besides assumption H_2, we make the following assumption:

H_3: A is empty, i.e. for any $x \in R^d$, T'_x is of dimension $d-1$, or equivalently $X_o \notin T'$.

This assumption is assumption (P) in Ishihara and Kunita [20].

By the Frobenius theorem [18] I.2, the involutive distribution T' determines a foliation of R^d by connected disjoint maximal leaves which are integral submanifolds of T', i.e.

a) R^d is the union of the leaves L_a. For each $x \in R^d$, there exists a local system of coordinates $(u, y^1, \ldots y^{d-1})$ around x so that the foliation is given by $(u = cst)$.

b) For $x \in L_a$, considering L_a as an immersed submanifold, $T_x(L_a)$ is exactly T'_x.

In general the L_a are not embedded as submanifolds of R^d, i.e. the natural topology of the L_a is stronger than the topology induced by R^d.

To avoid tricky problems related to the possible non closedness of some of the leaves, we also do the following assumption:

H_4: The leaves L_a are closed submanifolds of R^d.

For $x \in R^d$, L_x denotes the unique leaf containing x. We consider the differential equation

(5.11) $dy = X_o(y)dt$

$y(0) = y$

and the associated flow of diffeomorphisms of R^d $\psi_t: y \to y_t$.

We then have

<u>Proposition 5.4</u>: For each $t \geq 0$, the foliation L_a is stable by ψ_t, i.e. the image of a leaf by ψ_t is still a leaf. In particular, for $t \geq 0$, $x \in R^d$,

(5.12) $\psi^{*}{}_t^{-1} T'_{\psi_t(x)} = T'_x$

<u>Proof</u>: Let $Y_1 \ldots Y_{d-1}$ be a family of brackets which generate T'

at $x \in R^d$. They are free at x , and then free on a neighborhood. We have

(5.13) $\psi^{*}{}_t^{-1} Y_i(x) = Y_i(x) + \int_0^t \psi^{*}{}_s^{-1} [X_o, Y_i](x)ds$

Since $[X_o, Y_i] \in T'$, for t small enough, we may write

(5.14) $[X_o, Y_i] (\psi_t(x)) = c^j_{it} Y_j(\psi_t(x))$

where c^j_i are continuous functions. It is not hard to conclude that for t small enough , $\psi^{*}{}_t^{-1} T'_{\psi_t(x)} = T'_x$. Let S be defined by

(5.15) $S = \inf\{t \geq 0, \psi^{*}{}_t^{-1} T'_{\psi_t(x)} \neq T'_x\}$

If $S < +\infty$, $\psi^{*}{}_S^{-1} T'_{\psi_S(x)} = T'_x$. Now use the semi-group property of ψ_t to see that for $t - S > 0$ small enough, we still have $\psi^{*}{}_t^{-1} T'_{\psi_t(x)} = T'_x$, which contradicts the definition of S. Hence $S = +\infty$. (5.12) is then proved, and the Proposition follows easily [22] .

We finally have

<u>Theorem 5.5</u>: Consider the stochastic differential equation

(5.16) $dz = (\psi^{*}{}_t^{-1} X_i)(z_t) . dw^i$

$z(0) = z$

and the associated flow $\tau_t : z \to \tau_t(\omega, z) = z_t$. A.s., for any $z \in R^d$, $t \geq 0$, $\tau_t(\omega, z) \in L_z$. Moreover, a.s., for any $(t,x) \in R^+ \times R^d$,

(5.17) $\varphi_t(\omega, x) = \psi_t(\tau_t(\omega, x))$

<u>Proof</u>: By [3][4][10], we know that the flow $\tau_.(\omega, .)$ is the uniform limit in probability on the compact sets of $R^+ \times R^d$ of the flows $\tau^n_.(\omega, .)$ associated to the differential equations

(5.18) $dz^n = (\psi^{*}{}_t^{-1} X_i)(z^n)\, w^{i,n} dt$

where

(5.19) $w^{i,n} = 2^n(w^i(k+1/2^n) - w^i(k/2^n)) \qquad k/2^n \leq t < (k+1)/2^n$

By Proposition 5.4, since z^n is an integral curve of T', if $z^n(0) = z$, z^n stays in L^z. Since L_z is closed in R^d, the first part of the Theorem is proved. Using the Ito-Stratonovitch formula, the second part is trivial.

The diffusion $\varphi_t(\omega,x)$ has then been factored as the product of a diffusion in a fixed leaf L_x and a deterministic motion in the direction X_o. This is closely related to Ishihara and Kunita [20] .

It is proved in [22] that by restricting the flow $\tau_t(\omega,\cdot)$ to a fixed leaf L_x, it verifies— at least in a generalized sense, since $\tau_t(\omega,\cdot)$ is associated to a time-dependent stochastic differential equation--the assumptions of section 4, so that the distribution of $\tau_t(\omega,x)$ has a density in the leaf L_x. It is then clear that the resolvent operators of the diffusion (1.1) have a density in R^d, in particular because the mapping $(t,y) \in R^+ \times L_x \to \psi_t(y)$ is non singular.

c) <u>The general case</u>

We now sketch the proof of the existence of a density for the resolvent operators, when only H_2 is verified.

Using the results of section 4, we know we may assume that $x \notin A$(i.e. T'_x is not full at x). By a result of Rothschild and Stein[17](see also [16]), we know that if the brackets of $X_o \ldots X_m$ of length$\leq$ r span $T_x(R^d)$, there exists p and C^∞ vector fields on R^{d+p} $\tilde{X}_o \ldots \tilde{X}_m$ such that

a) If π is the projection of R^{d+p} on R^d, $\pi^* \tilde{X}_i = X_i$.

b)The brackets $\tilde{X}_{[I]}$ of $\tilde{X}_o \ldots \tilde{X}_m$ of length $\leq$ r(i.e. $|I| \leq r$)span $T_{(x,0)}(R^{d+p})$, and moreover they are free of order r at (x,0), i.e. the only algebraic dependences between the various $\tilde{X}_{[I]}(x)$ are the algebraic relations which exist in any formal Lie algebra, like the Jacobi identities.

Since T'_x is not full, $T'_{(x,0)}$ is not full.

<u>**Proposition 5.6**</u> : There exists a uniformly positive function u

on R^{d+p} which is bounded, C^{∞} with bounded differentials of any order such that if T'^{u} is the distribution associated to the vector fields $(u\tilde{X}_o,\tilde{X}_1,\ldots,\tilde{X}_m)$, T'^{u} is full at $(x,0)$.

Proof: By Proposition 2 in [16], we know there exists u C^{∞} and bounded such that if for a sequence $I = (i_1\ldots i_n)$ $\tilde{X}_I$ is the differential operator $\tilde{X}_{i_1}\tilde{X}_{i_2}\ldots\tilde{X}_{i_n}$, then

$$(5.20)\quad (\tilde{X}_I u)(x,0) = 0 \qquad |I| \le r-1$$
$$(\tilde{X}_1^r u)(x,0) = 1$$

By adding a constant we may assume that u is uniformly positive. Similarly we may assume that u is constant out of a compact set. An easy computation shows that if $\tilde{X}^u_{[I]}$ are the brackets corresponding to the distribution $(u\tilde{X}_o,\tilde{X}_1\ldots\tilde{X}_m)$, for $|I| \le r$, the vectors $\tilde{X}^u_{[I]}(x,0)$ are non zero multiples of the $\tilde{X}_{[I]}(x,0)$. We now calculate the bracket $[\tilde{X}_1, [\tilde{X}_1,\ldots[\tilde{X}_1,u\tilde{X}_o]]]$, where $\tilde{X}_1$ is repeated r times. It is trivial to see that it is equal to

$$(5.21)\quad C_r^j(\tilde{X}_1^j u)\ \tilde{Y}_{r-j}$$

where $\tilde{Y}_{r-j}$ is the bracket calculated with u=1 which contains r-j 1. Using (5.20), we find that this bracket is exactly equal to $u\tilde{Y}_r(x,0) + \tilde{X}_o(x,0)$. Since $\tilde{X}_o\ldots\tilde{X}_m$ are free of order r at (x,0) $\tilde{X}_o(x,0)$ does not belong to the vector space spanned by the other brackets of length $\le r$, which is necessarily of codimension 1. Since $\tilde{T}'_{(x,0)}$ is not full, it is exactly equal to this space. $\tilde{Y}_r(x,0)$ may then be written as a linear combination of $\tilde{X}_{[I]}(x,0)$ with $|I| \le r$, $I \ne 0$. $\tilde{X}_o(x,0)$ is then a linear combination of the $\tilde{X}^u_{[I]}(x,0)$ ($|I| \le r+1$, $I \ne 0$). The proposition is proved.

We now give the final key result:

Theorem 5.7: Under H_2, for $x \in R^d$, $T \ge 0$, the measure

$$(5.22)\quad f \to E\int_0^T f(\varphi_t(\omega,x))dt$$

has a density.

Proof: We may assume that $x \notin A$. Take $\tilde{X}_o\ldots\tilde{X}_m$, u as in Proposition

5.6. Let v be the positive function such that $v^2 = u$, and let $\tilde{\varphi}_s(\omega,.)$ be the flow on R^{d+p} associated to $(\tilde{X}_0 \ldots \tilde{X}_m)$. s_t is the change of time

(5.23) $$s_t = \inf\{s \geq 0 \; ; \int_0^s \frac{du}{v^2(\tilde{\varphi}_u(\omega,(x,0)))} = t\}$$

If w'^i_t is the process defined by

(5.24) $$w'^i_t = \int_0^{s_t} \frac{\delta w^i}{v(\tilde{\varphi}_u(\omega,(x,0)))}$$

w' is trivially a m-dimensional brownian motion. Moreover, $y_t = \tilde{\varphi}_{s_t}(\omega,(x,0))$ is the solution of

(5.25) $$dy = (v^2\tilde{X}_0 - \tfrac{1}{2}v(\tilde{X}_i v)\tilde{X}_i)(y)dt + (v\tilde{X}_i) . dw'^i$$

$$y(0) = (x,0)$$

It is not hard to see that the distribution T' associated to (5.25) is still full at $(x,0)$, so that by the results of section 4, the measure

(5.26) $$g \to E\int_0^T g(y_t)dt$$

has a density on R^{d+p}. Now

(5.27) $$E\int_0^T g(\tilde{\varphi}_t(\omega,(x,0)))\,dt = E\int_0^{+\infty} 1_{s_t \leq T}\, gv^2(y_t)dt$$

so that the measure (5.27) has a density.

Since $\pi\tilde{\varphi}_t(\omega,(x,0)) = \varphi_t(\omega,x)$, the result is proved.

<u>Remark</u>: The detailed proof may be found in [22] .

REFERENCES

[1] BAXENDALE P., Wiener processes on Manifolds of maps, J. of Diff. Geometry, to appear.

[2] BISMUT J.M., Principes de mécanique aléatoire, to appear.

[3] BISMUT J.M., Flots stochastiques et formule de Ito-Stratonovitch généralisée, CRAS 290, 483-486 (1980).

[4] BISMUT J.M., A generalized formula of Ito on stochastic flows, to appear.

[5] BISMUT J.M., An introductory approach to duality in optimal Stochastic control, SIAM Review 20 (1978), 62-78.

[6] CLARK J.M.C., The representation of functionals of Brownian motion by stochastic integrals, Ann. Math. Stat. 41 (1970), 1282-1295, 42 (1971), 1778.

[7] DELLACHERIE C., MEYER P.A., Probabilités et Potentiels, chap. I-IV, Paris, Hermann 1975, chap. V-VIII, Paris, Hermann 1980.

[8] ELWORTHY K.D., Stochastic dynamical systems and their flows, Stochastic analysis, A. Friedman and M. Pinsky ed. pp 79-95, New York Acad. Press 1978.

[9] HAUSSMANN U., Functionals of Ito processes as stochastic integrals, SIAM J. Control and Opt. 16 (1978), 252-269.

[10] MALLIAVIN P. Stochastic calculus of variations and hypoelliptic operators, Proceedings of the International Conference on Stochastic differential equations of Kyoto 1976), pp 195-263, Tokyo : Kinokuniya and New-York : Wiley 1978.

[11] MALLIAVIN P., C^k-hypoellipticity with degeneracy, Stochastic Analysis, A. Friedman and M. Pinsky ed., pp 199-214, New-York and London Acad. Press 1978.

[12] STROOCK D., The Malliavin calculus and its application to second order parabolic differential equations, Preprint 1980.

[13] STROOCK D.W. and VARADHAN S.R.S., Multidimensional diffusion processes, Grundlehren der Mathematischen Wissenschaften, Berlin-Heidelberg-New York, Springer 1979.

[14] JACOD J. and YOR M., Etude des solutions extrémales et représentation intégrale des solutions pour certains problèmes de martingales, Zeitschrift Wahrscheinlich keitstheorie verw. Gebiete 38 (1977) 83-125.

[15] HÖRMANDER L., Hypoelliptic second order differential equations, Acta Math. 119 (1967), 147-171.

[16] HÖRMANDER L. and MELIN A., Free systems of vector fields, Arkiv for Math. 16 (1978), 83-88.

[17] ROTHSCHILD L.P. and STEIN E.M., Hypoelliptic differential operators and nilpotent groups, Acta Math. 137 (1976), 247-320.

[18] ABRAHAM R. and MARSDEN J., Foundations of mechanics, Reading : Benjamin 1978.

[19] HAUSSMANN U., On the integral representation of functionals of Ito processes, Stochastics 3 (1979), 17-27.

[20] ICHIHARA K. and KUNITA H., A classification of second order degenerate elliptic operators and its probabilistic characterization, Zeitschrift Wahrscheinlich keitstheorie verw. Gebiete 30 (1974), 235-254.

[21] DAVIS M.H.A., Functionals of diffusion processes as stochastic integrals, Math. Proc. Camb. Phil. Soc. 87 (1980), 157-166.

22 BISMUT J.M.: Martingales, the Malliavin calculus and hypoellipticity under general Hörmander's conditions. To appear.

ON A REPRESENTATION OF LOCAL MARTINGALE ADDITIVE FUNCTIONALS OF SYMMETRIC DIFFUSIONS

M. Fukushima *

College of General Education
Osaka University
Toyonaka, Osaka, Japan

§ 1 Introduction

In studying the absolute continuity of diffusions with respect to Brownian motion, a very important role is played by the following representation of the positive continuous local martingale multiplicative functional L_t of the Brownian motion (Wentzell [7]):

$$L_t = \exp \{ \sum_{i=1}^{d} \int_0^t f_i(X_s)dB_s - \frac{1}{2} \sum_{i=1}^{d} \int_0^t f_i(X_s)^2 ds \},$$

where f_i are measurable functions with

$$P_x(\int_0^t f_i(X_s)^2 ds < \infty) = 1 \qquad \forall \; x \in R^d.$$

If we relax the above finiteness condition for f_i by requiring it only for q.e. $x \in R^d$ instead of " $\forall x \in R^d$", then this condition turns out to be equivalent to the quite simple analytical condition that

$$f_i \in L^2(\{K_n\})$$

for some nest $\{K_n\}$ (see § 3 for the precise definitions). Furthermore the preceding expression of L_t then provides us with the most general representation of the functional with the mentioned properties but admitting exceptional set of zero capacity.

In fact, we show in Theorem 3 that the assertion made in the above paragraph is true not only for the Brownian motion but also for a more general m-symmetric

* This work was done while the author was at Fakultät für Physik, Universität Bielefeld, West Germany.

diffusion process possessing the Dirichlet form (3.1). Brownian motion is merely a special case that m and ν_i are the Lebesgue measure.

Theorem 3 is acutually a corollary of the representation theorem (Theorem 2) of the local martingale additive functional as the stochastic integral. The author has obtained such a representation for the martingale additive functional with finite energy ([2]). So what we need is to show that any local martingale additive functional is locally of finite energy. This will be done in a rather general context (Theorem 1).

An application of Theorem 3 to distorted Brownian motions ([1]) will be treated elsewhere.

§ 2. Local martingale additive functionals are locally of finite energy

We start with a rather general question concerning an additive functional $A_t(\omega)$ of a Markov process $\underset{\sim}{M}$. A is said to be a local martingale additive functional if there exists a sequence of stopping times σ_n increasing to ζ (the killing time) such that $A_{t\wedge\sigma_n}$ is a square integrable martingale for each n. It is not clear however that this condition is equivalent to the following stronger and yet more useful property; there exists a sequence of stopping times σ_n increasing to ζ and a sequence $A_t^{(n)}(\omega)$ of square integrable martingale additive functionals of $\underset{\sim}{M}$ such that $A_t(\omega) = A_t^{(n)}(\omega)$, $\forall\ t < \sigma_n(\omega)$, for each n.

We show in this section that this is in fact true for a general symmetric Markov process. Moreover we prove that each $A^{(n)}$ may be choosen to be a martingale additive functional of finite energy in the sense of [2].

Let X be a locally compact separable Hausdorff space and m be an everywhere dense positive Radon measure on X. We consider a Hunt process $\underset{\sim}{M} = (\Omega, \mathcal{M}, X_t, P_x)$ on X whose transition function is m-symmetric and we assume that the associated Dirichlet form $\mathcal{E}$ on $L^2(X;m)$ is regular. See [2 ; Chap. 4] for the meanings of those notions.

For the notion of additive functionals of $\underset{\sim}{M}$ and the equivalence of additive functionals, we also refer to [2 ; Chap. 5]. Note that we admit exceptional sets in the state space X in the definition of additive functionals. The set of all positive continuous additive functionals (PCAF's) is denoted by $\underset{\sim}{A}_c^+$. We let

$$\mathcal{M} = \{M: M \text{ is an AF},\ \forall t > 0,\ E_x(M_t^2) < \infty,\ E_x(M_t) = 0 \quad \text{q.e.}\}$$

and call the elements of $\mathcal{M}$ a martingale additive functional (MAF). Here q.e. means "except for a set of capacity zero". The energy e(A) of an AF A is defined by $e(A) = \lim_{t\downarrow 0} \frac{1}{2t} E_m(A_t^2)$. The space of those elements of $\mathcal{M}$ with finite energy is denoted by $\overset{\circ}{\mathcal{M}}$.

An increasing sequence of compact sets $K_n \subset X$ is said to be a nest if

(2.1) $P_x(\lim_{n\to\infty} \tau_{K_n} < \zeta) = 0 \qquad$ q.e. $x \in X$,

τ_K being the first exit time from a set K. Note that condition (2.1) is equivalent to the following analytical condition ([2 ; Lemma 5.1.6]):

(2.2) $\lim_{n\to\infty} \text{cap}(K-K_n) = 0$ for any compact set K.

We say that an AF M is locally in $\overset{\circ}{\mathcal{M}}$ if there exist a nest $\{K_n\}$ and a sequence $M^{(n)} \in \overset{\circ}{\mathcal{M}}$, n=1,2,..., such that

(2.3) $P_x(M_t = M_t^{(n)}, \forall\, t < \tau_{K_n}) = 1$, q.e. $x \in X$.

The space of those AF's locally in $\overset{\circ}{\mathcal{M}}$ is denoted by $\overset{\circ}{\mathcal{M}}_{loc}$. (The present definition of the space $\overset{\circ}{\mathcal{M}}_{loc}$ is different from and a bit more general than the corresponding one introduced in [2 ; § 5.4].

Theorem 1. An AF M is local martingale if and only if $M \in \overset{\circ}{\mathcal{M}}_{loc}$.

Lemma 1. Let M be a local martingale additive functional.

(i) The quadratic variation $\langle M \rangle$ belongs to A_c^+.

(ii) If a Borel function f on X satisfies

$$E_x\left(\int_0^t f(X_s)^2 d\langle M\rangle_s\right) < \infty \qquad \text{q.e. } x,$$

then there exists a unique $f \cdot M \in \overset{\circ}{\mathcal{M}}$ such that

(2.4) $\langle f\cdot M, N\rangle = f \cdot \langle M,N\rangle$

for any local martingale additive functional N.

The MAF $f \cdot M$ is called the stochastic integral. Note that our additive functional is nothing but a usual perfect AF of the Hunt process $\underset{\sim}{M}/X-B$, B being a properly exceptional set depending on the functional. Therefore Lemma 1 is implied in the corresponding statements by P.A. Meyer ([4], [5]).

Proof of Theorem 1. Theorem 1 follows from Lemma 1 and the one-to-one correspondence between the space $\underset{\sim}{A}_c^+$ of PCAF's and the space of smooth measures ([2 ; Theorem 5.1.3]). Let M be a local martingale additive functional and let μ be the smooth measure associated with the PCAF $\langle M\rangle$ of Lemma 1. By the definition of the smooth measure, there exists a nest $\{K_n\}$ such that $\mu(K_n) < \infty$, n=1,2,... . Since the AF $I_{K_n} \cdot \langle M \rangle$ corresponds to the finite measure $I_{K_n} \cdot \mu$, [; Lemma 5.1.9] implies

$$E_x\left(\int_0^t I_{K_n}(X_s)\, d\langle M\rangle_s\right) < \infty \qquad \text{q.e. } x.$$

Then by Lemma 1 the stochastic integral $M^{(n)} = I_{K_n} \cdot M$ is well defined as an element of $\mathcal{M}$. It is easy to see $e(M^{(n)}) = \frac{1}{2}\mu(K_n) < \infty$, namely, $M^{(n)} \in \mathcal{M}$. Property (2.3) is also obvious. q.e.d.

§ 3. Local martingale additive functionals are stochastic integrals

Theorem 1 is now applied to the special case that $X = R^d$ d-dimensional Euclidean space, m is a positive Radon measure on R^d and $\underset{\sim}{M} = (\Omega, \mathcal{M}, X_t, P_x)$ is an m-symmetric diffusion process on R^d whose Dirichlet form on $L^2(R^d; m)$ possesses the space $C_o^1(R^d)$ of continuous differentiable function as its core and takes the expression

$$(3.1)\quad \mathcal{E}(u,v) = \frac{1}{2}\sum_{i=1}^{d}\int_{R^d} u_{x_i} v_{x_i}\, d\nu_i, \quad u,v \in C_o^1(R^d).$$

Here ν_i, $1 \leq i \leq d$, are positive Radon measures on R^d.

Denote the sample path as $X_t(\omega) = (X_t^1(\omega), X_t^2(\omega), \ldots, X_t^d(\omega))$. Since the coordinate function x_i is locally in the domain of $\mathcal{E}$, the process X_t^i admits a unique decomposition

$$(3.2)\quad X_t^i(\omega) - X_o^i(\omega) = M_t^i(\omega) + N_t^i(\omega), \quad 1 \leq i \leq d,$$

, where $M^i \in \mathring{\mathcal{M}}_{loc}$ and N^i is a continuous additive functional locally of zero energy. The smooth measure corresponding to the quadratic variation $\langle M^i \rangle$ of M^i is just ν_i appearing in the expression (3.1). (cf.[2 ; § 5.4]).

Furthermore we have in this case a complete description of the space $\mathring{\mathcal{M}}$ by means of stochastic integrals ([2 ; (5.4.36)]):

$$(3.3)\quad \begin{cases} \overset{\circ}{\mathcal{M}} = \{ \sum_{i=1}^{d} f_i \cdot M^i : f_i \in L^2(R^d; \nu_i),\ 1 \le i \le d\} \\ e(\sum_{i=1}^{d} f_i \cdot M^i) = \frac{1}{2} \sum_{i=1}^{d} \|f_i\|^2_{L^2(\nu_i)} \end{cases}.$$

Combining this with Theorem 1, we can get the next theorem.

Theorem 2. The totality of local martingale additive functionals of the diffusion $\underset{\sim}{M}$ coincides with the following family of stochastic integrals

$$(3.4)\ \{M = \sum_{i=1}^{d} f_i \cdot M^i : f_i \in L^2(\{K_n\}; \nu_i) \text{ for some nest } \{K_n\}\},$$

where $L^2(\{K_n\}; \nu_i)$ denotes the space of Borel functions f such that $I_{K_n} \cdot f \in L^2(R^d; \nu_i)$, n=1,2,... .

Proof. Take $M \in \overset{\circ}{\mathcal{M}}_{loc}$. There exist a nest $\{K_n\}$ and MAF's $M^{(n)} \in \overset{\circ}{\mathcal{M}}$ such that (2.3) holds. According to (3.3), each $M^{(n)}$ can be expressed as $M^{(n)} = \sum_{i=1}^{d} f_i^{(n)} \cdot M^i$ with some $f_i^{(n)} \in L^2(R^d; \nu_i)$, $1 \le i \le d$.
Let us prove

$$(3.5)\ f_i^{(n)} = f_i^{(n+1)} \qquad \nu_i\text{-a.e. on } K_n^o,$$

K_n^o being the fine interior of K_n. If we let $g_i = f_i^{(n)} - f_i^{(n+1)}$, then $\sum_{i=1}^{d} (g_i \cdot M^i)_t = 0$, $t < \tau_{K_n}$, and hence

$$(3.6)\ \sum_{i=1}^{d} \int_0^t g_i(X_s)^2 d\langle M^i\rangle_s = 0,\ t < \tau_{K_n}.$$

We may assume without loss of generality that $I_{K_n} \cdot \nu_i$ is of finite energy integral ([2 ; Theorem 3.2.3]). Then [2 ; Lemma 5.1.5] applies and we conclude from (3.6) that $g_i^2 \cdot I_{K_n} \cdot \nu_i$ vanishes on K_n^o, which means (3.5).

Define a function f_i on $\bigcup_{n=1}^{\infty} K_n^o$ by $f_i(x) = f_i^{(n)}(x)$, $x \in K_n^o$. But by virtue of the property (1.1) of the nest $\{K_n\}$, we see that $\bigcup_{n=1}^{\infty} K_n - \bigcup_{n=1}^{\infty} K_n^o$ is exceptional and consequently ν_i-negligible. It is now clear that M has the expression with this f_i.

Conversely take M from the space (3.4). Then $M^{(n)} = \sum_{i=1}^{d} I_{K_n} \cdot f_i \cdot M^i \in \overset{\circ}{\mathcal{M}}$

by (3.3) and we have the identity (2.3) as well.
Hence $M \in \overset{\circ}{\mathcal{M}}_{loc}$. q.e.d.

The condition in (3.4) for the function f_i can be simply stated as follows:

(3.7) $f_i^2 d\nu_i$ is a smooth measure.

This condition is in turn equivalent to the following probabilistic one:

(3.8) $P_x(\int_0^t f_i(X_s)^2 d\langle M^i\rangle_s < \infty,\ \forall\, t>0) = 1$ q.e. $x \in R^d$.

The analytic condition (3.7) is easier to be understood than (3.8). For instance any function in $L^2_{loc}(R^d;\ \nu_i)$ is readily seen to satisfy (3.7).

Example 1. The d-dimensional Brownian motion is the case when m and ν_i are the d-dimensional Lebesgue measure. In this case $M_t = (M_t^1, M_t^2, \ldots, M_t^d)$ is the Brownian motion $X_t - X_0$ starting from the origin and $\langle M^i\rangle_t = t$, $1 \leq i \leq d$.
By Theorem 2, the local martingale additive functional of the Brownian motion is just the stochastic integral $\sum_{i=1}^{d} f_i \cdot M^i$ with some f_i such that $f_i^2 dx$ is smooth.

An analogous assertion for the local martingale additive functional in the strict sense (admitting no exceptional set) was made by A.D. Wentzell [7] (a detailed proof is given in a Japanese article by H. Tanaka and M. Hasegawa [6]), where the integrand f_i is characterized by the condition

$$P_x(\int_0^t f_i(X_s)^2 ds < \infty,\ \forall\, t>0) = 1 \quad \forall\, x \in R^d.$$

This is of course a counterpart to (3.8).
Apparently no counterpart to (3.7) has appeared in the literature.

As an immediate consequence of Theorem 2, the following theorem holds:

Theorem 3. The next two conditions are equivalent to each other for the diffusion process $\underset{\sim}{M}$:

(i) L is a local martingale multiplicative functional (admitting exceptional set) such that

$$P_x(L_o = 1, L_t > 0, \forall t > 0) = 1 \qquad \text{q.e. } x \in R^d.$$

(ii) L is expressed as

$$L_t = \exp \{ \sum_{i=1}^{d} (f_i \cdot M^i)_t - \frac{1}{2} \sum_{i=1}^{d} f_i^2 \cdot \langle M^i \rangle)_t \}$$

with functions f_i satisfying (3.7).

It is in fact well known that such a functional as in (i) can be written as

$$L_t = \exp \{M_t - \frac{1}{2} \langle M \rangle_t \}$$

by some local martingale additive functional M_t (cf. [3]).

Acknowledgements Thanks are due to Professor D. Williams for his interest in the related topics, and to Professors S.Albeverio and L. Streit for their kind help in preparing the paper.

References

[1] S. Albeverio, R. Hoegh-Krohn and L. Streit, Energy forms, Hamiltonians, and distorted Brownian paths, J. Math. Phys., 18(1977), 907-917.

[2] M. Fukushima, Dirichlet forms and Markov processes, North-Holland Publ. Co. and Kodansha, 1980.

[3] H. Kunita and S. Watanabe, On square integrable martingales, Nagoya Math. J., 30.(1967), 209-245.

[4] P.A. Meyer, Intégrales stochastiques, III, Séminaire de Probabilité I, Lecture Notes in Math., Vol. 39, Springer-Verlag, 1967.

[5] P. A. Meyer, Martingales local fonctionneles additives I, Sémnaire de Probabilités XII, Lecture Notes in Math., Vol. 649, Springer-Verlag.

[6] H. Tanaka and M. Hasegawa, Stochastic differential equations, Seminar on Probability, Vol. 19 (in Japanese), 1964.

[7] A. D. Wentzell, Additive functionals of multidimensional Wiener process, D.A.H. CCCP, 139 (1961), 13-16.

SET-PARAMETERED MARTINGALES AND MULTIPLE STOCHASTIC INTEGRATION

Bruce Hajek and Eugene Wong

Coordinated Sciences Laboratory
University of Illinois at Urbana
and
Electronics Research Laboratory
University of California at Berkeley

Abstract

The starting point of this paper is the problem of representing square-integrable functionals of a multiparameter Wiener process. By embedding the problem in that of representing set-parameter martingales, we show that multiple stochastic integrals of various order arise naturally. Such integrals are defined relative to a collection of sets that satisfies certain regularity conditions. The classic cases of multiple Wiener integral and Ito integral (as well as its generalization by Wong-Zakai-Yor) are recovered by specializing the collection of sets appropriately.

Using the multiple stochastic integrals, we obtain a martingale representation theorem of considerable generality. An exponential formula and its application to the representation of likelihood ratios are also studied.

Research sponsored by the U.S. Army Research Office under Contract DAAG29-79-G-0186.

1. Introduction

Let R^n denote the collection of all Borel sets in $\mathbb{R}^n$ with finite Lebesgue measure (denoted by μ). Define a Wiener process $\{W(A), A \in R^n\}$ as a family of Gaussian random variables with zero mean and

$$(1.1)\qquad EW(A)W(B) = \mu(A \cap B)$$

As a set-parameter process, $W(A)$ is additive, i.e.,

$$(1.2)\qquad W(A+B) = W(A) + W(B) \quad , \text{ a.s.}$$

where $A + B$ denotes the union of disjoint sets, and intuitively, we can view $W(A)$ as the integral over A of a Gaussian white noise.

The connection with white noise renders the Wiener process important in applications as well as theory. Consider for example, the following signal detection problem.

A process ξ_t is observed on $t \in T \subset \mathbb{R}^n$, and we have to decide between the possibilities: (a) ξ_t contains a random signal Z_t plus an additive Gaussian white noise and (b) ξ_t contains only noise.

Formulated so as to avoid the pathologies of "white noise," the problem can be stated as follows: Let $\{W(A), A \in R^n(T)\}$ be a set-parameter process, with parameter space $R^n(T) = \{\text{Borel subsets of } T\}$, and defined on a fixed measurable space (Ω,F). Let P' and P be two probability measures such that (a) under P' $W(A) - \int_A Z_t dt$ is a Wiener process independent of $\{Z_t, t \in T\}$, (b) under P $W(A)$ is a Wiener process.

Now, let F_W denote the σ-field generated by the process W, and let P_W and P'_W denote the respective probability measures restricted to F_W. If $\int_T Z_t^2 dt < \infty$; a.s., then $P'_W \ll P_W$ and the detection problem in most cases reduces to one of computing the likelihood ratio

$$(1.3) \qquad \Lambda = \frac{dP'_W}{dP_W}$$

in terms of the observed process W.

With respect to the probability space (Ω, F, P) $\{W(A), A \in R^n(T)\}$ is a Wiener process. Hence, Λ is a positive integrable functional of a Wiener process. Computing Λ in terms of W is a problem that can be embedded in a more general one of finding representations of a Wiener functional, which in turn can be embedded (and illuminated in the process) in a still more general problem of representing martingales generated by a Wiener process.

For a random variable Y that is a square-integrable functional of a Wiener process $\{W(A), A \in R^n(T)\}$, several representations already exist. The first is the Hermite-Wiener series of Cameron and Martin [1]. The second is in terms of the multiple Wiener integrals as defined by Ito [5]. The third is in terms of the Ito integral [4], and its generalization as defined by Wong and Zakai [8] and Yor [10]. In the last representation the concept of martingales plays a crucial role.

For processes with a multidimensional parameter, it is both more natural and more general to define <u>martingales</u> for processes parameterized by sets rather than points in $\mathbb{R}^n$. Let $C \subset R^n(T)$ be a collection of closed sets. Let $\{F(A), A \in C\}$ be a family of σ-fields such that $A \supseteq B \Rightarrow F(A) \supseteq F(B)$. Let $\{M(A), A \in C\}$ be a set-parameter process. We say that $\{M(A), F(A), A \in C\}$ is a <u>martingale</u> if

$$E(M(A)|F(B)) = M(B) \quad \text{a.s.}$$

whenever $A \supset B$. Let $\{W(A), A \in R^n(T)\}$ be a Wiener process and denote

$$F_W(A) = \sigma(\{W(B), B \subset A\})$$

The main object of this paper is to show that under very general conditions on C, there is a canonical representation of all square-integrable martingales with respect to $\{F_W(A), A \in C\}$, and hence representation for square integrable Wiener functionals. For C = {all closed sets} the representation reduces to that of multiple Wiener integrals. For C = {all closed rectangles in $\mathbb{R}^n_+$ with the origin as one corner} the representations of Ito, Wong-Zakai, and Yor are recovered. These two are in a sense limiting cases, and between them lies a vast spectrum of choices for C, giving rise to an equally large array of representations for C-martingales and Wiener functionals.

The key to these representations is to define multiple stochastic integrals of the form

$$\int_{T^m} \phi(t_1, t_2, \ldots, t_m)\, W(dt_1) \ldots\ldots W(dt_m)$$

where ϕ are (in general) random integrands C-adapted in a suitable sense to be defined later.

The basic ideas underlying this paper were first introduced in the dissertation [3].

2. Multiple Stochastic Integrals

Let C be a collection of Borel subsets of a fixed rectangle T. Given sets $A_1, A_2, \ldots, A_m \in R^n(T)$, we shall define their support in C by

$$(2.1) \qquad S_{A_1, A_2, \ldots, A_m} = \cap \{B = B \in C \text{ and } B \cap A_i \neq \phi \text{ for every } i\}$$

If $t_1,t_2,..,t_m$ are points in T, their support will be written as $S_{t,t_2..t_m}$. We say $t_1,t_2,..,t_m$ are C-independent if no point is contained in the support of the remaining ones.

For $C = R^n(T)$, $S_{t_1,t_2..t_m}$ is just $\{t_1,t_2,..,t_m\}$ so that C-independent mean "distinct". For $C = \{T_t,\ t \in T \subset \mathbb{R}^n_+\}$ where T_t denotes the rectangle bounded by the origin and t, $S_{t_1,t_2..t_m}$ is the smallest T_t that contains $t_1,t_2,..,t_m$, and C-independent means "pairwise unordered". For C = {all convex sets in T}, $S_{t_1,t_2..t_m}$ is the convex hull of $\{t_1,t_2,..,t_m\}$, and C-independent means $t_1,t_2,..,t_m$ are extreme points of their convex hull. More examples will be given later.

Let $\hat{T}^m$ denote the subset of C-independent points in $T^m \subset \mathbb{R}^{mn}$. For a fixed n and C, $\hat{T}^m$ may be vacuous for sufficiently large m. For example, if $C = \{T_t\}$ is the collection of rectangles bounded by the origin and $t \in T \subset \mathbb{R}^n_+$, then $\hat{T}^m$ is empty for $m > n$. That is, no more than n points can be C-independent.

Let (Ω,F,P) be a fixed probability space. Let $\{F(A),\ A \in C\}$ be a family of σ-subfields parameterized by sets in $C \subset R^n(T)$. Let $\{W(A),\ A \in R^n(T)\}$ be a Wiener process such that: (a) $A \subset B \Rightarrow W(A)$ is F(B)-measurable, and (b) $A \cap B = \phi \Rightarrow \{W(A'),\ A' \subset A\}$ is F(B)-independent. We shall assume the following conditions on C:

(c_1) For every collection of rectangles $A_1,A_2,\ldots,A_m$ such that

$$\prod_{i=1}^{m} A_i \subset \hat{T}^m$$

$$\mu(A_i \cap S_{A_1A_2..A_m}) = 0, \quad i = 1,2,\ldots,m$$

(c_2) For each $m \geq 1$, the mapping

$$t = (t_1, t_2, \ldots, t_m) \rightsquigarrow S_t$$

is a continuous map from T^m to the collection of sets that are compact under the metric

(2.2) $$\rho(A,B) = (\max_{x \in A} \min_{y \in B} |x-y| + \max_{x \in B} \min_{y \in A} |x-y|)$$

(c_3) For each $m \geq 1$ and for almost all $t \in T^m$

$$\mu(S_t - \bigcup_{\varepsilon>0} S_{B(\varepsilon,t_1),B(\varepsilon,t_2)..,B(\varepsilon,t_m)}) = 0$$

when $B(\varepsilon,t_i)$ denotes the ball with radius ε centered at t_i.

For a C satisfying conditions c_1 - c_3, we shall define multiple stochastic integrals of order m

(2.3) $$\phi \circ W^m = \int_{\hat{T}^m} \phi_t \, W(dt_1)..W(dt_m)$$

for integrands $\phi(t,\omega)$, $(t,\omega) \in \hat{T}^m \times \Omega$, statifying

(h_1) ϕ is $F \times \mu^m$-measurable

(h_2) For each $t \in \hat{T}^m$ ϕ_t is $F(S_t)$-measurable.

(h_3) $$\int_{\hat{T}^m} E\phi_t^2 \, dt < \infty$$

The space of functions satisfying h_1 - h_3 will be denoted by $L_a^2(\hat{T}^m x\Omega)$.

Call ϕ _atomic_ if $\phi(t,\omega) = \alpha(\omega) I_A(t)$ where I_A is the indication function of a product of rectangles $A = \prod_{i=1}^{m} A_i$ such that $A \subset \hat{T}^m$. Two atomic functions

(2.4) $\phi(t,\omega) = \alpha(\omega)\ I_A(t) \quad , A \subset \hat{T}^m$

$\theta(t,\omega) = \beta(\omega)\ I_B(t) \quad , B \subset \hat{T}^p$

are said to be comparable if each pair (A_i,B_j) is either equal or disjoint modulo sets of zero Lebesgue measure, and similar if $m = p$ and $(B_1,B_2,\ldots,B_m)$ is a permutation of $(A_1,A_2,\ldots,A_m)$. Call ϕ simple if $\phi = \sum_{k=1}^{K} \phi_k$ and each ϕ_k is atomic.

For an atomic function ϕ define

(2.5) $$\phi \circ W^m = \alpha \prod_{i=1}^{m} W(A_i)$$

So define, $\phi \circ W^m$ has the following property:

Lemma 2.1. Let ϕ and θ be comparable atomic functions in $L_a^2(\hat{T}^m x\Omega)$ and $L_a^2(T^p x\Omega)$ of the form (2.4). Then

(2.6) $$E(\phi \circ W^m)\ (\theta \circ W^p) = 0$$

unless ϕ and θ are similar. In the latter case,

(2.7) $$E(\phi \circ W^m)\ (\theta \circ W^m) = \int_{\hat{T}^m} E\tilde{\theta}_t \tilde{\theta}_t\ dt \overset{\text{def.}}{=} \langle \tilde{\phi},\tilde{\theta} \rangle$$

where $\tilde{\phi}$ denotes the symmetrization of ϕ, i.e.,

(2.8) $$\tilde{\phi}_t = \frac{1}{m!} \sum_{\Pi} \phi_{\Pi(t)} \quad , \Pi(t) = \text{permutation of } t$$

Proof: First, assume ϕ and θ to be similar. Then,

$$(\phi \circ W^m)\ (\theta \circ W^m) = \alpha\beta \prod_{i=1}^{m} W^2(A_i)$$

and $\alpha\beta$ is measurable with respect to $F(S_{A_1A_2\ldots A_m})$. Therefore, condition c_1 implies that

$$E[(\phi\circ W^m)(\theta\circ W^m)|F(S_{A_1A_2..A_m})]$$

$$= \alpha\beta \prod_{i=1}^{m} E\, W^2(A_i)$$

$$= \alpha\beta \prod_{i=1}^{m} \mu(A_i)$$

and (2.7) follows.

Next, suppose that ϕ and θ are comparable but not similar. With no loss of generality assume $m \geq p$. Consider two possibilities:

(a) There exists a B_j (say B_1) such that

$$\mu(B_1 \cap [\bigcup_{i=1}^{m} A_i \cup S_{A_1A_2..A_m}]) = 0$$

(b) For every $j \leq p$

$$\mu(B_j \cap [\bigcup_{i=1}^{m} A_i \cup S_{A_1A_2..A_m}]) \neq 0$$

For case (a), let

$$D = \bigcup_{i=1}^{m} A_i \bigcup_{j=2}^{p} B_j \cup S_{A_1A_2..A_m} \cup S_{B_1B_2\ldots B_p}$$

Then, with probability 1

$$E[(\phi\circ W^m)(\theta\circ W^p)|F(D)] = \alpha\beta \prod_{i=1}^{m} W(A_i) \prod_{j=2}^{p} W(B_j)[EW(B_1)] = 0$$

and (2.6) is verified.

For case (b) we shall prove that $S_{A_1A_2..A_m} \supset S_{B_1B_2..B_p}$. Since ϕ and θ are comparable but not similar and $m \geq p$, there must exist an A_i(say A_1) such that $\mu(A_1 \cap B_j) = 0$ for every j. Hence, $W(A_1)$ is independent of $\alpha\beta \prod_{i=2}^{m} W(A_i) \prod_{j=1}^{p} W(B_j)$ and (2.6) is again proved.

To prove $S_{A_1A_2..A_m} \supset S_{B_1B_2..B_p}$ for case (b), let $D \in C$ be any set such that

$$D \cap A_i \neq \phi \text{ for every } i$$

then, $D \supset S_{A_1A_2..A_m}$ by definition. The defining condition for case (b) implies that for each j

either $B_j \cap \bigcup_i A_i \neq \phi$

which implies $B_j = A_i$ for some i

which in turn implies $D \cap B_j \neq \phi$

or $B_j \cap S_{A_1A_2..A_m} = \phi$

which implies $D \cap B_j \neq \phi$

Therefore,

$$D \cap A_i \neq \phi \text{ for every } i \Rightarrow D \cap B_j \neq \phi \text{ for every } j$$

and $S_{A_1A_2..A_m} \supset S_{B_1B_2...B_p}$ ¤

<u>Lemma 2.2</u>. For atomic functions ϕ and θ that are not necessarily comparable, we can write

$$(2.9) \qquad \phi = \sum_{k=1}^{K} \phi_k$$

$$\phi = \sum_{\lambda=1}^{L} \theta_\lambda$$

where ϕ_k, θ_λ are atomic and the set $\{\phi_k,\theta_\lambda\}$ is pairwise comparable. For any atomic ϕ and θ in L_a^2 the isometry

$$(2.10) \qquad E(\phi\circ W^m)\,(\theta\circ W^p) = \delta_{mp}\,\langle\tilde{\phi},\tilde{\theta}\rangle$$

holds.

Proof: ϕ and θ, being atomic, are of the form

$$\phi = \alpha\, I_{A_1 x A_2 x \ldots x A_m}$$

$$\theta = \beta\, I_{B_1 x B_2 x \ldots x B_p}$$

where $A_1, A_2, \ldots A_m, B_1, \ldots, B_p$ are rectangles in T. Since a union of rectangles is always a union of disjoint rectangles, there exist disjoint rectangles $D_1, D_2, \ldots, D_q$ such that each A_i or B_j is the union of some of the D_ν's. Hence (2.9) follows, with

$$\phi_k = \alpha\, I_{D_{k1} x D_{k2} x \ldots x D_{km}}$$

$$\theta_\lambda = \beta\, I_{D_{\lambda 1} x D_{\lambda 2} x \ldots x D_{\lambda p}}$$

where $D_{ki} \subset A_i$ and $D_{\lambda j} \subset B_j$ for every i and j. It follows that α is $F(S_{D_{k1}D_{k2}\ldots D_{km}})$-measurable and β is $F(S_{D_{\lambda 1}D_{\lambda 2}\ldots D_{\lambda m}})$-measurable for each k and λ. From lemma 2.1 we have

$$E(\phi_k\circ W^m)\,(\phi_\lambda\circ W^p) = \delta_{mp}\,\langle\tilde{\phi}_k,\tilde{\theta}_\lambda\rangle$$

and (2.10) follows from the bilinearity of $\langle \quad \rangle$. ¤

Lemma 2.3. Under conditions c_2 and c_3 the subset of simple functions is dense in $L_a^2(\hat{T}^m x\Omega)$.

A proof of this result is given in the appendix A.

Theorem 2.1. There is a unique linear map denoted by $\phi \circ W^m$ of $\phi \in L_a^2(\hat{T}^m x\Omega)$ into the space of square-integrable random variables such that

(a) For an atomic function $\phi = \alpha I_A$

$$\phi \circ W^m = \alpha \prod_i W(A_i)$$

(b) Symmetry:

$$\phi \circ W^m = \tilde{\phi} \circ W^m$$

(c) Isometry:

$$E(\phi \circ W^m)(\theta \circ W^p) = \langle \tilde{\phi}, \tilde{\theta} \rangle \delta_{mp}$$

Proof: First, any simple function ϕ is by definition of the form $\phi = \sum_{k=1}^{K} \phi_k$, where ϕ_k are atomic. Bilinearity of $\langle \quad \rangle$ then implies the isometry (2.10) for simple functions ϕ and θ. Let ϕ be any function from $L_a^2(\hat{T}^m x\Omega)$. Lemma 3.2 implies that there exists a sequence $\{\phi^{(n)}\}$ of simple functions such that

$$\phi^{(n)} \xrightarrow[n\to\infty]{L_a^2} \phi$$

Hence, $\{\phi^{(n)}\}$ is Cauchy. The isometry (2.10) then implies that $\{\phi^{(n)} \circ W^m\}$ is mean-square convergent as a sequence of random variables, and we take the limit to be $\phi \circ W^m$. Verification of the properties follows from the isometric property in a striaghtforward way. ¤

Remark: Observe that the isometry property of the multiple stochastic integral implies uniqueness up to equivalence of integrand. That is, if $\phi \circ W^m = \theta \circ W^m$ then

$$\|\phi-\theta\|^2 = \int_{\hat{T}^m} E(\tilde{\phi}_t-\tilde{\theta}_t)^2 \, dt = 0$$

Theorem 2.2. (Projection) For any $B \in \mathcal{R}^n(T)$

$$E(\phi \circ W^m | F(B)) = E(\phi | F(B)) \, I_{B^m} \circ W^m \tag{2.12}$$

Proof: It is enough to prove this for an atomic ϕ. Let $\phi = \alpha \, I_{A_1 \times A_2 \times \ldots \times A_m}$. Then

$$E(\phi \circ W^m | F(B)) = E(\alpha \prod_{i=1}^{m} W(A_i) | F(B))$$

$$= E\{\alpha \, E[\prod_{i=1}^{m} W(A_i) | F(B \cup S_{A_1 \ldots A_m})] | F(B)\}$$

$$= E\{\alpha \prod_{i=1}^{m} W(A_i \cap B) | F(B)\}$$

$$= E(\alpha | F(B)) \prod_{i=1}^{m} W(A_i \cap B)$$

$$= E(\phi | F(B)) \, I_{B^m} \circ W^m$$

¤

Corollary. If $B \in C$ then

$$E(\phi \circ W^m | F(B)) = \phi \, I_{B^m} \circ W^m$$

Proof: If $B \in C$ then $t_i \in B$ for each i implies $B \supset S_{t_1 t_2 \ldots t_m}$. Hence, $t \in B^m \Rightarrow \phi_t$ is $F(B)$-measurable and $E(\phi | F(B)) \, I_{B^m} = \phi \, I_{B^m}$ a.s. ¤

Let $\{(\phi \circ W^m)_B, B \in C\}$ be the set-parameterized process defined by

$$(\phi \circ W^m)_B = \phi \, I_{B^m} \circ W^m$$

Then the corollary to Theorem 2.2 implies that $\{(\phi \circ W^m)_B, B \in C\}$ is a martingale. We shall call $(\phi \circ W^m)_B$ the <u>indefinite integral</u> of $\phi \circ W^m$.

3. Relationship with Multiple Wiener Integrals and Representation of Wiener Functionals

Let $\tilde{T}^m$ denote the set of m-tuples of distinct points in T. Let $\theta(t)$, $t \in \tilde{T}^m$ satisfy

$$\int_{\tilde{T}^m} \theta^2(t)\, dt < \infty$$

Let $\theta \,\square\, W^m$ denote a multiple Wiener integral of order m.

<u>Theorem 3.1</u>. For a given C satisfying condition $c_1 - c_3$, a multiple Wiener integral can be represented as

$$(3.1) \qquad \theta \,\square\, W^m = \sum_{k=1}^{m} \binom{m}{k} \theta_k \circ W^k$$

where

$$(3.2) \qquad \theta_k(t_1, t_2, \cdots, t_k, \omega) = (\tilde{\theta}(t_1, t_2, \cdots, t_k, \cdot)\, I_{S^{m-k}_{t_1 t_2 \cdots t_k}}(\cdot) \,\square\, W^{m-k})(\omega)$$

and $\theta_k \circ W^k$ is a multiple stochastic integral defined relative to C.

Proof: Let $\Pi\theta$ denote the transformation of θ by a premutation of its arguments. Suppose for some permutation Π

$$\Pi\theta = I_{A_1 \times A_2 \times \cdots \times A_m}$$

where $A_1,\ldots\ldots,A_k$ are C-independent rectangles and $A_{k+1},\ldots,A_m$ are distinct rectangles contained in $S_{A_1A_2..A_k}$. Then, symmetry implies that

$$\theta \,\square\, W^m = \Pi\theta \,\square\, W^m = [\prod_{i=k+1}^{m} W(A_i)] \prod_{i=1}^{k} W(A_i)$$

$$= h_k \circ W^k$$

when

$$h_k(t_1,\ldots,t_k,\omega) = I_{A_1x\ldots xA_k}(t_1,t_2,\ldots,t_k)[I_{A_{k+1}x..xA_m} \,\square\, W^{m-k}](\omega)$$

$$= \Pi\theta(t_1,t_2,\ldots,t_k,\cdot) \,\square\, W^{m-k}$$

The isometry of multiple stochastic integrals implies that both k and the two sets $\{A_1,A_2,..,A_k\}$ and $\{A_{k+1},A_{k+2},\ldots A_m\}$ are unique. The integer k is unique because otherwise we would have

$$E(\theta \,\square\, W^m)^2 = E(h_k \circ W^k)\ (h_{k'} \circ W^{k'}) = 0$$

The collection $\{A_1,A_2,..,A_k\}$ is unique because otherwise we would have

$$\theta \,\square\, W^m = h_k \circ W^k = g_k \circ W^k$$

and $h_k g_k \equiv 0$. It follows that

$$\sum_{\text{all } \Pi} [(\Pi\theta)(t_1,t_2,..,t_k,\cdot)\ I_{S^{m-k}_{t_1t_2..t_k}}(\cdot) \,\square\, W^{m-k}] \circ W^k$$

$$= k!(m-k)!\ \theta \,\square\, W^m$$

$$= m!\ \theta_k \circ W^k$$

where

$$\theta_k(t_1,t_2,\ldots,t_k,\omega) = [\tilde{\theta}(t_1,t_2,\ldots,t_k,\cdot)\ I_{S^{m-k}_{t_1 t_2 \ldots t_k}}(\cdot) \,\square\, W^{m-k}](\omega)$$

Hence,

$$\theta \,\square\, W^m = \binom{m}{k} \quad \theta_k \circ W^k$$

In appendix B, it is proved that linear combinations of such θ's are dense in $L^2(T^m)$. Thus, the theorem is proved. ¤

Corollary 1. Let $F_W(A)$ denote the σ-field generated by $\{W(B),\ B \subset A\}$. Then, every square-integrable $F_W(T)$-measurable random variable Z has a representation of the form

$$(3.3) \qquad Z = EZ + \sum_{m=1}^{\infty} Z_m \circ W^m$$

where $Z_m \circ W^m$ are stochastic integrals defined with respect to the same C that satisfies conditions $c_1 - c_3$.

Proof: This corollary follows immediately from the main theorem and the well known result [5] that Z has a representation in a series of multiple Wiener integrals.

Corollary 2. For $f \in L^2(T)$, define

$$(3.4) \qquad \hat{f}^m(t_1,\ldots,t_m) = \prod_{i=1}^{m} f(t_i)$$

and set

$$(3.5) \qquad W_m(f,A) = (\hat{f}^m \square\, W^m)_A$$

Then, for $A \in C$

$$(3.6) \qquad W_m(f,A) = \sum_{k=1}^{m} \binom{m}{k} [\hat{f}^k(\cdot)\, W_{m-k}(f,S_{\cdot}) \circ W^k]_A$$

Proof: Observe that f^k is symmetric and

$$\hat{f}^m(t_1,t_2,\ldots,t_m) = \hat{f}^k(t_1,t_2,\ldots,t_k)\, \hat{f}^{m-k}(t_{k+1},\ldots,t_m)$$

Hence, (3.1) yields (3.6) for $A = T$, and the rest follows from the projection property (Theorem 2.2).

<u>Corollary 3</u>. For $f \in L^2(T)$ define

$$(3.7) \qquad L(f,A) = \exp\{(f \square W)_A - \frac{1}{2} \int_A f^2(t)\, dt\}$$

Then, for $A \in C$

$$(3.8) \qquad L(f,A) = 1 + \sum_{m=1}^{\infty} \frac{1}{m!} [\hat{f}^m(\cdot) L(f,S_{\cdot}) \circ W^m]_A$$

Proof: For multiple Wiener integrals (C = {all closed sets }}(3.8) reduces to

$$(3.9) \qquad L(f,A) = 1 + \sum_{m=1}^{\infty} \frac{1}{m!} W_m(f,A)$$

which is well known [5]. For the general case, we use (3.6) in (3.9) and write

$$L(f,A) = 1 + \sum_{m=1}^{\infty} \frac{1}{m!} \sum_{k=1}^{m} \binom{m}{k} [\hat{f}^k W_{m-k}(f,S_{\cdot}) \circ W^k]_A$$

$$= 1 + \sum_{k=1}^{\infty} \frac{1}{k!} [\hat{f}^k \sum_{j=0}^{\infty} \frac{1}{j!} W_j(f,S.) \circ W^k]_A$$

$$= 1 + \sum_{k=1}^{\infty} \frac{1}{k!} [\hat{f}^k L(f,S.) \circ W^k]_A \qquad \square$$

The expansion formula (3.8) for exponentials of the form (3.7) can be extended with the Wiener integral $f \,\square\, W$ in the exponent being replaced by a stochastic integral $f \circ W$. The result can be stated as follows:

Proposition 3.2. Equation (3.8) remains valid for $f \in L_a^2(T \times \Omega)$ such that f is bounded.

Proof: Define f to be a discrete simple function if f is a simple function

$$f(t,\omega) = \sum_{i=1}^{k} \alpha_i(\omega)\; I_{A_i}(t)$$

such that $P(\alpha_i \in J) = 1$ for some finite set J. Such a function may be written as $f(t,\omega) = g(t,\alpha(\omega))$ where $\alpha = (\alpha_1,\ldots,\alpha_k)$ and

$$g(t,c) = \sum_{i=1}^{k} c_i I_{A_i}(t) \text{ for } c \in J^k.$$

Then $g(\cdot,c) \in L^2(T)$ for each $c \in J^k$ so by Corollary 3 of Theorem 3.1,

$$(3.10) \qquad L(g(\cdot,c),A) = 1 + \sum_{m=1}^{\infty} \frac{1}{m!} [\hat{g}^m(\cdot,c) L(g(\cdot,c),S.) \circ W^m]_A.$$

This equality holds in $L^2(\Omega,F,P)$ for each $c \in J^k$ and hence it continues to hold in $L^2(\Omega,F,P)$ if c is replaced by the random vector $\alpha(\omega)$. By proposition C in appendix C, replacing c by $\alpha(\omega)$ in the stochastic integrals is equivalent to replacing c by $\alpha(\omega)$ in each of the integrands and then forming the stochastic integrals. (To apply propositon C to the $\underline{m}$th term on the right of (3.10), let $B_i = S_{A_i}$). This verifies equation (3.8) if f is a discrete simple function.

Conclude that $E[L(f,A)] = 1$ if f is discrete and simple. Moreover, if $p \geq 1$ and $|f(\cdot,\cdot)| \leq \Gamma$ for some constant Γ, then

$$L(f,A)^p = L(pf,A)\exp(\tfrac{1}{2}(p^2-p)\int_A f(t)^2dt)$$
$$\leq L(pf,A)\exp(\tfrac{1}{2}(p^2-p)\Gamma^2\mu(T)$$

so that

$$(3.11) \qquad E[L(f,A)^p] \leq \exp(\tfrac{1}{2}(p^2-p)\Gamma^2\mu(T))$$

Now choose any $f \in L_a^2(T\times\Omega)$ with $|f(\omega,t)| \leq \Gamma$. Then there is a sequence of discrete simple functions $f_j \to f$ in $L_a^2(T\times\Omega)$ such that $|f_j(\omega,t)| \leq \Gamma$ for each j. Hence $(f_j\circ W)_A \to (f\circ W)_A$ a.s. in $L^2(\Omega)$ so that taking a subsequence if necessary, we can assume that $(f_j\circ W)_A \to (f\circ W)_A$ with probability one. Thus $L(f_j,A) \to L(f,A)$ with probability one. By the estimate (3.11), the collection of random variables $\{L(f_j,A)^p : p \geq 1\}$ is uniformly integrable for each $p \geq 1$ so that $L(f_j,A) \to L(f,A)$ in $L^p(\Omega)$ for each $p > 1$. Moreover, $\hat{f}_j^m \to \hat{f}^m$ in $L_a^p(\hat{T}^m \times \Omega)$ for each $p \geq 1$ since these functions are uniformly bounded. Now (3.8) is true for f replaced by f_j, and it is then easily verified for f by taking the limit in $L^2(\Omega)$ term by term as $j \to +\infty$. ¤

4. A Likelihood Ratio Formula

Let $\{Z_t, t \in T\}$ be a bounded process defined on (Ω,F,P) and let $\{W(A), A \subset T\}$ be a Wiener process defined on the same space. Let $F(A) = \sigma(\{W(B), B \subset A\}, \{Z_t, t \in A\})$. We assume that $A \cap A' = \phi \Rightarrow W(A')$ in $F(A)$-independent. For any collection C the support

S_t contains t. Hence, Z_t is $F(S_t)$ measurable. For any C satisfying $c_1 - c_3$, the stochastic integral $Z \circ W$ is well-defined.

Now, let P' be a measure on (Ω, F) defined by:

$$\frac{dP'}{dP} = \exp\{Z \circ W - \frac{1}{2} Z^2 \circ \mu\} \tag{4.1}$$

and set

$$L(Z,A) = \exp\{(Z \circ W)_A - \frac{1}{2}(Z^2 \circ \mu)_A\} \tag{4.2}$$

For any C satisfying $c_1 - c_3$, proposition 3.2 yields

$$L(Z,A) = 1 + \sum_{m=1}^{\infty} \frac{1}{m!} [\hat{Z}^m(\cdot) L(Z,S_{\cdot}) \circ W^m]_A \tag{4.3}$$

It follows that

$$L(Z,A) = E(\frac{dP'}{dP} | F(A)) \tag{4.4}$$

and P' is a probability measure.

Next, let $F_W(A) = \sigma(\{W(B), B \subset A\})$, and define the likelihood ratio by

$$\Lambda(A) = E(\frac{dP'}{dP} | F_W(A)) \tag{4.5}$$

We shall use (4.3) to derive an expression for $\Lambda(A)$.

Proposition 4.1. Let $t \in \hat{T}^m$ and define

$$\tilde{Z}_m(t) = E'(Z(t_1)Z(t_2)..Z(t_m) | F_W(S_{t_1 t_2 \ldots t_m})) \tag{4.6}$$

Then the likelihood ratio is given by

$$(4.7) \qquad \Lambda(A) = 1 + \sum_{m=1}^{\infty} \frac{1}{m!} [\tilde{Z}_m(\cdot)\Lambda(S_.) \circ W^m]$$

Proof: We begin by writing

$$\Lambda(A) = E[L(Z,A)|F_W(A)]$$

and using (4.3). Observe that with P-measure 1,

$$E\{[\hat{Z}^m(\cdot)L(Z,S_.) \circ W^m]_A|F_W(A)\} = E[\hat{Z}^m(\cdot)L(Z,S_.)|F_W(A)] \circ W^m$$

Now, for $t = (t_1,t_2,\ldots,t_m) \in A^m$,

$$E[\hat{Z}^m(t)L(Z,S_t)|F_W(A)]$$

$$= E[Z(t_1)Z(t_2)\ldots Z(t_m)L(Z,S_{t_1t_2\ldots t_m})|F_W(A)]$$

$$= E[Z(t_1)Z(t_2)\ldots Z(t_m)L(Z,S_{t_1t_2\ldots t_m})|F_W(S_{t_1t_2\ldots t_m})]$$

$$= \Lambda(S_{t_1\ldots t_m})E'[Z(t_1)\ldots Z(t_m)|F_W(S_{t_1\ldots t_m})]$$

$$= \tilde{Z}_m(t)\,\Lambda\,(S_t)$$

and (4.7) follows. ¤

Two special cases are of particular interest. First, let $a \in \mathbb{R}^n$ be a fixed unit vector (i.e., $\|a\| = 1$) and let H_α denote the half space $\{t \in \mathbb{R}^n : (t,a) \geq \alpha\}$. Then, the collection $C = \{H_\alpha \cap T\}$ is a one-parameter family of sets such that $\hat{T}^m$ is vacuous for $m > 1$. That is, two or more points are always C-dependent. In this case the likelihood ratio formula reduces to

$$\Lambda(A) = 1 + [\tilde{Z}_1(\cdot)\Lambda(S_.) \circ W]_A \quad , \quad A \in C$$

and an application of (3.8) yields

(4.8) $$\Lambda(A) = L(\tilde{Z}_1,A) = \exp\{(\tilde{Z}_1 \circ W - \frac{1}{2}\tilde{Z}_1^2 \circ \mu)_A\}$$

where

$$\tilde{Z}_1(t) = E'(Z(t)|F_W(S_t))$$

$$= E'(Z(t)|F_W(H_{(t,a)} \cap T))$$

In this case we see that the likelihood ratio is expressible as an exponential of the conditional mean.

The second case of special interest results from taking $C = \{\text{all closed sets in } T\}$. For this case

$$S_{t_1 t_2 \ldots t_m} = \{t_1, t_2, \ldots, t_m\}$$

Hence, with P-measure 1

$$\Lambda(S_{t_1 t_2 \ldots t_m}) = 1$$

and

$$\tilde{Z}_m(t) = E' \; Z(t_1) \ldots Z(t_m)]$$

Furthermore, if we assume that Z and W are independent processes under P then Z is identically distributed under P'. Hence, for that case we can write

(4.9) $$\Lambda(A) = 1 + \sum_{m=1}^{\infty} \frac{1}{m!} (\rho_m \square W^m)_A$$

where ρ_m is the <u>m</u>th moment

(4.10) $$\rho_m(t_1, t_2, \ldots, t_m) = E[Z(t_1) \ldots Z(t_m)] \quad .$$

Equation (4.9) provides a martingale representation of the likelihood ratio for the "additive white Gaussian noise" model under very general conditions. In the one-dimensional case, it was recently obtained in [7].

Equation (4.7) is an integral equation in that Λ occurs on both sides. In special cases [2,6,9] the equation can be converted to yield an exponential formula for Λ in terms of conditional moments.

References

1. Cameron, R. H., Martin, W. T.: The orthogonal development of non-linear functionals in a series of Fourier-Hermite functions. Ann. of Math. 48, 385-392 (1947).
2. Duncan, T. E.: Likelihood functions for stochastic signals in white noise. Inform. Contr. 16, 303-310 (1970).
3. Hajek, B. E.: Stochastic Integration, Markov Property and Measure Transformation of Random Fields. Ph.D. dissertation, Berkeley, 1979.
4. Ito, K.: Stochastic integrals. Proc. Imp. Acad. Tokyo 20, 519-524 (1944).
5. Ito, K.: Multiple Wiener Integral. J. Math. Soc. Japan 3, 157-169 (1951).
6. Kailath, T.: A general likelihood-ratio formula for random signals in Gaussian noise. IEEE Trans. Inform. Th. 15, 350-361 (1969).
7. Mitter, S. K., Ocone, D.: Multiple integral expansion for nonlinear filtering. Proc. 18th IEEE Conference on Decision and Control, 1979.
8. Wong, E., Zakai, M.: Martingales and Stochastic integrals for processes with a multi-dimensional parameter. Z. Wahrscheinlichkeitstheorie 29, 109-122 (1974).
9. Wong, E., Zakai, M.: Likelihood ratios and transormation of probability associated with two-parameter Wiener processes. Z. Wahrscheinlichkeitstheorie 40, 283-309 (1977).
10. Yor, M.: Representation des martingales de carré integrable relative aux processus de Wiener et de Poisson à n paramétres. Z. Wahrscheinlichkeitstheorie 35, 121-129 (1976).

Appendix A: Proof that Simple Functions are Dense

The purpose of this appendix is to prove the following proposition:

Proposition A. Conditions c_2 and c_3 imply that the space of simple functions is dense in $L_a^2(\hat{T}^m x\Omega)$ for each $m \geq 1$.

Proof: We begin by introducing some additional notation. For $\varepsilon > 0$ and $t = (t_1,\ldots,t_m) \in T^m$, define the ε-support of t by

$$S_t^\varepsilon = S_{B(\varepsilon,t_1)B(\varepsilon,t_2)\ldots B(\varepsilon,t_m)} \quad ,$$

where $B(\varepsilon,t_i)$ denotes a ball with radius ε and center t_i, and define $S_t^{(-)} = \bigcup_{\varepsilon>0} S_t^\varepsilon$. Define $L_\varepsilon^2(\hat{T}^m x\Omega)$ the same way as $L_a^2(\hat{T}^m x\Omega)$ but with condition h_2 replaced by the stronger condition: (h_2^ε) for each $t \in \hat{T}^m$, ϕ_t is $F(S_t^\varepsilon)$-measurable. Finally, let $C_\varepsilon(\hat{T}^m x\Omega)$ be the subspace of $L_\varepsilon^2(\hat{T}^m x\Omega)$ consisting of $\phi \in L_\varepsilon^2(\hat{T}^m x\Omega)$ such that $\phi(\cdot,\omega)$ is continuous on $\hat{T}^m$ with probability one.

Proposition A is a consequence of the following sequence of lemmas.

$\square$

Lemma A.1. $\bigcup_{\varepsilon>0} L_\varepsilon^2(\hat{T}^m x\Omega)$ is dense in $L_a^2(\hat{T}^m x\Omega)$ under conditions c_2 and c_3.

Proof: Let $f \in L_a^2(\hat{T}^m x\Omega)$ be bounded by a constant $\Gamma > 0$. For any $\varepsilon > 0$, there is a Borel measurable mapping $u(\cdot,\varepsilon)$ of the open set $\hat{T}^m$ into a finite subset of $\hat{T}^m$ such that $|u(\underline{x},\varepsilon) - \underline{x}| < \varepsilon$ for all $\underline{x} \in \hat{T}^m$. Define

$$f^\varepsilon(S) = E[f(s)|F(S_{u(s,\varepsilon)}^{2\varepsilon})] \qquad s \in \hat{T}^m$$

A version of $f^\varepsilon(s)$ can be chosen for each s so that f^ε is a jointly measurable function of (s,ω). Indeed, for each fixed $t \in \hat{T}^m$ there exist versions of

$$g^{\varepsilon}(s,t) = E[f^{\varepsilon}(s)|F(S_t^{2\varepsilon})]$$

which are jointly measurable functions of (s,ε), and then $g^{\varepsilon}(s,u(s,\varepsilon))$ is a jointly measurable version of $f^{\varepsilon}(x)$. Also, f^{ε} can be assumed to be bounded by Γ. For each $s \in \hat{T}^m$, $f^{\varepsilon}(s)$ is measurable with respect to

$$F(S^{2\varepsilon}_{u(x,\varepsilon)}) \subset F(S_s^{\varepsilon})$$

so that $f^{\varepsilon} \in L^2_{\varepsilon}(\hat{T}^m x\Omega)$.

Since $S_s^{3\varepsilon} \subset S^{2\varepsilon}_{u(s,\varepsilon)} \subset S_s^{\varepsilon}$, $\lim_{\varepsilon\downarrow 0} S^{2\varepsilon}_{u(s,\varepsilon)} = \lim_{\varepsilon\downarrow 0} S_s^{\varepsilon} = S_s^{(-)}$ for each $\underline{s} \in \hat{T}^m$. By the continuity of σ-fields generated by the Wiener process, $\lim_{\varepsilon\downarrow 0} F(S_s^{\varepsilon}) = F(S_s^{(-)})$. Then, by L^2-martingale convergence, for each $s \in \hat{T}^m$,

$$E[(f(s)-f^{\varepsilon}(s))^2] = E\,(E[f(s)|F(R_s)] - E[f(s)|F(R^{2\varepsilon}_{u(s,\varepsilon)})])$$

$$\underset{\varepsilon\downarrow 0}{\rightarrow} E\,(E[f(s)|F(R_s)] - E[f(s)|F(R_s^{(-)})])$$

By condition c_3, $\mu(R_\varepsilon - R_s^{(-)}) = 0$ and so also $E[(f(s)-f^{\varepsilon}(s))^2] \to 0$, for a.e. $s \in \hat{T}^m$. Since $(f(s)-f^{\varepsilon}(s))^2 \leq 4\Gamma^2$,

$$\|f-f^{\varepsilon}\|^2 = \int_{\hat{T}^m} E[(f(s)-f^{\varepsilon}(s))^2\, ds \to 0$$

by the Lebesgue Dominated Convergence Theorem. Thus, any bounded function $f \in L^2_a(\hat{T}^m x\Omega)$ is the limit of functions in $\bigcup_{\varepsilon>0} L^2_{\varepsilon}(\hat{T}^m x\Omega)$. Since the bounded functions in $L^2_a(\hat{T}^m x\Omega)$ are dense in $L^2_a(\hat{T}^m x\Omega)$, the lemma is established. ¤

<u>Lemma A.2</u>. $\bigcup_{\varepsilon>0} C_{\varepsilon}(\hat{T}^m x\Omega)$ is dense in $\bigcup_{\varepsilon>0} L^2_{\varepsilon}(\hat{T}^m x\Omega)$.

Proof: Let $f \in L^2_{2\varepsilon}(\hat{T}^m x\Omega)$ be bounded by some constant $\Gamma > 0$. Choose $V \in C^\infty(\mathbb{R}^{mn})$ such that $V \geq 0$, $V(x) = 0$ if $|x| \geq 1$, and $\int_{\mathbb{R}^n} V(x)dx = 1$.

For $\delta > 0$, define $V^\delta \in C^\infty(\mathbb{R}^{mn})$ by $V^\delta(x) = (\frac{1}{\delta})^{mn} V(\frac{x}{\delta})$ and define a function f^δ on $\hat{T}^m$ by the convolution: $f^\delta(\cdot,\omega) = V^\delta * f(\cdot,\omega)$ for each fixed ω. Here the function $f(\cdot,\omega)$, which is a priori defined on $\hat{T}^m \subset T^m \subset (\mathbb{R}^n)^m \simeq \mathbb{R}^{mn}$, is extended to a function on all of $\mathbb{R}^{mn}$ by the convention $f(\underline{s},\omega) = 0$ if $\underline{s} \notin \hat{T}^m$. Note that f^δ is bounded by Γ and sample continuous, and since $V(x) = 0$ for $|x| \geq \delta$, $f^\delta \in C_{2\varepsilon-\delta}(\hat{T}^m x\Omega)$.

Observe that

$$
\begin{aligned}
\|f-f^\delta\|^2 &= E[\int_{\hat{T}^m} |f(s) - f^\delta(s)|^2 \, ds] \\
&\leq E\int_{\mathbb{R}^{mn}} |f(s) - V^\delta * f(s)|^2 \, ds] \\
&\leq \int_{\mathbb{R}^{mn}} V\ (x)\ E[\int_{\mathbb{R}^{mn}} |f(s) - f(s-x)\ ^2 \, ds] \, dx \qquad \text{(A.1)}
\end{aligned}
$$

Now $\int_{\mathbb{R}^{mn}} |f(s) - f(s-x)|^2 \, dx \to 0$ as $s \to 0$ for all ω since translations are continuous in $L^2(\mathbb{R}^{mn})$. Hence, the expertation in (A.1) converges to zero as $x \to 0$ by Lebesgues Bounded Convergence Theorem and so also $\|f-f^\delta\| \to 0$ as $\delta \to 0$. ¤

<u>Lemma A.3.</u> If $f \in C\ (\hat{T}^m x\Omega)$ for some $\varepsilon > 0$, then there is a sequence f^δ of simple function which converge to f in $L^2(\hat{T}^m x\Omega)$.

Proof: It suffices to prove the lemma under the additional assumption that f is bounded uniformly in (t,ω). Recall that under Condition

c_3, $\hat{T}^m$ is naturally identified with an open subset of $\mathbb{R}^{mn}$. For $\delta > 0$, let $\mathcal{I}_\delta$ denote sets of the form

$$(I_{11} \times \ldots \times I_{1n}) \times \ldots \times (I_{m1} \times \ldots \times I_{mn})$$

where each I_{ij} is an interval of the form $(k\delta, (k+1)\delta]$, and let $\hat{\mathcal{I}}_\delta$ consist of $A \in \mathcal{I}_\delta$ such that $A \subset \hat{T}^m$. Let $u(\cdot,\delta)$ be a function from $\hat{T}^m$ to $\hat{T}^m$ such that $u(x,\delta) = u(x',\delta) \in J$ whenever $x, x' \in J$ for some $J \in \mathcal{I}_\delta$. Define $f^\delta(\underline{s}) = f(u(\underline{s},\delta))$ if $\underline{s} \in J$ for some $J \in \hat{\mathcal{I}}_\delta$, and define $f^\delta(\underline{s}) = 0$ otherwise. For $\delta < \varepsilon/\sqrt{n}$, each of the m rectangles in T of a set in $\hat{\mathcal{I}}_\delta$ has diameter less than ε so that $f^\delta \in S_a(\hat{T}^m \times \Omega)$ for $\delta < \varepsilon/\sqrt{n}$. Furthermore, f^δ is bounded by the same constant that f is, and $f^\delta(s,\omega) \to f(s,\omega)$ as $\delta \to 0$ for each $(s,\omega) \in \hat{T}^m \times \Omega$ by the sample continuity of f. So $f^\delta \to f$ in $L^2(\hat{T}^m \times \Omega)$ as $\delta \to 0$ by dominated convergence. ¤

Appendix B

Let $\mathcal{I}_m$ denote the collection of subsets of T^m of the form $A_1 x \ldots x A_m$ such that each $A_i \in \mathcal{R}^n(T)$ and for some permutation Π,

1) $A_{\Pi(1)}, \ldots, A_{\Pi(k)}$ are C-independent, and

2) $A_{\Pi(k+1)}, \ldots, A_{\Pi(m)} \subset S_{A_{\Pi(1)} A_{\Pi(2)} \ldots A_{\Pi(k)}}$.

The purpose of this appendix is to prove the following proposition:

Proposition B. The linear span of $\{1_A : A \in \mathcal{I}_m\}$ is dense in $L^2(T^m)$ for each $m \geq 1$.

Proof: Consider the following two conditions on C:

(b_1) There is a countable subcollection of $\mathcal{I}_m$ which covers T^m a.e.

(b_2) There is a countable subcollection $\mathcal{I}_m^d$ of disjoint sets in $\mathcal{I}_m$ which covers T^m a.e.

By a sequence of lemmas it is shown below that conditions c_2 and $c_3 \Rightarrow$ condition $b_1 \Rightarrow$ condition $b_2 \Rightarrow$ the conclusion of Proposition B.

□

Lemma B.1.

$$\bigcup_{\ell=1}^{m} \bigcup_{\Pi \in \mathcal{P}(m)} \Pi \circ \{(\underline{x}, \underline{y}) : \underline{x} \in \hat{E}^{\ell},\ \underline{y} \in (S_{\underline{x}})^{m-\ell}\} = T^m \qquad (*)$$

Proof: Let $\underline{q} = (q_1, \ldots, q_m) \in T^m$. Choose a permutation $\underline{p} = (p_1, \ldots p_m) = \Pi(q_1, \ldots, q_m)$ so that for some ℓ with $1 \leq \ell \leq m$,

$$S_{\underline{q}} = S_{p_1, \ldots, p} \neq S_{p_1, \ldots, \hat{p}_1, \ldots, p_\ell} \quad \text{for } 1 \leq i \leq \ell$$

where "$\hat{p}_i$" denotes that p_i is to be omitted. That is, the permutation is choosen so that $p_1,\ldots,p_\ell$ is a minimal set from $q_1,\ldots,q_m$ with the same support as $p_1,\ldots,p_m$. Now $p_{\ell+1},\ldots,p_m \in S_{p_1,\ldots,p_\ell}$ since $q_1,\ldots q_m \in S_{\underline{q}} = S_{p_1,\ldots,p}$.

To show that $\underline{q}$ is contained in the left side of (*), it remains to show that $p_1,\ldots,p_\ell$ are C-independent. Now, if $p_1,\ldots,p_\ell$ were not C-independent, then $p_i \in S_{p_1,\ldots,\hat{p}_i,\ldots,p_\ell}$ for some i. Then

$$\{A\in C:\ p_1,\ldots,p \in A\} = \{A\in C:\ p_1,\ldots,\hat{p}_i,\ldots,p_\ell \in A\}.$$

Intersecting all the sets contained in this collection of sets yields that

$$S_{p_1,\ldots,p_\ell} = S_{p_1,\ldots,\hat{p}_i,\ldots,p_\ell}$$

which contradicts our choice of $p_1,\ldots,p_\ell$. Thus $p_1,\ldots,p_\ell$ are C-independent so that $\underline{p}$, and hence $\underline{q}$, is contained in the left side of (*). ¤

Lemma B.2. Conditions c_2 and c_3 imply Condition b_1.

Proof: Let I_m^0 denote the subsets of T^m of the form $A_1 x\ldots xA_m$ such that, for some $\Pi \in P(m)$ and some $\ell > 0$,

a) $A_\Pi,\ldots,A_{\Pi_\ell}$ are C-independent, closed rectangles whose vertices have rational coordinates in $T \subset \mathbb{R}^n$, and

b) $A_{\Pi(\ell+1)} = \ldots = A_{\Pi(m)} = S_{A_{\Pi(1)}A_{\Pi(2)}\cdots A_{\Pi(\ell)}}$

Then I_m^0 is a countable subset of I_m and

$$\bigcup_{A\in I_m^0} A \supset \bigcup_{\ell=1}^{m} \bigcup_{\Pi\in P(m)} \Pi\circ\{(\underline{x},\underline{y}):\underline{x}\in\hat{T}^{\ell},\ \underline{y}\in(S^{(-)})^{m-\ell}\}$$

$$= \bigcup_{\ell=1}^{m} \bigcup_{\Pi\in P(m)} \Pi\circ\{(\underline{x},\underline{y}):\underline{x}\in\hat{T}^{\ell},\ \underline{y}\in(S_{\underline{x}})^{m-\ell}\} \qquad (B.1)$$

$$- \bigcup_{\ell=1}^{m} \bigcup_{\Pi\in P(m)} \Pi\circ S_{m,\ell}$$

where

$$S_{m,\ell} = \{(\underline{x},\underline{y}):\underline{x}\in\hat{T}^{\ell},\ \underline{y}\in(R_{\underline{x}})^{m-\ell} - (R_{\underline{x}}^{(-)})^{m-\ell}\}\ .$$

The first term on the right hand side of (B.1) is equal to T^m by Lemma B.1. Thus, to complete the proof it must be shown that $\mu^m(S_{m,\ell}) = 0$ for all $m \geq 1$ and $1 \leq \ell \leq m$.

By Condition c_2,

$$F_{\varepsilon} = \{(\underline{x},\underline{y}):\underline{x}\in\hat{T}^{\ell},\ \underline{y}\in(S_{\underline{x}}^{\varepsilon})^{m-\ell}\}$$

is a closed subset of $\hat{T}^{\ell} \times T^{m-\ell}$ which increases as ε decreases to zero. Since $S_{m,\ell} = F_0 - \bigcup_{\varepsilon>0} F_{\varepsilon}$, it follows that $S_{m,\ell}$ is a Borel subset of T^m. By Condition c_3, the section

$$\{\underline{y}:(\underline{x},\underline{y}) \in S_{m,\ell}\} \subset T^{m-\ell}$$

of $S_{m,\ell}$ at $\underline{x}$ has Lebesgue measure zero for a.e. $\underline{y} \in \hat{T}^m$. Hence, by Fubini's theorem, $\mu^m(S_{m,\ell}) = 0$ for $1 \leq \ell \leq m$. ¤

<u>Lemma B.3</u>. Condition b_1 implies condition b_2.

Proof: Let $F_1, F_2, \ldots$ be a countable subcollection of I_m which covers T^m a.e.. Then the disjoint sets $D_i = F_i - \bigcup_{j=1}^{i-1} F_j$ $i \geq 1$ cover T^m a.e.. We claim that for each $i \geq 1$ there is a finite collection of disjoint sets $D_{i1}, \ldots, D_{in_i}$ in I_m such that $D_i = \bigcup_{j=1}^{n_i} D_{ij}$. Condition b is then satisfied with $I_m^d = \{D_{ij} : i \geq 1, 1 \leq j \leq n_i\}$. It remains to prove the claim.

By induction, it suffices to establish the cliam for $i = 2$. Now $F_1 = A_1 x \ldots x A_m$ for some Borel sets $A_1, \ldots, A_m \subset T$. Thus, $F_1^i = \bigcup_{j=1}^{r} K_j$ where $K_1, \ldots, K_r$ are disjoint and each K_j is the product of m Borel subsets of T. In fact, F_1^i is the union of all sets of the form $B_1 x \ldots x B_m$ such that $B_i = A_i$ or $B_i = A_i^c$ for each i and such that $B_i = A_i^c$ for at least one i, and these sets are disjoint. So $D_2 = \bigcup_{j=1}^{k} K_j \cap F_2$. The sets $K_j \cap F_2$. The sets $K_j \cap F_2$ are disjoint sets in I_m as required so the claim is established. ¤

<u>Lemma B.4</u>. Condition b_2 implies that the linear span of $\{1_A : A \in I_m\}$ is dense in $L^2(T^m)$.

Proof: Let $F = F_1 x \ldots x F_m$ where each $F_i \in R^n(T)$. Then $A \cap F \in I_m$ for any $A \in I_m$ and by Condition b_2,

$$1_F = \sum_{A \in I_m^d} 1_{A \cap F} \quad \text{a.e. in } T^m.$$

Since the linear span of functions of the form 1_F is dense in $L^2(T^m)$, the lemma is established. ¤

Appendix C

Proposition C. Assume Conditions $c_1 - c_3$. Let $B_1,\ldots,B_k$ be closed subsets of T and suppose that $\alpha_i(\omega)$ is an $F(B_i)$ measurable random variable with values in a finite set J for $1 \le i \le k$. Suppose for each $c \in J^k$ that $h(\cdot,\cdot,c) \in L_a^2(\hat{T}^m x\Omega)$ and that

$$h(t,\cdot,c) = h(t,\cdot,c') \text{ a.s.}$$

whenever $c_i = c_i'$ for all i such that $B_i \not\subset S_t$. Then $h(\cdot,\cdot,\alpha(\cdot)) \in L_a^2(\hat{T}^m x\Omega)$ and

$$h(\cdot,\cdot,\alpha(\cdot))\circ W^m = h(\cdot,\cdot,c)\circ W^m|_{c=\alpha(\cdot)} \quad \text{a.s.}$$

Proof: For each $\theta \in \{0,1\}^k$, define

$$\hat{T}^m_\theta = \{t\in\hat{T}^m : B_i\subset S_t \Leftrightarrow \theta_i = 1 \text{ for } 1 \le i \le k\}$$

By condition c_2, the set $\{t:B\subset S_t\}$ is open for each i so that $\hat{T}^m_\theta$ is Borel for each θ. Since $\bigcup_\theta \hat{T}^m_\theta = \hat{T}^m$ it suffices to prove the lemma when

$$h(t,\cdot,c) = h(t,\cdot,c)I_{\hat{T}^m_\theta}(t)$$

for all t,c. Now, for definiteness, suppose that $\theta_i = 1$ for $1 \le i \le \ell$ and $\theta_i = 0$ for $\ell \le i \le k$. Let $\Pi:\mathbb{R}^k \to \mathbb{R}^\ell$ denote projection onto the first ℓ coordinates. Then for all $c \in J^k$,

$$h(t,\omega,c) = \tilde{h}(t,\omega,\Pi(c))$$

where $\tilde{h}(t,\omega,c) = h(t,\omega,(\Pi(c),j_0,\ldots,j_0))$ for some fixed $j_0 \in J$.

Thus,

$$h(\cdot,\cdot,\alpha(\cdot))\circ W^m = \tilde{h}(\cdot,\cdot,\Pi(\alpha(\cdot)))\circ W^m$$

$$= \sum_{b\in J^\ell} [\tilde{h}(\cdot,\cdot,b)I_{(\Pi(\alpha(\cdot))=b}] \circ W^m$$

$$= \sum_{b\in J^\ell} I_{(\Pi(\alpha(\cdot))=b)}(\tilde{h}(\cdot,\cdot,b)\circ W^m)$$

$$= (\tilde{h}(\cdot,\cdot,b)\circ W^m)|_{b=\Pi(\alpha(\cdot))}$$

$$= (h(\cdot,\cdot,c)\circ W^m)|_{c=\alpha(\cdot)}$$

The second equality is easily proven by approximating $\tilde{h}(\cdot,\cdot,b)$ in $L_a^2(\hat{T}^m \times \Omega)$ for each b by simple functions which vanish off the open set $\{t\in\hat{T}^m : B_i \subset S_t \text{ for } 1 \le i \le \ell\}$. ¤

Generalized Ornstein - Uhlenbeck Processes as Limits of Interacting Systems

by

R. Holley [1] and D. Stroock [1]

0. Introduction.

Generalized random fields arise in several contexts. Notably in quantum field theory [10] , as limits of classical statistical mechanical systems [2] and as limits of models in population genetics [1] . In the latter two cases they come about by rescaling the equilibrium states for some infinite system. If one has a stochastic process consisting of infinitely many components it is possible to rescale the system even if it is not in equilibrium. The first instance of this of which the authors are aware is due to A. Martin - Lof [9] . When a stochastic process with infinitely many components is rescaled one would expect to get a generalized stochastic process with a stationary measure which coincides with the generalized random field that one gets by rescaling the stationary measure for the infinite system. If this happens it is sometimes possible to learn something about the generalized random field by studying the generalized stochastic process for which it is a stationary measure (see [3]).

We concentrate here on situations for which the limiting process is Markovian. This is not typically the case. In fact if the rescaling used is the usual central limit scaling in space, and time is left

1) Research partially supported by N.S.F. Grant MCS 77-14881 A 01 .

unchanged, then, at least for systems consisting of two state components, we have a good idea of what is needed for the limit to be Markovian [4]. If the rescaling used is the usual central limit scaling then, even if the limit is not Markovian, it is often possible to say something about the limit by the method of moments [5]. Our main interest here is in section 2 where we consider a more delicate rescaling involving both space and time. Our understanding of this procedure is not nearly as good as for the central limit rescaling and our methods work only when the limit is Markovian. At the moment all we have is a short list of examples for which this rescaling works; however, they all have certain features in common. In section two we work through one of these examples and point out what it has in common with other examples for which the same rescaling leads to the same limit.

In section one we introduce the basic examples and techniques and apply them to the less delicate situation in which space is rescaled by the usual central limit theorem rescaling and time is left unchanged.

1. <u>Rescaling Space</u>.

We begin with a simple example that has no interactions. Let $E = R^{Z^d}$. We denote the elements of E by η, thinking of η as a function from Z^d into R. Let $\tilde{\mu}$ be a probability measure on R with $\int x\tilde{\mu}(dx) = 0$ and $\tilde{\mu}(\{|x| > L\}) = 0$ for some finite L. Now the process η_t with state space E evolves as follows. At each $k \in Z^d$ there is a Poisson process with rate one. Whenever the Poisson process at k jumps then the value of $\eta(k)$ changes, the new value being chosen independently of everything else and having distribution $\tilde{\mu}$. We can characterize the process η_t in the following way. Give

E the product topology and let $\mathcal{D}$ be the set of bounded continuous functions on E for which there is a finite set $\Lambda \subset Z^d$ (depending on $f \in \mathcal{D}$) such that $f(\eta) = f(\bar{\eta})$ if $\eta(k) = \bar{\eta}(k)$ for all $k \in \Lambda$. Now define $\mathcal{A} : \mathcal{D} \to C(E)$ ($\equiv$ bounded continuous functions on E) by

$$\mathcal{A}f(\eta) = \sum_{k \in Z^d} \int (f(\eta^{k,x}) - f(\eta))\tilde{\mu}(dx) ,$$

where
$$\eta^{k,x}(j) = \begin{cases} \eta(j) & \text{if } j \neq k \\ x & \text{if } j = k \end{cases} .$$

Let $\Omega = D((0,\infty),E)$; the right continuous functions with left limits from $[0,\infty)$ into E. For each $\eta \in E$ there is a unique measure Q_η on Ω such that $Q_\eta(\eta_0 = \eta) = 1$ and for all $f \in \mathcal{D}$, $f(\eta_t) - \int_0^t \mathcal{A}f(\eta_s)ds$ is a Q_η - martingale. The measure Q_η is the distribution of the process η_t described above starting from η. Clearly the product measure $\mu = \prod_{k \in Z^d} \tilde{\mu}_k$, where each $\tilde{\mu}_k = \mu$, is the stationary measure for Q_η and $Q_\mu \equiv \int Q_\eta \, \mu(d\eta)$ is the distribution of the process when started in equilibrium. It is Q_μ that we want to rescale. For $\Lambda \subset R^d$ a bounded Borel set let $\alpha\Lambda = \{\alpha x : x \in \Lambda\}$ and define $\eta_t^{(\alpha)}(\Lambda) = \alpha^{-d/2} \sum_{k \in \alpha\Lambda} \eta_t(k)$. For each finite $\alpha \geq 1$, $\eta_t^{(\alpha)}(\cdot)$ is a signed measure; however when we let $\alpha \to \infty$ the limit is no longer a countably additive signed measure, and it will be convenient for us to think of $\eta_t^{(\alpha)}(\cdot)$ as a tempered distribution. Thus for $\varphi \in \mathcal{S}(R^d)$ (the Schwartz functions) we set $\eta_t^{(\alpha)}(\varphi) = \alpha^{-d/2} \sum_{k \in Z^d} \varphi(k/\alpha)\eta_t(k)$. Let $Q^{(\alpha)}$ be the distribution on $D([0,\infty),\mathcal{S}'(R^d))$ of $\eta^{(\alpha)}(\cdot)$ under Q_μ. We now want to let $\alpha \to \infty$ and try to identify what the weak limit of the $Q^{(\alpha)}$'s, if it exists, must be. To do this note that for every $\varphi \in \mathcal{S}(R^d)$ and $F \in C_0^\infty(R)$,

$$(1.1)\quad F(\eta_t^{(\alpha)}(\varphi)) - \int_0^t \sum_{k\in Z^d} \int [F(\eta_s^{(\alpha)}(\varphi) + \alpha^{-d/2}\varphi(k/\alpha)(x - \eta_s(k)) - F(\eta_s^{(\alpha)}(\varphi))]\,\tilde{\mu}(dx)ds$$

$$= F(\eta_t^{(\alpha)}(\varphi)) + \int_0^t F'(\eta_t^{(\alpha)}(\varphi))\alpha^{-d/2} \sum_k \varphi(k/\alpha)\eta_s(k)ds$$

$$- \int_0^t \frac{1}{2}F''(\eta_t^{(\alpha)}(\varphi)) \sum_{k\in Z^d} \alpha^{-d}\varphi^2(k/\alpha)\int (x - \eta_s(k))^2\tilde{\mu}(dx)ds$$

$$+ O(t\alpha^{-3d/2} \sum_k |\varphi(k/\alpha)|^3)$$

is a Q_μ martingale. The second term on the right side of (1.1) is just $\int_0^t F'(\eta_s^{(\alpha)}(\varphi))\eta_s^{(\alpha)}(\varphi)ds$, the third term converges in $L_2(Q_\mu)$ to $-\|\varphi\|_2^2 m_2\int_0^t F''(\eta_s^{(\alpha)}(\varphi))ds$ (here $m_2 = \int x^2\tilde{\mu}(dx)$), and the fourth term converges to zero. Thus any weak limit point of $\{Q^{(\alpha)}\ ;\ \alpha \geq 1\}$ as $\alpha \to \infty$ must be such that for all $\varphi \in \mathscr{S}(R^d)$ and $F \in C_0^\infty(R)$

$$F(\eta_t(\varphi)) + \int_0^t F'(\eta_s(\varphi))\eta_s(\varphi)ds - \|\varphi\|_2^2 m_2 \int_0^t F''(\eta_s(\varphi))ds$$

is a martingale. Also by the central limit theorem the distribution of $\eta_0(\varphi)$ is normal with mean 0 and variance $m_2\|\varphi\|_2^2$ under any weak limit of the $Q^{(\alpha)}$'s. This determines the weak limit, $Q^{(\infty)}$, uniquely (see [6]). In fact, for each $\varphi \in \mathscr{S}(R^d)$, $\eta_t(\varphi)$ is a one dimensional Ornstein - Uhlenbeck process under $Q^{(\alpha)}$.

We still must prove that $\{Q^{(\alpha)};\alpha \geq 1\}$ is relatively compact. Before stating the relevant theorem we introduce some notation. If $n = (n_1,\ldots,n_d)$ and $a = (a_1,\ldots,a_d)$ are multi-indices and $x = (x_1,\ldots,x_d) \in R^d$, then $x^n = x_1^{n_1},\ldots,x_d^{n_d}$ and $D^a = \dfrac{\partial^{|a|}}{\partial x_1^{a_1},\ldots,\partial x_d^{a_d}}$, where $|a| = \sum_{i=1}^d a_i$. For $\varphi \in \mathscr{S}(R^d)$ set $\|\varphi\|_{a,n} = \sup_{x\in R^d} |x^n D^a\varphi(x)|$.

(1.2) Theorem: Let $(P^{(\alpha)};\alpha \geq 1)$ be a family of probability measures on $D([0,\infty),\mathscr{S}'(R^d))$. Assume that there are a finite number of multi-indices $a^{(1)} \dots a^{(k)}, n^{(1)} \dots n^{(k)}$ and for each $T < \infty$ there is a constant $C(T) < \infty$ such that for all $\varphi \in \mathscr{S}(R^d)$

$$(1.3)\qquad \sup_{\alpha\geq 1} E^{P^{(\alpha)}}[\sup_{0\leq t\leq T} (\eta_t(\varphi))^2] \leq C(T) \sum_{j=1}^{k} \|\varphi\|_{a^{(j)},n^{(j)}} .$$

Suppose moreover that for all $\varepsilon > 0$ and all $\varphi \in \mathscr{S}(R^d)$

$$(1.4)\qquad \lim_{\alpha\to\infty} P^{(\alpha)}(\sup_{0\leq t<\infty} |\eta_{t+}(\varphi) - \eta_t(\varphi)| > \varepsilon) = 0 .$$

Finally suppose that for each $\varphi \in \mathscr{S}(R^d)$ and $\alpha \geq 1$ there are non-anticipating functions $\gamma^{(1)}_{\alpha,\varphi}(\cdot)$ and $\gamma^{(2)}_{\alpha,\varphi}(\cdot)$ such that $\eta_t(\varphi) - \int_0^t \gamma^{(1)}_{\alpha,\varphi}(s)ds$ and $(\eta_t(\varphi) - \int_0^t \gamma^{(1)}_{\alpha,\varphi}(s)ds)^2 - \int_0^t \gamma^{(2)}_{\alpha,\varphi}(s)ds$ are $P^{(\alpha)}$-martingales and that for all $T < \infty$

$$(1.5)\qquad \sup_{\alpha\geq 1} E^{P^{(\alpha)}}[\sup_{0\leq t\leq T} |\gamma^{(1)}_{\alpha,\varphi}(t)|^2] \vee E^{P^{(\alpha)}}[\sup_{0\leq t\leq T} |\gamma^{(2)}_{\alpha,\varphi}(t)|^2] < \infty .$$

Then $[P^{(\alpha)} : \alpha \geq 1]$ is relatively weakly compact and any weak limit is supported by $C([0,\infty),\mathscr{S}'(R^d))$.

For a proof of this theorem see section 4 of [6]. It is not difficult to check that $\{Q^{(\alpha)};\alpha \geq 1\}$ satisfies the hypotheses of Theorem (1.2). This is done in section 1 of [4] for a similar though more complicated process.

We now introduce a simple interaction into the underlying stochastic process with infinitely many components. Let E and $\mathfrak{D}$ be as above and let $p(k,j)$ be the transition function of a random walk on Z^d with $p(k,j) = 0$ if $|k - j| \geq L$ for some finite L and

(1.6) $$\sum_k kp(0,k) = 0 \quad\text{and}\quad \sum_k kk^t p(0,k) = \sigma^2 I \ .$$

(The assumption that the covariance matrix of p is $\sigma^2 I$ is not critical. If the covariance were A then the operator $\sigma^2\Delta$ in section 2 would be replaced by $\sum_{i,j} A_{i,j}\frac{\partial^2}{\partial x_i \partial x_j}$). Define an operator $\mathcal{L} : \mathcal{D} \to C(E)$ by

$$\mathcal{L}f(\eta) = \sum_k \sum_j p(k,j)[f(\eta^{(k,j)}) - f(\eta)] ,$$

where
$$\eta^{(k,j)}(\ell) = \begin{cases} \eta(\ell) & \text{if } \ell \neq k \\ \eta(j) & \text{if } \ell = k \ . \end{cases}$$

For each $\eta \in E$, let P_η be the unique (see [8]) measure on Ω such that $P_\eta(\eta_0 = \eta) = 1$ and for all $f \in \mathcal{D}$,

$$f(\eta_t) - \int_0^t \mathcal{L}f(\eta_s)ds \quad \text{is a } P_\eta\text{-martingale} .$$

The family $\{P_\eta : \eta \in E\}$ is a strong Markov family, and we call the corresponding process the voter model. If $\eta_t(k)$ represents the opinion of the person at k at time t, then we may form the following intuitive picture of the process. Each person has an independent Poisson process with rate one. Whenever person k's Poisson process jumps he changes his opinion to agree with person j with probability $p(k,j)$. The entire collection of people is doing this simultaneously, but with probability one the jumps only occur one at a time.

We want to use the same rescaling on the voter model that we used in the first example. Thus let μ be product measure as above and let $P_\mu = \int P_\eta \mu(d\eta)$. Just as before for $\varphi \in \mathcal{S}(R^d)$ and $\alpha \geq 1$ we define

$$\eta_t^{(\alpha)}(\varphi) = \alpha^{-d/2}\sum_k \varphi(k/\alpha)\eta_t(k)$$

and let $P^{(\alpha)}$ be the distribution of $\eta_\cdot^{(\alpha)}(\cdot)$ on $D([0,\infty),\mathcal{S}'(R^d))$. Again (see [4]) it is not difficult to check the hypotheses of Theorem (1.2), and thus the family $\{P^{(\alpha)} : \alpha \geq 1\}$ is relatively compact. We proceed as before to identify the limit. Let $\varphi \in \mathcal{S}(R^d)$ and $F \in C_0^\infty(R)$. Then

$$(1.7)\quad F(\eta_t^{(\alpha)}(\varphi)) - \int_0^t \sum_k \sum_j p(k,j)[F(\eta_s^{(\alpha)}(\varphi) + \alpha^{-d/2}\varphi(k/\alpha)(\eta_s(j) - \eta_s(k))) - F(\eta_s^{(\alpha)}(\varphi))]ds$$

$$= F(\eta_t^{(\alpha)}(\varphi)) - \int_0^t F'(\eta_s^{(\alpha)}(\varphi))\sum_k \sum_j p(k,j)\alpha^{-d/2}(\varphi(k/\alpha) - \varphi(j/\alpha))\eta_s(j)ds$$

$$- \frac{1}{2}\int_0^t F''(\eta_s^{(\alpha)}(\varphi))\sum_k \sum_j p(k,j)\alpha^{-d}\varphi^2(k/\alpha)(\eta_s(j) - \eta_s(k))^2 ds$$

$$- O(t\alpha^{-3d/2}\sum_k |\varphi(k/\alpha)|^3) .$$

Letting $\psi_\alpha(x) = \sum_j p(0,j)(\varphi(x - j/\alpha) - \varphi(x))$ we see from (1.6) that $\psi_\alpha \in \mathcal{S}(R^d)$ and there is a constant $C_\varphi < \infty$ such that $|\psi_\alpha(x)| \leq \alpha^{-2}C_\varphi(1 + |x|)^{-d-1}$. Thus the $L_2(P_\mu)$ norm of the second term on the right side of (1.7) is bounded by

$$(1.8)\quad t^{1/2}\|F'\|_\infty \alpha^{-2}C_\varphi\left(\int_0^t E[\alpha^{-d}\sum_k\sum_j (1 + |j/\alpha|)^{-d-1}(1 + |k/\alpha|)^{-d-1}\eta_s(j)\eta_s(k)]ds\right)^{1/2}.$$

Since μ is product measure with mean zero and $p(k,j) = 0$ if $|k - j| > L$ one can see without explicit computations that $\sum_j |E[\eta_s(k)\eta_s(j)]|$ is bounded for s in bounded sets (see Theorem 2.16 in [7]). This of course implies that the expression in (1.8) goes to zero as α goes to infinity. However, since we are going to need to know $E[\eta_s(k)\eta_s(j)]$ in the next section anyway, we will go ahead and compute it here. To do this let

$$\rho_t(k,j) = E^{P_\mu}(\eta_t(k)\eta_t(j)] .$$

Then

(1.9)
$$\frac{\partial\rho_t(k,j)}{\partial t} = \begin{cases} \sum_\ell p(k,\ell)(\rho_t(\ell,j) - \rho_t(k,j)) + \sum_\ell p(j,\ell)(\rho_t(k,\ell) - \rho_t(k,j)) & \text{if } k \neq j \\ 0 & \text{if } k = j . \end{cases}$$

Also

(1.10)
$$\rho_0(k,j) = \begin{cases} 0 & \text{if } k \neq j \\ m_2 & \text{if } k = j . \end{cases}$$

Equation (1.9) with initial conditions (1.10) can be solved in terms of two independent random walks $X_1(t)$ and $X_2(t)$ each having transition function $\rho^{t(p-I)}$ with one starting from k and the other from j. In fact if $\tau = \inf\{t : X_1(t) = X_2(t)\}$ then

(1.11)
$$\rho_t(k,j) = m_2 P^{(k,j)}(\tau \le t) .$$

We will also need the fourth moments of $\eta_t(k)$ in the next section. By a computation similar to but messier than the one above we get

(1.12)
$$E^{P_\mu}[\eta_t(i)\eta_t(j)\eta_t(k)\eta_t(\ell)] =$$
$$m_2^2 P^{(i,j,k,\ell)}(Y_t(1) = Y_t(2) \neq Y_t(3) = Y_t(4) \text{ or } Y_t(1) = Y_t(3) \neq Y_t(2) = Y_t(4) \text{ or } Y_t(1) = Y_t(4) \neq Y_t(2) = Y_t(3))$$

$$+ m_4 P^{(i,j,k,\ell)}(Y_t(1) = Y_t(2) = Y_t(3) = Y_t(4)) ,$$

where $m_4 = \int x^4 \tilde{\mu}(dx)$ and $Y_t(1)$, $Y_t(2)$, $Y_t(3)$, $Y_t(4)$ are random walks starting from i,j,k and ℓ which have transition function $e^{t(p-I)}$ and move independently until they collide at which time they stick together from then on.

Returning to (1.7) we see from (1.11) that $\sum_j E^{P_\mu}[\eta_s(k)\eta_s(j)] \le (1 + 2t)m_2$ and thus (1.8) goes to zero as α goes to infinity. The fourth term on the right side of (1.7) also goes to zero. This leaves only the third term. But by using (1.11) and (1.12) it is easy to see that

$$(1.13) \quad \frac{1}{2}\int_0^t F''(\eta_s^{(\alpha)}(\varphi))\sum_k\sum_j p(k,j)\alpha^{-d}\varphi^2(k/\alpha)(\eta_s(j) - \eta_s(k))^2 ds$$
$$- \frac{1}{2}\int_0^t F''(\eta_s^{(\alpha)}(\varphi))\sum_k\sum_j p(k,j)\alpha^{-d}\varphi^2(k/\alpha)2m_2 P^{(k,j)}(\tau > t)$$

converges to zero in $L_2(P_\mu)$. Also

$$(1.13) \quad \sum_k\sum_j p(k,j)\alpha^{-d}\varphi^2(k/\alpha)P^{(k,j)}(\tau > t) \to \|\varphi\|^2\sum_j p(0,j)P^{(0,j)}(\tau > t) .$$

Thus the third term on the right side of (1.7) converges in $L_2(P_\mu)$ to

$$(1.15) \quad -m_2\|\varphi\|_2^2 \int_0^t F''(\eta_s^{(\alpha)}(\varphi))\sum_j p(0,j)P^{(0,j)}(\tau > s)ds .$$

Thus the weak limit, $P^{(\infty)}$, of the $P^{(\alpha)}$'s is the unique measure on $C([0,\infty),\mathcal{S}'(R^d))$ such that for all $\varphi \in \mathcal{S}(R^d)$ $\eta_0(\varphi)$ has a normal distribution with mean zero and variance $m_2\|\varphi\|_2^2$ and for all $F \in C_0^\infty(R)$

$$F(\eta_t(\varphi)) - m_2\|\varphi\|_2^2 \int_0^t F''(\eta_s(\varphi))\sum_j p(0,j)P^{(0,j)}(\tau > s)ds$$

is a $P^{(\infty)}$ - martingale. Note that for $\varphi_1,\ldots,\varphi_n \in \mathcal{S}(R^d)$, $(\eta_t(\varphi_1),\ldots,\eta_t(\varphi_n))$ is just an n - dimensional time-inhomogeneous Brownian motion with covariance $\rho(t)(((\varphi_i,\varphi_j)_{L^2}))_{1\le i,j\le n}$ where $\rho(t)$ is non-increasing. In one and two dimensions the diffusion coefficient converges to zero as the time gets large and in three or more dimensions it has a strictly positive limit.

We want to emphasize again that even though the limit in each of these examples was Markovian, this is not the usual situation. For more complicated interactions the limit is hardly ever Markovian. For example if the interacting system is the stochastic Ising model and we hold the potential fixed except for the external field, then as we vary the external field there is at most one value of the external field for which the limiting process is Markovian. The Markovian limits obtained above resulted from the linear nature of the interaction. This linearity is one of the clues that the rescaling used in the next section may result in the more interesting limit obtained there. Another clue that the rescaling in the next section may work is contained in the second moment computation. From (1.11) we see that, if $d \ge 3$,

$$(1.16) \qquad \lim_{t\to\infty} E^{P_\mu}[\eta_t(k)\eta_t(j)] = m_2 P^{(k,j)}(\tau < \infty) \sim \frac{\text{constant}}{|k-j|^{d-2}}$$

$$\text{as } |k-j| \to \infty .$$

Every example that we have for which the rescaling in the next section yields an interesting limit satisfies (1.16) .

2. Rescaling Space and Time.

In this section we again consider the voter model but use a different rescaling. Thus P_μ will be the same as it was in section 1. Since we are going to speed up the time we can discover how to renormalize by considering the asymptotic behavior of

$$(2.1) \qquad \lim_{t\to\infty} E^{P_\mu}\left[\left(\sum_k \varphi(k/\alpha)\eta_t(k)\right)^2\right]$$

as α goes to infinity. From (1.11) we see that the asymptotic behavior of (2.1) depends on the dimension. If $d = 1$ or 2 then (2.1) is asymptotically $m_2\|\varphi\|_2^2\alpha^{2d}$ whereas if $d \geq 3$ then (2.1) is asymptotically constant $\int_{R^d}\int_{R^d}\varphi(x)\varphi(y)\frac{1}{|x - y|^{d-2}}\,dxdy\alpha^{d+2}$. This indicates that in 1 and 2 dimensions we should divide by α^d and in 3 or more dimensions we should divide by $\alpha^{\frac{d+2}{2}}$. If $d = 1$ or 2 this yields the uninteresting limit which is identically zero, thus we concentrate on the case $d \geq 3$. For $\varphi \in \mathscr{S}(R^d)$ and $\alpha \geq 1$ let $\eta_t^{(\alpha)}(\varphi) = \alpha^{-\frac{d+2}{2}}\sum_k \varphi(k/\alpha)\eta_{\beta(\alpha)t}(k)$, where $\beta(\alpha)$ is a function of α to be determined later. Let $P^{(\alpha)}$ be the distribution of $\eta^{(\alpha)}(\cdot)$ on $D([0,\infty),\mathscr{S}'(R^d))$ under P_μ. We use the same notation, $\eta^{(\alpha)}$ and $P^{(\alpha)}$, here that we used in section 1 even though they are different. The meanings that they had in section 1 will not be used in this section.

Again it is not difficult to show that $\{P^{(\alpha)} : \alpha \geq 1\}$ is relatively compact by checking the hypotheses of Theorem (1.2), so we will concentrate on identifying the limit. Fix $\varphi \in \mathscr{S}(R^d)$ and $F \in C_0^\infty(R)$. Then

$$(2.2)\quad F(\eta_t^{(\alpha)}(\varphi)) - \int_0^{\beta(\alpha)t}\sum_k\sum_j p(k,j)[F(\eta_{s/\beta(\alpha)}^{(\alpha)}(\varphi) + \alpha^{-\frac{d+2}{2}}\varphi(k/\alpha)(\eta_s(j) - \eta_s(k))) - F(\eta_{s/\beta(\alpha)}^{(\alpha)}(\varphi))]ds$$

$$= F(\eta_t^{(\alpha)}(\varphi)) - \int_0^t\sum_k\sum_j p(k,j)F'(\eta_s^{(\alpha)}(\varphi))(\eta_{\beta(\alpha)s}(j) - \eta_{\beta(\alpha)s}(k))\varphi(\tfrac{k}{\alpha})\beta(\alpha)\alpha^{-\frac{d+2}{2}}ds$$

$$- \frac{1}{2}\int_0^t F''(\eta_s^{(\alpha)}(\varphi))\sum_k\sum_j(\eta_{\beta(\alpha)s}(j) - \eta_{\beta(\alpha)s}(k))^2\varphi^2(k/\alpha)\alpha^{-(d+2)}\beta(\alpha)ds$$

$$- O(\alpha^{-\frac{d}{2}-3}\beta(\alpha)t)$$

$$= F(\eta_t^{(\alpha)}(\varphi)) - \int_0^t F'(\eta_s^{(\alpha)}(\varphi))\sum_k\sum_j p(k,j)(\varphi(\tfrac{k}{\alpha}) - \varphi(\tfrac{j}{\alpha}))\eta_{\beta(\alpha)s}(j)\beta(\alpha)\alpha^{-\frac{d+2}{2}}ds$$

$$- \frac{1}{2}\int_0^t F''(\eta_s^{(\alpha)}(\varphi))\sum_k\sum_j p(k,j)(\eta_{\beta(\alpha)s}(j) - \eta_{\beta(\alpha)s}(k))^2\varphi^2(\tfrac{k}{\alpha})\alpha^{-d}\alpha^{-2}\beta(\alpha)ds$$

$$- O(\alpha^{-\frac{d}{2}-3}\beta(\alpha)t)$$

$$= F(\eta_t^{(\alpha)}(\varphi)) - \int_0^t F'(\eta_s^{(\alpha)}(\varphi))\eta_s^{(\alpha)}(\tfrac{\sigma^2}{2}\Delta\varphi)\alpha^{-2}\beta(\alpha)ds$$

$$- \frac{1}{2}\int_0^t F''(\eta_s^{(\alpha)}(\varphi))\|\varphi\|_2^2\sum_j p(0,j)E^{P_\mu}[(\eta_{\beta(\alpha)s}(j) - \eta_{\beta(\alpha)s}(0))^2]\alpha^{-2}\beta(\alpha)ds$$

$$- \frac{1}{2}\int_0^t F''(\eta_s^{(\alpha)}(\varphi))[\sum_k\sum_j p(k,j)(\eta_{\beta(\alpha)s}(j) - \eta_{\beta(\alpha)s}(k))^2\varphi^2(k/\alpha)\alpha^{-d}$$

$$- \sum_k\sum_j p(k,j)E^{P_\mu}[(\eta_{\beta(\alpha)s}(j) - \eta_{\beta(\alpha)s}(k))^2]\varphi^2(k/\alpha)\alpha^{-d}]\alpha^{-2}\beta(\alpha)ds$$

$$- o(\alpha^{-2}\beta(\alpha)t)$$

is a P_μ - martingale. At this point it is clear that we should take $\beta(\alpha) = \alpha^2$. Now consider the third term on the right side of (2.2) .

$$(2.3)\quad \sum_j p(0,j)E^{P_\mu}[(\eta_{\alpha^2 s}(j)-\eta_{\alpha^2 s}(0))^2] = \sum_j p(0,j)(2m_2 - 2m_2 P^{(0,j)}(\tau \le \alpha^2 s))$$

$$\to 2m_2\rho \quad \text{as} \quad \alpha \to \infty ,$$

where ρ is the probability that a random walk with transition function $\frac{1}{2}[p(k,j) + p(j,k)]$ starting from the origin never returns to the origin. Thus for large α the third term is essentially

$$m_2\rho\|\varphi\|_2^2 \int_0^t F''(\eta_s^{(\alpha)}(\varphi))ds .$$

Finally we come to the fourth term on the right side of (2.2). In all of the examples which we have of this rescaling the computations all have the same ingredients and this term is always the most difficult to handle. We want to show that it converges to zero in $L_1(P_\mu)$. In this particular case it is not too difficult to show that in fact it converges in $L_2(P_\mu)$ because we have tractable expressions for the second and fourth moments of $\eta_s(k)$ given in (1.11) and (1.12). On the other hand, even with (1.11) and (1.12), the computation is tedious and we will leave it to the reader to check that the fourth term does indeed tend to zero in $L_2(P_\mu)$ as $\alpha_2 \to \infty$.

Note also that $E[(\eta_0^{(\alpha)}(\varphi))^2] \to 0$ as $\alpha \to \infty$. We have thus proved the following theorem.

(2.4) Theorem: *Let* $d \geq 3$. *As* α *goes to infinity* $P^{(\alpha)}$ *converges weakly to the measure* P *on* $C([0,\infty),\mathscr{S}'(R^d))$ *such that* $P(\eta_0 = 0) = 1$ *and*

$$(2.5)\quad F(\eta_t(\varphi)) - \int_0^t F'(\eta_s(\varphi))\eta_s(\frac{\sigma^2}{2}\Delta\varphi)ds - m_2\rho\|\varphi\|^2\int_0^t F''(\eta_s(\varphi))ds$$

is a P - *martingale for all* $\varphi \in \mathscr{S}(R^d)$ *and* $F \in C_0^\infty(R)$.

The initial conditions $P(\eta_0 = 0) = 1$ and (2.5) determine P uniquely (see [6]). P is a Gaussian process, in fact it is a generalized Ornstein-Uhlenbeck process with the drift for the φ'th coordinate given by $\frac{\sigma^2}{2}$ times the value at the $\Delta\varphi$ th coordinate. For more information about this process we refer the reader to [6].

We have chosen this example to illustrate a phenomenon which we do not fully understand. Namely, the voter model admits two different rescalings which yield different Gaussian Markov processes in the limit. This seems to be unusual. The first example given, while simpler, admits only one rescaling with a non-trivial limit. We know of a few other examples which also admit two different rescalings leading to the same limits as the voter model, and we have tried to point out what they have in common with the voter model; however, we do not have a theory explaining this phenomenon.

Finally we would like to point out that there are many other ways to rescale the voter model and get a non-trivial limit; however, these limits are closely related to the one in which space is rescaled by the usual central limit scaling. In particular for each $\beta \geq 0$ let

$$\eta_t^{(\alpha,\beta)}(\varphi) = \alpha^{-\frac{d+\beta}{2}} \sum_k \varphi(k/\alpha)\eta_{\alpha^\beta t}(k) .$$

$\eta_t^{(\alpha,0)}(\varphi)$ is equal to the $\eta_t^{(\alpha)}(\varphi)$ of section one and $\eta_t^{(\alpha,2)}(\varphi)$ is equal to the $\eta_t^{(\alpha)}(\varphi)$ of section two. For $0 < \beta < 2$ the limit $P^{(\beta)}$ satisfies

$$P^{(\beta)}(\eta_0 = 0) = 1 \quad \text{and for all} \quad \varphi \in \mathcal{S}(R^d) \quad \text{and} \quad F \in C_0^\infty(R) ,$$

$$F(\eta_t(\varphi)) - m_2\rho\|\varphi\|_2^2 \int_0^t F''(\eta_s(\varphi))ds \quad \text{is a} \quad P^{(\beta)}\text{-martingale.}$$

Thus for $0 < \beta < 2$ the limit does not depend on β, and for each $\varphi \in \mathcal{S}(R^d)$, $\eta_t(\varphi)$ is a one dimensional Brownian motion under $P^{(\beta)}$ (cf. the end of section 1). For $\beta > 2$ the limit satisfies $P^{(\beta)}(\eta_t \equiv 0) = 1$. Obviously, when $d \geq 3$, $\beta = 2$ is the interacting case.

References

[1] Dawson, D.A., Critical Measure Diffusion Processes. Z. Wahr. verw Geb. 40 (1977), 125-145.

[2] Gallavotti, G. and Jona-Lasinia, G., Limit Theorems for Multi-dimensional Markovian Processes, Commun. Math. Phys. 41 (1975), 301-307.

[3] Holley, R. and Stroock, D.W., The D.L.R. Conditions for Translation Invariant Gaussian Measures on $\mathcal{S}'(R^d)$, to appear.

[4] ________________, Central Limit Phenomena of Various Interacting Systems, Annals of Math. 110 (1979), 333-393.

[5] ________________, Rescaling Short Range Interacting Stochastic Processes in Higher Dimensions, to appear.

[6] ________________, Generalized Ornstein-Uhlenbeck Processes and Infinite Particle Branching Brownian Motions, Research Institute for Mathematical Sciences Kyoto University, 14 (1978), 741-788.

[7] ________________, L_2 Theory for the Stochastic Ising Model, Z. Wahr. verw. Gebiete, 35 (1976), 87-101.

[8] Liggett, T.M., Existence Theorems for Infinite Particle Systems, Trans. Amer. Math. Soc., 165 (1972), 471-481.

[9] Martin-Löf, A., Limit Theorems for Motion of a Poisson System of Independent Markovian Particles with High Density, Z. Wahr. Verw. Geb., 34 (1976), 205-223.

[10] Nelson, E., Construction of Quantum Fields from Markov Fields, J. Functional Anal., 12 (1973), 97-112.

WEAK AND STRONG SOLUTIONS OF STOCHASTIC DIFFERENTIAL EQUATIONS : EXISTENCE AND STABILITY

Jean JACOD and Jean MEMIN

1 - INTRODUCTION

We consider the following stochastic differential equation:

(1.1)
$$X_t = K_t + \int_0^t g_s(.,X_.(.))\, dZ_s$$

(Doléans-Dade and Protter's equation), where the driving process Z is an m-dimensional semimartingale, the solution X is a d-dimensional process, the coefficient g is a predictable process which depends on the path of X, and K is a d-dimensional process which plays the role of the initial condition.

In order to give a precise meaning to this equation, we introduce the following:

(1.2) A filtered probability space $(\Omega,\underline{\underline{F}},\underline{F}=(\underline{\underline{F}}_t)_{t\geq 0},P)$ equipped with an m-dimensional semimartingale $Z=(Z^j)_{j\leq m}$ with $Z_0=0$, and a d-dimensional adapted process $K=(K^j)_{j\leq m}$ with right-continuous paths with left-hand limits.

(1.3) $\mathcal{X}=D([0,\infty);\mathbb{R}^d)$, the Skorokhod space equipped with the canonical process X (i.e. $X_t(x)=x(t)$) and the canonical σ-field $\underline{\underline{\mathcal{X}}}$ and filtration $\underline{\mathcal{X}}=(\underline{\underline{\mathcal{X}}}_t)_{t\geq 0}$.

(1.4) The product space: $\bar{\Omega}=\Omega\times\mathcal{X}$, $\bar{\underline{\underline{F}}}=\underline{\underline{F}}\otimes\underline{\underline{\mathcal{X}}}$, $\bar{\underline{\underline{F}}}_t=\bigcap_{s>t}(\underline{\underline{F}}_s\otimes\underline{\underline{\mathcal{X}}}_s)$, on which is defined a predictable $\mathbb{R}^d\otimes\mathbb{R}^m$-valued process $g=(g^{jk})_{j\leq d,k\leq m}$.

Our notations and terminology will follows [9] and [20]. In particular, the (stochastic or Stieltjes) integral of a process U with respect to a process V will be denoted by $U\cdot V$. We recall that there exists a "maximal" set of predictable m-dimensional processes that are integrable with respect to Z: this set is denoted by $L(Z;\Omega,\underline{F},P)$ and contains all locally bounded predictable processes $H=(H^j)_{j\leq m}$, in which case one has $H\cdot Z=\sum_{j\leq m} H^j\cdot Z^j$ (cf. [10]; the useful properties of this set will be recalled later on).

If a function f is defined on a set E, we will systematically use the same symbol f to denote its natural extension to any product of the form $E\times F$. For instance Z and K (resp. X) are defined on $\overline{\Omega}$ as well as on Ω (resp. $\mathfrak{X}$).

An <u>extension</u> of $(\Omega,\underline{\underline{F}},\underline{F},P)$ is a filtered probability space $(\tilde{\Omega},\tilde{\underline{\underline{F}}},\tilde{\underline{F}},\tilde{P})$ such that

(i) $\tilde{\Omega}=\Omega\times\Omega'$, Ω' an auxiliary space;

(ii) $\underline{\underline{F}}\subset\tilde{\underline{\underline{F}}}$, $\underline{\underline{F}}_t\subset\tilde{\underline{\underline{F}}}_t$ (i.e.: $A\times\Omega'\in\tilde{\underline{\underline{F}}}$ (resp. $\tilde{\underline{\underline{F}}}_t$) if $A\in\underline{\underline{F}}$ (resp. $\underline{\underline{F}}_t$));

(iii) $\tilde{P}_{|\Omega}=P$, where $\tilde{P}_{|\Omega}$ denotes the "Ω-marginal" of $\tilde{P}$.

We are ready now to state the two possible definitions of a "solution" to Equation (1.1).

(1.5) DEFINITION: Let $(\tilde{\Omega},\tilde{\underline{\underline{F}}},\tilde{\underline{F}},\tilde{P})$ be an extension of $(\Omega,\underline{\underline{F}},\underline{F},P)$. <u>A mapping $\tilde{X}:\tilde{\Omega}\longrightarrow\mathfrak{X}$ is a solution-process of (1.1) if</u>

(i) $\tilde{X}$ is $\tilde{\underline{F}}$-adapted;

(ii) Z is a semimartingale on $(\tilde{\Omega},\tilde{\underline{\underline{F}}},\tilde{\underline{F}},\tilde{P})$;

(iii) if $g(\tilde{X})$ is defined by: $(\omega,\omega',t)\rightsquigarrow g(\tilde{X})_t(\omega,\omega')=g_t(\omega,\tilde{X}(\omega,\omega'))$, one has $g(\tilde{X})\in L(Z;\tilde{\Omega},\tilde{\underline{F}},\tilde{P})$ and $\tilde{X}=K+g(\tilde{X})\cdot Z$

(this equation must be read componentwise: $g(\tilde{X})^{j\cdot}\in L(Z;\tilde{\Omega},\tilde{\underline{F}},\tilde{P})$ and $\tilde{X}^j=K^j+g(\tilde{X})^{j\cdot}\cdot Z$ for every $j\leq d$; note that (i) and (1.4) imply that $g(\tilde{X})$ is predictable for $\tilde{\underline{F}}$). ∎

(1.6) DEFINITION: <u>A probability measure</u> $\overline{P}$ <u>on</u> $(\overline{\Omega},\overline{\underline{\underline{F}}})$ <u>is a solution-measure</u> (or, a weak solution) <u>if</u> $(\overline{\Omega},\overline{\underline{\underline{F}}},\overline{\underline{F}},\overline{P})$ <u>is an extension of</u> $(\Omega,\underline{\underline{F}},\underline{F},P)$, <u>on which the process</u> X <u>is a solution-process of (1.1)</u>: since $g(X)=g$, this amounts to saying that $g\in L(Z;\overline{\Omega},\overline{\underline{F}},\overline{P})$ and that $X=K+g\cdot Z$ (an equality that obviously holds up to a $\overline{P}$-null set). ∎

Of course, Ito's equations are a particular case of equation (1.1): take g not depending on ω, and Z with $Z^1,\dots,Z^{m-1}$ independent Brownian motions and $Z^m_t=t$. However, Definitions (1.5) and (1.6) are slightly different from the ordinary ones: for instance in (1.5), if Z^1 is a Brownian motion on $(\Omega,\underline{\underline{F}},\underline{F},P)$, it is a semimartingale but not necessarily a Brownian motion on $(\tilde{\Omega},\tilde{\underline{\underline{F}}},\tilde{\underline{F}},\tilde{P})$. We are thus led to the following:

(1.7) DEFINITION: a) A solution-process on $(\tilde{\Omega},\tilde{\underline{\underline{F}}},\tilde{\underline{F}},\tilde{P})$ (resp. a solution-measure $\overline{P}$) is good if Z is a semimartingale with the same local characteristics (cf. [9], this notion will be recalled below) on $(\Omega,\underline{\underline{F}},\underline{F},P)$ and on $(\tilde{\Omega},\tilde{\underline{\underline{F}}},\tilde{\underline{F}},\tilde{P})$ (resp. $(\overline{\Omega},\overline{\underline{\underline{F}}},\overline{\underline{F}},\overline{P})$).

b) A solution-process on $(\tilde{\Omega},\tilde{\underline{\underline{F}}},\tilde{\underline{F}},\tilde{P})$ (resp. a solution-measure $\overline{P}$) is very good if every martingale on $(\Omega,\underline{\underline{F}},\underline{F},P)$ is also a martingale on $(\tilde{\Omega},\tilde{\underline{\underline{F}}},\tilde{\underline{F}},\tilde{P})$ (resp. $(\overline{\Omega},\overline{\underline{\underline{F}}},\overline{\underline{F}},\overline{P})$). ∎

Notice that this statement is a property of the extension rather than of the processes $\tilde{X}$ or $\overline{X}$. Notice also that a very good solution is a-fortiori good ! Now, the usual notion of a weak solution of an Ito's equation corresponds here to the concept of a good solution-measure.

Since many papers have been devoted to studying existence and uniqueness of the solution-process on the (non-extended) space $(\Omega,\underline{\underline{F}},\underline{F},P)$, see e.g. Protter [24], Doléans-Dade [3], Doléans-Dade and Meyer [4], Métivier and Pellaumail [18], Jacod [9], we will write very little on this subject: see section 4. We will rather concentrate on three main topics:

1) The relationships between solution-processes, solution-measures, and martingale problems; we will see in particular that the famous result of Yamada and Watanabe [29] on the links between uniqueness of solution-processes (pathwise uniqueness) and of solution-measures, still holds in our general case. All this is studied in section 2, which follows rather closely [11] in a slightly different setting.

2) The existence of a solution-measure; the simplest result in that direction is the

(1.8) THEOREM: The following assumptions imply the existence of at least one very good solution-measure:

(i) we have identically $|g_t(\omega,x)| \leq \gamma(1+\sup_{s<t}|x(s)|)$, where $\gamma\in\mathbb{R}_+$;

(ii) for all $\omega\in\Omega$, $t\geq 0$, the mapping: $x \rightsquigarrow g_t(\omega,x)$ is continuous on $\mathscr{X}$ endowed with the uniform topology.

When (i) is replaced by: g is bounded, similar results have been proved by Lebedev [16](when g has the form $g_t(\omega,x) = \overline{g}_t(\omega,x(t-))$, in which case (ii) reduces to the continuity of $\overline{g}_t(\omega,.)$ over $\mathbb{R}^d$; actually Lebedev uses a slightly different notion of a solution-measure, and he studies a more general equation than (1.1), involving random measures), by Pellaumail [23] (under a rather more stringent condition than (ii), including in particular that $g_t(\ ,.)$ is continuous on $\mathscr{X}$ endowed with Skorokhod topology), and by ourselves [13].

3) Rather than repeating the proof of [13], we will get (1.8) (even under weaker, but more difficult to state, assumptions) as a corollary of some stability results that we obtain in section 3: if $(Z^n,K^n,g^n,P^n)_{n\geq 1}$ is a sequence converging in some sense towards (Z,K,g,P), then the corresponding sequence $(\overline{P}^n)$ of solution-measures admits limit points for a suitable topology, and these limit points are solution-measures of (1.1). We will also obtain some strong stability results, about solution-processes (section 3-f), to be compared with similar results of Emery [6] and of Métivier and Pellaumail [18].

At last in section 4 we go back to existence and uniqueness of a solution-process on $(\Omega,\underline{\underline{F}},\underline{F},P)$: taking advantage of the results of the previous sections, we show that in many cases only uniqueness needs to be proved, and in particular we apply this to the monotonicity condition of [12].

2 - SOLUTION-PROCESSES AND SOLUTION-MEASURES

§2-a is not needed for the subsequent paragraphs of this section (except for the definition of local characteristics). Contrarywise, only §2-a is needed for section 3.

2-a. SOLUTION-MEASURES AND MARTINGALE PROBLEMS.

One of the main advantages of using weak solutions of an Ito's equation is that they are the solutions of a given martingale problem. We will see that the same holds for good solution-measures of equation (1.1).

If $y,y' \in \mathbb{R}^p$ we denote by $\langle y|y'\rangle$ their scalar product and by $|y|$ the Euclidian norm. If h is a $p\times q$-matrix, we denote by $|h|$ its norm as an operator: $\mathbb{R}^q \longrightarrow \mathbb{R}^p$; for instance, if g is as in (1.4),

$$|g_t| = \sup_{z\in\mathbb{R}^m,|z|\leq 1} |g_t z| = \sup_{z\in\mathbb{R}^m,|z|\leq 1} \Big(\sum_{j=1}^{d}\Big(\sum_{k=1}^{m} g_t^{jk} z^k\Big)^2\Big)^{1/2}.$$

The local characteristics (B,C,ν) of the semimartingale $Y=(Y^j)_{j\leq p}$ on $(\Omega,\underline{\underline{F}},\underline{F},P)$ consist in:

$$(2.1)\quad \begin{cases} B=(B^j)_{j\leq p}, \text{ an } \mathbb{R}^p\text{-valued predictable process with finite variation;} \\ C=(C^{jk})_{j,k\leq p}, \text{ a continuous adapted process, } C_0=0,\ C_t-C_s \text{ is a} \\ \quad p\times p \text{ symmetric nonnegative matrix if } s\leq t; \\ \nu=\nu(\omega;dt,dz), \text{ a positive predictable random measure on } \mathbb{R}_+\times\mathbb{R}^p \text{ with} \\ \quad \nu(\omega;\mathbb{R}_+\times\{0\})=\nu(\omega;\{0\}\times\mathbb{R}^p)=0,\ \int\nu(\omega;[0,t]\times dz)(|z|^2\wedge 1)<\infty, \end{cases}$$

and they are characterized as follows: first, Y has the unique decomposition:

$$Y_t = Y_0 + B_t + \sum_{0<s\leq t} \Delta Y_s I_{\{|\Delta Y_s|>1\}} + Y_t^c + Y_t^d$$

where B is like in (2.1), Y^c (resp. Y^d) is a continuous (resp. purely discontinuous) local martingale. Next, C is $C^{jk}=\langle (Y^c)^j,(Y^c)^k\rangle$. At last, ν is the "dual predictable projection" of the jump measure of Y, that is for all Borel subsets $A\subset\mathbb{R}^p$ at a positive distance of 0, $\nu([0,t]\times A)$ is the unique increasing predictable process such that $\sum_{s\leq t} I_A(\Delta Y_s) - \nu([0,t]\times A)$ is a local martingale.

There exists another nice characterization of (B,C,ν): if $u\in\mathbb{R}^p$ we define a predictable $\mathbb{C}$-valued process with finite variation by setting

$$(2.2)\quad \Phi_t^u = \int_0^t e^{i\langle u|Y_{s-}\rangle}\Big[i\sum_{j\leq p} u^j dB_s^j - \frac{1}{2}\sum_{j,k\leq p} u^j u^k dC_s^{jk}\Big] + \int_0^t\int_{\mathbb{R}^p} e^{i\langle u|Y_{s-}\rangle}\big(e^{i\langle u|z\rangle}-1-i\langle u|z\rangle I_{\{|z|\leq 1\}}\big)\nu(ds,dz).$$

Then (cf. [9, proof of (3.51)], or [7]):

(2.3) LEMMA: <u>An adapted, right-continuous process</u> Y <u>with left-hand limits is a semimartingale with local characteristics</u> (B,C,ν) <u>if and only if, for all</u> $u\in\mathbb{R}^p$, <u>the process</u> $e^{i\langle u|Y\rangle}-\Phi^u$ <u>is a local martingale</u>.

Now we go back to equation (1.1). In the following we assume (1.2), (1.3), (1.4), and (B,C,ν) denote the local characteristics of Z over $(\Omega,\underline{\underline{F}},\underline{F},P)$. The following is an increasing predictable process:

$$(2.4) \qquad A_t = \sum_{j\le m}\Big[\int_0^t |dB_s^j| + C_t^{jj}\Big] + \int_0^t\int_{\mathbb{R}^m}\nu(ds,dz)(|z|^2\wedge 1)\,.$$

We have a factorization

$$(2.5) \qquad B = b\bullet A\,, \quad C = c\bullet A\,, \quad \nu(dt,dz) = dA_t\times N_t(dz)\,,$$

$$\begin{cases} b=(b^j)_{j\le m}\,, \text{ an } \mathbb{R}^m\text{-valued predictable process,} \\ c=(c^{jk})_{j,k\le m}\,, \text{ an } m\times m \text{ nonnegative symmetric matrix-valued predictable process,} \\ N_t(\omega,dz)\,, \text{ a predictable transition measure on } \mathbb{R}^m\,. \end{cases}$$

From (2.4) it is easy to see that one may choose b,c,N such that

$$(2.6) \qquad |b| + \sum_{j\le m} c^{jj} + \int_{\mathbb{R}^m} N(dz)(|z|^2\wedge 1) \le 1\,.$$

Now, define the following collection of predictable processes on $(\overline{\Omega},\overline{\underline{\underline{F}}},\overline{\underline{F}})$, with $u\in\mathbb{R}^{m+d}$, and with this additional piece of notation: if $z\in\mathbb{R}^m$ and $x\in\mathbb{R}^d$, we denote by $y=(z,x)$ the vector of $\mathbb{R}^{m+d}$ whose components are $y^j=z^j$ if $j\le m$, $y^j=x^{j-m}$ otherwise.

$$(2.7) \qquad c_t^{g,jk} = \begin{cases} c_t^{jk} & \text{if } j,k\le m \\ \sum_{l\le m} g_t^{j-m,l}\, c_t^{lk} & \text{if } k\le m<j\le m+d \\ \sum_{l\le m} c_t^{jl}\, g_t^{k-m,l} & \text{if } j\le m<k\le m+d \\ \sum_{l,q\le m} g_t^{j-m,l}\, c_t^{lq}\, g_t^{k-m,q} & \text{if } m<j,k\le m+d \end{cases}$$

$$(2.8) \qquad \begin{aligned} v_t^{g,u} &= i\langle u|(b_t,g_tb_t)\rangle - \tfrac{1}{2}\sum_{j,k\le m+d} u^j\, c_t^{g,jk}\, u^k \\ &\quad + \int N_t(dz)\big[e^{i\langle u|(z,g_tz)\rangle} - 1 - i\langle u|(z,g_tz)\rangle I_{\{|z|\le 1\}}\big] \end{aligned}$$

$$(2.9) \qquad \Phi_t^{g,u} = \int_0^t \big[v_s^{g,u}\exp i\langle u|(Z_{s-},X_{s-}-K_{s-})\rangle\big]\, dA_s$$

(putting $+\infty$ if this integral is not well-defined).

(2.10) THEOREM: <u>A probability measure</u> $\overline{P}$ <u>on</u> $(\overline{\Omega},\overline{\underline{\underline{F}}})$ <u>is a good solution-measure of (1.1) if and only if</u>:

(i) $\overline{P}_{|\Omega} = P$ <u>and</u> $\overline{P}(X_0=K_0)=1$.

(ii) $\exp i\langle u|(Z,X-K)\rangle - \Phi^{g,u}$ <u>is a complex-valued local martingale on</u> $(\overline{\Omega},\overline{\underline{\underline{F}}},\overline{\underline{F}},\overline{P})$ <u>for all</u> $u\in\mathbb{R}^{m+d}$.

This result exactly generalizes the result for Ito's equations (cf. for instance [28]).

Proof. At first, we complete our collection of processes by setting

$$\hat{b}_t^j = b_t^j - \int N_t(dz)\, z^j\, I_{\{|z|\le 1 < |(z,g_t z)|\}} \qquad (j\le m)$$

$$b_t^{g,j} = \begin{cases} \hat{b}_t^j & \text{if } j\le m \\ \sum_{k\le m} g_t^{j-m,k}\, \hat{b}_t^k & \text{if } m<j\le m+d \end{cases}$$

$$N_t^g(G) = \int N_t(dz)\, I_G(z,g_t z) \qquad (\text{ G Borel set of } \mathbb{R}^{m+d})$$

(2.11) $$B^g = b^g{\scriptstyle\bullet}A, \quad C^g = c^g{\scriptstyle\bullet}A, \quad \nu^g(dt,dy) = dA_t \times N_t^g(dy).$$

Now, a proof exactly similar to that of Theorem (6.3) of [11], and too lenghty to be reproduced here, shows that if $Y=(Z,X-K)$, the two following assertions are equivalent:

a) Y is a semimartingale with local characteristics (B^g,C^g,ν^g) on $(\bar{\Omega},\bar{\underline{F}},\bar{\underline{F}},\bar{P})$.

b) Z is a semimartingale with local characteristics (B,C,ν) on $(\bar{\Omega},\bar{\underline{F}},\bar{\underline{F}},\bar{P})$, and on this space one has $\bar{P}$-almost surely

$$X-K = X_0 - K_0 + g{\scriptstyle\bullet}Z.$$

Therefore $\bar{P}$ is a good solution-measure if and only if one has (i), and a) above. Hence the result follows from Lemma (2.3) and from the fact that the process Φ^u associated to Y and (B^g,C^g,ν^g) by (2.2) is equal to $\Phi^{g,u}$, a fact that is easily verified once noticed that the process $v^{g,u}$ defined by (2.8) is equal to

$$v_t^{g,u} = i\langle u|(\hat{b}_t,g_t\hat{b}_t)\rangle - \frac{1}{2}\sum_{j,k\le m+d} u^j c_t^{g,jk} u^k + \int_{\mathbb{R}^{m+d}} N_t^g(dy)\left[e^{i\langle u|y\rangle} - 1 - i\langle u|y\rangle I_{\{|y|\le 1\}}\right]. \quad \blacksquare$$

Because of (2.6), the process $v^{g,u}$ is bounded whenever g is so, and is "Lipschitz in g". More precisely an easy (but rather tedious) computation shows that there exists a constant α, not depending on (b,c,N) satisfying (2.6), such that for all $u\in\mathbb{R}^{m+d}$:

(2.12) $$|v^{g,u}| \le \alpha(1+|u|^2)(1+|g|^2)$$

(2.13) $$|v^{g,u} - v^{g',u}| \le \alpha(1+|u|^2)(1+|g|+|g'|)\,|g-g'|.$$

2-b. COMPARISON BETWEEN SOLUTION-MEASURES AND SOLUTION-PROCESSES.

Let us begin with some technical lemmas, starting with the essential one:

(2.14) LEMMA: *Let* Y *be a* p-*dimensional semimartingale on* $(\Omega,\underline{\underline{F}},\underline{F},P)$. *Let* $\underline{G}$ *be a subfiltration of* $\underline{F}$ (i.e. $\underline{\underline{G}}_t \subset \underline{\underline{F}}_t$) *and assume* Y *is* $\underline{G}$-*adapted*.

(a) Y *is a* $\underline{G}$-*semimartingale*.

(b) *a* $\underline{G}$-*predictable* p-*dimensional process* H *belonging to* $L(Y;\Omega,\underline{F},P)$ *also belongs to* $L(Y;\Omega,\underline{G},P)$, *and the stochastic integral* $H\bullet Y$ *is the same, relative to* $\underline{F}$ *and to* $\underline{G}$.

(c) *If the* $\underline{F}$-*local characteristics of* Y *are* $\underline{G}$-*predictable, then they are the* G-*local characteristics of* Y, *and one has* $L(Y;\Omega,\underline{G},P) \subset L(Y;\Omega,\underline{F},P)$.

(a) is due to Stricker [27]; (b) and (c) may be found in [10]: notice that the last inclusion in (c) is not necessarily true for every $\underline{G}$-adapted $\underline{F}$-semimartingale.

We turn now to studying extensions of $(\Omega,\underline{\underline{F}},\underline{F},P)$. The following lemma is an easy corollary of the previous one, and of the rules governing changes of probability spaces (see e.g. [9,§X-2]).

(2.15) LEMMA: *Let* Y *be a* p-*dimensional semimartingale on* $(\Omega,\underline{\underline{F}},\underline{F},P)$, H *be a* p-*dimensional predictable process on this space, and* $(\tilde\Omega,\tilde{\underline{\underline{F}}},\tilde{\underline{F}},\tilde P)$ *be an extension of this space*.

(a) *If* Y *is a semimartingale on this extension, and if* $H\in L(Y;\tilde\Omega,\tilde{\underline{F}},\tilde P)$, *then* $H\in L(Y;\Omega,\underline{F},P)$ *and the stochastic integral* $H\bullet Y$ *is the same on both spaces*.

(b) *If* Y *is a semimartingale with the same local characteristics on both spaces and if* $H\in L(Y;\Omega,\underline{F},P)$, *then* $H\in L(Y;\tilde\Omega,\tilde{\underline{F}},\tilde P)$.

Suppose now that the extension has the following special form, where $(\Omega',\underline{\underline{F}}',\underline{F}')$ is an auxiliary filtered space and Q is a transition probability of $(\Omega,\underline{\underline{F}})$ into $(\Omega',\underline{\underline{F}}')$:

$$
(2.16)\quad \begin{cases} \tilde\Omega = \Omega\times\Omega', \quad \tilde{\underline{\underline{F}}} = \underline{\underline{F}}\otimes\underline{\underline{F}}', \quad \tilde{\underline{\underline{F}}}_t = \bigcap_{s>t}(\underline{\underline{F}}_s\otimes\underline{\underline{F}}'_s) \\ \tilde P(d\omega,d\omega') = P(d\omega)Q(\omega,d\omega'). \end{cases}
$$

Then, if $\underline{F}^P$ denotes the usual P-completion of the filtration $\underline{F}$, one has:

(2.17) LEMMA: *The extension is very good* (that is, according to (1.7), if any martingale on $(\Omega,\underline{\underline{F}},\underline{F},P)$ is a martingale on $(\tilde\Omega,\tilde{\underline{\underline{F}}},\tilde{\underline{F}},\tilde P)$) *if and*

<u>only if</u> $Q(.,G')\in \underline{\underline{F}}_t^P$ <u>for all</u> $G'\in \underline{\underline{F}}'_t$, $t\geq 0$.

<u>Proof</u>. Suppose the extension is very good. Let $G'\in \underline{\underline{F}}'_t$. For any $F\in \underline{\underline{F}}$ we have $P(F|\underline{\underline{F}}_t)=\tilde{P}(F|\tilde{\underline{\underline{F}}}_t)$ by hypothesis, so

$$E[I_F Q(G')] = \tilde{E}(I_F I_{G'}) = \tilde{E}[I_{G'}\; P(F|\underline{\underline{F}}_t)] = E[P(F|\underline{\underline{F}}_t)\, Q(G')],$$

which clearly implies that $Q(.,G')\in \underline{\underline{F}}_t^P$.

Suppose conversely that $Q(.,G')\in \underline{\underline{F}}_t^P$ for all $G'\in \underline{\underline{F}}'_t$, $t\geq 0$. Let V be a right-continuous martingale on $(\Omega,\underline{\underline{F}},\underline{F},P)$. Let $s\geq t\geq 0$, $G\in \underline{\underline{F}}_t$, $G'\in \underline{\underline{F}}'_t$. Then

$$\tilde{E}(V_t I_{G\times G'}) = E[V_t I_G Q(G')] = E[V_s I_G Q(G')] = \tilde{E}(V_s I_{G\times G'}).$$

Thus $V_t=\tilde{E}(V_s|\underline{\underline{F}}_t\otimes\underline{\underline{F}}'_t)$. Since V is right-continuous, it follows that $V_t=\tilde{E}(V_s|\tilde{\underline{\underline{F}}}_t)$, thus V is a martingale over $(\tilde{\Omega},\tilde{\underline{\underline{F}}},\tilde{\underline{F}},\tilde{P})$. ∎

We are ready now to compare solution-processes and solution-measures. To every solution-measure $\overline{P}$ it corresponds a solution-process, namely X itself, on $(\overline{\Omega},\overline{\underline{\underline{F}}},\overline{\underline{F}},\overline{P})$. Conversely, one has:

(2.18) THEOREM: <u>Let</u> $\tilde{X}$ <u>be a solution-process on an extension</u> $(\tilde{\Omega},\tilde{\underline{\underline{F}}},\tilde{\underline{F}},\tilde{P})$. <u>Let</u> $\varphi:\tilde{\Omega}=\Omega\times\Omega'\longrightarrow\overline{\Omega}=\Omega\times\mathcal{X}$ <u>be defined by</u> $\varphi(\omega,\omega')=(\omega,\tilde{X}(\omega,\omega'))$. <u>Then</u> $\overline{P}=\tilde{P}\circ\varphi^{-1}$ <u>is a solution-measure. Moreover if</u> $\tilde{X}$ <u>is a good (resp. very good) solution-process, then</u> $\overline{P}$ <u>is good (resp. very good)</u>.

<u>Proof</u>. Let $\tilde{\underline{\underline{F}}}'_t=\varphi^{-1}(\overline{\underline{\underline{F}}}_t)$. Z is still a semimartingale over $(\tilde{\Omega},\tilde{\underline{\underline{F}}},\tilde{\underline{F}}',\tilde{P})$ and $g(\tilde{X})$ is $\tilde{\underline{F}}'$-predictable, so (2.14,b) implies that $g(\tilde{X})\in L(Z;\tilde{\Omega},\tilde{\underline{F}}',\tilde{P})$ and $\tilde{X}$ is still a solution process on $(\tilde{\Omega},\tilde{\underline{\underline{F}}},\tilde{\underline{F}}',\tilde{P})$. Moreover it is obvious that this solution-process is good (resp. very good) on $(\tilde{\Omega},\tilde{\underline{\underline{F}}},\tilde{\underline{F}}',\tilde{P})$ if it is so on $(\tilde{\Omega},\tilde{\underline{\underline{F}}},\tilde{\underline{F}},\tilde{P})$. Since $\tilde{X}=X\circ\varphi$ by definition of φ, while $Z=Z\circ\varphi$ and $(B,C,\nu)=(B\circ\varphi,C\circ\varphi,\nu\circ\varphi)$ and $g(\tilde{X})=g\circ\varphi$, the result immediately follows from the usual rules of change of probability spaces (see e.g. [9,§X-2]). ∎

Any solution-measure $\overline{P}$ factorizes as

$$\overline{P}(d\omega,dx) = P(d\omega)\,Q(\omega,dx). \tag{2.19}$$

We obtain as an immediate corollary to (2.17) and (2.18):

(2.20) COROLLARY: <u>For any solution-measure</u> $\overline{P}$, <u>there is equivalence between:</u>

(a) $\overline{P}$ <u>is a very good solution-measure</u>.

(b) In factorization (2.19) of $\overline{P}$, one has $Q(.,G)\in \underline{\underline{F}}_t^P$ for all $G\in \underline{\underline{\mathscr{X}}}_t$, $t\geq 0$.

(c) $\overline{P}$ is realized (in the sense of (2.18)) by a very good solution-process.

2-c. STRONG SOLUTIONS.

Here again we shall see that the notion of a strong solution for an Ito's equation generalizes with almost no change, and with the same properties.

(2.21) DEFINITION: A solution-measure is called strong if it can be realized (in the sense of (2.18)) by a solution-process on the space $(\Omega,\underline{\underline{F}},\underline{F}^P,P)$.

(2.22) THEOREM: For any solution-measure $\overline{P}$ there is equivalence between:

(a) $\overline{P}$ is strong.

(b) There exists an $\underline{F}^P$-adapted mapping $\tilde{X}:\Omega\longrightarrow\mathscr{X}$ such that

$$\overline{P}(d\omega,dx) = P(d\omega)\,\varepsilon_{\tilde{X}(\omega)}(dx)\,. \tag{2.23}$$

(c) $\overline{P}$ is realized by a solution-process $\tilde{X}$ over an extension $(\tilde{\Omega},\tilde{\underline{\underline{F}}},\tilde{\underline{F}},\tilde{P})$, on which $\tilde{X}$ is $\underline{F}^{\tilde{P}}$-adapted (where here, $\underline{F}_t=\{G\times\Omega' : G\in\underline{\underline{F}}_t\}$).

Moreover, these properties imply:

(d) $\overline{P}$ is a very good solution-measure.

(e) In (b), $\tilde{X}$ is the unique (up to a P-null set) solution-process over $(\Omega,\underline{\underline{F}},\underline{F},P)$ that realizes $\overline{P}$.

(f) On any good extension of $(\Omega,\underline{\underline{F}},\underline{F},P)$ there exists exactly one (up to a null set) solution-process which realizes $\overline{P}$, and this solution-process is the process $\tilde{X}$ introduced in (b).

Proof. We obviously have: (a) $\Longrightarrow$ (b) $\Longrightarrow$ (d) (use (2.20)), and: (f) $\longrightarrow$ (e) $\Longrightarrow$ (a) $\Longrightarrow$ (c). It remains to prove that: (b) $\Longrightarrow$ (f) and (c) $\longrightarrow$ (b).

Suppose (b). Let $(\tilde{\Omega},\tilde{\underline{\underline{F}}},\tilde{\underline{F}},\tilde{P})$ be a good extension of $(\Omega,\underline{\underline{F}},\underline{F},P)$. Since $\tilde{X}=X$ $\overline{P}$-a.s. on $\overline{\Omega}$, $\tilde{X}$ is a solution-process on $(\overline{\Omega},\overline{\underline{\underline{F}}},\overline{\underline{F}},\overline{P})$, thus also on $(\Omega,\underline{\underline{F}},\underline{F}^P,P)$ by using (2.14,b). Using again the rules of change of probability spaces, it follows that $\tilde{X}$ is still a solution-process on $(\tilde{\Omega},\tilde{\underline{\underline{F}}},\underline{F}^{\tilde{P}},\tilde{P})$. Since our extension is good, we deduce from (2.14,c) that $\tilde{X}$ is also a solution-process on $(\tilde{\Omega},\tilde{\underline{\underline{F}}},\tilde{\underline{F}},\tilde{P})$. Uniqueness in (f) follows immediately from (2.18) and (2.23).

Suppose (c). Let $\tilde{X}$ be an $\underline{F}^{\tilde{P}}$-adapted solution-process over the extension $(\tilde{\Omega},\tilde{\underline{\underline{F}}},\tilde{\underline{F}},\tilde{P})$. By changing $\tilde{X}$ on a $\tilde{P}$-null set, we may actually assume that $\tilde{X}$ is defined on Ω and is $\underline{F}^{P}$-adapted. Then (b) immediately follows from (2.18).∎

(2.24) DEFINITION: We say that pathwise (resp. good pathwise, resp. very good pathwise) uniqueness holds if, on any extension (resp. good extension, resp. very good extension) of $(\Omega,\underline{\underline{F}},\underline{F},P)$ there is at most one solution-process, up to a null set. ∎

Of course, we have the implications: pathwise uniqueness $\Longrightarrow$ good pathwise uniqueness $\Longrightarrow$ very good pathwise uniqueness.

The results of Yamada and Watanabe [29] and of Zvonkin and Krylov [30] generalize as follows, with a slightly complicated situation here due to our three kinds of solutions.

(2.25) THEOREM: (a) If $\overline{P}$ is a very good solution-measure, there is equivalence between:

(a-i) very good pathwise uniqueness;

(a-ii) $\overline{P}$ is strong and is the unique very good solution-measure.

(b) If $\overline{P}$ is a strong solution-measure, there is equivalence between:

(b-i) good pathwise uniqueness;

(b-ii) $\overline{P}$ is the unique good solution-measure.

(c) If $\overline{P}$ is a strong solution-measure and if the process g is locally bounded on $(\overline{\Omega},\overline{\underline{\underline{F}}},\overline{\underline{F}})$ (i.e.: there exists a sequence (T_n) of $\underline{F}$-stopping times going to $+\infty$ and such that g is bounded on each interval $[\![0,T_n]\!]$), there is equivalence between:

(c-i) pathwise uniqueness;

(c-ii) $\overline{P}$ is the unique solution-measure.

Notice that the restriction on g in (c) is due to the bad situation encountered in Lemma (2.14), in which the last inclusion is not always true.

Proof. α) Let $\overline{P}$ be a strong solution-measure, to which a solution-process $\tilde{X}$ on $(\Omega,\underline{\underline{F}},\underline{F},P)$ is associated by (2.22). Let $\tilde{X}^1$ and $\tilde{X}^2$ be two solution-processes (resp. good solution-processes, resp. very good solution-processes) on the same extension $(\tilde{\Omega},\tilde{\underline{\underline{F}}},\tilde{\underline{F}},\tilde{P})$. Assume (c-ii) (resp. (b-ii), resp. (a-ii)). Then $\tilde{X}^1$ and $\tilde{X}^2$ also realize $\overline{P}$. Thus, due to (2.23), it is clear that $\tilde{X}^i = \tilde{X}$ $\tilde{P}$-a.s. for $i = 1,2$. Hence we have proved that: (c-ii) $\Longrightarrow$ (c-i); (b-ii) $\Longrightarrow$ (b-i); (a-ii) $\Longrightarrow$ (a-i).

β) Let again $\overline{P}$ be a strong solution-measure, and $\tilde{X}$ the corresponding solution-process on $(\Omega, \underline{\underline{F}}, \underline{F}, P)$. Let $\overline{P}'$ be a good solution-measure (resp. a solution-measure, and assume g is locally bounded). Applying Lemma (2.15,c) (resp. the local boundedness of g) we obtain that $g(\tilde{X}) \in L(Z;\overline{\Omega},\overline{\underline{F}},\overline{P}')$ and that $\tilde{X}$ is again a solution-process on $(\overline{\Omega},\overline{\underline{\underline{F}}},\overline{\underline{F}},\overline{P}')$, as well as X itself. Under (b-i) (resp. (c-i)) it follows that $\tilde{X} = X$ $\overline{P}'$-a.s., which implies by (2.18) that $\overline{P}' = \overline{P}$. Hence we have proved that: (b-i) $\longrightarrow$ (b-ii); (c-i) $\longrightarrow$ (c-ii).

γ) Suppose (a-i) holds. Let $\overline{P}_1, \overline{P}_2$ be two very good solution-measures, with their factorizations $\overline{P}_i(d\omega, dx) = P(d\omega) Q_i(\omega, dx)$. Set

$$\tilde{\Omega} = \Omega \times \mathcal{X} \times \mathcal{X}, \quad \tilde{\underline{\underline{F}}} = \underline{\underline{F}} \otimes \underline{\underline{\mathcal{X}}} \otimes \underline{\underline{\mathcal{X}}}, \quad \tilde{\underline{\underline{F}}}_t = \bigcap_{s>t} (\underline{\underline{F}}_s \otimes \underline{\underline{\mathcal{X}}}_s \otimes \underline{\underline{\mathcal{X}}}_s)$$

$$\tilde{P}(d\omega, dx_1, dx_2) = P(d\omega)\, Q_1(\omega, dx_1)\, Q_2(\omega, dx_2)$$

$$\tilde{X}^1(\omega, x_1, x_2) = X(x_1), \qquad \tilde{X}^2(\omega, x_1, x_2) = X(x_2).$$

According to (2.17), $(\tilde{\Omega},\tilde{\underline{\underline{F}}},\tilde{\underline{F}},\tilde{P})$ is a very good extension of $(\overline{\Omega},\overline{\underline{\underline{F}}},\overline{\underline{F}},\overline{P}_i)$, so Lemma (2.15,c) implies that $\tilde{X}^i$ is a solution-process on $(\tilde{\Omega},\tilde{\underline{\underline{F}}},\tilde{\underline{F}},\tilde{P})$. Thus (a-i) implies that $\tilde{X}^1 = \tilde{X}^2$ $\tilde{P}$-a.s., that is $\tilde{P}(D) = 0$ if $D = \{(x_1, x_2) : x_1 \neq x_2\}$. From the definition of $\tilde{P}$, that fact implies the existence of a mapping $\tilde{X} : \Omega \longrightarrow \mathcal{X}$ such that $Q_i(\omega, dx) = \varepsilon_{\tilde{X}(\omega)}(dx)$ P-a.s. for $i = 1,2$. Moreover (2.17) implies that $\tilde{X}$ is $\underline{F}^P$-adapted, and we have

$$\overline{P}_i(d\omega, dx) = P(d\omega)\varepsilon_{X(\omega)}(dx), \quad i = 1,2.$$

Thus $\overline{P}_1 = \overline{P}_2$ is strong, because of (2.22), and we have (a-ii).∎

(2.26) COROLLARY: a) <u>If</u> $\overline{P}$ <u>is a very good solution-measure, and if good pathwise uniqueness holds, then</u> $\overline{P}$ <u>is strong and is the only good solution-measure</u>.

b) <u>If</u> g <u>is locally bounded, if</u> $\overline{P}$ <u>is a very good solution-measure, and if pathwise uniqueness holds, then</u> $\overline{P}$ <u>is strong and is the only solution-measure</u>.

In particular, under the conditions of Theorem (1.8), in order to obtain existence and uniqueness of a solution-process on $(\Omega, \underline{\underline{F}}, \underline{F}, P)$, it is sufficient to show that good pathwise uniqueness holds, a fact that is often easy to check: we will take advantage of this remark in section 4.

2-d. ABSOLUTELY CONTINUOUS CHANGE OF MEASURE.

This short paragraph is aimed to show that, although in Theorem (1.8) we get a very good solution-measure, one may very well encounter solution-measures which are neither very good, nor even good.

Let $P' \ll P$ and $L_\infty = \frac{dP'}{dP}$ be the Radon-Nikodym derivative. Let $L_t = E(L_\infty | \underline{\underline{F}}_t)$. We know that Z is a semimartingale over $(\Omega, \underline{\underline{F}}, \underline{F}, P')$ and we shall call (1.1)' the stochastic differential equation based upon $(\Omega, \underline{\underline{F}}, \underline{F}, P')$, Z, K, g.

(2.27) PROPOSITION: Let $\overline{P}$ be a solution-measure of (1.1). Any $\overline{P}'$ such that $\overline{P}' \ll \overline{P}$ and that $\overline{P}'_{|\Omega} = P'$ is a solution-measure of (1.1)'.

Proof. Immediate, since Z is a semimartingale on $(\overline{\Omega}, \underline{\underline{\overline{F}}}, \underline{\overline{F}}, \overline{P}')$ and since $L(Z;\overline{\Omega}, \underline{\overline{F}}, \overline{P}) \subset L(Z;\overline{\Omega}, \underline{\overline{F}}, \overline{P}')$: cf. [9]. ∎

In particular we can examinate the measure $\overline{P}' = L_\infty \cdot \overline{P}$, which is by (2.27) a solution-measure of (1.1)'. One may ask whether this solution is good (resp. very good) when $\overline{P}$ is such. In this direction, let us state two results, without proof:

(2.28) If $\overline{P}$ is very good, then $\overline{P}'$ is very good.

(2.29) If $\overline{P}$ is good, then $\overline{P}'$ will also be good, provided L belongs to the stable subspace of martingale of $(\Omega, \underline{\underline{F}}, \underline{F}, P)$ generated by Z (that is, generated by Z^c and by the jump random measure of Z : cf. [9, ch. IV]).

3 - EXISTENCE AND STABILITY FOR SOLUTION-MEASURES

3-a. A TOPOLOGY FOR PROBABILITY MEASURES ON $(\overline{\Omega}, \underline{\underline{\overline{F}}})$.

We wish to study "weak stability" of solution-measures of equation (1.1). To do so, our first task is to introduce a reasonnable topology on the space of all probability measures on $(\overline{\Omega}, \underline{\underline{\overline{F}}})$. This can be achieved through somehow "combining" reasonnable topologies on the spaces of probability measures on $(\Omega, \underline{\underline{F}})$ and on $(\mathfrak{X}, \underline{\underline{\mathfrak{X}}})$.

1) We denote by $\underline{\underline{M}}_m(\Omega)$ the set of all probability measures on $(\Omega, \underline{\underline{F}})$ endowed with the coarsest topology for which all mappings: $P \rightsquigarrow P(F)$,

$F \in \underline{\underline{F}}$, are continuous.

2) Let $\mathcal{T}_s$ be the Sorokhod J_1 topology on $\mathcal{X}$, for which one knows that $(\mathcal{X},\underline{\underline{\mathcal{X}}})$ is a Polish space with its Borel σ-field. Ve denote by $\underline{\underline{M}}_c(\mathcal{X})$ the set of all probability measures on $(\mathcal{X},\underline{\underline{\mathcal{X}}})$ endowed with the <u>weak topology</u>, that is the coarsest one for which all mappings: $\mu \rightsquigarrow \mu(V)$, V bounded continuous function on $\mathcal{X}$, are continuous. See [2],[26].

Now, let us denote by $B_{mc}(\bar{\Omega})$ the set of all bounded measurable functions $V : \bar{\Omega} \longrightarrow \mathbb{R}$ such that $V(\omega,.)$ is $\mathcal{T}_s$-continuous on $\mathcal{X}$ for all $\omega \in \Omega$.

(3.1) DEFINITION: We denote by $\underline{\underline{M}}_{mc}(\bar{\Omega})$ the set of <u>all probability measures on</u> $(\bar{\Omega},\bar{\underline{\underline{F}}})$ endowed with the <u>coarsest topology</u> for which all mappings: $\mu \rightsquigarrow \mu(V)$, $V \in B_{mc}(\bar{\Omega})$, are continuous. ■

This topology has been introduced by many authors in quite different contexts. All the results recalled below may be found in [15], but some of them already appear in [13] and [21].

(3.2) LEMMA: <u>The topology on</u> $\underline{\underline{M}}_{mc}(\bar{\Omega})$ <u>is also the coarsest one for which all mappings</u>: $V \rightsquigarrow \mu(V)$ <u>with</u> $V(\omega,x) = U(\omega)W(x)$, U <u>bounded measurable on</u> $(\Omega,\underline{\underline{F}})$ <u>and</u> W <u>bounded uniformly continuous on</u> $\mathcal{X}$, <u>are continuous</u>.

If $\bar{P} \in \underline{\underline{M}}_{mc}(\bar{\Omega})$ we will denote by $\bar{P}_{|\Omega}$ and $\bar{P}_{|\mathcal{X}}$ its marginals on Ω and on $\mathcal{X}$. Of course, the mappings: $\bar{P} \rightsquigarrow \bar{P}_{|\Omega}$ (resp. $\bar{P} \rightsquigarrow \bar{P}_{|\mathcal{X}}$) are continuous from $\underline{\underline{M}}_{mc}(\bar{\Omega})$ onto $\underline{\underline{M}}_m(\Omega)$ (resp. $\underline{\underline{M}}_c(\mathcal{X})$).

(3.3) THEOREM: <u>A subset</u> $\mathcal{M}$ <u>of</u> $\underline{\underline{M}}_{mc}(\bar{\Omega})$ <u>is relatively compact if and only if both</u> $\mathcal{M}_\Omega = \{\bar{P}_{|\Omega} : \bar{P} \in \mathcal{M}\}$ <u>and</u> $\mathcal{M}_{\mathcal{X}} = \{\bar{P}_{|\mathcal{X}} : \bar{P} \in \mathcal{M}\}$ <u>are relatively compact in</u> $\underline{\underline{M}}_m(\Omega)$ <u>and</u> $\underline{\underline{M}}_c(\mathcal{X})$, <u>respectively</u>.

(3.4) THEOREM: <u>Let</u> $(\bar{P}^n)$ <u>be a sequence converging to</u> $\bar{P}$ <u>in</u> $\underline{\underline{M}}_{mc}(\bar{\Omega})$. <u>Let</u> $F \in \bar{\underline{\underline{F}}}$ <u>be such that each section</u> $F_\omega = \{x \in \mathcal{X} : (\omega,x) \in F\}$ <u>is</u> $\mathcal{T}_s$<u>-closed in</u> $\mathcal{X}$ <u>and that</u>: $\limsup_{(n)} \bar{P}^n(F) = 1$. <u>Let</u> V <u>be a bounded measurable function on</u> $(\bar{\Omega},\bar{\underline{\underline{F}}})$ <u>such that for each</u> $\omega \in \Omega$ <u>the restriction of</u> $V(\omega,.)$ <u>to</u> F_ω <u>is continuous for the topology induced by</u> $\mathcal{T}_s$ <u>on</u> F_ω. <u>Then</u> $\bar{E}^n(V) \longrightarrow \bar{E}(V)$.

Let us now introduce $\mathcal{T}_u$, the topology of uniform convergence on compact sets, on the space $\mathcal{X}$, with the metric $\delta_u(x,x') = \sum_{n\geq 1} 2^{-n}[1 \wedge \sup_{s\leq n} |x(s) - x'(s)|]$. One knows that $\mathcal{T}_u$ is stronger than $\mathcal{T}_s$. However, let $k : \mathbb{R}_+ \longrightarrow \mathbb{R}_+$ be such that

(3.5) for all $a>0$, the set $\{t : k(t) \geq a\}$ is discrete (i.e.: locally finite),

and set

$$\mathcal{X}_k = \{x \in \mathcal{X} : |\Delta x(t)| \leq k(t)(1 + \sup_{s<t} |x(s)|) \text{ for all } t>0\}$$

(where $\Delta x(t) = x(t) - x(t-)$). Then:

(3.6) LEMMA: <u>$\mathcal{X}_k$ is a closed set for $\mathcal{T}_u$ and $\mathcal{T}_s$, and both $\mathcal{T}_u$ and $\mathcal{T}_s$ induce the same topology on $\mathcal{X}_k$.</u>

<u>Proof</u>. Let (x^n) be a sequence of points in $\mathcal{X}_k$, converging to $x \in \mathcal{X}$ for $\mathcal{T}_s$. Let $t>0$ be such that $|\Delta x(t)| > 0$. From classical properties of $\mathcal{T}_s$ there exists a sequence (t_n) converging to t, such that $\Delta x^n(t_n) \longrightarrow \Delta x(t)$. We have

$$\sup_{(n)} \sup_{s\leq t+1} |x^n(s)| < \infty$$

$$\lim_{\varepsilon\to 0} \sup_{s\neq t, t-\varepsilon\leq s\leq t+\varepsilon} k(s) = 0 .$$

Since $x^n \in \mathcal{X}_k$, $t_n \longrightarrow t$ and $|\Delta x^n(t_n)| \longrightarrow |\Delta x(t)| > 0$, it follows that $t_n = t$ for n large enough. Therefore $\Delta x^n(t) \longrightarrow \Delta x(t)$. If $\Delta x(t) = 0$ one also know that $\Delta x^n(t) \longrightarrow 0$. Hence $\Delta x^n(t) \longrightarrow \Delta x(t)$ for all $t>0$, which implies that $x^n \longrightarrow x$ in $\mathcal{T}_u$ as well as in $\mathcal{T}_s$ (use Theorem 2.6.2 of [26]). The lemma follows from that, and from the obvious fact that $\mathcal{X}_k$ is $\mathcal{T}_u$-closed. ∎

3-b. STATEMENT OF STABILITY RESULTS.

In order to state our stability results, we unhappily have to introduce a huge number of notations and of conditions, due to the fact that we want everything in (1.1) to change: P, Z, K, g.

The filtered spaces $(\Omega,\underline{\underline{F}},\underline{F})$, $(\mathcal{X},\underline{\underline{\mathcal{X}}},\underline{\mathcal{X}})$, $(\bar{\Omega},\bar{\underline{\underline{F}}},\bar{\underline{F}})$ are fixed as before. But instead of (1.2) and (1.4) we are given for each $n \in \overline{\mathbb{N}} = \mathbb{N} \cup \{\infty\}$:

(3.7) A probability measure P^n on $(\Omega,\underline{\underline{F}})$; a semimartingale $Z^n = (Z^{n,j})_{j\leq m}$ with $Z^n_0 = 0$ on $(\Omega,\underline{\underline{F}},\underline{F},P^n)$; a right-continuous $\underline{F}$-adapted process $K^n = (K^{n,j})_{j\leq d}$ with left-hand limits on $(\Omega,\underline{\underline{F}})$. ∎

(3.8) A predictable process $g^n = (g^{n,jk})_{j\leq d, k\leq m}$ on $(\bar{\Omega}, \bar{\underline{\underline{F}}}, \bar{\underline{F}})$. ∎

We will call (1.1,n) the following equation:

$$(1.1,n) \qquad X_t = K_t^n + \int_0^t g_s^n(.,X_.(.))dZ_s^n \quad \text{based upon } (\Omega, \underline{\underline{F}}, \underline{F}, P^n).$$

We denote by (B^n, C^n, ν^n) the local characteristics of Z^n over $(\Omega, \underline{\underline{F}}, \underline{F}, P^n)$, and to which we associate A^n, (b^n, c^n, N^n) by (2.4) and (2.5). To $u \in \mathbb{R}^{m+d}$ we also associate with g^n, (b^n, c^n, N^n), the process $v^{n,g^n,u}$ defined by (2.8), and we set

$$(3.9) \qquad \Phi_t^{n,g^n,u} = \int_0^t \left[v_s^{n,g^n,u} \exp i\langle u|(Z_{s-}^n, X_{s-} - K_{s-}^n)\rangle\right] dA_s^n .$$

We turn now to a first set of conditions.

(3.10) Condition on P^n: (P^n) converges to P^∞ in $\underline{\underline{M}}_m(\Omega)$.

(3.11) Condition on K^n: $\lim_{(n)} P^n[\sup_{s\leq t} |K_s^n - K_s^\infty| > \varepsilon] = 0$, all $t \geq 0, \varepsilon > 0$.

(3.12) Condition on Z^n: $\lim_{(n)} P^n[|Z_T^n - Z_T^\infty| > \varepsilon] = 0$ for all $\varepsilon > 0$ and all bounded $\underline{F}$-stopping time T on Ω.

(3.13) Tightness and linear growth condition: for each $n \in \bar{\mathbb{N}}$ there exists a predictable process $\gamma^n \geq 0$ and a predictable increasing process $\tilde{A}^n$ on $(\Omega, \underline{\underline{F}}, \underline{F})$, such that $\tilde{A}_t^n - \int_0^t (1 \vee (\gamma_s^n)^2)\, dA_s^n$ is increasing and that:

(i) $|g_t^n(\omega,x)| \leq \gamma_t^n(\omega)(1 + \sup_{s<t} |x(s)|)$, all $t \geq 0, (\omega,x) \in \bar{\Omega}$, $n \in \bar{\mathbb{N}}$;

(ii) $\lim_{(n)} P^n(\delta_s(\tilde{A}^n, \tilde{A}^\infty) > \varepsilon) = 0$ for all $\varepsilon > 0$, where δ_s denotes the Skorokhod distance on the space $D([0,\infty), \mathbb{R}_+)$;

(iii) there exists a measurable process $V \geq 0$ on $(\Omega, \underline{\underline{F}})$ such that for all $a > 0$, $\omega \in \Omega$ the set $\{t : V_t(\omega) > a\}$ is discrete, and such that for all $t \geq 0$ we have

$$\lim_{(n)} P^n[(1 + \gamma_s^n)|\Delta Z_s^n| + |\Delta K_s^n| \leq V_s \text{ for all } s \in (0,t]] = 1. ∎$$

(3.14) Convergence condition on solution-measures: this is a condition on a sequence $(\bar{P}^n)_{n\in\mathbb{N}}$, each $\bar{P}^n$ being a solution-measure of (1.1,n). We have: $\lim_{(n)} \bar{P}^n(|\Phi_T^{n,g^n,u} - \Phi_T^{\infty,g^\infty,u}| > \varepsilon) = 0$ for all $\varepsilon > 0$, $u \in \mathbb{R}^{m+d}$, and all bounded $\bar{\underline{F}}$-stopping time T on $\bar{\Omega}$.

(3.15) Continuity of g^∞: for all $\omega \in \Omega$, $t \geq 0$, $g_t^\infty(\omega,.)$ is continuous on $\mathcal{X}$ endowed with the uniform topology (or equivalently with $\mathcal{C}_u$, since $g_t^{\infty}(\omega,x)$ depends upon x only through the values $x(s)$, $s<t$).

Then an ugly but quite general theorem goes as follows:

(3.16) THEOREM: For each $n\in\mathbb{N}$ let $\bar{P}^n$ be a good solution-measure of (1.1,n).

(a) Under (3.10), (3.11) and (3.13) the sequence $(\bar{P}^n)_{n\in\mathbb{N}}$ is relatively compact in $\underline{\underline{M}}_{mc}(\bar{\Omega})$.

(b) If moreover one has (3.12), (3.14) and (3.15), then all limit points of this sequence are good solution-measures of $(1.1,\infty)$.

(c) Assume the sequence $(\bar{P}^n)_{n\in\mathbb{N}}$ is relatively compact and all its limit points are good solution-measures of $(1.1,\infty)$; assume that $(1.1,\infty)$ admits at most one solution-measure. Then it admits exactly one solution-measure $\bar{P}^{\infty}$ and $(\bar{P}^n)$ tends to $\bar{P}^{\infty}$. Moreover if for all $n\in\bar{\mathbb{N}}$, $\bar{P}^n$ corresponds to a solution-process $\tilde{X}^n$ on $(\Omega,\underline{\underline{F}},\underline{F},P^n)$, then $\lim_{(n)} P^n(\sup_{s\leq t}|\tilde{X}^n_s-\tilde{X}^{\infty}_s|>\varepsilon)=0$ for all $t\geq 0$, $\varepsilon>0$.

Some of the previous conditions look rather bad, for instance (3.13,ii, iii). Still worse: condition (3.14) involves the solution measures $\bar{P}^n$ themselves! so, let us state another, perhaps better to look at, theorem; for it, we need again some new notations.

If F is an increasing process, we write $V(F)$ for the process of its variation, that is $V(F)_t=\int_0^t|dF_s|$. Let $f:\mathbb{R}^m\longrightarrow\mathbb{R}^m$ denote a function, fixed throughout, that is continuous, bounded by 1, with compact support, and such that $f(z)=z$ if $|z|\leq 1$. Define

$$(3.17)\quad\begin{cases}\hat{B}^n_t = B^n_t-\int\nu^n([0,t]\times dz)\,[f(z)-z\,I_{\{|z|\leq 1\}}]\\ \hat{C}^{n,jk}_t = C^{n,jk}_t+\int\nu^n([0,t]\times dz)\,f^j(z)\,f^k(z)\\ \hat{\nu}^n_t(h) = \int\nu^n([0,t]\times dz)h(z)\,.\end{cases}$$

$\hat{\nu}^n_t(h)$ is well-defined whenever $h:\mathbb{R}^m\longrightarrow\mathbb{R}$ is Borel and $h(z)/(|z|^2\wedge 1)$ is bounded.

(3.18) Linear growth: there exists a locally bounded nonnegative process γ on $(\Omega,\underline{\underline{F}},\underline{F})$ such that $|g^n_t(\omega,x)|\leq\gamma_t(\omega)(1+\sup_{s<t}|x(s)|)$ for all $t\geq 0$, $(\omega,x)\in\bar{\Omega}$, $n\in\bar{\mathbb{N}}$.

(3.19) <u>Convergence of</u> g^n : $g^n_t(\omega,.)$ converges towards $g^\infty_t(\omega,.)$ uniformly on each γ_s-compact subset of $\mathcal{X}$, for all $t \geq 0$, $\omega \in \Omega$.

(3.20) <u>Conditions on</u> Z^n :

(i) $\lim_{(n)} P^n(\sup_{s \leq t} |Z^n_s - Z^\infty_s| \geq \varepsilon) = 0$ for all $t \geq 0$, $\varepsilon > 0$;

(ii) $V(\hat{B}^{n,j} - \hat{B}^{\infty,j})_t \xrightarrow{P^\infty} 0$ for all $t \geq 0$, $j \leq m$;

(iii) $V(\hat{C}^{n,jk} - \hat{C}^{\infty,jk})_t \xrightarrow{P^\infty} 0$ for all $t \geq 0$, $j,k \leq m$;

(iv) $V(\hat{\nu}^n(h) - \hat{\nu}^\infty(h))_t \xrightarrow{P^\infty} 0$ for all $t \geq 0$ and all functions $h : \mathbb{R}^m \longrightarrow \mathbb{R}$ that are continuous and with $h(z)/(|z|^3 \wedge 1)$ bounded.■

(3.21) <u>Condition on</u> P^n : $\lim_{(n)} \sup_{F \in \underline{\underline{F}}} |P^n(F) - P^\infty(F)| = 0$ (we have of course (3.21) $\Longrightarrow$(3.10)).

(3.22) REMARKS: 1) We will see in section 3-f that, when $P^n = P$ for all $n \in \overline{\mathbb{N}}$, then condition (3.20) is implied by the convergence of Z^n towards Z^∞ for Emery's topology of semimartingales.

2) Conditions (3.20,ii,iii,iv) may appear particularly artificial ! in fact, it is quite the opposite; for example if each Z^n is a process with independent increment for P^n, then $\hat{B}^n$, $\hat{C}^n$, $\hat{\nu}^n$ are deterministic, and (3.20,ii,iii,iv) is equivalent to the fact that the distribution of Z^n under P^n converges towards the distribution of Z^∞ under P^∞.■

(3.23) REMARK: When (3.21) holds, it is immediate to check that

$$(3.11) \Longleftrightarrow \sup_{s \leq t} |K^n_s - K^\infty_s| \xrightarrow{P^\infty} 0$$

$$(3.20,i) \Longleftrightarrow \sup_{s \leq t} |Z^n_s - Z^\infty_s| \xrightarrow{P^\infty} 0 .$$

When (3.10) holds, but not (3.21), we will see later (Lemma (3.55)) that the above implications $\Longleftarrow$ hold, but not necessarily the implications $\Longrightarrow$.■

(3.24) THEOREM: <u>For each</u> $n \in \mathbb{N}$ <u>let</u> $\overline{P}^n$ <u>be a good solution-measure of (1.1,n). Under (3.11), (3.15), (3.18), (3.19), (3.20), and either (3.21) or (3.10) and (3.13,iii) with</u> $\gamma^n = 0$, <u>then the sequence</u> $(\overline{P}^n)_{n \in \mathbb{N}}$ <u>is relatively compact in</u> $\underline{\underline{M}}_{mc}(\overline{\Omega})$ <u>and all its limit points are good solution-measures of (1.1,∞)</u>.

We will prove (3.16) in §3-d and (3.24) in §3-e. In §3-f we shall give a "strong" stability result, to be compared with [6], [18].

3-c. EXISTENCE OF A SOLUTION-MEASURE.

In this paragraph we go back to the setting of section 1. We want to prove the following slight improvement of Theorem (1.8).

(3.25) THEOREM: The following assumptions insure the existence of at least one very good solution:

(i) there exists a predictable process $\gamma \geq 0$ on $(\Omega, \underline{\underline{F}}, \underline{F})$ such that $\int_0^t (\gamma_s)^2 dA_s < \infty$ P-a.s. for all $t \geq 0$, and that we have identically:

$$|g_t(\omega,x)| \leq \gamma_t(\omega)(1 + \sup_{s<t} |x(s)|) .$$

(ii) for all $\omega \in \Omega$, $t \geq 0$, the mapping: $x \rightsquigarrow g_t(\omega,x)$ is continuous on $\mathcal{X}$ endowed with the uniform topology (or equivalently, with $\mathcal{E}_u$).

For all $s \geq 0$, $x \in \mathcal{X}$, we define $s_n \geq 0$, $x^s \in \mathcal{X}$, $x^{s-} \in \mathcal{X}$ by

$$s_n = k/n \quad \text{if} \quad \frac{k}{n} < s \leq \frac{k+1}{n}, \qquad s_n = 0 \quad \text{if} \quad s = 0$$

$$x^s(t) = x(s \wedge t)$$

$$x^{s-}(t) = \begin{cases} x(t) & \text{if} \quad t < s \\ x(s-) & \text{if} \quad t \geq s \quad (\text{with} \quad x(0-) = x(0)). \end{cases}$$

For each $n \in \mathbb{N}$ we define a new predictable process g^n on $(\overline{\Omega}, \overline{\underline{\underline{F}}}, \overline{\underline{F}})$ by

(3.26)
$$g_t^n(\omega,x) = g_t(\omega, x^{t_n}) .$$

This is the choice made by Stroock and Varadhan [28] for proving Theorem (3.25) for Ito's equations. We will first prove the existence of a solution-process with the coefficient g^n, which is "strictly non-anticipating", then we will apply Theorem (3.16).

(3.27) LEMMA: Under (3.25,i), the equation

(3.28)
$$X_t = K_t + \int_0^t g_s^n(.,X_.(.)) \, dZ_s$$

admits a unique solution-process $\tilde{X}^n$ on the space $(\Omega, \underline{\underline{F}}, \underline{F}, P)$.

Proof. It is sufficient to prove existence and uniqueness on each interval $[0,\frac{k}{n}]$, and this will be proved by induction on k. This is trivial for $k=0$. Assume now that it is true for k. Put $r = \frac{k}{n}$, $r' = \frac{k+1}{n}$, and denote by $\tilde{X}$ the unique solution-process on $[0,r]$. If $t \leq r'$ we have by (3.25,i): $|g_t^n(\tilde{X})| \leq \gamma_r(1 + \sup_{s<r} |\tilde{X}_s|)$, so $I_{(r,r']} g^n(\tilde{X})$ is locally

bounded and we can set

$$\tilde{X}'_t = \begin{cases} \tilde{X}_t & \text{if } t \leq r \\ \tilde{X}_r + K_t - K_r + \int_r^t g^n_s(\tilde{X}) dZ_s & \text{if } r < t \leq r'. \end{cases}$$

Since $\tilde{X}' = \tilde{X}$ on $[0,r]$, (3.26) implies that $g^n(\tilde{X}') = g^n(\tilde{X})$ on $[0,r']$. One deduces that $\tilde{X}'$ is a solution-process of (3.28) on $[0,r']$.

Any other solution-process $\tilde{X}''$ on $[0,r']$ must coincide with $\tilde{X}$ and $\tilde{X}'$ on $[0,r]$, so $g^n(\tilde{X}'') = g^n(\tilde{X}')$ on $[0,r']$, so $\tilde{X}'' = \tilde{X}'$ on $[0,r']$. ∎

For each $\omega \in \Omega$, define the following subset of $\mathcal{X}$:

$$\mathcal{X}(\omega) = \{x \in \mathcal{X} : |\Delta x(t)| \leq \gamma_t(\omega) |\Delta Z_t(\omega)| (1 + \sup_{s<t} |x(s)|) \text{ for all } t > 0\}.$$

(3.29) LEMMA: <u>For</u> P-<u>almost all</u> ω, $\mathcal{X}(\omega)$ <u>is</u> $\mathcal{C}_s$- <u>and</u> $\mathcal{C}_u$-<u>closed, and both</u> $\mathcal{C}_s$ <u>and</u> $\mathcal{C}_u$ <u>induce the same topology on</u> $\mathcal{X}(\omega)$.

<u>Proof</u>. Due to (3.6), it suffices to prove that for P-almost all ω, the function: $t \rightsquigarrow \gamma_t(\omega)|\Delta Z_t(\omega)|$ satisfies (3.5). But $\int_0^t (\gamma_s)^2 dA_s < \infty$ P-a.s. for all $t \geq 0$, which implies that $\gamma \in L(Z^j; \Omega, \underline{F}, P)$ for all $j \leq m$ (cf. [10]). Hence $\gamma_t \Delta Z_t$ is the "jump process" of the m-dimensional semimartingale $\gamma \cdot Z$, and the results immediately follows. ∎

By throwing out a null set, we may suppose that the properties (3.29) of $\mathcal{X}(\omega)$ hold everywhere.

(3.30) LEMMA: <u>For every</u> $\mathcal{C}_s$-<u>compact set</u> H <u>of</u> $\mathcal{X}$, <u>for all</u> $\omega \in \Omega$, $t \geq 0$, <u>we have</u>: $\lim_{(n)} \sup_{x \in H \cap \mathcal{X}(\omega)} \int_0^t |g^n_s(\omega,x) - g_s(\omega,x)| (1 + \gamma_s(\omega)) dA_s(\omega) = 0$.

<u>Proof</u>. Let us pick an $\omega \in \Omega$. Since every $\mathcal{C}_s$-compact subset of $\mathcal{X}$ is contained in a $\mathcal{C}_s$-compact set which is stable by stopping and strict stopping (i.e.: $x \in H$, $s \geq 0 \Longrightarrow x^s, x^{s-} \in H$), we may assume that H itself enjoys those properties. From (3.29), $H \cap \mathcal{X}(\omega)$ is a $\mathcal{C}_u$-compact subset of $\mathcal{X}$, so $g_s(\omega,.)$ is $\mathcal{C}_u$-uniformly continuous in restriction to $H \cap \mathcal{X}(\omega)$. Thus if $\varepsilon > 0$ and if $\delta(x,x') = \sup_{s \geq 0} |x(s) - x'(s)|$, there exists $\eta_s(\omega) > 0$ such that

(3.31) $\quad s \leq t,\ x,x' \in H \cap \mathcal{X}(\omega),\ \delta(x,x') \leq \eta_s(\omega) \Longrightarrow |g_s(\omega,x) - g_s(\omega,x')| \leq \varepsilon$.

Since H is $\mathcal{C}_s$-compact, there exists $a > 0$ such that

$$x \in H \Longrightarrow \sup_{s \leq t} |x(s)| \leq a.$$

We choose $\eta(\omega)>0$ such that

(3.32)
$$\int_0^t I_{\{s:\eta_s(\omega)<\eta(\omega)\}}(1+\gamma_s(\omega))\, dA_s(\omega) \leq \varepsilon .$$

H being $\mathcal{C}_s$-compact, there exists $\theta(\omega)>0$ such that [2]:

(3.33)
$$x\in H,\ s\leq t,\ s<r<s'\leq s+\theta(\omega) \Longrightarrow |x(r)-x(s)| \wedge |x(s'-)-x(r)| \leq \frac{\eta(\omega)}{4}$$

Let $t_1<\ldots<t_p$ be the points of $[0,t]$ where $\gamma_t(\omega)|\Delta Z_t(\omega)|(1+a) > \frac{\eta(\omega)}{4}$. If $x\in\mathcal{X}(\omega)$ and if (s,s') does not intersect the set $\{t_1,\ldots,t_p\}$, we have $|\Delta x(r)|\leq\frac{\eta(\omega)}{4}$ if $r\in(s,s')$. Therefore if $\sup_{s<r\leq s'}|x(r)-x(s)|>\eta(\omega)$ there certainly exists a point $r\in(s,s')$ such that both $|x(r)-x(s)|$ and $|x(s'-)-x(r)|$ are bigger than $\frac{\eta(\omega)}{4}$. Thus (3.33) yields

(3.35)
$$x\in H\cap\mathcal{X}(\omega),\ s\leq t,\ s<s'\leq s+\theta(\omega),\ (s,s')\cap\{t_1,\ldots,t_p\}=\emptyset \Longrightarrow \delta(x^{s'-},x^s)\leq\eta(\omega).$$

We can find $\rho(\omega)>0$ such that

(3.36)
$$\sum_{i\leq p}\int_{t_i}^{t_i+\rho(\omega)}(1+\gamma_s(\omega))(1+a)\,dA_s(\omega) \leq \varepsilon .$$

If $D(\omega)=[\bigcup_{i\leq p}(t_i,t_i+\rho(\omega))]\bigcup\{s:\eta_s(\omega)<\eta(\omega)\}\cap[0,t]$, (3.32) and (3.36) imply

(3.37)
$$\int(1+\gamma_s(\omega))(1+a)I_{D(\omega)}(s)\, dA_s(\omega) \leq 2\varepsilon.$$

Let us recall that $g_s(\omega,x)=g_s(\omega,x^{s-})$ and $g_s^n(\omega,x)=g_s(\omega,x^{s_n})$. If $n(\omega)$ is an integer bigger than $1/\theta(\omega)$ and than $1/\rho(\omega)$, (3.31) and (3.35) and the fact that H and $H\cap\mathcal{X}(\omega)$ are stable by stopping and strict stopping, yield

$$n\geq n(\omega),\ x\in H\cap\mathcal{X}(\omega),\ s\in[0,t]\cap D(\omega)^c \Longrightarrow |g_s^n(\omega,x)-g_s(\omega,x)| \leq \varepsilon.$$

Since $|g_s^n(\omega,x)-g_s(\omega,x)|\leq 2\gamma_s(\omega)(1+a)$ if $s\leq t,\ x\in H$, the above implication and (3.37) yield

$$n\geq n(\omega),\ x\in H\cap\mathcal{X}(\omega) \Longrightarrow \int_0^t(1+\gamma_s(\omega))|g_s^n(\omega,x)-g_s(\omega,x)|\, dA_s(\omega)\leq 4\varepsilon+\varepsilon A_t(\omega)$$

and the result follows from the arbitrariness of $\varepsilon>0$. ∎

<u>Proof of (3.25)</u>. We will apply (3.16) to the following: $P^n=P$, $Z^n=Z$, $K^n=K$ for all $n\in\overline{\mathbb{N}}$, g^n is given by (3.26) for $n\in\mathbb{N}$ and $g^\infty=g$. We have obviously (3.10), (3.11), (3.12), (3.15), and also (3.13): take $\gamma^n=\gamma$, $\tilde{A}_t^n=\int_0^t(1\vee\gamma_s^2)dA_s$ and $V=(1+\gamma)|\Delta Z|+|\Delta K|$. We denote by $\overline{P}^n$

the very good solution-measure of (3.28) = (1.1,n) associated to the process $\tilde{X}^n$ introduced in (3.27). We will now prove that (3.14) holds.

Let T be a bounded stopping time, and $u \in \mathbb{R}^{m+d}$. With the notations (2.9) and (3.9) we have $\phi^{n,g^n,u} = \phi^{g^n,u}$ and $\phi^{\infty,g^\infty,u} = \phi^{g,u}$. Let $t \in \mathbb{R}_+$ such that $T \leq t$. By (3.16,a), the sequence $(\overline{P}^n)_{n\in\mathbb{N}}$ is relatively compact in $\underline{\underline{M}}_{mc}(\overline{\Omega})$, so $(\overline{P}^n_{|\mathcal{X}})_{n\in\mathbb{N}}$ is tight. Hence if $\eta > 0$ there exists a $\mathcal{C}_s$-compact subset H of $\mathcal{X}$ such that $\overline{P}^n_{|\mathcal{X}}(H) = P(\tilde{X}^n \in H) \geq 1 - \eta$ for all $n \in \mathbb{N}$. There exists $a > 0$ such that $x \in H \longrightarrow \sup_{s\leq t} |x(s)| \leq a$. Since $\tilde{X}^n$ is a solution of (3.28), we also have $\tilde{X}^n(\omega) \in \mathcal{X}(\omega)$ for P-almost all $\omega \in \Omega$ and all $n \in \mathbb{N}$. If $\varepsilon > 0$, we obtain

$$\overline{P}^n(|\phi_T^{g^n,u} - \phi_T^{g,u}| > \varepsilon) \leq \eta + \overline{P}^n(|\phi_T^{g^n,u} - \phi_T^{g,u}| > \varepsilon,\ X(.) \in H \cap \mathcal{X}(.))$$

$$\leq \eta + P\Big\{ \Big| \int_0^T [(v_s^{g^n(\tilde{X}^n),u} - v_s^{g(\tilde{X}^n),u}) \exp i\langle u|(Z_{s-}, \tilde{X}^n_{s-} - K_{s-})\rangle]\, dA_s \Big| > \varepsilon, \text{ and } \tilde{X}^n(.) \in H \cap \mathcal{X}(.)\Big\}$$

$$(3.38) \leq \eta + P\Big\{\tilde{X}^n(.) \in H \cap \mathcal{X}(.),\ 3\alpha(1+|u|^2)(1+a)\int_0^t (1+\gamma_s)|g_s^n(\tilde{X}^n) - g_s(\tilde{X}^n)|\, dA_s > \varepsilon\Big\}$$

where for the last inequality we have used (2.13), (3.25,i) and the fact that $\sup_{s\leq t} |x(s)| \leq a$ if $x \in H$. Now, lemma (3.30) implies that the last term in (3.38) tends to 0 when $n \uparrow \infty$; since $\eta > 0$ is arbitrary, we obtain (3.14).

It remains to prove that if $\overline{P}$ is a limit point of $(\overline{P}^n)_{n\in\mathbb{N}}$, then $(\overline{\Omega}, \overline{\underline{\underline{F}}}, \overline{\underline{F}}, \overline{P})$ is a very good extension of $(\Omega, \underline{\underline{F}}, \underline{F}, P)$, and the theorem will follow from (3.16). We may assume that $(\overline{P}^n)_{n\in\mathbb{N}}$ tends to $\overline{P}$ in $\underline{\underline{M}}_{mc}(\overline{\Omega})$. Let M be a bounded martingale on $(\Omega, \underline{\underline{F}}, \underline{F}, P)$, $s \leq t$, and $U \in B_{mc}(\overline{\Omega})$ that is $\overline{\underline{\underline{F}}}_s$-measurable. Then

$$\overline{E}[U(M_t - M_s)] = \lim_{(n)} \overline{E}^n[U(M_t - M_s)] = 0,$$

because each $(\overline{\Omega}, \overline{\underline{\underline{F}}}, \overline{\underline{F}}, \overline{P}^n)$ is a very good extension. The set of all $\overline{\underline{\underline{F}}}_s$-measurable $U \in B_{mc}(\overline{\Omega})$ generates a σ-field that is in between $\overline{\underline{\underline{F}}}_{s-}$ and $\overline{\underline{\underline{F}}}_s$, so we have proved that $\overline{E}(M_t - M_s | \overline{\underline{\underline{F}}}_{s-}) = 0$. The right-continuity of M implies that $\overline{E}(M_t - M_s | \overline{\underline{\underline{F}}}_s) = 0$, and the result is proved. ■

3-d. PROOF OF THE MAIN THEOREM (3.16).

Let us begin with a "Gronwall lemma", whose proof is reproduced here for the sake of completeness (see [12]).

(3.39) LEMMA: <u>There exists a mapping</u> $k: \mathbb{R}_+^3 \longrightarrow \mathbb{R}_+$ <u>with the following property: if</u> F <u>(resp.</u> Φ<u>) is a right-continuous nonnegative predictable (resp. adapted) increasing process on some filtered probability space</u> $(\Omega,\underline{\underline{F}},\underline{F},P)$, <u>if</u> $F_0=0$, <u>if</u> $F_T\le\eta$ <u>for some stopping time</u> T, <u>and if</u>

$$E(\Phi_{S-}) \le \alpha + \beta E[(\Phi_-\bullet F)_S] \tag{3.40}$$

<u>for all stopping times</u> $S\le T$ (with the convention $\Phi_{0-}=0$), <u>then</u> $E(\Phi_{T-})\le k(\alpha,\beta,\eta)$. <u>Moreover</u> $k(0,\beta,\eta)=0$.

<u>Proof</u>. Set $U_0=0$, $U_{n+1}=\inf(t>U_n: \Delta F_t\ge 1/4\beta)$. Each stopping time U_n is predictable, and hence is announced by a sequence $(U(n,j))_{j\in\mathbb{N}}$ of stopping times: we have $U(n,j)<U_n$ and $\lim_{(j)}\uparrow U(n,j)=U_n$ for $n\ge 1$. Set

$$V_n = U_n\wedge T, \qquad V(n,j) = [U_{n-1}\vee U(n,j)]\wedge T;$$

we have

$$V_{n-1}\le V(n,j)\le T, \quad \lim_{(j)}\uparrow V(n,j)=V_n, \quad \{U_{n-1}<T\}\subset\{V(n,j)<U_n\}. \tag{3.41}$$

Let us keep n, j fixed for the moment. Set $S_0=V_{n-1}$ and

$$S_{k+1} = V(n,j)\wedge \inf(t>S_k: F_t-F_{S_k}\ge 1/4\beta), \qquad x_k = E(\Phi_{(S_k)-}).$$

Since $V(n,j)<U_n$ if $V_{n-1}<T$, and $V(n,j)=T=S_k$ if $V_{n-1}=T$, we have $F_{S_{k+1}}\le F_{S_k}+1/2\beta$. Thus (3.40) implies:

$$x_{k+1} \le \alpha+\beta E(\Phi_-\bullet F_{S_k}) + \beta\frac{1}{2\beta}x_{k+1} \le \alpha+\beta\eta x_k + x_{k+1}/2$$

$$x_{k+1} \le 2\alpha + 2\beta\eta x_k,$$

which yields

$$x_{k+1} \le 2\alpha\sum_{0\le i\le k}(2\beta\eta)^i + x_0(2\beta\eta)^{k+1}.$$

We have $x_0=E(\Phi_{(V_{n-1})-})$, and if m denotes the integer part of $4\beta\eta$ we have $S_{m+1}=V(n,j)$, because $F_T\le\eta$. Therefore we have proved that

$$E(\Phi_{V(n,j)-}) \le 2\alpha\sum_{0\le i\le m}(2\beta\eta)^i + (2\beta\eta)^{m+1}E(\Phi_{(V_{n-1})-}).$$

But (3.41) implies that: $\lim_{(j)}\uparrow\Phi_{V(n,j)-}=\Phi_{(V_n)-}$. Thus, since $\Phi_{0-}=0$,

$$E(\Phi_{(V_n)-}) \le 2\alpha\sum_{0\le i\le m}(2\beta\eta)^i + (2\beta\eta)^{m+1}E(\Phi_{(V_{n-1})-})$$

$$\le \Big[2\alpha\sum_{0\le i\le m}(2\beta\eta)^i\Big]\sum_{0\le j\le n-1}(2\beta\eta)^{j(m+1)}.$$

Finally, if m' is the integer part of $2\beta\eta$, we have $V_{m'+1}=T$, so the lemma is proved with

$$k(\alpha,\beta,\eta) = \Big[2\alpha\sum_{0\le i\le m}(2\beta\eta)^i\Big]\Big[\sum_{0\le j\le m'}(2\beta\eta)^{j(m+1)}\Big]. \blacksquare$$

We turn now to the situation of (3.16,a): we assume (3.10), (3.11) and (3.13). For all $a \geq 1$, $n \in \overline{\mathbb{N}}$, we set

(3.42) $$\begin{cases} \tau^n(a) = \inf(t : |\Delta Z^n_t| > a) \\ Z^n(a)_t = Z^n_t - \sum_{s\leq t} \Delta Z^n_s \, I_{\{|\Delta Z^n_s| > a\}} . \end{cases}$$

$Z^n(a)$ is a semimartingale with bounded jumps, so it is special and we denote by $Z^n(a) = M^n(a) + B^n(a)$ its canonical decomposition over $(\Omega, \underline{\underline{F}}, \underline{F}, P^n)$, where $M^n(a)$ is a locally bounded local martingale and $B^n(a)$ is a predictable process with finite variation. From the local characteristics (B^n, C^n, ν^n) of Z^n and from (2.5), it is easy to compute that

(3.43) $$\begin{cases} B^n(a)_t = B^n_t + \int_0^t dA^n_s \int N^n_s(dz)\, z\, I_{\{1<|z|\leq a\}} \\ \langle M^{n,j}(a), M^{n,k}(a)\rangle_t = C^{n,jk}_t + \int_0^t dA^n_s \int N^n_s(dz)\, z^j z^k I_{\{|z|\leq a\}} \\ \qquad - \sum_{s\leq t} (\Delta A^n_s)^2 [\int N^n_s(dz) z^j I_{\{|z|\leq a\}}][\int N^n_s(dz) z^k I_{\{|z|\leq a\}}]. \end{cases}$$

Now, $\overline{P}^n$ is a good solution-measure of (1.1,n), so the same canonical decomposition of $Z^n(a)$ holds on $(\overline{\Omega}, \overline{\underline{\underline{F}}}, \overline{\underline{F}}, \overline{P}^n)$, as well as (3.43). Let

(3.44) $$X^*_t = \sup_{s\leq t} |X_s| .$$

If F, F' are two increasing processes, we will write $F \prec F'$ if $F' - F$ is also an increasing process: for instance in (3.13), we have $(1 \vee (\gamma^n)^2) \cdot A^n \prec \tilde{A}^n$.

(3.45) LEMMA: (a) <u>On</u> $(\overline{\Omega}, \overline{\underline{\underline{F}}}, \overline{\underline{F}}, \overline{P}^n)$, $g^n \cdot Z^n(a)$ <u>is a special semimartingale whose canonical decomposition</u> $g^n \cdot Z^n(a) = \tilde{M}^n(a) + \tilde{B}^n(a)$ <u>is</u>

$$\tilde{M}^n(a) = g^n \cdot M^n(a) , \qquad \tilde{B}^n(a) = g^n \cdot B^n(a) .$$

(b) <u>We have</u>:

$$\sum_{j\leq m} [\langle \tilde{M}^{n,j}(a), \tilde{M}^{n,j}(a)\rangle + V(\tilde{B}^{n,j}(a))] \prec 4d(1+a^2)(1+X^{*2}_-) \cdot \tilde{A}^n .$$

(c) <u>For every stopping time</u> T <u>we have</u>

$$\overline{E}^n(\sup_{s\leq T} |g^n \cdot Z^n(a)_s|^2) \leq 16(1+a^3)\, \overline{E}^n[(\tilde{A}^n_T \vee 1) \int_0^T (1+X^{*2}_{s-}) d\tilde{A}^n_s].$$

<u>Proof</u>. Using (2.5), (2.6), and the second formula in (3.43), in which we obtain a majoration by deleting the last term, we have:

$$\sum_{j\leq d} \sum_{k,l\leq m} (g^{n,jk} g^{n,jl}) \cdot \langle M^{n,k}(a), M^{n,l}(a)\rangle \prec (1+a^2)\, |g^n|^2 \cdot A^n ,$$

and this expression is $\overline{P}^n$-a.s. finite. Thus $\tilde{M}^n(a) = g^n \cdot M^n(a)$ is well defined, is a locally square-integrable local martingale, and $\sum_{j\leq d}\langle\tilde{M}^{n,j}(a),\tilde{M}^{n,j}(a)\rangle \prec 2(1+a^2)(1+X_-^{*2})\cdot\tilde{A}^n$ because of (3.13,i). Now, Doob's inequality implies that for every stopping time T,

$$\text{(3.46)}\qquad \bar{E}^n(\sup_{s\leq T}|g^n\cdot M^n(a)_s|^2) \leq 8(1+a^2)\,\bar{E}^n[(1+X_-^{*2})\cdot\tilde{A}_T^n].$$

Using (2.5) and (2.6) first, then (3.13,i), we obtain

$$\int_0^t dA_s^n\Big|\sum_{k\leq m} g_s^{n,jk}\Big(b_s^{n,k}+\int N_s^n(dz)z^k I_{\{1<|z|\leq a\}}\Big)\Big| \prec \int_0^t (1+a)|g_s^n|\,dA_s^n \prec (1+a)(1+X_-^*)\cdot\tilde{A}_t^n.$$

This implies first that $\tilde{B}^n(a) = g^n\cdot B^n(a)$ is well-defined and is predictable with finite variation, so (a) holds. Secondly, put together with the previous results, it proves (b). At last, we also have

$$\begin{aligned}\sup_{s\leq T}|g^n\cdot B^n(a)_s|^2 &\leq \Big\{\int_0^T dA_s^n\Big[|g_s^n b_s^n| + \int N_s^n(dz)|g_s^n z| I_{\{1<|z|\leq a\}}\Big]\Big\}^2\\ &\leq 2A_T^n\int_0^T dA_s^n\Big[|g_s^n b_s^n|^2 + \int N_s^n(dz)I_{\{1<|z|\leq a\}}\int N_s^n(dz)|g_s^n z|^2 I_{\{1<|z|\leq a\}}\Big]\\ &\leq 2A_T^n(1+a^3)\int_0^T(\gamma_s^n)^2(1+X_{s-}^{*2})\,dA_s^n.\end{aligned}$$

Putting together with (3.46), we obtain (c). ∎

(3.47) LEMMA: <u>There exists a sequence</u> $(T_q^n)_{n\in\bar{\mathbb{N}},q\in\mathbb{N}}$ <u>of bounded</u> $\underline{F}$<u>-stopping times on</u> Ω <u>such that</u>

(i) $\lim_{(q)}\uparrow T_q^\infty = \infty$ P^∞-<u>a.s.</u>

(ii) $\lim_{(n)} P^n(T_q^n < T_q^\infty) = 0$ <u>for all</u> $q\in\mathbb{N}$, <u>and</u> $T_q^n \leq T_q^\infty$;

(iii) <u>we have identically</u> $\tilde{A}^n_{T_q^n} \leq q$.

<u>Proof</u>. Since $\tilde{A}^\infty$ is predictable and finite-valued, the existence of a sequence $(T_q^\infty)_{q\geq 1}$ of $\underline{F}$-stopping times satisfying (i) and $\tilde{A}^\infty_{T_q^\infty}\leq q/2$ is well-known ([9],[20]). Put $S_q^n = \inf(t:\tilde{A}_t^n \geq q)$. Condition (3.13,ii) implies that: $\lim_{(n)} P^n(S_q^n \leq T_q^\infty) = 0$. Since S_q^n is predictable, there exists another stopping time $S_q'^n$ such that $S_q'^n < S_q^n$ and $P^n(S_q'^n < T_q^\infty) \leq P^n(S_q^n \leq T_q^\infty) + 1/n$. It remains to put $T_q^n = S_q'^n \wedge T_q^\infty$. ∎

Let $\varphi_n : \bar{\Omega} \longrightarrow \bar{\Omega}$ be defined by $\varphi_n(\omega,x) = (\omega, x-K^n(\omega))$. Then, if $\bar{P}'^n = \bar{P}^n\circ\varphi_n^{-1}$, for all $n\in\mathbb{N}$ the measure $\bar{P}'^n_{|\mathcal{X}}$ is the distribution of the process $X-K^n$ over $(\bar{\Omega},\bar{\underline{F}},\bar{P}^n)$.

(3.48) LEMMA: <u>The sequence</u> $(\bar{P}'^n_{|\mathcal{X}})_{n\in\mathbb{N}}$ <u>is tight</u> (i.e., relatively compact in $\underline{M}_c(\mathcal{X})$).

Proof. $\tau^n(a)$ and T_q^n are defined in (3.42) and (3.47). If $a>0$, put

$$\theta^n(a) = \inf(t : |K_t^n| > a), \qquad \sigma(a) = \inf(t : |X_t| > a).$$

Now, if $a, a', a'' \geq 1$, $q \in \mathbb{N}$, we put

$$R^n(a,a',a'',q) = \tau^n(a) \wedge \theta^n(a') \wedge \sigma(a'') \wedge T_q^n$$

$$Y^n(a,a',a'',q)_t = \int_0^{t \wedge R^n(a,a',a'',q)} g_s^n \, dZ^n(a)_s .$$

From Lemma (3.45), $Y^n(a,a',a'',q)$ is a special semimartingale whose canonical decomposition $\hat{M}^n + \hat{B}^n$ satisfies

$$\sum_{j \leq d} [<\hat{M}^{n,j}, \hat{M}^{n,j}> + V(\hat{B}^{n,j})] \prec 4d(1+a^2)(1+a''^2)\, \tilde{A}^n .$$

Then if $\mathcal{L}(Y^n(a,a',a'',q))$ denotes the image of $\overline{P}^n$ by the mapping: $\overline{\Omega} \longrightarrow \mathcal{X}$ defined by $(\omega,x) \rightsquigarrow Y^n(a,a',a'',q)(\omega,x)$, conditions (3.10) and (3.13,ii) imply that the sequence $(\mathcal{L}(Y^n(a,a',a'',q)) : n \in \mathbb{N})$ is tight: apply Theorem (7.1) of [14] with condition (C5), using the fact that $\overline{P}^n|_{\Omega} = P^n$ and that all $\tilde{A}^n$ $(n \in \overline{\mathbb{N}})$ are $\underline{F}$-predictable.

Now, we have $Y^n(a,a',a'',q) = X - K^n$ $\overline{P}^n$-a.s. on $[0,R^n(a,a',a'',q))$, because $Z^n = Z^n(a)$ on this stochastic interval. Hence, to prove that $(\mathcal{L}(X - K^n) : n \in \mathbb{N})$ is tight, it is sufficient to prove that for all $\varepsilon > 0$, $N > 0$, there exist $a \geq 1$, $a' > 0$, $a'' > 0$, $q \in \mathbb{N}$, $n_o \in \mathbb{N}$ with

(3.49) $$n \geq n_o \implies \overline{P}^n(R^n(a,a',a'',q) < N) \leq \varepsilon .$$

Let $\varepsilon > 0$, $N > 0$. Condition (3.13,iii) implies: firstly the existence of $a \geq 1$ such that $P^{\infty}(\sup_{s \leq N} V_s > a) \leq \varepsilon/16$, secondly the existence of n_1' such that: $n \geq n_1' \implies P^n(|\Delta Z_s^n| > V_s$ for at least one $s \leq N) \leq \varepsilon/8$. Condition (3.10) implies the existence of $n_1 \geq n_1'$ such that: $n \geq n_1$ $P^n(\sup_{s \leq N} V_s > a) \leq \varepsilon/8$. Then

(3.50) $$n \geq n_1 \implies P^n(\tau^n(a) \leq N) = P^n(|\Delta Z_s^n| > a \text{ for at least one } s \leq N) \leq \frac{\varepsilon}{4}.$$

There exists $a' > 1$ such that $P^{\infty}(\sup_{s \leq N} |K_s^{\infty}| > a' - 1) \leq \varepsilon/16$. From (3.11) there exists n_2' such that: $n \geq n_2' \implies P^n(\sup_{s \leq N} |K_s^n - K_s^{\infty}| \geq 1) \leq \varepsilon/8$. From (3.10) there exists $n_2 \geq n_2'$ such that: $n \geq n_2 \implies P^n(\sup_{s \leq N} |K_s^{\infty}| > a' - 1) \leq \varepsilon/8$. Then

(3.51) $$n \geq n_2 \longrightarrow P^n(\theta^n(a') \leq N) = P^n(\sup_{s \leq N} |K_s^n| > a') \leq \frac{\varepsilon}{4}.$$

From (3.47) there exists $q \in \mathbb{N}$, $n_3' \in \mathbb{N}$ such that $P^{\infty}(T_q^{\infty} \leq N) \leq \varepsilon/16$ and that: $n \geq n_3' \implies P^n(T_q^n < T_q^{\infty}) \leq \varepsilon/8$. From (3.10) there exists $n_3 \geq n_3'$ such that: $n \geq n_3 \implies P^n(T_q^{\infty} \leq N) \leq \varepsilon/8$. Then

$$(3.52) \qquad n \geq n_3 \Longrightarrow P^n(T^n_q \leq N) \leq \frac{\varepsilon}{4} .$$

Let $S^n = \tau^n(a) \wedge \theta^n(a') \wedge T^n_q$. We have $X = K^n + g^n \cdot Z^n(a)$ on $[0,S^n)$ for $\overline{P}^n$, and $|K^n| \leq a'$ on $[0,S^n)$, and $\tilde{A}^n_{S^n} \leq q$. Then for every stopping time $S \leq S^n$ we have

$$\overline{E}^n(X^{*2}_{S-}) \leq 2a'^2 + 32(1+a^3)q^2 + 32(1+a^3)q\,\overline{E}^n(X^{*2}_- \cdot A^n_S)$$

because of Lemma (3.45,c). Using again $\tilde{A}^n_{S^n} \leq q$, Lemma (3.39) implies that

$$\overline{E}^n(X^{*2}_{(S^n)-}) \leq \delta := k(2a'^2 + 32(1+a^3)q^2,\ 32(1+a^3)q\ ,\ q) .$$

Then if $a'' = 2(\delta/\varepsilon)^{1/2}$, we obtain

$$\overline{P}^n(\sigma(a'') < S^n) \leq \frac{1}{a''^2}\,\overline{E}^n(X^{*2}_{(S^n)-}) \leq \frac{\varepsilon}{4} .$$

This inequality, with (3.50), (3.51) and (3.52), yields (3.49) with $n_0 = n_1 \vee n_2 \vee n_3$. ∎

<u>Proof of (3.16,a)</u>. Each φ_n $(n \in \overline{\mathbb{N}})$ is bijective and bi-measurable. It follows that since $\overline{P}'^n = \overline{P}^n \circ \varphi_n^{-1}$, we have $\overline{P}^n = \overline{P}'^n \circ \varphi_n$ for $n \in \mathbb{N}$; it also follows that: $U \in B_{mc}(\overline{\Omega}) \Longleftrightarrow U \circ \varphi_\infty \in B_{mc}(\overline{\Omega})$, therefore the correspondance: $\overline{P} \rightsquigarrow \overline{P} \circ \varphi_\infty^{-1}$ is bi-continuous on $\underline{\underline{M}}_{mc}(\overline{\Omega})$. We will show that a subsequence $(\overline{P}^{n'})$ converges to a limit $\overline{P}$ if and only if $(\overline{P}'^{n'})$ converges to a limit $\overline{P}'$, and then $\overline{P}' = \overline{P} \circ \varphi_\infty^{-1}$. From these facts it will follow that the closure of $(\overline{P}^n)_{n \in \mathbb{N}}$ and the closure of $(\overline{P}'^n)_{n \in \mathbb{N}}$ in $\underline{\underline{M}}_{mc}(\overline{\Omega})$ are isomorphic and, since the sequence $(\overline{P}'^n)_{n \in \mathbb{N}}$ is relatively compact in $\underline{\underline{M}}_{mc}(\overline{\Omega})$ by Theorem (3.3) (because of Lemma (3.48), of $\overline{P}'^n_{|\Omega} = P^n$ and of (3.10)), the sequence $(\overline{P}^n)_{n \in \mathbb{N}}$ will also be relatively compact. <u>Added in proof</u>: See Note 2 on page 212.

Thus we are left to prove the following (up to a relabelling of sequences): (i) $\overline{P}^n \longrightarrow \overline{P}$ $\Longleftrightarrow$ (ii) $\overline{P}^n \circ \varphi_n^{-1} \longrightarrow \overline{P}' = \overline{P} \circ \varphi_\infty^{-1}$. Let $U = VW$, V bounded measurable on $(\Omega, \underline{\underline{F}})$, W bounded uniformly continuous on $\mathscr{X}$. Since $U \circ \varphi_\infty \in B_{mc}(\overline{\Omega})$, in case (i) (resp. (ii)) we have: $\overline{E}^n(U \circ \varphi_\infty) \longrightarrow \overline{E}(U \circ \varphi_\infty)$ (resp. $\overline{E}^n(U \circ \varphi_\infty \circ \varphi_n^{-1}) \longrightarrow \overline{E}(U \circ \varphi_\infty \circ \varphi_\infty^{-1}) = \overline{E}(U)$). Henceforth, by using Lemma (3.2) we see that it suffices to proves the following:

$$(3.53) \qquad \overline{E}^n(|U \circ \varphi_n - U \circ \varphi_\infty|) \longrightarrow 0 , \qquad \overline{E}^n(|U - U \circ \varphi_\infty \circ \varphi_n^{-1}|) \longrightarrow 0 .$$

We have $(U \circ \varphi_n - U \circ \varphi_\infty)(\omega,x) = V(\omega)[W(x - K^n(\omega) - W(x - K^\infty(\omega))]$, and $(U - U \circ \varphi_\infty \circ \varphi_n^{-1})(\omega,x) = V(\omega)[W(x) - W(x - K^\infty(\omega) + K^n(\omega))]$. Let α be a bound for $|V|$ and for $|W|$. If $\varepsilon > 0$ there exists $\eta > 0$ such that: $\delta_u(x,x') \leq \eta \Longrightarrow |W(x) - W(x')| \leq \varepsilon$. Thus

$$\left.\begin{array}{l}\overline{E}^n(|U\circ\varphi_n - U\circ\varphi_\infty|)\\ \overline{E}^n(|U - U\circ\varphi_\infty\circ\varphi_n^{-1}|)\end{array}\right\} \leq \varepsilon\alpha + \alpha^2 P^n(\delta_u(K^n,K^\infty)>\eta).$$

Since (3.11) can also be read as follows: $P^n(\delta_u(K^n,K^\infty)>\eta) \longrightarrow 0$ for all $\eta > 0$, and since $\varepsilon > 0$ is arbitrary, we obtain (3.53). ∎

Proof of (3.16,b). Using (3.16,a), we may assume that the sequence $(\overline{P}^n)$ itself converges to a limit $\overline{P}^\infty$. It is obvious, by (3.10), that $\overline{P}^\infty|_\Omega = P^\infty$.

Let $U(\omega,x) = 1 \wedge |K_0^\infty(\omega) - x(0)|$, which belongs to $B_{mc}(\overline{\Omega})$, so $\lim_{(n)} \overline{E}^n(U) = \overline{E}^\infty(U)$. But $X_0 = K_0^n$ $\overline{P}^n$-a.s., so (3.11) implies that $\overline{E}^n(U) \longrightarrow 0$. Hence $\overline{P}^\infty(X_0 = K_0^\infty) = 1$.

We will apply the characterization of Theorem (2.10): to obtain that $\overline{P}^\infty$ is a good solution of $(1.1,\infty)$ it suffices to prove that if Y^n is the (m+d)-dimensional process $Y^n = (Z^n, X - K^n)$, and if $M^n = \exp i\langle u|Y^n\rangle - \phi^{n,g^n,u}$, then M^n is a local martingale on $(\overline{\Omega},\overline{\underline{F}},\overline{\underline{F}},\overline{P}^n)$ for $n = \infty$, while one knows that it is true for $n \in \mathbb{N}$.

Let $\sigma(a) = \inf(t : |X_t| > a)$, and consider the T_q^n's defined in (3.47). We have: $\lim_{a\uparrow\infty} \sigma(a) = \infty$, and: $\lim_{(q)}\uparrow T_q^\infty = \infty$ P^∞-a.s., thus $\overline{P}^\infty$-a.s. Therefore it suffices to prove that, for each fixed a, q and if $T^n = T_q^n \wedge \sigma(a)$, then $(M^\infty_{t\wedge T^\infty})_{t\geq 0}$ is a $\overline{P}^\infty$-local martingale, while knowing that $(M^n_{t\wedge T^n})_{t\geq 0}$ is a $\overline{P}^n$-local martingale for each $n \in \mathbb{N}$. Actually, one may replace "local martingale" by "martingale", since M^n is bounded by $1 + 4\alpha(1 + |u|^2)(1 + a^2)q$ on $[0,T^n]$ for all $n \in \overline{\mathbb{N}}$ (apply (2.12), (3.47,iii), (3.13,i) and the definition of $\sigma(a)$).

Let us assume that for all $t \geq 0$, all U in $B_{mc}(\overline{\Omega})$ and $\overline{\underline{F}}_t$-measurable, we have:

$$(3.54) \qquad \overline{E}^n(U M^n_{t\wedge T^n}) \longrightarrow \overline{E}^\infty(U M^\infty_{t\wedge T^\infty}).$$

Let $s \leq t$. If $U \in B_{mc}(\overline{\Omega})$ is $\overline{\underline{F}}_s$-measurable, and since $(M^n_{t\wedge T^n})_{t\geq 0}$ is a $\overline{P}^n$-martingale, (3.54) implies that $\overline{E}^\infty(U(M^\infty_{t\wedge T^\infty} - M^\infty_{s\wedge T^\infty})) = 0$. But the set of all $\overline{\underline{F}}_s$-measurable U belonging to $B_{mc}(\overline{\Omega})$ generates a σ-field which is in between $\overline{\underline{F}}_{s-}$ and $\overline{\underline{F}}_s$. Therefore $\overline{E}^\infty(M^\infty_{t\wedge T^\infty} - M^\infty_{s\wedge T^\infty}|\overline{\underline{F}}_{s-}) = 0$. Since M^∞ is right-continuous, it follows that $(M^\infty_{t\wedge T^\infty})_{t\geq 0}$ is a $\overline{P}^\infty$-martingale.

It remains to prove (3.54); we may write the left-hand side of (3.54) as $\beta_1^n + \beta_2^n + \beta_3^n$, where

$$\beta_1^n = \overline{E}^n[U(\exp i\langle u|Y^n_{t\wedge T^n}\rangle - \exp i\langle u|Y^\infty_{t\wedge T^\infty}\rangle)]$$

$$\beta_2^n = \overline{E}^n[U(\exp i\langle u|Y^\infty_{t\wedge T^\infty}\rangle - \phi^{\infty,g^\infty,u}_{t\wedge T^\infty})] = \overline{E}^n(U M^\infty_{t\wedge T^\infty})$$

$$\beta_3^n = \overline{E}^n[U(\phi^{\infty,g^\infty,u}_{t\wedge T^\infty} - \phi^{n,g^n,u}_{t\wedge T^n})].$$

By (3.47) we have: $\lim_{(n)} \overline{P}^n(T^n \neq T^\infty) = 0$. Using (3.14) and the fact that $|\phi^{n,g^n,u}_{t\wedge T^n}|$ is bounded, uniformly in $n\in\overline{\mathbb{N}}$, it easily follows that $\beta_3^n \longrightarrow 0$. (3.11) implies that: $\lim_{(n)} P^n(|K^n_{t\wedge T^\infty} - K^\infty_{t\wedge T^\infty}| > \varepsilon) = 0$ for all $\varepsilon > 0$. Using this fact, with (3.12) and again that $\lim_{(n)} \overline{P}^n(T^n \neq T^\infty) = 0$, we easily obtain that: $\beta_1^n \longrightarrow 0$.

Let $F^t = \{(\omega,x) : x\in F^t_\omega\}$, where

$$F^t_\omega = \{x\in \mathcal{X}:\ |\Delta x(s)| \leq V_s(\omega)(1+\sup_{r<s}|x(r)|) \ \text{ for } \ s\in(0,t]\}.$$

Let $x_n \in F^t_\omega$ be such that $x_\infty = \mathcal{T}_s\text{-}\lim_{(n)} x_n$ exists. One proves exactly like in (3.6) that $\Delta x_n(t) \longrightarrow \Delta x_\infty(t)$. Then a classical result about Skorokhod topology implies that if $x^t_n(s) = x_n(t\wedge s)$, then $x^t_\infty = \mathcal{T}_s\text{-}\lim_{(n)} x^t_n$. But $x^t_n \in F^\infty_\omega$ for all $n\in\mathbb{N}$, so (3.6) implies that $x_\infty \in F^\infty_\omega$ and $x^t_\infty = \mathcal{T}_u\text{-}\lim_{(n)} x^t_n$. It follows that $x_\infty \in F^t_\omega$, that is F^t_ω is $\mathcal{T}_s$-closed. Now if $W = UM^\infty_{t\wedge T^\infty}$, then $W(\omega,.)$ is $\underline{\underline{\mathcal{X}}}_t$-measurable, so $W(\omega,x_n) = W(\omega,x^t_n)$. Using that $U\in B_{mc}(\overline{\Omega})$, (2.13), and (3.15), it follows that $W(\omega,.)$ is $\mathcal{T}_u$-continuous, so $W(\omega,x_n) \longrightarrow W(\omega,x_\infty)$ and the restriction of $W(\omega,.)$ to F^t_ω is $\mathcal{T}_s$-continuous.

$\overline{P}^n$ being a solution-measure of (1.1,n) for $n\in\mathbb{N}$, we have

$$|\Delta X_s| \leq |\Delta K^n_s| + \gamma^n_s|\Delta Z^n_s|(1+X^*_{s-}), \quad s\in(0,t], \ \overline{P}^n\text{-a.s.}$$

Then (3.13,iii) yields: $\lim_{(n)} \overline{P}^n(F^t) = 1$. Theorem (3.4) yields: $\beta_2^n \longrightarrow \beta_2^\infty$, and we deduce that (3.54) holds. ∎

<u>Proof of (3.16,c)</u>. The first assertion is trivial. The second assertion follows from a general property of convergence in $\underline{\underline{M}}_{mc}(\overline{\Omega})$, when each $\overline{P}^n$ has the form $\overline{P}^n(d\omega,dx) = P^n(d\omega)\varepsilon_{\tilde{X}^n(\omega)}(dx)$ for $n\in\overline{\mathbb{N}}$: see [15]. ∎

<u>3-e. PROOF OF (3.24).</u>

Let us first prove the lemma announced in Remark (3.23).

(3.55) LEMMA: <u>If</u> (V^n) <u>is a sequence of random variables on</u> $(\Omega,\underline{F})$, <u>which converges to</u> 0 <u>in</u> P^∞<u>-measure. If (3.10) holds, then for all</u> $\varepsilon > 0$ <u>we</u>

have: $\lim_{(n)} P^n(|V^n|>\varepsilon) = 0$.

Proof. Let $\mathbb{N}'\subset\mathbb{N}$. One may find a sequence (n_k) in $\mathbb{N}'$ such that $V^{n(k)}\longrightarrow 0$ P^{∞}-a.s. Let $\varepsilon>0$, $\eta>0$. There exists $k_0(\omega)\in\overline{\mathbb{N}}$ such that: $k\geq k_0(\omega) \Longrightarrow |V^{n(k)}(\omega)|\leq\varepsilon$, and $k_0<\infty$ P^{∞}-a.s. There exists $k_1\in\mathbb{N}$ such that $P^{\infty}(\omega: k_1<k_0(\omega))\leq\eta$. By (3.10) there exists $k_2\geq k_1$ such that: $k\geq k_2 \Longrightarrow P^{n(k)}(\omega: k_1<k_0(\omega))\leq 2\eta$. Then

$$k\geq k_2 \quad\Longrightarrow\quad P^{n(k)}(|V^{n(k)}|>\varepsilon) \leq 2\eta.$$

Hence we have proved that for all $\varepsilon>0$ and from any subsequence $\mathbb{N}'$ one can find a sub-subsequence $n(k)$ such that $P^{n(k)}(|V^{n(k)}|>\varepsilon)\longrightarrow 0$, thus obtaining the result. ∎

(3.56) LEMMA: Under the assumptions of (3.24), the sequence $(\overline{P}^n)_{n\in\mathbb{N}}$ is relatively compact in $\underline{\underline{M}}_{mc}(\overline{\Omega})$.

Proof. We have (3.10), (3.11), (3.13,i) with $\gamma^n=\gamma$. Let $h_0(z)=|z|^3\wedge 1$. Using (2.4) and (3.17), it is easy to check that

$$(1\vee\gamma^2)\cdot A^n \prec \tilde{A}^n := (1\vee\gamma^2)\cdot\left[\sum_{j\leq m}(\hat{C}^{n,jj}+V(\hat{B}^{n,j})) + (d+1)\hat{\nu}^n(h_0)\right].$$

Since γ is locally bounded, (3.20,ii,iii,iv) and (3.55) imply that for all $t\geq 0$, $\varepsilon>0$,

$$\lim_{(n)} P^n[V(\tilde{A}^n-\tilde{A}^{\infty})_t>\varepsilon] = 0.$$

This property implies (3.13,ii), because the topology of convergence in variation is stronger than $\mathcal{E}_s$. We could then apply (3.16,a), except that (3.13,iii) does not necessarily hold. But if we come back to the proof of (3.16,a), we see that condition (3.13,iii) is used only once, in the course of the proof of (3.48), for proving (3.50).

However, with the notations of that proof, we have: first, there exists $a>1$ with $P^{\infty}(|\Delta Z_s^{\infty}|>a-1$ for at least one $s\leq N)\leq\varepsilon/16$. Secondly, there exists $n_1'\in\mathbb{N}$ such that $n\geq n_1' \longrightarrow P^n(|\Delta Z_s^{\infty}|>a-1$ for at least one $s\leq N)\leq\varepsilon/8$, because of (3.10). Thirdly, there exists $n_1\geq n_1'$ such that: $n\geq n_1 \longrightarrow P^n(\sup_{s\leq N}|Z_s^n-Z_s^{\infty}|>1)\leq\varepsilon/8$, because of (3.20,i). Then, we obtain that (3.50) holds. ∎

(3.57) LEMMA: Under the assumptions of (3.24), there exists a subsequence satisfying (3.13,iii).

Proof. Let us first assume (3.21), and put $W_t^n=\sup_{s\leq t}[|Z_s^n-Z_s^{\infty}|+|K_s^n-K_s^{\infty}|]$. Then (3.11) and (3.20,i) yield that (W_t^n) goes to 0 in P^{∞}-measure.

We can find a subsequence $\mathbb{N}'\subset\mathbb{N}$ such that $\Omega_0 = \{\omega : \lim_{\mathbb{N}'} W^n_t(\omega) = 0$ for all $t \geq 0\}$ satisfies $P^\infty(\Omega_0) = 1$. Define

$$V_t(\omega) = \begin{cases} \sup_{n\in\mathbb{N}'}[|\Delta Z^n_t(\omega)| + |\Delta K^n_t(\omega)|] & \text{if } \omega\in\Omega_0 \\ 0 & \text{if } \omega\notin\Omega_0 . \end{cases}$$

By definition of Ω_0, each set $\{t : V_t(\omega) > a\}$ is discrete for every $a > 0$, $\omega\in\Omega$, and V is measurable. Moreover by (3.21) there exists a subsequence satisfying (3.13,iii) with $\gamma^n = 0$ and some process V. Now, since γ is locally bounded, for each $\omega\in\Omega$, $a > 0$, the set $\{t : (\gamma_t(\omega) + 1)V_t(\omega) > a\}$ is discrete, and our subsequence satisfies (3.13,iii) with $\gamma^n = \gamma$ and $V' = (1+\gamma)V$. ∎

<u>Proof of (3.24)</u>. 1) Due to (3.56), it remains to prove that all limit points of the sequence $(\overline{P}^n)_{n\in\mathbb{N}}$ are good solution-measures of (1.1,∞); and for this we can assume that, up to a relabelling, $(\overline{P}^n)_{n\in\mathbb{N}}$ converges to a limit $\overline{P}^\infty$. <u>Added in proof</u>: See Note 2 on page 212.

We have (3.10), (3.11), (3.12) (by (3.20,i)), (3.15), and (3.13,i,ii) as seen in the proof of (3.56). Let C be the set of all continuous functions: $\mathbb{R}^m \longrightarrow \mathbb{R}$ such that $h(z)/|z|^3 \wedge 1$ is bounded, and which have a limit when $|z| \longrightarrow \infty$. This set is separable for the uniform convergence topology, and we denote by $\mathcal{H}$ a dense countable subset of C, containing the function h_0 occuring in the proof of (3.56). By taking again a subsequence, still denoted by $(\overline{P}^n)_{n\in\mathbb{N}}$, we can assume by (3.57) that (3.13,iii) holds, and by (3.20,ii,iii,iv) that there exists a P^∞-full set Ω_0 such that:

$$(3.58)\quad \begin{cases} \omega\in\Omega_0,\ t\geq 0,\ h\in\mathcal{H},\ j,k\leq m \Longrightarrow V(\hat{B}^{n,j} - \hat{B}^{\infty,j})_t(\omega) \longrightarrow 0 , \\ V(\hat{C}^{n,jk} - \hat{C}^{\infty,jk})_t(\omega) \longrightarrow 0,\quad V(\hat{\nu}^n(h) - \hat{\nu}^\infty(h))_t(\omega) \longrightarrow 0 , \end{cases}$$

which also implies: $V(\tilde{A}^n - \tilde{A}^\infty)_t(\omega) \longrightarrow 0$, where $\tilde{A}^n$ is defined in (3.56).

2) <u>It remains to prove (3.14)</u>. Let T be an $\overline{\underline{F}}$-stopping time on $(\overline{\Omega},\overline{\underline{F}})$, bounded by N. Let (S_p) be a sequence of $\underline{F}$-stopping times increasing to $+\infty$, such that $\gamma_s \leq p$ if $s \leq S_p$. From (3.10) it is easy to see that

$$\lim_{p\uparrow\infty} \sup_{(n)} P^n(S_p < N) = 0 .$$

On the other hand, one may apply (3.47) and (3.48). In particular, with the notations of these lemmas, we have from (3.49):

$$\lim_{q\uparrow\infty, a\uparrow\infty} \sup_{(n)} \overline{P}^n(\sigma(a) \wedge T_q^n < N) = 0$$

$$\lim_{(n)} P^n(T_q^n < T_q^\infty) = 0 .$$

Therefore it is sufficient to prove that for all $\varepsilon > 0$, $a > 0$, $q \in \mathbb{N}$, $p \in \mathbb{N}$, $u \in \mathbb{R}^{m+d}$,

(3.59) $$\lim_{(n)} \overline{P}^n(|\Phi_T^{n,g^n,u} - \Phi_T^{\infty,g^\infty,u}| > \varepsilon,\ T \leqslant \sigma(a) \wedge S_p \wedge T_q^n) = 0 ,$$

or, in other words, we may suppose that $T \leqslant \sigma(a) \wedge S_p \wedge T_q^n \wedge N$. Then by (2.12) and (2.13), we get on $[0,T]$:

(3.60) $$\begin{cases} |v_s^{n,g^n,u}| \leqslant \rho := (1+|u|^2)[1+p^2(1+a)^2] \\ |v_s^{\infty,g^\infty,u} - v_s^{\infty,g^n,u}| \leqslant 2\rho \wedge \rho' |g_s^n - g_s^\infty|, \text{ with} \\ \qquad \rho' = (1+|u|^2)(1+2p(1+a)) . \end{cases}$$

Let $Y^n = (Z^n, X - K^n)$. Using the definition (3.9), we easily get a majoration $|\Phi_T^{n,g^n,u} - \Phi_T^{\infty,g^\infty,u}| \leqslant \beta_1^n + \beta_2^n + \beta_3^n$, with

$$\beta_1^n = \int_0^T |\exp i\langle u|Y_{s-}^n\rangle - \exp i\langle u|Y_{s-}^\infty\rangle|\, |v_s^{n,g^n,u}|\, dA_s^n$$

$$\beta_2^n = \int_0^T |v_s^{\infty,g^\infty,u} - v_s^{\infty,g^n,u}|\, dA_s^\infty$$

$$\beta_3^n = \left|\int_0^T \exp i\langle u|Y_{s-}^\infty\rangle\, v_s^{n,g^n,u}\, dA_s^n - \int_0^T \exp i\langle u|Y_{s-}^\infty\rangle\, v_s^{\infty,g^n,u}\, dA_s^\infty\right| .$$

Because of (3.60) and of $A_T^n \leqslant q$ (recall that $T \leqslant T_q^n$), we have

$$\overline{P}^n(\beta_1^n \geqslant \varepsilon) \leqslant \overline{P}^n(\sup_{s\leqslant N} |Y_s^n - Y_s^\infty| \geqslant \frac{\varepsilon}{2\rho q |u|}) ,$$

which goes to 0 because of (3.11) and (3.20,i).

Let H be a $\mathcal{C}_s$-compact subset of $\mathcal{X}$, and let $h_H^n(\omega,s) = \operatorname{ess\,sup}_{x\in H} |g_s^n(\omega,x) - g_s^\infty(\omega,x)|$, this "ess sup" being taken with respect to the measure $P^\infty(d\omega)dA_s^\infty(\omega)$. Then $\beta_2^n(\omega,x) \leqslant \overline{\beta}_H^n(\omega) := \int_{[0,T_q^n(\omega)]} (2\rho \wedge \rho' h_H^n(s,\omega))dA_s^\infty(\omega)$ for all $x \in H$, P^∞-a.s. in ω (use again (3.60)). By (3.15) and Lebesgue convergence Theorem, we have $\overline{\beta}_H^n \longrightarrow 0$ P^∞-a.s., and (3.55) implies that $P^n(\overline{\beta}_H^n \geqslant \varepsilon) \longrightarrow 0$. But

(3.61) $$\overline{P}^n(\beta_2^n \geqslant \varepsilon) \leqslant P^n(\overline{\beta}_H^n \geqslant \varepsilon) + \overline{P}^n(\overline{\Omega} \setminus (\Omega \times H)) .$$

By (3.56) and (2.3), the sequence $(\overline{P}^n|_{\mathcal{X}})$ is tight, so one may find H, $\mathcal{C}_s$-compact subset of $\mathcal{X}$, such that $\overline{P}^n(\overline{\Omega} \setminus (\Omega \times H))$ is as small as we want, uniformly in n. Then we can make $P^n(\overline{\beta}_H^n \geqslant \varepsilon)$ as small as we want for n large enough, and (3.61) implies that $\overline{P}^n(\beta_2^n \geqslant \varepsilon) \longrightarrow 0$.

3) <u>It remains to prove that</u> $\overline{P}^n(\beta_3^n \geq \varepsilon) \longrightarrow 0$ <u>for all</u> $\varepsilon > 0$. Let F be an increasing process such that $dA_t^n \ll dF_t$ for all $n \in \overline{\mathbb{N}}$. Let H^n be a nonnegative predictable process on $(\Omega, \underline{\underline{F}}, \underline{F})$ such that $A^n = H^n \cdot F$. Set

$$\begin{aligned}\hat{b}_t^n &= H_t^n \left[b_t^n - \int N_t^n(dz)(f(z) - z I_{\{|z| \leq 1\}})\right] \\ \hat{c}_t^{n,jk} &= H_t^n \left[c_t^{n,jk} + \int N_t^n(dz) f^j(z) f^k(z)\right] \\ \hat{N}_t^n(dz) &= H_t^n N_t^n(dz).\end{aligned}$$

(3.17) implies that $\hat{B}^n = \hat{b}^n \cdot F$, $\hat{C}^n = \hat{c}^n \cdot F$, $\hat{\nu}^n(h) = \hat{N}^n(h) \cdot F$. If $\hat{u} \in \mathbb{R}^m$ we set

$$(3.63) \quad w_t^{n,\hat{u}} = i\langle \hat{u} | b_t^n \rangle - \frac{1}{2} \sum_{j,k \leq m} \hat{u}^j c_t^{n,jk} \hat{u}^k + \int N_t^n(dz)(e^{i\langle \hat{u} | z \rangle} - 1 - i\langle \hat{u} | z \rangle I_{\{|z| \leq 1\}})$$

hence if $\varphi_{\hat{u}}(z) = e^{i\langle \hat{u} | z \rangle} - 1 - i\langle \hat{u} | f(z) \rangle + \frac{1}{2}(\langle \hat{u} | f(z) \rangle)^2$, we have

$$(3.64) \quad H_t^n w_t^{n,\hat{u}} = i\langle \hat{u}, \hat{b}_t^n \rangle - \frac{1}{2} \sum_{j,k \leq m} \hat{u}^j \hat{c}_t^{n,jk} \hat{u}^k + \hat{N}_t^n(\varphi_{\hat{u}}).$$

We pick $\omega \in \Omega_0$. (3.58) implies:

$$(3.65) \quad t \geq 0,\ h \in \mathcal{H} \Longrightarrow \int_0^t \left[|\hat{b}_s^n(\omega) - \hat{b}_s^\infty(\omega)| + |\hat{c}_s^n(\omega) - \hat{c}_s^\infty(\omega)| + |\hat{N}_s^n(\omega,h) - \hat{N}_s^\infty(\omega,h)|\right] dF_s(\omega) \longrightarrow 0.$$

Consider an infinite subset $\mathbb{N}' \subset \mathbb{N}$. (3.65) implies that there exists an infinite subset $\mathbb{N}''(\omega) \subset \mathbb{N}'$ and a $dF_s(\omega)$-full subset $D(\omega) \subset \mathbb{R}_+$, with

$$(3.66) \quad s \in D(\omega),\ h \in \mathcal{H} \Longrightarrow \lim_{n \in \mathbb{N}''(\omega)} \hat{b}_s^n(\omega) = \hat{b}_s^\infty(\omega), \quad \lim_{n \in \mathbb{N}''(\omega)} \hat{c}_s^n(\omega) = \hat{c}_s^\infty(\omega),\ \lim_{n \in \mathbb{N}''(\omega)} \hat{N}_s^n(\omega,h) = \hat{N}_s^\infty(\omega,h)$$

In particular, $\lim_{n \in \mathbb{N}''(\omega)} \hat{N}_s^n(\omega,h_0) = \hat{N}_s^\infty(\omega,h_0)$, which yields: $\lim_{a \uparrow \infty} \sup_{n \in \mathbb{N}''(\omega)} \hat{N}_s^n(\omega,\{z : |z| > a\}) = 0$. Hence, since $\mathcal{H}$ is dense in C, it is easy to check that (3.66) holds for all $h \in C$, and in particular for $h = \varphi_{\hat{u}}$ (recall that $f(z) = z$ for z small enough). Then (3.64) and (3.66) imply that: $\lim_{n \in \mathbb{N}''(\omega)} H_s^n(\omega) w_s^{n,\hat{u}}(\omega) = H_s^\infty(\omega) w_s^{\infty,\hat{u}}(\omega)$ if $s \in D(\omega)$. Now by (3.63) we see that the function: $\hat{u} \rightsquigarrow \exp(H_s^n(\omega) w_s^{n,\hat{u}}(\omega))$ is for each n the characteristic function of an infinitely divisible distribution. Therefore the convergence is uniform in $\hat{u}$, on each compact subset of $\mathbb{R}^m$. In particular, if

$$\alpha_s^n(\omega) = \sup_{\hat{u} \in \mathbb{R}^m,\ |\hat{u}| \leq |u|(1 + p(1+a))} |w_s^{n,\hat{u}}(\omega) H_s^n(\omega) - w_s^{\infty,\hat{u}}(\omega) H_s^\infty(\omega)|,$$

then $\alpha_s^n(\omega)$ is measurable in (ω, s) (because $w_s^{n,\hat{u}}$ is continuous in $\hat{u}$) and

$$(3.67) \quad s \in D(\omega) \Longrightarrow \lim_{n \in \mathbb{N}''(\omega)} \alpha_s^n(\omega) = 0.$$

Define an $\mathbb{R}^m$-valued process $\tilde{u}(t,\omega,x)$ by $\tilde{u}^j(t,\omega,x) =$

$u^j + \sum_{k\leq m} u^{m+k} g_t^{\infty,jk}(\omega,x)$. Then an easy computation shows that $v_t^{n,g^\infty,u}(\omega,x) = w_t^{n,\tilde{u}(t,\omega,x)}(\omega)$. If $t \leq T(\omega,x)$, we have $|g_t^\infty(\omega,x)| \leq p(1+a)$, and hence $|\tilde{u}(t,\omega,x)| \leq |u|(1+p(1+a))$. Then

$$\beta_3^n(\omega,x) = \left|\int_0^{T(\omega,x)} \exp i\langle u|Y_{s-}^\infty(\omega,x)\rangle (v_s^{n,g^n,u}(\omega,x)H_s^n(\omega) - v_s^{\infty,g^n,u}(\omega,x)H_s^\infty(\omega))\, dF_s(\omega)\right|$$

$$\leq \overline{\beta}_3^n(\omega) := \int_0^N \alpha_s^n(\omega)dF_s(\omega)$$

(recall that $T \leq N$). Now, the same computation than for (2.12) shows that $|w^{n,\hat{u}}| \leq \alpha(1+|u|^2)$, hence if $\alpha' = \alpha[1+|u|^2(1+p(1+a))^2]$, we have $\alpha_s^n \leq \alpha'(H_s^n + H_s^\infty)$. Moreover, the same computation than in the proof of (3.56) shows that $H^n \leq \overline{H}^n := \sum_{j\leq m}(\hat{b}^{n,j} + \hat{c}^{n,jj}) + (d+1)\hat{\nu}^n(h_0)$, while (3.65) implies that $\int_0^N |\overline{H}_s^n(\omega) - \overline{H}_s^\infty(\omega)|\, dF_s(\omega) \longrightarrow 0$. Then

$$\sup_{(n)} \overline{\beta}_3^n(\omega) \leq \alpha' \sup_{(n)} \int_0^N [\overline{H}_s^n(\omega) + \overline{H}_s^\infty(\omega)]\, dF_s(\omega) < \infty \tag{3.68}$$

$$\overline{\beta}_3^n(\omega) \leq \varepsilon F_N(\omega) + 2\alpha' \int_0^N I_{\{s:\alpha_s^n(\omega)>\varepsilon\}} \overline{H}_s^\infty(\omega)\, dF_s(\omega) + \alpha' \int_0^N |\overline{H}_s^n(\omega) - \overline{H}_s^\infty(\omega)|\, dF_s(\omega).$$

Since $\varepsilon > 0$ above is arbitrary, and using (3.67) again, we obtain that $\lim_{n\in\mathbb{N}''(\omega)} \overline{\beta}_3^n(\omega) = 0$. Since from all infinite subset $\mathbb{N}' \subset \mathbb{N}$ we can extract a subsequence $\mathbb{N}''(\omega) \subset \mathbb{N}'$ with this property, and since we have (3.68), it follows that $\overline{\beta}_3^n(\omega) \longrightarrow 0$ when $n \uparrow \infty$, for all $\omega \in \Omega_0$, that is $\overline{\beta}_3^n \longrightarrow 0$ P^∞-a.s.

Then the proof of (3.55) shows that: $P^n(\overline{\beta}_3^n \geq \varepsilon) \longrightarrow 0$, implying that: $\overline{P}^n(\overline{\beta}_3^n \geq \varepsilon) \longrightarrow 0$, for all $\varepsilon > 0$. ∎

3-f. STRONG STABILITY.

In this paragraph we consider a sequence $(Z^n)_{n\in\overline{\mathbb{N}}}$ of semimartingales with $Z_0^n = 0$ and a sequence of right-continuous adapted processes $(K^n)_{n\in\overline{\mathbb{N}}}$ with left-hand limits, on the same filtered probability space $(\Omega, \underline{\underline{F}}, \underline{F}, P)$. We consider the coefficients $(g^n)_{n\in\overline{\mathbb{N}}}$, and we suppose that each equation

$$X_t = K_t^n + \int_0^t g_s^n(.,X_{.}(.))dZ_s^n \tag{1.1,n}$$

admits a solution-process $\tilde{X}^n$ on $(\Omega, \underline{\underline{F}}, \underline{F}, P)$, for $n \in \overline{\mathbb{N}}$.

Of course one could apply Theorems (3.16) and (3.24) to this situation: in particular, if the solution-measure associated to $\tilde{X}^\infty$ is the only one good solution-measure of (1.1,∞), under the assumptions of (3.16) or of (3.24) one would obtain: for all $t \geq 0$, then $\sup_{s\leq t} |\tilde{X}_s^n - \tilde{X}_s^\infty| \xrightarrow{P} 0$.

However, we wish to obtain a better form of convergence of $\tilde{X}^n$ to $\tilde{X}^\infty$, or rather of $\tilde{X}^n - K^n$ to $\tilde{X}^\infty - K^\infty$. For this purpose, let us first recall some facts about Emery's topology of semimartingales [5]. We denote by $\underline{\underline{S}}$ the vector space of all real-valued semimartingales on $(\Omega,\underline{\underline{F}},\underline{F},P)$ with the topology generated by the distance

$$\delta(Y,Y') = \sum_{n\geq 1} 2^{-n} \sup\{E(|H\cdot(Y-Y')_n|\wedge 1): H \text{ predictable}, |H|\leq 1\}.$$

Here are some properties of this topology:

(3.69) If $Y^n \longrightarrow Y^\infty$ in $\underline{\underline{S}}$, then $\sup_{s\leq t}|Y^n_s - Y^\infty_s| \xrightarrow{P} 0$ [5].

(3.70) If Y, Y' have finite variation, then $\delta(Y,Y') \leq \sum_{n\geq 1} 2^{-n} E[1\wedge V(Y-Y')_n]$ and if Y and Y' are predictable, this inequality is an equality.

(3.71) If $Y^n \longrightarrow Y^\infty$ and $Y'^n \longrightarrow Y'^\infty$ in $\underline{\underline{S}}$, then $V([Y^n,Y'^n]-[Y^\infty,Y'^\infty])_t \xrightarrow{P} 0$ (use Lemma II-4 in [17] and Kunita and Watanabe inequality).

(3.72) If $Y^n \longrightarrow Y^\infty$ in $\underline{\underline{S}}$, if for each $n\in\overline{\mathbb{N}}$, Y^n is a special semimartingale whose canonical decomposition is $Y^n = M^n + A^n$, and if $\sup_{n,\omega,s}|\Delta Y^n_s(\omega)| < \infty$, then $V(A^n - A^\infty)_t \xrightarrow{P} 0$ ([17,IV-3] and (3.70)).

(3.73) If $Y^n \longrightarrow Y^\infty$ in $\underline{\underline{S}}$, and if $(H^n)_{n\in\overline{\mathbb{N}}}$ is a sequence of locally bounded predictable processes such that: $\sup_{s\leq t}|H^n_s - H^\infty_s| \xrightarrow{P} 0$ for all $t\geq 0$, then: $H^n\cdot Y^n \longrightarrow H^\infty\cdot Y^\infty$ in $\underline{\underline{S}}$ [17,III.13].

We denote by $\underline{\underline{S}}^m$ the space of all $\mathbb{R}^m$-valued semimartingales, with the product topology.

(3.74) THEOREM: <u>Assume the following</u>:

(i) $\sup_{s\leq t}|K^n_s - K^\infty_s| \xrightarrow{P} 0$ <u>for all</u> $t\geq 0$;

(ii) $g^n_t(\omega,.)$ <u>converges to</u> $g^\infty_t(\omega,.)$ <u>uniformly on each</u> $\mathcal{C}_s$<u>-compact set, for all</u> $t\geq 0$, $\omega\in\Omega$;

(iii) $g^\infty_t(\omega,.)$ <u>is</u> $\mathcal{C}_u$<u>-continuous on</u> $\mathcal{X}$, <u>for all</u> $t\geq 0$, $\omega\in\Omega$;

(iv) $|g^n_t(\omega,x)| \leq \gamma_t(\omega)(1 + \sup_{s<t}|x(s)|)$ <u>for all</u> $t\geq 0$, $\omega\in\Omega$, $x\in\mathcal{X}$, $n\in\overline{\mathbb{N}}$, <u>where</u> γ <u>is a locally bounded predictable process</u>;

(v) $Z^n \longrightarrow Z^\infty$ <u>in</u> $\underline{\underline{S}}^m$;

(vi) <u>for each</u> $n\in\overline{\mathbb{N}}$, $\tilde{X}^n$ <u>is a solution-process of (1.1,n) over</u> $(\Omega,\underline{\underline{F}},\underline{F},P)$, <u>and the solution-measure</u> $\overline{P}(d\omega,dx) = P(d\omega)\varepsilon_{\tilde{X}^\infty(\omega)}(dx)$ <u>is the unique good solution-measure of (1.1,∞)</u>.

Then, $\sup_{s\le t}|\tilde{X}^n_s - \tilde{X}^{\infty}_s| \xrightarrow{P} 0$ for all $t\ge 0$. If moreover

(ii') $g^n_s(\omega,x) \longrightarrow g^{\infty}_s(\omega,x)$ uniformly in (s,x) on each $[0,t]\times H$, H $\mathcal{C}_s$-compact set, for all $\omega\in\Omega$;

(iii') the family $(g^{\infty}_s(\omega,.): s\le t)$ is equicontinuous on $\mathcal{X}$ with the topology $\mathcal{C}_u$, for all $\omega\in\Omega$, $t\ge 0$,

then $(\tilde{X}^n - K^n)_{n\in\mathbb{N}}$ tends to $\tilde{X}^{\infty} - K^{\infty}$ in $\underline{\underline{S}}^d$.

(3.75) LEMMA: If $Z^n \longrightarrow Z^{\infty}$ in $\underline{\underline{S}}^m$, then (3.20) holds.

Proof. We have (3.20,i) by (3.69). f being the function showing in (3.17), define

$$\check{Z}^n_t = \sum_{s\le t}(\Delta Z^n_s - f(\Delta Z^n_s)), \qquad \tilde{Z}^n = Z^n - \check{Z}^n.$$

Since $f(z)=z$ for $|z|\le 1$, it is obvious that (3.20,i) implies that $V(\check{Z}^n - \check{Z}^{\infty})_t \xrightarrow{P} 0$ for all $t\ge 0$; hence $\check{Z}^n \longrightarrow \check{Z}^{\infty}$ in $\underline{\underline{S}}^m$ by (3.70), hence $\tilde{Z}^n \longrightarrow \tilde{Z}^{\infty}$ in $\underline{\underline{S}}^m$. A simple computation shows that the predictable process with finite variation in the canonical decomposition of $\tilde{Z}^n$ is $\hat{B}^n$, while $|\Delta\tilde{Z}^n|\le 1$ because of $|f|\le 1$. Thus (3.72) implies (3.20,ii). By (3.71) we have

(3.76) $$V([Z^{n,j},Z^{n,k}] - [Z^{\infty,j},Z^{\infty,k}])_t \xrightarrow{P} 0, \text{ all } t\ge 0.$$

Since $C^{n,jk}$ is the "continuous part" of $[Z^{n,j},Z^{n,k}]$, it follows that $V(C^{n,jk} - C^{\infty,jk})_t \xrightarrow{P} 0$. For each $h:\mathbb{R}^m \longrightarrow \mathbb{R}$ continuous, such that $h(z)/(|z|^2\wedge 1)$ is bounded, we set $\mu^n_t(h) = \sum_{s\le t} h(\Delta Z^n_s)$. Then from (3.76) and (3.20,i) it is easy to deduce that $V(\mu^n(h) - \mu^{\infty}(h))_t \xrightarrow{P} 0$. Thus $\mu^n(h) \longrightarrow \mu^{\infty}(h)$ in $\underline{\underline{S}}$ and, since $\Delta(\mu^n(h))$ is bounded uniformly in n, and because the dual predictable projection of $\mu^n(h)$ is $\hat{\nu}^n(h)$, (3.72) implies that $V(\hat{\nu}^n(h) - \hat{\nu}^{\infty}(h))_t \xrightarrow{P} 0$ for all $t\ge 0$. This gives (3.20,iv) (take h with h/h_0 bounded), and (3.20,iii) (take $h(z) = f^j(z)f^k(z)$ and combine with the previous results on $C^{n,jk}$). ■

(3.77) REMARK: The converse of this lemma, namely: (3.20) $\Longrightarrow$ $Z^n \longrightarrow Z^{\infty}$ in $\underline{\underline{S}}^m$, is not true. In fact, if $Z^n \longrightarrow Z^{\infty}$ in $\underline{\underline{S}}^m$, we have seen that $V(C^{n,jk} - C^{\infty,jk})_t \xrightarrow{P} 0$ (this result is due to Emery [5]), while this is not necessarily the case under (3.20). For instance we may approximate a Wiener process Z^{∞} by pure jump processes Z^n in such a way that (3.20) holds, but of course we have $C^{\infty}_t = t$ and $C^n_t = 0$ if $n\in\mathbb{N}$. ■

Added in proof : See Note 1 on page 212.

Proof of (3.74). Because of Lemma (3.75), we can apply (3.24) and (3.16,c), thus obtaining that: $\sup_{s\leq t}|\tilde{X}^n_s-\tilde{X}^\infty_s| \xrightarrow{P} 0$ for all $t\geq 0$.

We have $\tilde{X}^n - K^n = g^n(\tilde{X}^n)\cdot Z^n$, and each $g^n(\tilde{X}^n)$ is locally bounded by (iv). Since we have (3.73), it remains to prove that:
$\sup_{s\leq t}|g^n(\tilde{X}^n)_s - g^\infty(\tilde{X}^\infty)_s| \xrightarrow{P} 0$ for all $t\geq 0$ under (ii') and (iii').

For all $\eta>0$, there exists a $\mathcal{C}_s$-compact set H such that $P(\tilde{X}^n\in H) \geq 1-\eta$ for all $n\in\overline{\mathbb{N}}$. Then (ii') implies that

$$\lim_{(n)} P(\sup_{s\leq t}|g^n(\tilde{X}^n)_s - g^\infty(\tilde{X}^n)_s|\geq \varepsilon) = 0$$

for all $t\geq 0$, $\varepsilon>0$, while (iii') implies that

$$\lim_{(n)} P(\sup_{s\leq t}|g^\infty(\tilde{X}^n)_s - g^\infty(\tilde{X}^\infty)_s|\geq \varepsilon) = 0$$

for all $t\geq 0$, $\varepsilon>0$. Hence the result follows. ∎

Now we can compare this Theorem with Emery's results [6] (Métivier and Pellaumail in [18], §7.5, Corollary 2, have proved a slightly different result than Emery's one). In [6], it is proved that
$\tilde{X}^n - K^n \longrightarrow \tilde{X}^\infty - K^\infty$ in $\underline{\underline{S}}^d$ under the following assumptions:

(a) (3.74,i,v);

(b) each g^n_s is left-continuous in s;

(c) $|g^n_s(\omega,x) - g^n_s(\omega,x')| \leq a \sup_{s<t}|x(s)-x'(s)|$ for all $\omega\in\Omega$, $x,x'\in\mathcal{X}$, $n\in\overline{\mathbb{N}}$;

(d) if $\tilde{X}^\infty$ is the solution-process of (1.1,∞) on $(\Omega,\underline{\underline{F}},\underline{F},P)$ (this solution exists and is unique because of (b) and (c)), then:
$\sup_{s\leq t}|g^n_s(\tilde{X}^\infty) - g^\infty_s(\tilde{X}^\infty)| \xrightarrow{P} 0$ for all $t\geq 0$.

Of course (c) implies (iii'), and implies (iv) up to a minor technical detail, namely that γ is replaced by γ^n. (b) and (c) also imply (vi). But of course (d) does not imply (ii'), nor even (ii), so in some sense Emery's result is much stronger than ours. On the other hand, assumptions (iii), (iv), (vi) are much weaker than (b) and (c).

4 - EXISTENCE AND UNIQUENESS OF STRONG SOLUTIONS

In this section we consider again Equation (1.1), and we use the notations in force in section 2, especially A, (B,C,ν), as defined in (2.4). We will assume the following:

(4.1) There exists a nonnegative predictable process γ on $(\Omega,\underline{\underline{F}},\underline{F},P)$ such that $\int_0^t(\gamma_s)^2 dA_s < \infty$ P-a.s. for all $t \geq 0$, and that $|g_t(\omega,x)| \leq \gamma_t(\omega)(1+\sup_{s<t}|x(s)|)$, for all $t \geq 0$, $\omega \in \Omega$, $x \in \mathfrak{X}$.

(4.2) $g_t(\omega,.)$ is continuous on $\mathfrak{X}$ endowed with the uniform topology, for $P(d\omega)\times dA_t(\omega)$-almost all (ω,t).

If F is a $P(d\omega)\times dA_t(\omega)$-full set such that $g_t(\omega,.)$ is $\mathcal{C}_u$-continuous on $\mathfrak{X}$ for all $(\omega,t)\in F$, let $g'_t(\omega,x) = I_F(\omega,t)g_t(\omega,x)$. It is easy to see that any good solution-measure of (1.1) with coefficient g is a good solution-measure for the equation with coefficient g', and vice-versa. Therefore, Theorem (3.25) applies:

(4.3) Under (4.1) and (4.2) there exists at least one very good solution-measure.

Our purpose in this section is to show that (4.3) and (2.26) enable us to obtain easily existence and uniqueness of a solution-process on $(\Omega,\underline{\underline{F}},\underline{F},P)$ in some cases.

4-a. THE LOCALLY LIPSCHITZ CASE.

Let us make the following assumption:

(4.4) Local Lipschitz condition: for all $n\in\mathbb{N}$, $t\geq 0$, $\omega\in\Omega$, $x, x'\in\mathfrak{X}$ such that $\sup_{s<t}|x(s)|\leq n$, $\sup_{s\leq t}|x'(s)|< n$, we have $|g_t(\omega,x)-g_t(\omega,x')|\leq \beta^n_t(\omega)\sup_{s<t}|x(s)-x'(s)|$, where β^n is a nonnegative predictable process such that $\int_0^t(\beta^n_s)^2 dA_s < \infty$ P-a.s. for all $t\geq 0$.

Then, we have the following well-known result ([3], [4], [6], [9], [18], [24]). Of course, the method presented here for obtaining this result is more complicated than the usual ones, because it hinges on the rather difficult Theorem (3.25).

(4.5) THEOREM: Under (4.1) and (4.4), there exists one and only one solution-process of (1.1) on the space $(\Omega,\underline{\underline{F}},\underline{F},P)$.

Proof. Of course (4.4) $\Longrightarrow$ (4.2). Then by (2.26,a) it suffices to prove that if $(\tilde{\Omega},\tilde{\underline{\underline{F}}},\tilde{\underline{F}},\tilde{P})$ is a good extension of $(\Omega,\underline{\underline{F}},\underline{F},P)$ and if $\tilde{X}$ and $\tilde{X}'$ are two solution-processes on this extension, then $\tilde{X}=\tilde{X}'$ $\tilde{P}$-a.s.

Let $Y=\tilde{X}-\tilde{X}'$, $\sigma(n) = \inf(t : |\tilde{X}_t|\geq n$ or $|\tilde{X}'_t|\geq n)$, $\tau(a) = \inf(t: |\Delta Z_t|>a)$

We have $Y = [g(\tilde{X}) - g(\tilde{X}')] \cdot Z$ and, with the notation (3.42), we obtain

(4.6) $$Y = I_{[0,\sigma(n))}[g(\tilde{X}) - g(\tilde{X}')] \cdot Z(a) \quad \text{on} \quad [0,\sigma(n) \wedge \tau(a)) .$$

Let $\tilde{A}^n_t = \int_0^t (\beta^n_s)^2 dA_s$. Since $(\tilde{\Omega}, \tilde{\underline{\underline{F}}}, \tilde{\underline{F}}, \tilde{P})$ is a good extension, we can use (3.45) on this space. If $Y^*_t = \sup_{s\leq t} |Y_s|$, for any $\underline{F}$-stopping time T such that $T \leq \sigma(n) \wedge \tau(a)$ we have, by (4.4) and (4.6):

$$\tilde{E}(Y^{*2}_{T-}) \leq 16(1+a^3)\, \tilde{E}[(1 \vee \tilde{A}^n_T) \int_0^T Y^{*2}_{s-}\, d\tilde{A}^n_s];$$

(we use $|g(\tilde{X}) - g(\tilde{X}')| \leq \beta^n Y^*$ on $[0,T)$, instead of $|g^n| \leq \gamma(1+X^*)$ in (3.45), which explains the difference in the two expressions). If $(T^n_q)_{q\geq 1}$ is a sequence of stopping times increasing to $+\infty$, such that $\tilde{A}^n_{T^n_q} < q$, Lemma (3.39) implies that $\tilde{E}(Y^{*2}_{(\sigma(n)\wedge\tau(a)\wedge T^n_q)-}) = 0$ (we apply this lemma with $\alpha = 0$, $\phi = Y^{*2}_-$, $F = \tilde{A}^n$). Since $\lim_{(q)} T^n_q = \infty$, and $\lim_{(n)} \sigma(n) = \infty$ $\tilde{P}$-a.s., and $\lim_{a\uparrow\infty} \tau(a) = \infty$ $\tilde{P}$-a.s., we obtain that $Y = 0$ $\tilde{P}$-a.s., that is $\tilde{X} = \tilde{X}'$ $\tilde{P}$-a.s. ∎

4-b. THE MONOTONE CASE.

In this paragraph, we assume the following:

(4.7) Condition on Z: Z is a special semimartingale; let $Z = M + F$ be its canonical decomposition; then M is a locally square integrable, local martingale. ∎

Let us define

(4.8) $$\begin{cases} f_t = b_t + \int N_t(dz)\, z\, I_{\{|z|>1\}} \\ \sigma^{jk}_t = c^{jk}_t + \int N_t(dz)\, z^j z^k - \Delta A_t [\int N_t(dz)\, z^j][\int N_t(dz)\, z^k], \end{cases}$$

such as to have

(4.9) $$F = f \cdot A, \quad \langle M^j, M^k \rangle = \sigma^{jk} \cdot A .$$

(4.10) Monotonicity condition: for all $n \in \mathbb{N}$ there exists a predictable process $\delta^n \geq 0$ such that $\int_0^t \delta^n_s dA_s < \infty$ P-a.s. for all $t \geq 0$, and that:

$$\begin{aligned} &2 \langle x(t-) - x'(t-) | [g_t(\omega,x) - g_t(\omega,x')] f_t(\omega) \rangle \\ &+ \Delta A_t(\omega) \, |[g_t(\omega,x) - g_t(\omega,x')] f_t(\omega)|^2 \\ &+ \textstyle\sum_{j\leq d; k,l\leq m} [g^{jk}_t(\omega,x) - g^{jk}_t(\omega,x')] \sigma^{kl}_t(\omega) [g^{jl}_t(\omega,x) - g^{jl}_t(\omega,x')] \\ &\qquad\qquad \leq \delta^n_t(\omega) \, |x(t-) - x'(t-)|^2 \end{aligned}$$

for all $\omega \in \Omega$, $t \geq 0$, and $x, x' \in \mathcal{X}$ with: $\sup_{s \leq t} |x(s)| \leq n$, $\sup_{s \leq t} |x'(s)| \leq n$ (recall that gf is the vector of components $(gf)^j = \sum_{k \leq m} g^{jk} f^k$, and that $\langle . | . \rangle$ denotes the scalar product). ∎

(4.11) REMARKS: 1) On the set where $\Delta A > 0$, (4.10) implies that

$$\Delta A_t(\omega) \left| [g_t(\omega,x) - g_t(\omega,x')] f_t(\omega) \right| \leq (\sqrt{1 + \Delta A_t(\omega) \delta_t^n(\omega)} - 1) |x(t-) - x'(t-)|$$

(cf. [8]). Hence on this set, g satisfies a local Lipschtiz condition, and a very strong one since it involves $|x(t-) - x'(t-)|$ instead of $\sup_{s \leq t} |x(s) - x'(s)|$.

2) Contrarily, on the set where $\Delta A = 0$, this condition is weaker than local Lipschitz. For instance (cf. [25]) assume that $m = 2$, Z^1 is a Wiener process and $Z_t^2 = t$. Then $f^1 = 0$, $f^2 = 1$, $\sigma^{11} = 1$, $\sigma^{12} = \sigma^{21} = \sigma^{22} = 0$, if $A_t = t$. Assume that $g_t(\omega,x) = \bar{g}(x(t-))$, with

$$\left. \begin{aligned} \bar{g}^1(y) &= 8\,\frac{p-1}{p^2}\,|y|^{p/2} \\ \bar{g}^2(y) &= -|y|^{p-1}\,\mathrm{sign}(y) \end{aligned} \right\} \quad p \in (1,2).$$

Then g satisfies (4.10), but $\bar{g}(y)$ is not Lipschitz in y, so $g_t(\omega,x)$ is not Lipschitz in x. ∎

(4.12) THEOREM: <u>Under (4.1), (4.2), (4.7) and (4.10), there exists one and only one solution-process of (1.1) on the space</u> $(\Omega, \underline{\underline{F}}, \underline{F}, P)$.

This result has been proved by Rozovskii [25] for Ito's equations and $g_t(\omega,x) = \bar{g}_t(\omega,x(t-))$, extending ideas of Bensoussan and Temam [1], Métivier and Pistone [19] and Pardoux [22], then by Jacod [12] under (4.7) and (4.10) with δ^n not depending on n and again $g_t(\omega,x) = \bar{g}_t(\omega,x(t-))$, then again by Gyöngy and Krylov [8] for the same form of coefficients and under the "local" condition (4.10).

In fact all these proofs are rather complicated (the difficult part being the existence part) and work only for $g_t(\omega,x) = \bar{g}_t(\omega,x(t-))$. So, although Theorem (3.25) is not very simple, the proof that we present here is as simple as those given in [12] or [8], and it works in addition for "general" coefficients $g_t(\omega,x)$, depending on the whole past $x(s)$, $s < t$.

The key point is the following lemma ([12], ameliorated in [8]):

(4.13) LEMMA: <u>Let</u> Y <u>be a</u> d-<u>dimensional special semimartingale with</u> $Y_0 = 0$, *whose canonical decomposition* $Y = M' + F'$ *is such that* M' *is a locally square integrable local martingale, and such that* F' *and* $\langle M'^j, M'^k\rangle$ *admit the factorizations:*

(4.14) $$F' = f' \bullet A, \qquad \langle M'^j, M'^k\rangle = \sigma'^{jk} \bullet A.$$

Let H *be the unique increasing predictable process, solution of the equation:*

(4.15) $$H_t = 1 + \int_0^t H_{s-} \delta_s^n \, dA_s .$$

Then $\frac{|Y|^2}{H}$ *is a special semimartingale whose canonical decomposition* $N + G$ *satisfies*

(4.16) $$G_t = \int_0^t \frac{1}{H_s} (2\langle Y_{s-} | f'_s\rangle + \Delta A_s |f'_s|^2 + \sum_{j\le d} \sigma'^{jj}_s - \delta_s^n |Y_{s-}|^2) dA_s .$$

Proof. The proof is an application of Ito's formula. We shall write: $U \sim V$, if $U - V$ is a local martingale. We have

$$\frac{|Y|^2}{H} = 2\sum_{j\le d} (\frac{Y^j_-}{H_-}) \bullet Y^j - (\frac{Y_-}{H_-})^2 \bullet H + \sum_{j\le m} \frac{1}{H} \bullet \langle Y^{j,c}, Y^{j,c}\rangle$$
$$+ \sum_{s\le \cdot} (\frac{|Y_s|^2}{H_s} - \frac{|Y_{s-}|^2}{H_{s-}} - 2\sum_{j\le d} \frac{Y^j_{s-}}{H_{s-}} \Delta Y^j_s + (\frac{Y_{s-}}{H_{s-}})^2 \Delta H_s) .$$

We have $|Y|^2 = |Y_-|^2 + 2\langle Y_- | \Delta Y\rangle + |\Delta Y|^2$, and $H_s = H_{s-}(1 + \delta_s^n \Delta A_s)$. Then, using (4.14) and (4.15), we obtain

$$\frac{|Y|^2}{H} \sim 2(\frac{1}{H_-} \langle Y_- | f'\rangle) \bullet A - (\frac{|Y_-|^2}{H_-} \delta^n) \bullet A + \sum_{j\le m} \frac{1}{H} \bullet \langle Y^{j,c}, Y^{j,c}\rangle$$
$$+ \sum_{s\le \cdot} (\frac{|\Delta Y_s|^2}{H_s} - 2\frac{\langle Y_{s-} | \Delta Y_s\rangle}{H_s} \delta_s^n \Delta A_s + \frac{|Y_{s-}|^2}{H_s} (\delta_s^n \Delta A_s)^2) .$$

Now the dual predictable projection of $\sum_{s\le \cdot} |\Delta Y_s|^2 / H_s$ is

$$\frac{1}{H} (\sum_{s\le \cdot} |\Delta F'_s|^2 + \sum_{j\le d} \langle M'^j - M'^{j,c}, M'^j - M'^{j,c}\rangle) ,$$

because $|\Delta Y|^2 = |\Delta M'|^2 + |\Delta F'|^2 + 2\langle \Delta M' | \Delta F'\rangle$ and $\sum_{s\le \cdot} \langle \Delta M'_s | \Delta F'_s\rangle$ is a local martingale because F' is predictable. Similarly, $\sum_{s\le \cdot} \frac{1}{H_s} \langle Y_{s-} | \Delta M'_s\rangle \delta_s^n \Delta A_s$ is a local martingale. Thus

$$\frac{|Y|^2}{H} \sim 2(\frac{1}{H_-} \langle Y_- | f'\rangle) \bullet A - (\frac{|Y_-|^2}{H_-} \delta^n) \bullet A + (\frac{1}{H} \sum_{j\le d} \sigma'^{jj}) \bullet A$$
$$+ \sum_{s\le \cdot} (\frac{1}{H_s} |f'_s|^2 \Delta A_s^2 - \frac{2}{H_s} \langle Y_{s-} | f'_s\rangle \delta_s^n \Delta A_s^2 + \frac{|Y_{s-}|^2}{H_s} (\delta_s^n \Delta A_s)^2)$$
$$\sim \frac{1}{H} (2\langle Y_- | f'\rangle + |f'|^2 \Delta A + \sum_{j\le d} \sigma'^{jj} - \delta^n |Y_-|^2) \bullet A = G . \blacksquare$$

Proof of (4.12). By (2.26,a), it suffices to prove that if $(\tilde{\Omega},\tilde{\underline{F}},\tilde{F},\tilde{P})$ is a good extension of $(\Omega,\underline{F},F,P)$ and if $\tilde{X}$ and $\tilde{X}'$ are two solution-processes on this extension, then $\tilde{X}=\tilde{X}'$ $\tilde{P}$-a.s. Let $Y=\tilde{X}-\tilde{X}'$, $\sigma(n)=\inf(t: |\tilde{X}_t|\geq n$ or $|\tilde{X}'_t|\geq n)$, $\tilde{g}=g(\tilde{X})-g(\tilde{X}')$. Then we have $Y=\tilde{g}\cdot Z$.

Since $(\tilde{\Omega},\tilde{\underline{F}},\tilde{F},\tilde{P})$ is a good extension, the canonical decomposition of Z on this space is still $Z=M+F$, and since $\tilde{g}$ is locally bounded by (4.1), Y is a special semimartingale whose canonical decomposition is $Y=M'+F'$, with $M'=\tilde{g}\cdot M$ and $F'=\tilde{g}\cdot F$. We will apply (4.13) to this process Y, and in (4.14) we have

$$(4.17)\qquad f' = \tilde{g}f, \qquad \sigma'^{jj} = \Sigma_{k,l\leq m}\tilde{g}^{jk}\sigma^{kl}\tilde{g}^{jl}.$$

Let $n\in\mathbb{N}$ be fixed, and G (depending on n) be given by (4.16). We have that $N=|Y|^2/H-G$ is a local martingale. But (4.16), (4.17), the facts that $\tilde{g}=g(\tilde{X})-g(\tilde{X}')$ and that $Y=\tilde{X}-\tilde{X}'$, and (4.10), imply that $G_t\leq 0$ if $t\leq\sigma(n)$. Since $|Y|^2/H\geq 0$, it follows that $(N_{t\wedge\sigma(n)})_{t\geq 0}$ is a local martingale that is nonnegative. Hence it is a nonnegative supermartingale, and since $N_0=0$ we must have $N_{t\wedge\sigma(n)}=0$ $\tilde{P}$-a.s. for all $t\geq 0$. Hence $|Y|^2/H=G$ on $[0,\sigma(n)]$ and since $|Y|^2/H\geq 0$ and $G\leq 0$ on $[0,\sigma(n)]$ we have $|Y|^2=0$ on $[0,\sigma(n)]$ $\tilde{P}$-a.s. Now, $\lim_{(n)}\sigma(n)=\infty$ $\tilde{P}$-a.s., therefore $Y=0$ $\tilde{P}$-a.s. and the result is proved.∎

REFERENCES

1 A. BENSOUSSAN, R. TEMAM: Equations aux dérivées partielles stochastiques non linéaires. Israël J. Math. 11, 1972, 95-129.

2 P. BILLINGSLEY: Convergence of probability measures. Wiley and Sons: New-York, 1968.

3 C. DOLEANS-DADE: On the existence and unicity of solution of stochastic integral equations. Z. für Wahr. 34, 93-101, 1976.

4 C. DOLEANS-DADE, P.A. MEYER: Equations différentielles stochastiques. Sém. Probab. XI, Lect. Notes in Math. 581, 376-382, 1977.

5 M. EMERY: Une topologie sur l'espace des semimartingales. Sém. Probab. XIII, Lect. Notes in Math. 721, 260-281, 1979.

6 M. EMERY: Equations différentielles stochastiques lipschitziennes: étude de la stabilité. Sém. Probab. XIII, Lect. Notes in Math. 721 281-293, 1979.

7 B. GRIGELIONIS, R. MIKULEVICIUS: On weak convergence of semimartingales. Lit. Math. J. XXI, 1981.

8 I. GYONGY, N.V. KRYLOV: On stochastic equations with respect to semimartingales I, to appear, 1980.

9 J. JACOD: Calcul stochastique et problèmes de martingales. Lect. Notes in Math. 714, Springer Verlag: Berlin, 1979.

10 J. JACOD: Intégrales stochastiques par rapport à une semimartingale vectorielle et changements de filtration. Sém. Probab. XIV, Lect. Notes in Math. 784, 161-172, 1980.

11 J. JACOD: Weak and strong solutions of stochastic differential equations. Stochastics, 3, 171-191, 1980.

12 J. JACOD: Une condition d'existence et d'unicité pour les solutions fortes d'équations différentielles stochastiques. To appear in Stochastics, 1980.

13 J. JACOD, J. MEMIN: Existence of weak solutions for stochastic differential equations driven by semimartingales. To appear in Stochastics, 1980.

14 J. JACOD, J. MEMIN, M. METIVIER: Tightness and stopping times: some new conditions. To appear, 1980.

15 J. JACOD, J. MEMIN: Sur un type de convergence intermédiaire entre la convergence en probabilité et la convergence en loi. To appear in: Sém. de Probab. XV, 1980.

16 V.A. LEBEDEV: On the existence of a solution of the stochastic equation with respect to a martingale and a stochastic measure. Int. Symp. on Stoch. Diff. Equa., Vilnius, 65-69, 1978.

17 J. MEMIN: Espaces de semimartingales et changements de probabilité. Z. für Wahr. 52, 9-40, 1980.

18 M. METIVIER, J. PELLAUMAIL: Stochastic integration. To appear, 1980.

19 M. METIVIER, G. PISTONE: Sur une équation d'évolution stochastique. Bull. Soc. Math. France, 104, 65-85, 1976.

20 P.A. MEYER: Un cours sur les intégrales stochastiques. Sém. Probab. X, Lect. Notes in Math. 511, 245-400, 1976.

21 P.A. MEYER: Convergence faible et compacité des temps d'arrêt d'après Baxter et Chacon. Sém. Probab. XII, Lect. Notes in Math. 649, 411-423, 1978.

22 E. PARDOUX: Thèse, Univ. Paris-Sud, 1975.

23 J. PELLAUMAIL: Solutions faibles pour des processus discontinus. Comptes Rendus Acad. Sci. Paris (A) 290, 431-433, 1980.

24 P. PROTTER: On the existence, uniqueness, convergence and explosions of solutions of systems of stochastic differential equations. Ann. Probab. 5, 243-261, 1977.

25 B.L. ROZOVSKII: A note on strong solutions of stochastic differential equations with random coefficients. To appear: 1980.

26 A.V. SKOROKHOD: Limit theorems for stochastic processes. Theor. Probab. and Appl. 1, 261-290 (AMS Transl.) 1956.

27 C. STRICKER: Quasimartingales, martingales locales, semimartingales et filtrations. Z. für Wahr. 39, 55-63, 1977.

28 D.W. STROOCK, S.R.S. VARADHAN: Multidimensional diffusion processes. Springer Verlag: Berlin, 1979.

29 T. YAMADA, S. WATANABE: On the uniqueness of solutions of stochastic differential equations. J. Math. Kyoto Univ. 11, 156-167, 1971.

30 A.V. ZVONKIN, N.V. KRYLOV: On strong solutions of stochastic differential equations. School-Seminar (Druskininkai), Vilnius, Ac. Sci. Lit. SSR, II, 9-88, 1975.

Département de Mathématiques et Informatique
Université de Rennes
Campus de Beaulieu
35042 - RENNES Cedex

Notes added in proof.

1) The statement in Remark (3.77) is false: (3.20) is actually equivalent to the convergence $Z^n \to Z$ in $\underline{S}^m$ when all the P^n's are the same.

2) Of course, for a non-metrizable space, compactness does not imply sequential compactness. Hence, our proofs of (3.16,a), (3.16,b), and (3.24), are not quite right : instead of taking subsequences, one needs to take subnets (indexed by a directed set); nothing else is changed, and the statements of these theorems are (hopefully) still true!

On the decomposition of solutions of stochastic differential equations

Hiroshi Kunita

Department of Applied Science

Kyushu University, Fukuoka, Japan

We shall consider stochastic differential equation of the form

$$(0.1) \qquad d\xi_t = \sum_{j=1}^{r} X_j(\xi_t) \circ dM_t^j ,$$

where $X_1,\dots,X_r$ are smooth vector fields on a manifold M, and $M_t^1,\dots,M_t^r$ are continuous semimartingales. The symbol $\circ$ denotes the Stratonovich integral. We denote the solution with initial condition $\xi_0 = x$ as $\xi_t(x)$ or $\xi_t(x,\omega)$. Then $\xi_t(\cdot,\omega)$ defines a map from M into itself for each t and a.s. ω.

In the first part of the paper (Sections 1-3), we will show that under additional contitions on $X_1,\dots,X_r$ or M, the maps $\xi_t(\cdot,\omega)$ become a flow of diffeomorphisms a.s. ω. The property appears important in recent study of stochastic differential geometry, and has been studied by several authors, e.g. Elworthy [7], Malliavin [15], Ikeda-Watanabe [8], Bismut [1]. We will propose here still other method for the proof of diffeomorphism.

In Section 1 we consider Ito SDE's rather than Stratonovich SDE's on R^d. We will prove that the solution map $\xi_t(\cdot,\omega)$ of Ito SDE is a flow of homeomorphisms, provided that coefficients are Lipschitz continuous. (Theorem 1.2) Furthermore, if coefficients are of C^k-class, then $\xi_t(\cdot,\omega)$ is a flow of C^{k-1}-diffeomorphisms. This is proved in Theorem 2.3 of Section 2. In Section 3, we discuss a similar problem in case of SDE on manifold.

The second part of the paper is of Sections 4-6. We will apply results of the first part to the decomposition of the solution of SDE; namely, we decompose the equation (0.1) to two simpler equations such that ξ_t is decomposed as $\xi_t(x) = \zeta_t \circ \eta_t(x)$, where ζ_t and η_t are solutions of these two equations. There, the differentials ξ_{t*} of maps ξ_t are fully used. (Sections 4 and 5)

As an application of this, we shall consider in Section 6 the representation of the solution when the Lie algebra generated by $X_1,\ldots X_r$ is solvable. In [13], the author showed that under the same condition the solution is decomposed to those of several nilpotent equations, making use of Campbell-Hausdorff formula. Here we establish another and perhaps simpler representation, following the method of K. T. Chen [3].

Acknowledgment. In the proof of Theorem 1.2, the author owes much to S. R. S. Varadhan (by private communication). In particular, Lemma 1.4 and the onto property of the map ξ_t are due to him. Lemma 1.3 is motivated by a result of T. Yamada's, which will appear in [16]. The author expresses thanks to them.

1. Flow of homeomorphisms for the solution of SDE.

Let $X_1(x),\dots,X_r(x)$ be continuous mappings from R^d into itself and $M_t^1,\dots,M_t^r$ be continuous semimartingales difined on a probability space $(\Omega, F, P ; F_t)$. Here F_t, $0 \le t < \infty$ is an increasing family of sub σ-fields of F such that $\bigwedge_{\varepsilon>0} F_{t+\varepsilon} = F_t$ holds for each t. Consider an Ito stochastic differential equation (SDE) on R^d;

$$(1.1)\qquad d\xi_t = \sum_{j=1}^{r} X_j(\xi_t)dM_t^j .$$

A sample continuous F_t-adapted stochastic process ξ_t with values in R^d is called a solution of (1.1), if it satisfies

$$(1.2)\qquad \xi_t = \xi_0 + \sum_{j=1}^{r}\int_0^t X_j(\xi_s)dM_s^j ,$$

where the right hand side is the Ito integral.

Concerning coefficients of the equation, we will assume in this section that they are Lipschitz continuous, i.e., there is a positive constant L such that

$$|X_j^i(x) - X_j^i(y)| \le L|x - y|, \qquad \forall x,y \in R^d$$

holds for all indices i, j, where $X_j^i(x)$ is the i-th component of the vector function $X_j(x)$. Then for a given point x of R^d, the equation has a unique solution such that $\xi_0 = x$. We denote it as $\xi_t(x)$ or $\xi_t(x,\omega)$. It is continuous in (t,x) a.s. In fact, the following proposition is well known.

Proposition 1.1. (cf. Stroock-Varadhan [18]). $\xi_t(x,\omega)$ is continuous in $[0,\infty) \times R^d$ for almost all ω. Furthermore, for any $T > 0$ and $p \geq 2$, there is a positive constant $K^{(1)}_{p,T}$ such that

$$(1.3) \qquad E|\xi_t(x) - \xi_s(y)|^p \leq K^{(1)}_{p,T}(|x - y|^p + |t - s|^{\frac{p}{2}})$$

holds for all x, y of R^d and t, s of $[0,T]$.

We thus regard that for fixed t, $\xi_t(\cdot,\omega)$ is a continuous map from R^d into itself for almost all ω. The purpose of this section is to prove that the map $\xi_t(\cdot,\omega)$ is one to one and onto, and that the inverse map $\xi_t^{-1}(\cdot,\omega)$ is also continuous. Namely we will prove

Theorem 1.2. Suppose that $X_1,\ldots,X_r$ of equation (1.1) are Lipschitz continuous. Then the solution map $\xi_t(\cdot,\omega)$ is a homeomorphism of R^d for all t, a.s. ω.

Before the proof of the theorem, we shall mention a few remarks.

Remark 1. In case of one dimensional SDE, Ogura and Yamada [16] has shown the same result under a weaker condition, using a strong comparison theorem of solutions. In fact, if coefficients are Lipschitz continuous on any finite interval (local Lipschitzian) and if they are of linear growth, i.e., $|X_j(x)| \leq C(1 + |x|)$ holds for all x with some positive C, then the solution $\xi_t(\cdot,\omega)$ is a homeomorphism for any t a.s.

Remark 2. The (local) Lipschitz continuity of coefficients is crucial for the theorem. Ogura and Yamada [16] has given an example of one dimensional SDE with α-Hölder continuous coefficients $(\frac{1}{2} < \alpha < 1)$, which has a unique strong solution but does not have the "one to one" property.

Remark 3. It is enough to prove the theorem in case that M_t^i, $i = 1,\dots,r$ satisfy properties below: Let $M_t^j = N_t^j + A_t^j$ be the decomposition of semimartingale such that N_t^j is a continuous local martingale and A_t^j is a continuous process of bounded variation. Let $\langle N^j\rangle_t$ be the quadratic variation of N_t^j. Then it holds for each j,

$$(1.4)\qquad A_t^j - A_s^j \le t - s, \quad \langle N^j\rangle_t - \langle N^j\rangle_s \le t - s, \qquad \forall s < \forall t.$$

In fact if it is not satisfied, set

$$C_t = \sum_j (A_t^j + \langle N^j\rangle_t) + t$$

and consider the inverse function K_t. Define $\check{M}_t^j = M_{K_t}^j$ and $\check{\xi}_t = \xi_{K_t}$. Obviously $\check{M}_t^j$ is $\check{F}_t \equiv F_{K_t}$ adapted and is a continuous semimartingale such that $\check{A}_t^j$ and $\langle \check{N}^j\rangle_t$ satisfy (1.4). Furthermore $\check{\xi}_t$ is a solution of (1.1) replacing M_t^j by $\check{M}_t^j$. Hence if the theorem is valid for $\check{\xi}_t$, then the same is valid for ξ_t, since $\xi_t = \check{\xi}_{C_t}$.

In the following discussion, condition (1.4) is always assumed. We will first show the "one to one" property. Our approach is based on several elementary inequalities.

Lemma 1.3. Let $T > 0$ and p be any real number. Then there is a positive constant $K_{p,T}^{(2)}$ such that

$$(1.5)\qquad E|\xi_t(x) - \xi_t(y)|^p \le K_{p,T}^{(2)}|x - y|^p, \quad \forall x,y \in R^d, \quad \forall t \in [0,T].$$

Proof. If $x = y$, the inequality is clearly satisfied for any positive constant $K^{(2)}_{p,T}$. We shall assume $x \neq y$. Let ε be an arbitrary positive number and $\sigma_\varepsilon = \inf\{t > 0;\ |\xi_t(x) - \xi_t(y)| < \varepsilon\}$. We shall apply Ito's formula to the function $f(z) = |z|^p$. Then it holds for $t < \sigma_\varepsilon$,

$$|\xi_t(x) - \xi_t(y)|^p - |x - y|^p$$

$$= \sum_{i,j} \int_0^t \frac{\partial f}{\partial z_i}(\xi_s(x) - \xi_s(y))(X_j^i(\xi_s(x)) - X_j^i(\xi_s(y)))dM_s^j$$

$$+ \frac{1}{2} \sum_{i,j,k,\ell} \int_0^t \frac{\partial^2 f}{\partial z_i \partial z_j}(\xi_s(x) - \xi_s(y))(X_k^i(\xi_s(x)) - X_k^i(\xi_s(y)))$$

$$\times (X_\ell^j(\xi_s(x)) - X_\ell^j(\xi_s(y)))d\langle M^k, M^\ell\rangle_s$$

$$= I_t + J_t .$$

Note $\frac{\partial f}{\partial z_i} = p|z|^{p-2} z_i$ and apply Lipschitz inequality. Then

$$\sum_i \left|\frac{\partial f}{\partial z_i}(\xi_s(x) - \xi_s(y))(X_j^i(\xi_s(x)) - X_j^i(\xi_s(y)))\right|$$

$$\leq |p|\sqrt{d}\, L\, |\xi_s(x) - \xi_s(y)|^p.$$

Therefore we have

$$|EI_{t\wedge\sigma_\varepsilon}| \leq |p|r\sqrt{d}\, L \int_0^t E|\xi_{s\wedge\sigma_\varepsilon}(x) - \xi_{s\wedge\sigma_\varepsilon}(y)|^p ds.$$

Next, note that

$$\frac{\partial^2 f}{\partial z_i \partial z_j} = p|z|^{p-2}\delta_{ij} + p(p-2)|z|^{p-4} z_i z_j,$$

where δ_{ij} is the Kronecker's delta. Then

$$\Big|\sum_{i,j} \frac{\partial^2 f}{\partial z_i \partial z_j}(\xi_s(x) - \xi_s(y))(X_k^i(\xi_s(x)) - X_k^i(\xi_s(y)))(X_\ell^i(\xi_s(x)) - X_\ell^i(\xi_s(y)))\Big|$$

$$\leq |p|(|p-2| + d)L^2|\xi_s(x) - \xi_s(y)|^p.$$

Therefore

$$|E\, J_{t\wedge\sigma_\varepsilon}| \leq \frac{1}{2}r^2\, |p|(|p-2| + d)L^2\int_0^t E|\xi_{s\wedge\sigma_\varepsilon}(x) - \xi_{s\wedge\sigma_\varepsilon}(y)|^p ds.$$

Summing up these two inequalities, we obtain

$$E|\xi_{t\wedge\sigma_\varepsilon}(x) - \xi_{t\wedge\sigma_\varepsilon}(y)|^p \leq |x-y|^p + C_p\cdot\int_0^t E|\xi_{s\wedge\sigma_\varepsilon}(x) - \xi_{s\wedge\sigma_\varepsilon}(y)|^p ds,$$

where C_p is a positive constant. By Gronwall's inequality,

$$E|\xi_{t\wedge\sigma_\varepsilon}(x) - \xi_{t\wedge\sigma_\varepsilon}(y)|^p \leq K_{p,T}^{(2)}|x-y|^p, \qquad \forall t \in [0,T]$$

where $K_{p,T}^{(2)} = e^{C_p T}$. Letting ε tend to 0, we have

$$E|\xi_{t\wedge\sigma}(x) - \xi_{t\wedge\sigma}(y)|^p \leq K_{p,T}^{(2)}|x-y|^p,$$

where σ is the first time t such that $\xi_t(x) = \xi_t(y)$. However, it holds $\sigma = \infty$ a.s., since otherwise the left hand side would be infinity if $p < 0$. The proof is complete.

The above lemma shows that if $x \neq y$ then $\xi_t(x) \neq \xi_t(y)$ holds for all t a.s. But it does not conclude that $\xi_t(\cdot,\omega)$ is "one to one", since the exceptional null set $N_{x,y} = \{\omega | \xi_t(x) = \xi_t(y)$ for some $t\}$ depends on the pair (x, y). To overcome this point, we shall prove the following lemma.

Lemma 1.4. (Varadhan) Set

$$\eta_t(x,y) = \frac{1}{|\xi_t(x) - \xi_t(y)|}$$

Then $\eta_t(x,y)$ is continuous in $[0,\infty) \times \{(x,y) \in R^{2d} \mid x \neq y\}$.

Proof. Suppose $p > 2(2d+1)$. It holds

$$|\eta_t(x,y) - \eta_{t'}(x',y')|^p$$

$$\leq 2^p \eta_t(x,y)^p \eta_{t'}(x',y')^p \{|\xi_t(x) - \xi_{t'}(x')|^p + |\xi_t(y) - \xi_{t'}(y')|^p\}$$

By Hölder's inequality,

$$E|\eta_t(x,y) - \eta_{t'}(x',y')|^p$$

$$\leq 2^p \{E(\eta_t(x,y)^{4p}) E(\eta_{t'}(x',y')^{4p})\}^{\frac{1}{4}} \{(E|\xi_t(x) - \xi_{t'}(x')|^{2p})^{\frac{1}{2}} + (E|\xi_t(y) - \xi_{t'}(y')|^{2p})^{\frac{1}{2}}\}.$$

By Lemma 1.3 and Proposition 1.1, we have

$$(1.7)\qquad E|\eta_t(x,y) - \eta_{t'}(x',y')|^p$$

$$\le C_{p,T}|x-y|^{-p}|x'-y'|^{-p}\{|x-x'|^p + |y-y'|^p + 2|t-t'|^{\frac{p}{2}}\}$$

$$\le C_{p,T}\delta^{-2p}\{|x-x'|^p + |y-y'|^p + 2|t-t'|^{\frac{p}{2}}\}$$

if $|x-y| \ge \delta$ and $|x'-y'| \ge \delta$, where $C_{p,T}$ is a positive constant. Then by Kolmogorov's theorem, $\eta_t(x,y)$ is continuous in $[0,T] \times \{(x,y) \big| |x-y| \ge \delta\}$. Since T and δ are arbitrary positive numbers, we get the assertion. The proof is complete.

The above lemma leads immediately to the "one to one" property of the map $\xi_t(\cdot,\omega)$ for all t a.s.

We shall next consider the onto property. We first establish

Lemma 1.5. Let $T > 0$ and p be any real number. Then there is a positive constant $K^{(3)}_{p,T}$ such that

$$(1.8)\qquad E(1 + |\xi_t(x)|^2)^p \le K^{(3)}_{p,T}(1 + |x|^2)^p, \quad \forall x \in R^d, \quad \forall t \in [0,T].$$

Proof. We shall apply Ito's formula to the function $f(z) = (1 + |z|^2)^p$. It holds

$$f(\xi_t(x)) - f(x) = \sum_{i,j}\int_0^t \frac{\partial f}{\partial z_i}(\xi_s(x))X^i_j(\xi_s(x))dM^j_s$$

$$+ \frac{1}{2}\sum_{i,j,k,\ell}\int_0^t \frac{\partial^2 f}{\partial z_i \partial z_j}(\xi_s(x))X^i_k(\xi_s(x))X^j_\ell(\xi_s(y))d\langle M^k,M^\ell\rangle_s$$

$$= I_t + J_t .$$

Let K be a positive constant such that

$$|X_j^i(x)| \le K(1 + |x|^2)^{\frac{1}{2}}$$

holds for all i and j. Then,

$$|\sum_i \frac{\partial f}{\partial z_i}(\xi_s(x))X_j^i(\xi_s(x))| \le 2\sqrt{d}|p|K(1 + |\xi_s(x)|^2)^p .$$

Therefore,

$$|E\, I_t| \le 2r\sqrt{d}|p|K\int_0^t E(1 + |\xi_s(x)|^2)^p ds .$$

Similarly,

$$|\frac{1}{2}\sum_{i,j} \frac{\partial^2 f}{\partial z_i \partial z_j}(\xi_s(x))X_k^i(\xi_s(x))X_\ell^j(\xi_s(x))|$$

$$\le |p|(d + 2|p-1|)K^2(1 + |\xi_s(x)|^2)^p ,$$

so that

$$|E\, J_t| \le |p|r^2(d + 2|p-1|)K^2\int_0^t E(1 + |\xi_s(x)|^2)^p ds .$$

Therefore we have

$$E(1 + |\xi_t(x)|^2)^p \le (1 + |x|^2)^p + \text{const.}\int_0^t E(1 + |\xi_s(x)|^2)^p ds.$$

By Gronwall's inequality, we get the inequality of the lemma.

Remark. It holds $(1 + |x|^2) \le (1 + |x|)^2 \le 2(1 + |x|^2)$. Therefore, inequality (1.8) implies

$$(1.9) \qquad E(1 + |\xi_t(x)|)^{2p} \le 2^{|p|} K^{(3)}_{p,T} (1 + |x|)^{2p}.$$

Now taking negative p in the above lemma, we see that $|\xi_t(x)|$ tends to infinity in probability as x tends sequencially to infinity. We shall prove a stronger convergence. We claim

Lemma 1.6. Let $\hat{R}^d = R^d \cup \{\infty\}$ be the one point compactification of R^d. Set

$$\eta_t(x) = \begin{cases} \dfrac{1}{1 + |\xi_t(x)|} & \text{if } x \in R^d \\ 0 & \text{if } x = \infty \end{cases}$$

Then $\eta_t(x,\omega)$ is a continuous map from $[0,\infty) \times \hat{R}^d$ into R^1 a.s.

Proof. Obviously $\eta_t(x)$ is continuous in $[0,\infty) \times R^d$. Hence it is enough to prove the continuity in the neighborhood of infinity. Suppose $p > 2(2d+1)$. It holds

$$|\eta_t(x) - \eta_s(y)|^p \le \eta_t(x)^p \eta_s(y)^p |\xi_t(x) - \xi_s(y)|^p$$

By Hölder's inequality, Proposition 1.1 and Lemma 1.5, we have

$$E|\eta_t(x) - \eta_s(y)|^p \le (E\eta_t(x)^{4p})^{\frac{1}{4}}(E\eta_s(y)^{4p})^{\frac{1}{4}}(E|\xi_t(x) - \xi_s(y)|^{2p})^{\frac{1}{2}}$$

$$\le C_{p,T}(1 + |x|)^{-p}(1 + |y|)^{-p}(|x-y|^p + |t-s|^{\frac{p}{2}})$$

if $t,s \in [0,T]$ and $x,y \in R^d$, where $C_{p,T}$ is a positive constant. Set $\frac{1}{x} = (x_1^{-1},\dots,x_d^{-1})$. Since

$$\frac{|x-y|}{(1+|x|)(1+|y|)} \le \left|\frac{1}{x} - \frac{1}{y}\right|,$$

we get the inequality

$$E|\eta_t(x) - \eta_s(y)|^p \le C_{p,T}\left(\left|\frac{1}{x} - \frac{1}{y}\right|^p + |t-s|^{\frac{p}{2}}\right).$$

Define

$$\tilde{\eta}_t(x) = \begin{cases} \eta_t(\frac{1}{x}) & \text{if } x \neq 0 \\ 0 & \text{if } x = 0 \end{cases}$$

Then the above inequality implies

$$E|\tilde{\eta}_t(x) - \tilde{\eta}_s(y)|^p \le C_{p,T}(|x-y|^p + |t-s|^{\frac{p}{2}}), \quad x \neq 0,\ y \neq 0$$

In case $y = 0$, we have

$$E|\tilde{\eta}_t(x)|^p \leq C_{p,T}|x|^p.$$

Therefore $\tilde{\eta}_t(x)$ is continuous in $[0,\infty) \times R^d$ by Kolmogorov's theorem. This proves that $\eta_t(x)$ is continuous in $[0,\infty) \times$ neighborhood of infinity.

Proof of "onto" property (Varadhan). Define a stochastic process $\hat{\xi}_t$ on $\hat{R}^d = R^d \cup \{\infty\}$ by

$$\hat{\xi}_t(x) = \begin{cases} \xi_t(x) & \text{if } x \in R^d \\ \infty & \text{if } x = \infty \end{cases}$$

Then $\hat{\xi}_t(x)$ is continuous in $[0,\infty) \times \hat{R}^d$ by the previous lemma. Thus for each $t > 0$, the map $\hat{\xi}_t(\cdot,\omega)$ is homotopic to the identity map on $\hat{R}^d$, which is homeomorphic to d-dimensional sphere S^d. Then $\hat{\xi}_t(\cdot,\omega)$ is an onto map of $\hat{R}^d$ by a well known homotopic theory.

Now the map $\hat{\xi}_t$ is a homeomorphism of $\hat{R}^d$, since it is one to one, onto and continuous. Since ∞ is the invariant point of the map $\hat{\xi}_t$, we see that ξ_t is a homeomorphism of R^d. This completes the proof of Theorem 1.2.

2. Smoothness of the solution

In the previous section, we have seen that the solution $\xi_t(\cdot,\omega)$ of Ito SDE (1.1) is a homeomorphism for all t, provided that coefficients are Lipschitz continuous. We shall show in this section that if coefficients are of C^k-class, then $\xi_t(\cdot,\omega)$ is a C^{k-1}-diffeomorphism for any t. We first state

Proposition 2.1. (Blagoveščenskii and Freidlin [2]). Suppose that coefficients $X_1,\dots,X_r$ of equation (1.1) are of C^k-class and their derivatives are all bounded. Then $\xi_t(x,\omega)$ is of C^{k-1}-class for any t a.s. Furthermore Jacobian matrix $D\xi_t = (\frac{\partial \xi_t^i}{\partial x_j})$ satisfies

$$(2.1) \qquad D\xi_t(x) = I + \sum_{k=1}^{r}\int_0^t X_k'(\xi_s(x))D\xi_s(x)\,dM_s^k,$$

where I is the identity matrix and

$$X_k'(x) = (\frac{\partial X_k^i(x)}{\partial x_j})$$

Proof. Following [2], we will give the proof. Let $e_\ell = (0,\dots,0,1,0,\dots 0)$ (1 is the ℓ-th component) be a unit vector in R^d and let $x' = x + he_\ell$, where h is a non zero number. Set

$$\eta_t(x,x') = \frac{1}{h}(\xi_t(x') - \xi_t(x)).$$

Then it satisfies

$$(2.2) \qquad \eta_t(x,x') = e_\ell + \sum_k \int_0^t \frac{1}{h}(X_k(\xi_s(x')) - X_k(\xi_s(x)))\,dM_s^k$$

$$= e_\ell + \sum_k \int_0^t (\int_0^1 X_k'(\xi_s(x) + v(\xi_s(x') - \xi_s(x)))\,dv)\eta_s(x,x')\,dM_s^k$$

We may consider that $\zeta_t(x,x') = (\xi_t(x), \xi_t(x'), \eta_t(x,x'))$ is a R^{3d}-valued stochastic process with Lipschitz continuous coefficients. Then by Proposition 1.1, we have

$$E|\zeta_t(x,x') - \zeta_s(y,y')|^p \leq K_{p,T}^{(4)}(|x-y|^p + |x'-y'|^p + |t-s|^{\frac{p}{2}})$$

Therefore $\zeta_t(x,x')$ is continuous in (t,x,x') in $[0,T] \times \{(x,x')|x\neq x'\}$ and has a continuous extension to $[0,T] \times R^{2d}$. This proves that $\frac{\partial \xi_t}{\partial x_\ell}(x)$ exists and is continuous in (t,x) for almost all ω.

To get (2.1), make h tend to 0 in (2.2). Then we obtain

$$(2.3) \qquad \frac{\partial \xi_t(x)}{\partial x_\ell} = e_\ell + \sum_{k=1}^{r} \int_0^t X_k'(\xi_s(x)) \frac{\partial \xi_s(x)}{\partial x_\ell} \, dM_s^k .$$

This proves (2.1).

Consider next the SDE for the pair $(\xi_t(x), \frac{\partial \xi_t}{\partial x_\ell}(x))$. Coefficients of these equations are of C^{k-1}-class and their derivatives are bounded. We may apply the same argument to the pair. Then we see that $\frac{\partial \xi_t}{\partial x_\ell}$ have continuous derivatives $\frac{\partial^2 \xi_t}{\partial x_k \partial x_\ell}$ provided $k \geq 3$. Repeating this argument, we get the assertion of the proposition.

The smoothness of $\xi_t(\cdot,\omega)$ is a local property and boundedness of coefficients X_j and their derivatives are not needed. In fact if the boundedness is not satisfied, choose for each X_j a sequence of C^k-class functions $X_j^{(n)}$, n=1,2... such that $X_j^{(n)}(x) = X_j(x)$ if $|x| \leq n$ and derivatives of each $X_j^{(n)}$ are bounded. Then the solution $\xi_t^{(n)}(x,\omega)$ associated to $X_1^{(n)},\ldots,X_r^{(n)}$ coincides with $\xi_t(x,\omega)$ for

$t < \tau_n(x,\omega) \equiv \inf\{t > 0 : |\xi_t(x,\omega)| \geq n\}$. Therefore $\xi_t(x)$ is of C^{k-1}-class in $|x| \leq m < n$ if $\tau_n(x,\omega) > t$. Since $\tau_n(x)$ tend to infinity, we see that $\xi_t(x,\omega)$ is of C^{k-1}-class everywhere for any t.

It remains to prove the smoothness of the inverse map ξ_t^{-1}. We claim

Lemma 2.2. (c.f. Ikeda-Watanabe [8]) Matrix $D\xi_t(x)$ is non-singular for all $t > 0$ and x a.s.

Proof. We shall consider a matrix valued SDE for each x

$$(2.4) \qquad K_t = I - \sum_{k=1}^{r} \int_0^t K_s X_k'(\xi_s(x))dM_s^k$$
$$- \sum_{k,\ell} \int_0^t K_s X_k'(\xi_s(x))X_\ell'(\xi_s(x))d\langle M^k, M^\ell\rangle_s .$$

The solution K_t satisfies $K_t D\xi_t(x) = I$ for any t. In fact, by Ito's formula we have

$$(2.5) \qquad d(K_t D\xi_t(x)) = dK_t \cdot D\xi_t(x) + K_t dD\xi_t(x)$$
$$+ \sum_{k,\ell} K_s X_k' X_\ell' D\xi_s d\langle M^k, M^\ell\rangle_s .$$

Substitute (2.1) and (2.4) to (2.5), then we see that the right hand side of the above is 0. This proves $K_t D\xi_t(x) = I$, showing that $D\xi_t(x)$ is nonsingular.

Now the inverse mapping theorem states that $\xi_t(\cdot,\omega)^{-1}$ is again a C^{k-1}-class map. We have thus obtained the following theorem.

Theorem 2.3. Suppose that coefficients of equation (1.1) are of C^k-class and their first derivatives are bounded. Then the solution $\xi_t(\cdot,\omega)$ is a diffeomorphism of C^{k-1}-class for any t a.s.

In later discussion, we shall mainly concerned with Stratonovich SDE. Therefore it is convenient to get analogous results for Stratonovich SDE. Let us consider a SDE

$$(2.6) \qquad d\xi_t = \sum_{j=1}^{r} X_j(\xi_t)\circ dM_t^j ,$$

where the right hand side denotes the Stratonovich integral. If $X_1,\ldots,X_r$ are of C^2-class, the equation is written as the Ito SDE

$$(2.7) \qquad d\xi_t = \sum_{j=1}^{r} X_j(\xi_t)dM_t^j + \frac{1}{2}\sum_{\ell,j}\left(\sum_k X_\ell^k \frac{\partial}{\partial x_k} X_j\right)(\xi_t)d\langle M^\ell,M^j\rangle_t$$

Therefore if $X_1,\ldots,X_r$ together with their first and second derivatives are bounded, all coefficients of equation (2.7) are Lipschitz continuous. Then the solution $\xi_t(\cdot,\omega)$ is a homeomorphism for any t a.s. We then have

Theorem 2.4. Suppose that coefficients $X_1,\ldots,X_r$ of Stratonovich SDE (2,4) are of C^k-class $(k \geq 2)$. Suppose further that $X_1,\ldots,X_r$ together with their first and second derivatives are all bounded. Then the solution $\xi_t(\cdot,\omega)$ is a C^{k-2}-diffeomorphism for any t a.s. Here, C^0-diffeomorphism means a homeomorphism.

3. Case of Manifold

In this section, we shall consider SDE's on manifolds. Let M be a σ-compact, connected C^∞-manifold of dimension d. Let $X_1,\ldots,X_r$ be C^k-vector fields on M where $k \geq 2$ and $M_t^1,\ldots,M_t^r$ be continuous semimartingales. We shall consider SDE on the manifold M;

(3.1) $$d\xi_t = \sum_{j=1}^{r} X_j(\xi_t)\circ dM_t^j.$$

A sample continuous F_t-adapted process ξ_t, $0 \leq t < \tau$ with life time τ, taking values in M is called the solution of (3.1), if it satisfies

(3.2) $$f(\xi_t) = f(\xi_0) + \sum_{j=1}^{r}\int_0^t X_j f(\xi_s)\circ dM_s^j, \qquad 0 \leq t < \tau$$

for any C^∞-fucntion f on M. Here $\circ$ denotes the Stratonovich integral. Given a point x of M, the equation has a unique solution such that $\xi_0 = x$, up to the explosion time $\tau(x)$ (= life time). We denote it as $\xi_t(x)$ or $\xi_t(x,\omega)$ as before.

Let $(x^1,\dots,x^d)$ be a local coordinate in a coordinate neighborhood, and let $X_j = \sum_i X_j^i(x)\frac{\partial}{\partial x^i}$ be the representation of the vector field X_j. Then (3.1) is represented as

(3.3) $$d\xi_t^i = \sum_{j=1}^{r} X_j^i(\xi_t)\circ dM_t^j$$

Then by Theorem 2.4, $\xi_t(\cdot,\omega)$ is a C^{k-2}-map in the coordinate neighborhood.

We are interested in the global property of the map ξ_t. We first state

Lemma 3.1. Let $\tau(x,\omega)$ be the explosion time of $\xi_t(x,\omega)$.

Set

$$D_t(\omega) = \{x \mid \tau(x,\omega) > t\}.$$

Then it is an open set for any t a.s.

Proof. Let D_n, $n=1,2,\ldots$ be a sequence of domains in M with compact closure such that $\bigcup_n D_n = M$. Let $\tau_n(x,\omega)$ be the first leaving time of $\xi_t(x,\omega)$ from $\bar{D}_n$. Then for each t, the set $D_t^{(n)}(\omega) = \{x \mid \tau_n(x,\omega) > t\}$ is open a.s. In fact, take any point x_0 from the set. Then there is a neighborhood U of x_0 such that $\xi_t(x,\omega) \in D_n$ for all x of U, since $\xi_t(x,\omega)$ is continuous in x. Now, since

$$D_t(\omega) = \bigcup_n D_t^{(n)}(\omega),$$

we see that $D_t(\omega)$ is an open set.

Lemma 3.2. $\xi_t(\cdot,\omega)$ is an one to one map from $D_t(\omega)$ into M.

Proof. Let U_n, $n=1,2,\ldots$ be coordinate neighborhoods of M such that $\bigcup_n U_n = M$. Let S_m, $m=1,2,\ldots$ be a set of open time intervals generating all open sets in $[0,\infty)$. We denote by $N_{n,m}$ the set of all ω such that there are x, y $(x \neq y)$ of M and $\sigma(\omega) \in S_m$ such that $\xi_t(x) = \xi_t(y)$ for $t \geq \tau(\omega)$, $\xi_t(x) \neq \xi_t(y)$ for $t < \sigma(\omega)$ and $\xi_\sigma(x) = \xi_\sigma(y)$ is in U_n. In the coordinate neighborhood U_n, we see by Theorem 1.2 that $N_{n,m}$ is a null set. Therefore $\bigcup_{n,m} N_{n,m}$ is a null set. Note that if $\xi_t(\cdot,\omega)$ is not an one to one map for some t, then ω belongs to some $N_{n,m}$. The proof is complete.

Lemma 3.3. The map $\xi_t(\cdot,\omega)$ is a local C^{k-2}-diffeomorphism for any t a.s.

Proof. Consider a trajectory $\{\xi_s(x_0,\omega)\ ;\ 0 \le s \le t\}$ such that $\tau(x_0,\omega) > t$, where x_0, t and ω are fixed. We may choose coordinate neighborhoods $U_0,U_1,\ldots,U_n$ such that $x_0 \in U_0$ and for any $k = 1,\ldots,n$

$$\bigcup_{x\in U_0} \{\xi_s(x)\ :\ s \in [\tfrac{k-1}{n}t\ ,\ \tfrac{k}{n}t]\} \subset U_k .$$

We denote by $\xi_s^k(y)$, $s \ge \frac{k-1}{n}t$ the solution of (3.1) starting at y at time $\frac{k-1}{n}t$. Then it holds

$$\xi_t(x) = \xi_t^n \circ \xi^{n-1}_{\frac{n-1}{n}t} \circ \ldots \circ \xi^1_{\frac{1}{n}t}(x).$$

In each coordinate neighborhood U_k, Jacobian matrix of the map $\xi^k_{\frac{k}{n}t}$ is nonsingular by Lemma 2.2. Hence $\xi^k_{\frac{k}{n}t}$ is a local diffeomorphism for any k. Therefore ξ_t is again a local diffeomorphism.

Summing up these lemmas, we obtain

Theorem 3.4. The solution $\xi_t(\cdot,\omega)$ is a C^{k-2}-diffeomorphism form $D_t(\omega)$ into M for any t a.s.

The next problem we are concerned is to check that $D_t(\omega) = M$ and $\xi_t(\cdot,\omega)$ becomes an onto map. We shall consider three cases separately in the following Theorems 3.5, 3.6 and 3.8.

Theorem 3.5. (Elworthy [7]) If M is a compact manifold, tnen $\xi_t(\cdot,\omega)$ is a C^{k-2}-diffeomorphism of M for all t a.s.

Proof. We have $D_t(\omega) = M$ for all t a.s., since the life time of $\xi_t(\cdot,\omega)$ is infinite for all x a.s. Let $R_t(\omega)$ be the range of M by the map $\xi_t(\cdot,\omega)$. Since M is compact and $\xi_t(\cdot,\omega)$

is a continuous map, the set $R_t(\omega)$ is closed. Furthermore, $R_t(\omega)$ is open since Jacobian matrix of the map $\xi_t(x)$ is non singular for all t. Therefore $R_t(\omega)$ must be the whole space M because it is connected.

If the manifold is not compact, the problem appears complicated. The following theorem is comparable with Theorem 2.4 in Euclidean space.

Theorem 3.6. Let M be a non compact complete Riemannian manifold with non positive curvature. Suppose that $X_1,\ldots,X_r$ together with their first and second covariant derivatives along geodesics parameterized by the distance are bounded relative to the Riemannian norm. Then the solution $\xi_t(\cdot,\omega)$ is a C^{k-2}-diffeomorphism of M for any t a.s.

Proof. Let $\tilde{M}$ be the universal covering manifold of M. We may introduce a Riemannian metric on $\tilde{M}$, which is isometric to that of M. Let $\tilde{X}_1,\ldots,\tilde{X}_r$ be vector fields on $\tilde{M}$ such that $p_*\tilde{X}_j$[1)] $= X_j$, $j=1,\ldots,r$ holds, where p is the covering projection. Consider a SDE on $\tilde{M}$;

$$(3.4) \qquad d\tilde{\xi}_t = \sum_{j=1}^{r} \tilde{X}_j(\tilde{\xi}_t)\circ dM_t^1.$$

Then the projection $\xi_t = p(\tilde{\xi}_t)$ satisfies the SDE (3.1). Therefore if $\tilde{\xi}_t$ is a diffeomorphism of $\tilde{M}$, then ξ_t is a diffeomorphism of M. Thus it is enough to prove the theorem in case that M is a simply connected Riemannian manifold with non positive curvature.

Now let us introduce a global normal coordinate to the manifold mentioned above. Take a point x_0 in M and regard it as the origin.

1) p_* is the differential of the map p.

Let $T_{x_0}(M)$ be the tangent space at x_0 and let X be an element of $T_{x_0}(M)$. The geodesic $\gamma(t)$ such that $\gamma(0) = x_0$ and $\dot{\gamma}(t)|_{t=0} = X$ is denoted by $\exp_{x_0}(tX)$. Then $\gamma(1) = \exp_{x_0}(X)$ is a map from $T_{x_0}(M)$ into M. Under our assumption on the manifold M, it is known that $\exp_{x_0}$ is a diffeomorphism from $T_{x_0}(M)$ to M. Furthermore, $\exp_{x_0}$ increases the distance, i.e., $\|X\| \leq \|(\exp_{x_0})_* X\|$ holds for any tangent vector X of $T_{x_0}(M)$. (Kobayashi-Nomizu [12], p.103). Given an orthonormal basis $Y_1,\ldots,Y_d$ of $T_{x_0}(M)$, the associated Cartesian coordinate on $T_{x_0}(M)$ induces naturally a global coordinate on M, called a normal coordinate: The coordinate $(x^1,\ldots,x^d)$ of a point x of M is defined by the relation $x = \exp_{x_0}(\sum_i x^i Y_i)$.

Let $X_1,\ldots,X_r$ be vector fields of (3.1) and $X_j = \sum_i X_j^i(x)\frac{\partial}{\partial x^i}$ be the coordinate expression. Then it holds $(\sum_i X_j^i(x)^2)^{1/2} \leq \|X_j(x)\|$ since $\exp_{x_0}$ increases the distance. Therefore $X_j^i(x)$ are bounded in x. Furthermore, the next lemma shows that $\frac{\partial}{\partial x^k} X_j^i(x)$ and $\frac{\partial^2}{\partial x^\ell \partial x^k} X_j^i(x)$ are also bounded in x.

Now using the above normal coordinate, equation (3.1) is written as

$$(3.5) \qquad d\xi_t^i = \sum_{j=1}^{r} X_j^i(\xi_t) \circ dM_t^j$$

Then the assertion of the theorem follows from Theorem 2.4.

Lemma 3.7. Let X be a vector field on M such that the norm of the covariant derivatives along geodesics parameterized by the distance is bounded. Let $X = \sum X^i(x)\frac{\partial}{\partial x^i}$ be the expression relative to the normal coordinate. Then $\frac{\partial}{\partial x^k} X^i(x)$ are bounded in x for all i and k.

Proof. Let us fix an arbitrary point y_0 of M and introduce a normal coordinate $(y^1,\ldots,y^d)$ with the origin y_0 by the relation $x = \exp_{y_0}(\sum_i y^i(x)(\frac{\partial}{\partial x^i})_{y_0})$. Then it holds $(\frac{\partial}{\partial y^i})_{y_0} = (\frac{\partial}{\partial x^i})_{y_0}$. Furthermore we have $\| (\frac{\partial}{\partial y^i})_{y_0} \| \geq 1$ since $\exp_{y_0}$ increases the distance.

Now let $\gamma(t)$ be a geodesic such that $\gamma(0) = y_0$ and

$$\dot{\gamma}(t)|_{t=0} = \| (\frac{\partial}{\partial y^k})_{y_0} \|^{-1} (\frac{\partial}{\partial y^k})_{y_0}.$$

The covariant derivative $\nabla_{\dot{\gamma}(t)} X$ at y_0 is written as

$$\nabla_{\dot{\gamma}(t)} X(y_0) = \| (\frac{\partial}{\partial y^k})_{y_0} \| \sum_i X^i_{,k}(y_0) (\frac{\partial}{\partial y^i})_{y_0}$$

Since $\exp_{x_0}$ increases the distance, we have

$$(3.6) \qquad \|\nabla_{\dot{\gamma}(t)} X(y_0)\| \geq \| (\frac{\partial}{\partial y^k})_{y_0} \| (\sum_i X^i_{,k}(y_0)^2)^{\frac{1}{2}} \geq (\sum_i X^i_{,k}(y_0)^2)^{\frac{1}{2}}$$

Note that

$$X^i_{,k}(y_0) = \frac{\partial}{\partial y^k} X^i(y_0) - \sum_j \Gamma^i_{j\,k}(y_0) X^j(y_0)$$

and $\Gamma^i_{j\,k}(y_0) = 0$ since $(y^1,\ldots,y^d)$ is a normal coordinate with origin y_0. Then we have $X^i_{,k}(y_0) = \frac{\partial}{\partial x^k} X^i(y_0)$. Inequality (3.6) implies that $\frac{\partial}{\partial x^k} X^i(y_0)$ is bounded in y_0.

Remark. If the sectional curvature of a connected complete Riemannian manifold is greater than a positive number, then the manifold is compact. Hence the solution of (3.1) is always a flow of diffeomorphisms

We shall finally consider the equation (3.1) when the Lie algebra generated by vector fields $X_1,\dots,X_r$ is of finite dimension. We will not assume any condition to the manifold where the equation (3.1) is defined.

For two vector fields X, Y, we define the Lie bracket $[X,Y]$ as $XY - YX$. It is again a vector field. The Lie algebra generated by vector fields $X_1,\dots,X_r$ is the linear span of vector fields $[X_{i_1}[X_{i_2}[\dots [X_{i_{n-1}},X_{i_n}]\dots]$, $n=1,2,\dots$, where $i_1,\dots,i_n \in \{1,2,\dots,r\}$. We denote it as L.

Theorem 3.8. Suppose that $X_1,\dots,X_r$ are complete C^∞-vector fields and that the Lie algebra generated by them is of finite dimension. Then the solution $\xi_t(x,\omega)$ of SDE (3.1) is conservative and is a C^∞-diffeomorphism of M for any $t > 0$ a.s. ω.

Proof. We need a fact from differential geometry. It is known (e.g. Palais [17]) that any element of L is complete and that there exists a Lie group G with properties (i)-(iii) below: (i) G is a Lie transformation group of M, i.e. there esists a C^∞-map ϕ from the product manifold $G \times M$ into M such that (a) for each g $\phi(g,\cdot)$ is a diffeomorphism of M and (b) $\phi(e,\cdot) =$ identity, $\phi(gh,\cdot) = \phi(g,\phi(h,\cdot))$ for any g, h of G. (ii) The map $g \longrightarrow \phi(g,\cdot)$ is an isomorphism from G into the group of all diffeomorphisms of M. (iii) Let $\underline{\underline{G}}$ be the Lie algebra of G (= right invaraiant vector fields). For any X of L there exists $\hat{X}$ of $\underline{\underline{G}}$ such that

$$(3.7) \qquad \hat{X}(f\circ\phi_x)(g) = Xf(\phi(g,x))$$

holds for any C^∞-function f on M. Here $f\circ\phi_x$ is a C^∞-function on G such that $f\circ\phi_x(g) = f\circ\phi(g,x)$.

Now let $\hat{X}_j$ $(j=1,\ldots,r)$ be elements of $\underline{\underline{G}}$ relating to X_j by the formula (3.7). Consider SDE on G

$$(3.8) \qquad d\hat{\xi}_t = \sum_j \hat{X}_j(\hat{\xi}_t)\circ dM_t^j$$

If $(M_t^1,\ldots,M_t^r)$ is a Brownian motion, the solution $\hat{\xi}_t$ is so called a Brownian motion on Lie group G. Ito has shown that it is conservative [9]. His argument can be applied to the above (3.8), provided that M_t^j, $j=1,\ldots,r$ satisfies property (1.4). Then the conservativeness for general M_t^j, $j=1,\ldots,r$ can be proved by the method of time change, as we have stated in Section 1.

Set $\xi_t(x,\omega) = \phi(\hat{\xi}_t(e),x)$, where e is the unit of G. Then for each (t,ω), $\xi_t(\cdot,\omega)$ is a diffeomorphsim. We have

$$\begin{aligned} f(\xi_t(x)) &= f\circ\phi(\hat{\xi}_t(e),x) \\ &= f\circ\phi(e,x) + \sum_j \int_0^t \hat{X}_j(f\circ\phi_x)(\hat{\xi}_s(e))\circ dM_s^j \\ &= f(x) + \sum_j \int_0^t X_j f(\xi_s(x))\circ dM_s^j \end{aligned}$$

Therefore ξ_t is a solution of (3.1). The proof is complete.

4. Decomposition of solutions

Consider a Stratonovich SDE on a manifold M:

(4.1) $$d\xi_t = \sum_{j=1}^{r} X_j(\xi_t)\circ dM_t^j.$$

We shall assume from now that vector fields $X_1,\ldots,X_r$ are of C^∞ for simplicity. The solution $\xi_t(x)$ is a functional of vector fields $X_1,\ldots,X_r$ and paths $M_s^1,\ldots,M_s^r$, $0 \leq s \leq t$, obviously. We are interested how the functional is written explicitly. We begins with a simple case. The following proposition is more or less known.

Proposition 4.1. Suppose that $X_1,\ldots,X_r$ are complete vector fields and commutative each other. Then the solution of (4.1) is represented as

(4.2) $$\xi_t(x) = \mathrm{Exp}\, M_t^1 X_1\circ \ldots \circ \mathrm{Exp}\, M_t^r X_r(x),$$

where $\mathrm{Exp}\, sX_i$, $-\infty < s < \infty$ is the one parameter group of transformations generated by X_i. Here $\circ$ denotes the composition of maps.

Proof. For simplicity, we only consider the case $r = 2$. Let f be a C^∞-function on M. Set

$$F(t_1,t_2) = f(\mathrm{Exp}\, t_1X_1\circ \mathrm{Exp}\, t_2X_2(x)).$$

(x is fixed). By Ito's formula ralative to Stratonovich integral (c.f. [10]),

$$(4.3)\qquad F(M_t^1, M_t^2) = F(0, 0) + \int_0^t \frac{\partial F}{\partial t_1}(M_s^1, M_s^2)\circ dM_s^1 + \int_0^t \frac{\partial F}{\partial t_2}(M_s^1, M_s^2)\circ dM_s^2 .$$

Since $X_1X_2 = X_2X_1$, it holds $\mathrm{Exp}\ t_1X_1\circ\mathrm{Exp}\ t_2X_2 = \mathrm{Exp}\ t_2X_2\circ\mathrm{Exp}\ t_1X_1$. Therefore it holds

$$\frac{\partial F}{\partial t_1} = X_1 f(\mathrm{Exp}\ t_1X_1\circ\mathrm{Exp}\ t_2X_2(x))$$

$$\frac{\partial F}{\partial t_2} = X_2 f(\mathrm{Exp}\ t_1X_1\circ\mathrm{Exp}\ t_2X_2(x)).$$

Set $\xi_t(x) = \mathrm{Exp}\ M_t^1X_1\circ\mathrm{Exp}\ M_t^2X_2(x)$. Then the equality (4.3) is written as

$$f(\xi_t(x)) = f(x) + \int_0^t X_1 f(\xi_s)\circ dM_s^1 + \int_0^t X_2 f(\xi_s)\circ dM_s^2$$

Therefore, ξ_t is a solution of (4.1).

The proposition shows that the equation (4.1) is decomposed to r equations

$$d\xi_t^j = X_j(\xi_t)\circ dM_t^j , \qquad j=1,\dots,r$$

and the solution is the composition of solutions of these r equations. Obviously this is not the case if vector fields $X_1,\dots,X_r$ are not commutative. In the following, we shall discuss the decomposition problem in general settings. A similar decomposition has been discussed in [13] under some restricted framework. A basic tool is the differential of the solution map $\xi_t(\cdot,\omega)$.

Let $\xi_t(\cdot,\omega)$ be the solution of (4.1) with life time $\tau(x,\omega)$. Set $D_t(\omega) = \{x \mid \tau(x,\omega) > t\}$. It is the domain of the map $\xi_t(\cdot,\omega)$. Denote the range of the map $\xi_t(\cdot,\omega)$ as $R_t(\omega)$. Then ξ_t is a diffeomorphism from $D_t(\omega)$ onto $R_t(\omega)$. Given a point x of $D_t(\omega)$, the differential $(\xi_{t*})_x$ of the map ξ_t is defined as a linear map from $T_x(M)$ to $T_{\xi_t(x)}(M)$ such that

$$(\xi_{t*})_x X_x f = X_x(f\circ\xi_t) \qquad \forall X_x \in T_x(M).$$

Given a vector field X on M, we denote by X_x the restriction of X at the point $x \in M$. We define a new vector field $(\xi_{t*})(X)$ on R_t by

$$\xi_{t*}(X)_x = (\xi_{t*})_{\xi_t^{-1}(x)} X_{\xi_t^{-1}(x)}, \qquad x \in R_t$$

Then it holds

$$\xi_{t*}(X)f(x) = X(f\circ\xi_t)(\xi_t^{-1}(x))$$

for any C^∞-function f on M.

Let $(x^1,\dots,x^d)$ be a local coordinate. Taking $f(x) = x^i$ above, we see that the i-th component of $(\xi_{t*})(X)$ ralative to the coordinate is

$$\xi_{t*}(X)^i(x) = \sum_k X^k(\xi_t^{-1}(x))\left(\frac{\partial}{\partial x^k}\xi_t^i\right)(\xi_t^{-1}(x))$$

Hence, denoting Jacobian matrix $(\frac{\partial}{\partial x^k}\xi_t^i(x))$ as $D\xi_t(x)$, the vector $\xi_{t*}(X)(x)$ with components $\xi_{t*}(X)^i(x)$ is

$$\xi_{t*}(X)(x) = (D\xi_t \cdot X)(\xi_t^{-1}(x)).$$

Now let $(\xi_{t*})^{-1}$ be the inverse of ξ_{t*}. The vector field $(\xi_{t*})^{-1}(X)$ is then defined on D_t as

$$\xi_{t*}^{-1}(X)_x = (\xi_{t*})_x^{-1} X_{\xi_t(x)}, \qquad \forall x \in D_t.$$

Then it holds

$$\xi_{t*}^{-1}(X)f(x) = X(f\circ\xi_t^{-1})(\xi_t(x))$$

for any C^∞-function f on M. With a local coordinate $(x^1,\dots,x^d)$, we have

$$\xi_{t*}^{-1}(X)(x) = (D\xi_t(x))^{-1}X(\xi_t(x))$$

Remark. If X is commuting to all $X_1,\dots,X_r$, then $\xi_{t*}(X) = X$ and $\xi_{t*}^{-1}(X) = X$ hold. These properties follow from Proposition 5.2 and 5.3 of the next section.

Suppose now we are given other C^∞-vector fields $Y_1,\dots,Y_s$ on M and continuous semimartingales $N_t^1,\dots,N_t^s$. Consider SDE

$$d\zeta_t = \sum_k \xi_{t*}^{-1}(Y_k)(\zeta_t)\circ dN_t^k. \tag{4.4}$$

A sample continuous stochastic process $\zeta_t(x)$ on M with life time $\sigma(x)$ is called a solution of (4.4) if $\zeta_t(x)$ is in D_t for all $t < \sigma(x)$ and satisfies

$$f(\zeta_t(x)) = f(x) + \sum_{k=1}^{s}\int_0^t \xi_{u*}^{-1}(Y_k)f(\zeta_u(x))\circ dN_u^k$$

for all C^∞-function on M. Then we have

$$\begin{aligned} f(\zeta_t(x)) &= f(x) + \sum_{k=1}^{s}\int_0^t \xi_{u*}^{-1}(Y_k)f(\zeta_u(x))\circ dN_u^k \\ &= f(x) + \sum_{k=1}^{s}\int_0^t Y_k(f\circ\xi_u^{-1})(\xi_u\circ\zeta_u(x))\circ dN_u^k. \end{aligned}$$

We shall first obtain SDE governing the composition map $\xi_t\circ\zeta_t(x)$.

Proposition 4.2. The composition map $\eta_t(x) \equiv \xi_t\circ\zeta_t(x)$, $t \in [0,\sigma(x))$ satisfies SDE

$$\text{(4.5)} \qquad d\eta_t = \sum_j X_j(\eta_t)\circ dM_t^j + \sum_k Y_k(\eta_t)\circ dN_t^k .$$

Proof. We shall apply an extended Ito's formula [14]. Let f be a C^∞-function on M and let $F_t(x) = f\circ\xi_t(x)$. Using a local coordinate $(x^1,\dots,x^d)$, we shall write ζ_t as $(\zeta_t^1,\dots,\zeta_t^d)$. Since

$$dF_t(x) = \sum_j X_j f(\xi_t(x))\circ dM_t^j$$

we have by Theorem 1.2 of [14],

$$(4.6) \qquad dF_t(\zeta_t(x)) = \sum_j X_j f(\xi_t\circ\zeta_t(x))\circ dM_t^j + \sum_i \frac{\partial F_t}{\partial x^i}(\zeta_t(x))\circ d\zeta_t^i .$$

The second term of the right hand side equals

$$\sum_{i,k} \frac{\partial(f\circ\xi_t)}{\partial x^i}(\zeta_t(x))\xi_{t*}^{-1}(Y_k)^i(\zeta_t(x))\circ dN_t^k ,$$

where $\xi_{t*}^{-1}(Y_k)^i$ is the i-th component of the vector field $\xi_{t*}^{-1}(Y_k)$ relative to the local coordinate $(x^1,\dots,x^d)$. The above is equal to

$$\sum_k \xi_{t*}^{-1}(Y_k)(f\circ\xi_t)(\zeta_t(x))\circ dN_t^k$$

$$= \sum_k Y_k f(\xi_t\circ\zeta_t(x))\circ dN_t^k .$$

Hence (4.6) is written as

$$df(\eta_t(x)) = \sum_j X_j f(\eta_t(x))\circ dM_t^j + \sum_k Y_k f(\eta_t(x))\circ dN_t^k .$$

The proof is complete.

Remark. Instead of (4.4), consider

$$(4.7) \qquad d\kappa_t = \sum_k Y_k(\kappa_t)\circ dN_t^k .$$

Then the composition $\lambda_t \equiv \xi_t\circ\kappa_t$ satisfies the equation

$$(4.8) \qquad d\lambda_t = \sum_j X_j(\lambda_t)\circ dM_t^j + \sum_k \xi_{t*}(Y_k)(\lambda_t)\circ dN_t^k$$

This can be proved analogously as Proposition 4.2.

We can now get the decomposition of solution of (4.1)

Theorem 4.3. Consider two SDE's

$$(4.9)\qquad d\zeta_t = \sum_{j=1}^{r} Y_j(\zeta_t)\circ dM_t^j$$

$$(4.10)\qquad d\eta_t = \sum_{j=1}^{r} \zeta_{t*}^{-1}(Z_j)(\eta_t)\circ dM_t^j.$$

If $X_j = Y_j + Z_j$, $j=1,\ldots,r$ hold, the composition $\xi_t(x) \equiv \zeta_t\circ\eta_t(x)$, $0 \leq t < \sigma(x)$ is a solution of (4.1). Furthermore, if both of $\xi_t(x)$ and $\eta_t(x)$ are flows of diffeomorphisms, then so is $\eta_t(x)$ and the composition $\xi_t = \zeta_t\circ\eta_t$ is the solution of (4.1) for all $t \geq 0$ a.s.

Proof. The first half of the theorem is immediate from Proposition 4.2. The second half will be obvious.

Corollary. If Z_j, $j=1,\ldots,r$ are commutative to all $Y_1,\ldots,Y_r$ then η_t of the theorem is determined by

$$(4.11)\qquad d\eta_t = \sum_{j=1}^{r} Z_j(\eta_t)\circ dM_t^j .$$

Proof. Since $[Y_j,Z_k] = 0$, we have $\zeta_{t*}^{-1}(Z_k) = Z_k$ by Proposition 5.2, which will be established at the next section.

A typical example of the decomposition of the solution is that of linear SDE on R^d;

$$d\xi_t = A\xi_t dt + BdW_t,$$

where A is a d×d-matrix, B is a d×r-matrix and W_t is a r-dimensional Wiener process. The equation is decomposed to

$$d\zeta_t = A\zeta_t dt, \qquad d\eta_t = \zeta_{t*}^{-1}(B)dW_t$$

Clearly we have $\zeta_t(x) = e^{At}x$. Then

$$\zeta_{t*}^{-1}B = (D\zeta_t)^{-1}B = e^{-At}B.$$

Consequently, $\eta_t(x) = x + \int_0^t e^{-As}BdW_s$. We have thus the decomposition

$$\xi_t(x) = e^{At}(x + \int_0^t e^{-As}BdW_s).$$

Some other examples of decompositions are found in [13].

We will mention that the technique of the decomposition is used in filtering theory in order to get a "robust" solution (c.f. Doss [6], Clark [4] and Davis [5]). Consider a SDE on R^d

$$d\xi_t = X_0(\xi_t)dt + \sum_{j=1}^{r} X_j(\xi_t)dW_t^j,$$

where $W_t = (W_t^1,\dots,W_t^r)$ is a Wiener process. Suppose that $X_1,\dots,X_r$ are commuting each other, but they are not commuting with X_0. Decompose the equation to

$$d\zeta_t = \sum_{j=1}^{r} X_j(\zeta_t)\circ dM_t^j, \qquad d\eta_t = \zeta_{t*}^{-1}(X_0)(\eta_t)dt.$$

Then, it holds $\zeta_t(x) = \mathrm{Exp}\, W_t^1 X_1 \circ \dots \circ \mathrm{Exp}\, W_t^r X_r$.

Then $\zeta_t(x,\omega)$ and its Jacobian matrix $D\zeta_t(x,\omega)$ are locally Lipschitz continuous with respect to the Wiener process W_t, i.e., for each $T > 0$ and $N > 0$ there is a positive constant $K_{N,T}$ such that

$$|\zeta_t(x,\omega) - \zeta_t(x,\tilde{\omega})| \leq K_{N,T} \| W.(\omega) - W.(\tilde{\omega}) \|_T$$

if $\| W.(\omega)\|_T \leq N$ and $\| W.(\tilde{\omega})\|_T \leq N$, where $\| W.(\omega)\|_T = \sup_{0\leq s\leq T} |W_s(\omega)|$. Consequently the vector field $\zeta_{t*}^{-1}(X_0)(\omega)$ is also locally Lipschitz continuous with respect to $W_t(\omega)$. Then the solution $\eta_t(x)$ of the ordinary differential equation $\frac{d\eta_t}{dt} = \zeta_{t*}^{-1}(X_0)$ is also locally Lipschitz continuous with respect to the parameter $W.(\omega)$. This implies that the solution $\xi_t(x)$ itself is locally Lipschitz continuous with respect to $W.(\omega)$.

5. SDE for differentials ξ_{t*}^{-1} and ξ_{t*}

The decomposition theorem of the solution can be applied in some cases for getting the explicit expression of the solution. The problem will be considered at the next section in case that the Lie algebra generated by $X_1,\dots,X_r$ is solvable. There the exact form of the differential ξ_{t*}^{-1} will be required. In this section we shall obtain a SDE governing ξ_{t*}^{-1} and ξ_{t*}.

We assume in this section that the solution $\xi_t(\cdot, \omega)$ of equation (4.1) is a flow of diffeomorphisms. We first obtain the equation govering the inverse map ξ_t^{-1}.

Proposition 5.1. ξ_t^{-1} satisfies

(5.1) $$d\xi_t^{-1} = -\sum_j \xi_{t*}^{-1}(X_j)(\xi_t^{-1}) \circ dM_t^j$$

Proof. Set $Y_k = -X_k$ and $N_t^k = M_t^k$, $k=1,\ldots,r$ in (4.4). Then $\eta_t \equiv \xi_t \circ \zeta_t$ satisfies $d\eta_t = 0$ by Proposition 4.2. Therefore it holds $\xi_t \circ \zeta_t(x) = x$ for all x a.s. This proves $\zeta_t = \xi_t^{-1}$. Hence ξ_t^{-1} satisfies (5.1).

We shall next obtain SDE for differential ξ_{t*}^{-1}.

Proposition 5.2. It holds

(5.2) $$\xi_{t*}^{-1}(X) = X + \sum_j \int_0^t \xi_{s*}^{-1}([X_j, X]) \circ dM_s^j$$

or simply

(5.3) $$\xi_{t*}^{-1} = I + \sum_j \int_0^t \xi_{s*}^{-1} \mathrm{ad}_{X_j} \circ dM_s^j ,$$

where I is the identity and ad_{X_j} is a linear map defined by $\mathrm{ad}_{X_j} X = [X_j, X]$.

Proof. Let f be a C^∞-function on M. Since

$$f(\xi_t^{-1}(x)) = f(x) - \sum_{j=1}^r \int_0^t \xi_{s*}^{-1}(X_j) f(\xi_s^{-1}(x)) \circ dM_s^j$$

holds by (5.1), we have

$$X(f \circ \xi_t^{-1})(x) = Xf(x) - \sum_{j=1}^r \int_0^t X\{(\xi_{s*}^{-1}(X_j) f) \circ \xi_s^{-1}\}(x) \circ dM_s^j .$$

Set $F_t(x) = X(f\circ\xi_t^{-1})(x)$ and apply an extended Ito's formula in [14] to $F_t(\xi_t(x))$. Then

$$\xi_{t*}^{-1}(X)f(x) = Xf(x) - \sum_j \int_0^t \xi_{s*}^{-1}(X)\xi_{s*}^{-1}(X_j)f(x)\circ dM_s^j$$
$$+ \sum_i \int_0^t \frac{\partial F_s}{\partial x^i}\circ d\xi_s^i$$

The last member equals

$$\sum_{i,j}\int_0^t \frac{\partial X(f\circ\xi_s^{-1})}{\partial x_i}(\xi_s(x))\, X_j^i(\xi_s(x))\circ dM_s^j$$
$$= \sum_j \int_0^t \xi_{s*}^{-1}(X_j)\xi_s^{-1}(X)f(x)\circ dM_s^j$$

Note that $[\xi_{s*}^{-1}(X_j),\ \xi_{s*}^{-1}(X)] = \xi_{s*}^{-1}([X_j,\ X])$. Then we get

$$\xi_{t*}^{-1}(X)f(x) = Xf(x) + \sum_j \int_0^t \xi_{s*}^{-1}([X_j,\ X])f(x)\circ dM_s^j\ .$$

This proves (5.2) or (5.3). The proof is complete.

Corollary. Suppose that the Lie algebra L generated by $X_1,\ldots,X_r$ is of finite dimension. If X is in L, then $\xi_{t*}^{-1}(X)$ is in L for any t and a.s. ω.

Proof. The equation (5.3) may be considered as an equation for the linear map on L, since ad_{X_j} is a linear map on L. Then the solution ξ_{t*}^{-1} maps L into itself. The proof is complete.

Similarly as above, we may obtain an equation for ξ_{t*}

Proposition 5.3. It holds

$$(5.4) \qquad \xi_{t*} = I - \sum_j \int_0^t \mathrm{ad}_{X_j} \xi_{s*} \circ dM_s^j$$

Proof. Set $F_t(x) = X(f \circ \xi_t)(x)$. Then

$$F_t(x) = Xf(x) + \int_0^t \sum_j XX_j f(\xi_s(x)) \circ dM_s^j$$

Apply an extended Ito's formula in [14] to $F_t(\xi_t^{-1}(x))$ $(= \xi_{t*}(X)f(x))$. Then by a similar calculation, we get

$$\xi_{t*}(X)f(x) = Xf(x) + \sum_j \int_0^t [\xi_{s*}(X),\ X_j] \circ dM_s^j$$

This implies (5.4).

6. Representation of solution. Solvable case

As an application of the decomposition of the solution, we shall obtain an exact solution of the equation (4.1) when the Lie algebra generated by $X_1,\dots,X_r$ is solvable.

We shall define a chain of ideals of L by $L_1 = [L, L]$, $L_2 = [L_1, L_1],\dots, L_m = [L_{m-1}, L_{m-1}]$. The Lie algebra L is called solvable if $L_m = \{0\}$ for some m. If L is a finite dimensional solvable Lie algebra, then by Lie's theorem there is a basis of L denoted by $\{Y_1,\dots,Y_n\}$ with following property: There is a sequence of ideals

$$L = L_0 \supset L_1 \supset \cdots \supset L_n = \{0\}$$

such that $Y_{i+1},\ldots,Y_n$ is a basis of the ideal L_i for each i.

Theorem 6.1. Suppose that $X_1,\ldots,X_r$ are all complete vector fields and generate a finite dimensional solvable Lie algebra L. Let $\{Y_1,\ldots,Y_n\}$ be a basis of L mentioned above. Then the solution of (4.1) is represented as

$$(6.1) \qquad \xi_t(x) = \mathrm{Exp}\, N_t^1 Y_1 \circ \mathrm{Exp}\, N_t^2 Y_2 \circ \cdots \circ \mathrm{Exp}\, N_t^n Y_n(x),$$

where $N_t^1,\ldots,N_t^n$ are continuous semimartingales constructed from $M_t^1,\ldots,M_t^r$ through finite repetition of the following elementary calculations

(i) linear sums and products of $M_t^1,\ldots,M_t^r$.

(ii) Stratonovich-Fisk integrals based on $M_t^1,\ldots,M_t^r$.

(iii) substitution to the exponential function e^x.

Furthermore, if L is nilpotent, $N_t^1,\ldots,N_t^r$ are constructed via (i) and (ii) only.

Remark. The algorithm of calculating $N_t^1,\ldots,N_t^n$ will be found in the proof of the theorem. It is determined by the structure constants of Lie algebra L relative to the basis $\{Y_1,\ldots,Y_n\}$.

Before going to the proof, we remark a preliminary fact on solvable Lie algebra L. Let $\{Y_1,\ldots,Y_n\}$ be the basis of L mentioned above.

Let $Z \in L$. Then $ad_Z Y_i$ is in L_{i-1} so that it is written as $\sum_{j\geq i} c_{ij}Y_j$. Then $ad_Z = (c_{ij})$ is a triangular matrix such that $c_{ij} = 0$ if $j < i$. Furthermore if Z is in L_k, then $c_{ij} = 0$ if $j \leq k$. Hence

$$ad_{Y_k} = \left(\begin{array}{c|ccc} \overbrace{0}^{k} & \overbrace{* \quad\quad *}^{n-k} \\ \hline & * & & * \\ 0 & & \ddots & \\ & 0 & & * \end{array}\right) \begin{array}{l} \}\, k \\ \\ \}\, n-k \\ \\ \end{array}$$

Note that $\exp ad_{Y_k}$ is also a triangular matrix.

Proof of Theorem. Vector fields $X_1,\ldots,X_r$ are written as linear sums of $Y_1,\ldots,Y_n$, say $X_j = \sum_k a_{jk}Y_k$. Then SDE (4.1) is written as

$$d\xi_t = \sum_{k=1}^{n} Y_k(\xi_t)\circ d\tilde{M}_t^k ,$$

where $\tilde{M}_t^k = \sum_j a_{jk}M_t^j$. We write $\tilde{M}_t^k$ as M_t^k for simplicity.

Consider two SDE's

$$\text{(6.2)} \qquad d\zeta_t^{(1)} = Y_1(\zeta_t^{(1)})\circ dM_t^1 ,$$

$$\text{(6.3)} \qquad d\eta_t^{(1)} = \sum_{j=2}^{r} (\zeta_{t*}^{(1)})^{-1}(Y_j)\circ dM_t^j$$

The solution of (6.2) is written as $\zeta_t^{(1)}(x) = \mathrm{Exp}\, M_t^1 Y_1(x)$. By Theorem 4.3 it holds $\xi_t = \zeta_t^{(1)}\circ\eta_t^{(1)}$. Furthermore, we have by Proposition 5.2,

$$(\zeta_{t*}^{(1)})^{-1} = I + \int_0^t (\zeta_{s*}^{(1)})^{-1} ad_{Y_1} \circ dM_s^1 .$$

This leads to $(\zeta_{t*}^{(1)})^{-1} = \exp M_t^1 ad_{Y_1}$.

With vector notations $Y = (Y_1,\ldots,Y_n)$, $\hat{M}_t = (0, M_t^2,\ldots,M_t^n)$ and inner product $(\ ,\ )$, equation (6.3) is written as

$$(6.4) \qquad d\eta_t^{(1)} = ((\zeta_{t*}^{(1)})^{-1}(Y), \circ d\hat{M}_t) = (Y, \circ \exp M_t^1 ad'_{Y_1} \circ d\hat{M}_t),$$

where ad'_{Y_1} is the transpose of ad_{Y_1}. Define

$$(6.5) \qquad M_t^{(1)} = \int_0^t \exp M_s^1 ad'_{Y_1} \circ d\hat{M}_s .$$

Since $\exp M_s^1 ad'_{Y_1}$ is a triangular matrix, the first component of the vector $M_t^{(1)}$ is 0. Hence equation (6.4) becomes

$$(6.6) \qquad d\eta_t^{(1)} = (Y, \circ dM_t^{(1)}) = \sum_{j=2}^{n} Y_j(\eta_t^{(1)}) \circ dM_t^{(1)j} ,$$

where $M_t^{(1)} = (0, M_t^{(1)2},\ldots,M_t^{(1)n})$.

We shall define $\zeta_t^{(i)}$ and $\eta_t^{(i)}$ $i=2,\ldots,n$, by induction. Suppose that $\zeta_t^{(k)}$, $M_t^{(k)} = (0,\ldots, 0, M_t^{(k)k+1},\ldots,M_t^{(k)n})$ and η_t^k, $k=1,\ldots,i-1$ are well defined. Set

$$(6.7) \qquad \zeta_t^{(i)} = \operatorname{Exp} M_t^{(i-1)i} Y_i ,$$

$$(6.8) \qquad M_t^{(i)} = \int_0^t \exp M_s^{(i-1)i} ad'_{Y_i} \circ d\hat{M}_s^{(i-1)} ,$$

(6.9) $$d\eta_t^{(i)} = (Y,\circ dM_t^{(i)}),$$

where $\hat{M}_t^{(i-1)} = (0,\ldots, 0, M_t^{(i-1)i+1},\ldots,M_t^{(i)n})$. Then it holds $\eta_t^{(i-1)} = \zeta_t^{(i)} \circ \eta_t^{(i)}$. Since $\exp M_s^{(i-1)i} ad'_{Y_i}$ is triangular, the first i components of $M_t^{(i)}$ are 0. Hence (6.9) is written as

(6.10) $$d\eta_t^{(i)} = \sum_{\ell \geq i+1} Y_\ell \circ dM_t^{(i)\ell}$$

Consequently we obtain

(6.11) $$\xi_t = \zeta_t^{(1)} \circ \ldots \circ \zeta_t^{(n)}$$

$$= \text{Exp } M_t^1 Y_1 \circ \text{Exp } M_t^{(1)2} Y_2 \circ \ldots \circ \text{Exp } M_t^{(n-1)n} Y_n .$$

Clearly $M_t^1, M_t^{(1)2},\ldots,M_t^{(n-1)n}$ are constructed from $M_t^1,\ldots,M_t^r$ via (i) - (iii) of the theorem.

In case where L is a nilpotent Lie algebra, diagonal elements of matrix ad_{Y_i} are all 0. Therfore any component of matrix $\exp (M_s^{(i-1)i} ad'_{Y_i})$ is a polynomial of $M_s^{(i-1)i}$, not containing exponential functions. Therefore, $M_t^{(i)}$ are constructed from $M_t^1,\ldots,M_t^r$ via operations (i) and (ii) only. The proof is complete.

Remark. In case that the Lie algebra L is nilpotent, the assertion of the theorem states that $N_t^1,\ldots,N_t^r$ in the expression (6.1) are linear sums of multiple Wiener integrals of the form

$$\int_0^t \cdots \int_0^{t_{m-1}} \circ dM_{t_1}^{i_1} \circ \cdots \circ dM_{t_m}^{i_m}$$

This corresponds to a result by Yamato [19].

References

[1] J. M. Bismut; Flots stochastiques et formula de Ito-Stratonovich généralisée, C. R. Acad. Sci. Paris, t. 290 (10 mars 1980)

[2] Ju. N. Blagoveščenskii-M. I. Freidlin; Some properties of diffusion processes depending on a parameter, DAN 138 (1961), Soviet Math. 2 (1961), 633-636

[3] K. T. Chen; Decomposition of differential equations, Math. Annalen 146 (1962), 263-278

[4] J. M. C. Clark; The design of robust approximations to the stochastic differential equations of non linear filtering, in "Communication Systems and Random Process Theory" ed. by J. K. Skwirzynski, NATO Advanced Study Institute Series, Sijthoff and Noordhoff, Alphen aan den Rijn, 1978

[5] M. H. A. Davis; A pathwise solution of the equations of non linear filtering, 12th European Meeting of Statisticians, Varna, 1979

[6] H. Doss; Liens entre equations differentielles stochastiques et ordinaires, Ann. Inst. H. Poincare 13 (1979), 99-125

[7] K. D. Elworthy; Stochastic dynamical systems and their flows, Stochastic analysis ed. by A. Friedman and M. Pinsky, 79-95, Academic press, New York, 1978

[8] N. Ikeda-S. Watanabe; Stochastic differential equations and diffusion processes, forthcoming book.

[9] K. Itô; Brownian motions in a Lie group, Proc. Japan Acad., 8 (1950), 4-10

[10] K. Itô; Stochastic differentials, Apl. Math. Optimization 1, 374-384 (1975)

[11] N. Jacobson; Lie algebras, New York-London, Wiley, 1962

[12] S. Kobayashi-K.Nomizu; Fundations of differential geometry II, Interscience, 1969

[13] H. Kunita; On the representation of solutions of stochastic differential equations, Seminaire des Probabilites XIV, Lecture Note in Math., 784 (1980), 282-303

[14] H. Kunita; Some extensions of Ito's formula, Seminaire des Probabilites XV, Lecture Note in Math., to appear

[15] P. Malliavin; Stochastic calculus of variation and hypoelliptic operators, Kyoto, Conference, 1976, Wiley 1978, 195-263

[16] Y. Ogura-T. Yamada; On the strong comparison theorem of solutions of stochastic differential equations, to appear

[17] R. S. Palais; A global formulation of the Lie theory of transformation groups, Mem. Amer. Math. Soc. 22 (1957)

[18] D. W. Stroock-S. R. S. Varadhan; Multidimensional diffusion processes, 1979, Springer-Verlag, New York

[19] Y. Yamato; Stochastic differential equations and nilpotent Lie algebra, Z. W. 47 (1979), 213-229

A DIFFERENTIAL GEOMETRIC FORMALISM
FOR THE ITO CALCULUS

(P.A. Meyer)

This report doesn't contain any new result, just a partially new language to state old results. However, I think that this language is well adapted to the geometry of continuous semimartingales in manifolds, and may even have from time to time some interest for the non-probabilist.

We shall be concerned here only with continuous semimartingales and C^∞ manifolds, and everything below (except the semimartingales themselves !), all maps, fields, etc, will be assumed to be C^∞.

The Einstein convention that Σ_i over a diagonally repeated index i is omitted will be in force throughout this report.

The reader will notice at once the numerous references to related work by J.M. Bismut and L. Schwartz. My debt to them, particularly to Schwartz, is gratefully acknowledged. A set of lectures on the same material will appear (in French) in the volume Séminaire de Probabilités XV, edited by J. Azéma and M. Yor (Lecture Notes in Mathematics, 1981).

I. NOTATIONS

M is a manifold (C^∞) . To simplify things, I will even assume that M is an open set in $\mathbb{R}^n$ with its linear structure forgotten, and therefore has global coordinates (x^i). Given a point $a\in M$, $T_a(M)$ is the tangent space to M at a, and T(M) is the tangent bundle. So T(M) also has global coordinates (x^i,u^i) with $u^i=dx^i$. We denote by D_i the tangent vector $\partial/\partial x^i$, either at some fixed $a\in M$ or considered as a vector field on M.

We denote by $\tau_a(M)$ the space of all differential operators of order $\leqq 2$ at a, without a constant term. That is, every element of $\tau_a(M)$ is a mapping from $C^\infty(M)$ to $\mathbb{R}$

$$f \mapsto (\lambda^i D_i f + \lambda^{ij} D_{ij} f)_a$$

and this representation is unique if we assume – as we shall always do – that $\lambda^{ij}=\lambda^{ji}$. Since every differential operator of order 1 has order $\leqq 2$, we have $T_a(M) \subset \tau_a(M)$. Considering x^i,λ^i and λ^{ij} ($i\leqq j$) as coordinate mappings on $\tau(M)=\cup_a \tau_a(M)$, we turn $\tau(M)$ into a C^∞ bundle over M. Elements of $\tau_a(M)$ will be called <u>second order tangent vectors</u> to M at a.

If N is a manifold and $h : N \to M$ is a map, we have an obvious mapping $h_* : \tau(N) \to \tau(M)$, ($\tau_a(N) \to \tau_{h(a)}(M)$). For instance, let N be $\mathbb{R}$ and h be a <u>curve</u> in M . At each point t of $\mathbb{R}$ we have a second order tangent vector D^2, the second derivative. Then $h_*(D^2)$ is the second order tangent vector $\ddot{h}^i D_i + \dot{h}^i \dot{h}^j D_{ij}$ at h(t), which we denote by $\ddot{h}(t)$ and call the <u>acceleration</u> of the curve at time t.

2. THE PRINCIPLE OF SCHWARTZ

Let (X_t) be a (continuous) semimartingale with values in M - this just means that the real valued processes $X_t^i(\omega)$ are semimartingales, and this is invariant by a diffeomorphism of M . Consider also the canonical decompositions of the coordinate semimartingales

$$dX_t^i = dA_t^i + dM_t^i$$

where as usual A_t^i is a finite variation, adapted, continuous process, and M_t^i is a local martingale. Let f be a C^∞ function on M, and let Y_t be the real valued process $f(X_t)$. According to the Ito formula

(1) $$dY_t = D_i f(X_t) dX_t^i + \tfrac{1}{2} D_{ij} f(X_t) d\langle X^i, X^j\rangle_t$$

with decomposition $Y_t = B_t + N_t$ given by

(2) $$dB_t = D_i f(X_t) dA_t^i + \tfrac{1}{2} D_{ij} f(X_t) d\langle X^i, X^j\rangle_t$$

(3) $$dN_t = D_i f(X_t)(dX_t^i - dA_t^i)$$

The fundamental remark of Schwartz is the following : the left sides of these equalities are intrinsic, while the right sides apparently depend on the choice of coordinates. So the right sides are really invariant by a change of coordinates, meaning that formally

(4) $$dX_t^i D_i + \tfrac{1}{2} d\langle X^i, X^j\rangle_t D_{ij}$$

behaves like a second order tangent vector at X_t , the same is true for

(5) $$dA_t^i D_i + \tfrac{1}{2} d\langle X^i, X^j\rangle_t D_{ij}$$

and finally

(6) $$(dX_t^i - dA_t^i) D_i$$

behaves under changes of coordinates as a first order tangent vector. These are really mere shortcuts for computation, meaning that " any formula that you can prove in geometry for second order tangent vectors will have a probabilistic interpretation", since (4),(5),(6) aren't mathematically defined objects . However, (5) can be given a precise meaning : choose once and for all a scalar valued increasing process (C_t) - usually $C_t=t$ - and assume that $dA_t^i \ll dC_t$, $d\langle X^i, X^j\rangle_t \ll dC_t$ with densities respectively α_t^i, ξ_t^{ij} . Then you can consider the <u>true</u> second order tangent vectors $\alpha_t^i D_i + \frac{1}{2}\xi_t^{ij} D_{ij}$ at X_t . And now, if you consider another coordinate system, the same absolute continuity relations will take place, and your two systems of second order tangent vectors will be the same a.e.-dC_t .

It is convenient to call (5) the <u>local characteristics vector</u> of X at time t. I like to call (4) the <u>speed</u> (not the acceleration !) of X and to denote it by d^2X_t.

For instance, if (X_t) is a diffusion with generator $L=a^i D_i + a^{ij} D_{ij}$, the local characteristics vector at time t will be $L_{X_t}dt$. So this language amounts to considering all continuous semimartingales as diffusions, in a loose symbolic sense (but still helpful to intuition and computation).

3. MARTINGALES IN MANIFOLDS

The preceding section I have learnt from Schwartz, and this one I have learnt from Bismut. It turns out that the idea of a linear connection, possibly the most difficult to grasp for the ordinary student of differential geometry, has a simple probabilistic meaning .

Consider first a nice diffusion process X in $\mathbb{R}^n$, with generator L. Since $\mathbb{R}^n$ is a vector space, it has a meaning to ask whether X is a local martingale, and the easy answer is that X is a local martingale if and only if L is a pure second order operator, while the first order part represents a drift term. Now it isn't possible in a manifold to decompose intrinsically a second order operator into a "pure second order part" and a "drift term" : this requires a new structure, as follows :

DEFINITION. A linear connection at $a \in M$ is a linear mapping Γ from $\tau_a(M)$ (the second order operators) to $T_a(M)$ (the first order operators) which reduces to identity on $T_a(M)$. If $\lambda \in \tau_a(M)$, $\Gamma(\lambda)$ is called the drift of λ.

To know Γ, by linearity we need only know the values of $\Gamma(D_i)$ and $\Gamma(D_{ij})$. The first one is just D_i , and the second one can be written as

(7) $$\Gamma(D_{ij}) = \Gamma^k_{ij}(a)D_k$$

These coefficients are the Christoffel symbols of the connection. Since we have $D_{ij}=D_{ji}$, we also have $\Gamma^k_{ij}=\Gamma^k_{ji}$, and therefore our connections aren't quite the general ones considered by differential geometers : they are the torsion free connections. However, we shall just use the word connection without adjective.

A connection on M, of course, is a family of connections given at every point of M, with C^∞ Christoffel symbols. So if L is a differential operator on M (a field of second order tangent vectors), $\Gamma(L)$ is an ordinary vector field, and we have $\Gamma(fL)=f\Gamma(L)$ for $f \in C^\infty(M)$.

DEFINITION. A semimartingale X with values in M is a (local) martingale (w.r.to the connection Γ) if its local characteristics have no drift :

$$dA^i_t + \tfrac{1}{2}\Gamma^i_{jk}(X_t)d\langle X^j,X^k\rangle_t = 0 \quad , i=1,\dots,n$$

or equivalently

$$dX^i_t + \tfrac{1}{2}\Gamma^i_{jk}(X_t)d\langle X^j,X^k\rangle_t = \text{differential of a real valued local martingale.}$$

The word local is generally useless, since there are no martingales in the usual sense taking values in manifolds.

Example : The brownian motion of a riemannian manifold is the diffusion with generator $\frac{1}{2}\Delta$, where Δ is the Laplace-Beltrami operator. This operator is usually defined as "div grad ", but the following expression (which is just a little less well known) is more interesting for us

(10) $$\Delta = g^{ij}(D_{ij}-\Gamma^k_{ij}D_k)$$

So Δ has no drift, and brownian motion is a martingale. Brownian motion is more precisely an example of a conformal martingale, that is

(11) $$d\langle X^i,X^j\rangle_t = g^{ij}(X_t)dC_t$$

where (C_t) is a scalar valued increasing process.

COMMENT. The definition of connections we have given is known to differential geometers, but not currently taught (see Ambrose-Palais-Singer [1]). The standard definitions use either the covariant derivative point of view, or the horizontal subspace point of view. I will comment briefly on the first one, and at length on the second one (but this discussion will be for the most part deferred to section 8).

With our definition, the covariant derivative $\nabla_X Y$ of the field Y along the field X is just the field $\Gamma(XY)$, where XY is the second order operator arising from the composition of the first order operators X and Y.

To understand the horizontal subspace point of view, we must look at another "second order object", namely the second tangent bundle T(T(M)). Just as T(M) has 2n coordinates (x^i,u^i) with $u^i=dx^i$, TT(M) has 4n coordinates (x^i,u^i,v^i,w^i) with $v^i=\delta x^i$ and $w^i=\delta u^i$ (I distinguish by different letters the differentiation operators on functions on M and T(M) for clarity). Let now t be a tangent vector to T(M) at (x,u), with coordinates $(x^i,u^i\ ;\ v^i,w^i)$. We associate to it a second order tangent vector $\rho(t)$ at x as follows. We take a function f on M , and differentiate it, thus getting a function on T(M)

$$df = D_i f(x)u^i$$

Differentiate again and take the value at t :

$$\langle\delta df,t\rangle = D_i f(x)w^i + D_{ji}f(x)v^j u^i$$

Since the left side doesn't mention the coordinates, the right side is intrinsic too, and we set

$$\rho(t) = w^i D_i + u^i v^j D_{ji} \in \tau_x(M)$$

We say that t is horizontal if the drift of $\rho(t)$ is 0, that is

(12) $$(w^k + u^i v^j \Gamma^k_{ij})D_k = 0$$

Every tangent vector $v^i D_i$ to M at x has a unique horizontal lift at (x,u), namely the tangent vector with coordinates $(x^i,u^i,v^i,\ -u^j v^k \Gamma^i_{jk})$.

The horizontal subspace point of view lends itself to far reaching generalizations, with probabilistic significance. Let us interrupt the discussion for a while.

4. THE LANGEVIN EQUATION

Since I have been describing T(M) and TT(M), let me make a short digression. The best model for the real brownian motion isn't the Wiener process, but the Ornstein-Uhlenbeck process in its complete form, that is a process (X_t) whose particles have a continuous speed (V_t), but no acceleration, and satisfy the Langevin equation

$$\begin{aligned} dX_t &= V_t dt \\ m dV_t &= [K(X_t) - bV_t]dt + \sigma(X_t)dW_t \end{aligned} \tag{13}$$

Here W is a Wiener process, K is the driving force, b is a friction coefficient (usually σ is a constant, and written as $b\sqrt{D}$). If the driving force is 0, the second line stands alone as a stochastic differential equation of a simple kind, and its solution (for σ constant) is the <u>Ornstein-Uhlenbeck speed process</u>. See Nelson's well known book [1].

So it is a natural problem to investigate semimartingales Y_t with values in the tangent bundle T(M), and such that the projection of Y on M is a differentiable path. Setting $Y_t=(X_t,V_t)$, where V_t is a tangent vector at X_t, we may either assume that

$$\frac{dX_t}{dt} = V_t \tag{14}$$

or do the much more general assumption that (X_t) <u>is a continuous process with finite variation</u>, without even assuming the continuity of (V_t) - jumps in the tangent space can be efficiently handled, as I have learnt in a recent paper of T.E. Duncan [1]. Take now two systems of (global) coordinates (x^i,u^i) and $(\bar{x}^\alpha, \bar{u}^\alpha)$ on T(M), with the usual transformation rules

$u^i = dx^i$, $\bar{u}^\alpha = d\bar{x}^\alpha$, $\bar{u}^\alpha = p_i^\alpha(x)u^i$ where $p_i^\alpha = D_i\bar{x}^\alpha$ ($= \partial\bar{x}^\alpha/\partial x^i$)

therefore $\bar{V}_t^\alpha = p_i^\alpha(X_t)V_t^i$, and the integration by parts formula gives, since X is a continuous finite variation process

$$d\bar{V}_t^\alpha = p_i^\alpha(X_t)dV_t^i + D_k p_i^\alpha(X_t)V_{t-}^i dX_t^k \tag{15}$$

This means that the system $(X_t^i, V_{t-}^i, dX_t^i, dV_t^i)$ transforms like an element of TT(M), with the implication that

$$(dV_t^i + \Gamma_{jk}^i(X_t)V_{t-}^j dX_t^k)D_i \tag{16}$$

formally represents an intrinsic first order vector (note also, incidentally, that $\Sigma_{i<j} (V_{t-}^i dX_t^j - V_{t-}^j dX_t^i)D_i \wedge D_j$, $g_{ij}V_{t-}^i dX_t^j$ are intrinsic too).

It follows also from (15) that the brackets $[V^i,V^j]$ transform as

$$d[\bar{V}^\alpha,\bar{V}^\beta]_t = p_i^\alpha(X_t)p_j^\beta(X_t)d[V^i,V^j]_t \tag{17}$$

and the same is true for the brackets $\langle V^i,V^j \rangle$ assuming they exist.

It is now trivial to write the Langevin equation in a Riemannian manifold (note there is no more individual mass, since a point of M may represent a complicated system, also it is slightly unnatural to write the friction term as $-bV_t$ in such a set up, rather than to say vaguely that "K might depend also on V" (may be linearly ?). Here is the system :

$$dX_t^i = V_t dt$$
$$dV_t^i + \Gamma_{jk}^i(X_t)V_t^j dX_t^k = [K^i(X_t) - bV_t^i]dt + \sigma(X_t)dW_t^i$$

where W_t is the brownian motion of the Riemannian manifold M. This amounts to saying that V is continuous, and (omitting the friction for simplicity)

$$dV_t^i + \Gamma_{jk}^i(X_t)V_t^j dX_t^k - K^i(X_t)dt = \text{differential of a local martingale for } i=1,\dots,n$$
$$d\langle V^i, V^j\rangle_t = \sigma g^{ij}(X_t)dt \ .$$

But the situation studied above is much more general : it lends itself to the description of something like a random electromagnetic field fluctuating around a deterministic light ray.

5. INTEGRATION FORMALISM : PRELIMINARIES

We are going to give a geometric meaning to the Ito and Stratonovich integrals, and in this section we prepare the work, with the introduction of second order differential forms. This is a surprising story. Higher order differentials were quite familiar to mathematicians of the last century, and are mentioned in the classical "Cours d'Analyse Mathématique" (Goursat, Hadamard (1917), even Valiron (1942)) as a quite efficient method for computing higher order partial derivatives. On the other hand, they are now completely extinct. I hope to show here that second order forms are the natural objects to integrate along semimartingale paths.

DEFINITION. A second order differential form on M is a C^∞ function on $\tau(M)$ which is linear on each $\tau_a(M)$.

The basic example is the following

DEFINITION. Let f be a C^∞ function on M. Then its second differential d^2f is the second order differential form given by

$$\text{if } \lambda \in \tau_a(V), \quad \langle \lambda, d^2f \rangle = \lambda(f)_a$$

It is obvious that second order forms can be multiplied by C^∞ functions on M. So we may also define :

DEFINITION. Let f and g be C^∞ functions on M. Then df.dg is the second order differential form

$$df.dg = \tfrac{1}{2}(d^2(fg) - fd^2g - gd^2f).$$

We know that $T(M) \subset \tau(M)$, so each second order form has a restriction to T(M), which is an ordinary form. It is obvious that

$$d^2f|_{T(M)} = df \quad , \quad df.dg|_{T(M)} = 0$$

Remember now that we have global coordinates on M. We prove the intrinsic character of the "full second order differential of f" as it appears in the classical books :

THEOREM (trivial). $d^2f = D_i f\, d^2x^i + D_{ij} f\, dx^i . dx^j$.

Proof : fix $a \in M$. We don't change anything by replacing x^i by $x^i - a^i$, so we may assume that $x^i(a)=0$, in which case $dx^i.dx^j = \frac{1}{2} d^2(x^i x^j)$, and the formula reduces to the fact that $d^2 f|_a = d^2 g|_a$, where

$$g(x) = D_i f(a) x^i + \tfrac{1}{2} D_{ij} f(a) x^i x^j$$

This is just the Taylor formula of order 2 : $f(x)-f(a)-g(x)$ has a zero of order ≥ 3 at a, so all differential operators of order 2 at a vanish on it.

> Just out of curiosity, it is natural to wonder about forms of higher orders . It turns out that they exist, that one can define the d and . operations in a nice way, but there are deep differences between orders ≤ 2 and >2 : essentially, for $n>2$, they are dual to something larger than differential operators of order n. They seem to be quite useless and inoffensive.

COROLLARY. The coordinates λ^i, λ^{ij} on $\tau(M)$ are just the d^2x^i and $dx^i.dx^j$.

Our next step consists in extending the product . and the differentiation d to arbitrary 1-forms. This is now obvious. Given two 1-forms $\rho = a_i dx^i$ and $\sigma = b_j dx^j$, define

$$\rho . \sigma = a_i b_j dx^i . dx^j \quad , \quad d\rho = a_i d^2 x^i + D_j a_i dx^j . dx^i$$

We must check that the result doesn't depend on the coordinate system. Now we have the properties :

- the product is commutative, and bilinear w.r.to $C^\infty(M)$ multiplication
- $d(f\rho) = f d\rho + df.\rho$

which in turn characterize both operations uniquely. Note that

(18) $$d\rho|_{T(V)} = \rho \quad , \quad \rho.\sigma|_{T(V)} = 0 .$$

> There is another intrinsic characterization of the d operator (which doesn't extend to higher orders). Let ω be a form of order 1, and $h(t)$ be a curve in M. Then we have
>
> $$\frac{d}{dt} < \dot{h}(t), \omega > = < \ddot{h}(t), d\omega > .$$

Differentiation raises by one unit the order of forms. There is another such operation, deduced from a linear connection Γ . Since Γ maps linearly $\tau_a(M)$ into $T_a(M)$, its dual (which we denote by Γ too) maps forms of order 1 into forms of order 2, with the property that

(19) $$\Gamma(\rho)|_{T(V)} = \rho \quad , \quad \Gamma(f\rho) = f\Gamma(\rho) .$$

The Christoffel symbols appear in the expression of Γ as follows

(20) $$\Gamma(dx^k) = d^2x^k + \Gamma^k_{ij} dx^i . dx^j$$

Note that $d - \Gamma$ is a second order form whose restriction to $T(M)$ is zero, i.e. just a symmetric bilinear form. If $\omega = a_i dx^i$ is a form

$$(d-\Gamma)\omega = (D_j a_i - a_k\Gamma^k_{ij})dx^i.dx^j$$

and on a Riemannian manifold we may take the trace of this quadratic form w.r. to the metric to get the scalar function

(21) $$-\delta\omega = (D_j a_i - a_k\Gamma^k_{ij})g^{ij}$$

(just consider the left side as the definition of δ on forms, since we don't need any general theory of the δ operator, and remark that $-\delta df = \Delta f$ according to (10)).

6. ITO AND STRATONOVICH INTEGRALS

In the usual set-up of stochastic integration, the Ito integral has an awkward geometric status (it doesn't "behave well" under a change of coordinates), while the Stratonovich integral has an awkward analytic status (as Yor [1] shows, the approximation procedures which are traditionnally used to justify its use aren't valid for all semimartingales). The use of second order forms will clarify the situation. Roughly stated, the true stochastic integral is a second order object, like the semimartingale differentials themselves. To reduce it to first order, one may use two geometric procedures, which yield Ito or Stratonovich integrals.

Also, remember that forms shouldn't be integrated only on paths, but on chains, i.e. on paths provided with formal multipliers. Here our multipliers will be predictable (locally)bounded processes. For simplicity, we shall omit the multipliers most of the time.

DEFINITION. Let $X=(X^i_t)$ a semimartingale with values in M, and let $\Theta = a_i d^2x^i + a_{ij}dx^i.dx^j$ be a (C^∞) form of order 2 on M. Then we define the stochastic integral $\int_{X^t_0}\Theta$ of Θ along the path X^t_0 as the real valued process

$$Y_t = \int_{X^t_0}\Theta = \int_0^t a_i(X_s)dX^i_s + \frac{1}{2}\int_0^t a_{ij}(X_s)d\langle X^i,X^j\rangle_s \quad \text{(Einstein convention !)}$$

Let (K_t) be a (locally) bounded predictable process. Then the stochastic integral of Θ alors the chain $K\circ X^t_0$ is the process

$$\int_{K\circ X^t_0}\Theta = \int_0^t K_s dY_s = \int_0^t K_s a_i(X_s)dX^i_s + \frac{1}{2}\int_0^t K_s a_{ij}(X_s)d\langle X^i,X^j\rangle_s$$

PROPERTIES. 1) Probabilistic. Those of the usual stochastic integrals in $\mathbb{R}^n$: stochastic integrals are real valued semimartingales ; they remain unchanged if P is replaced by an equivalent law Q (more generally, if Q<<P the P-s.i. is a version of the Q-s.i.) ; they are local on Ω (if two semimartingales X and X' have the same path on some subset A of Ω, the corresponding s.i. have the same paths on A). Etc...

The use of multipliers is convenient at many places. For instance, if U is a coordinate patch, it is convenient to use the multiplier $I_{\{X\in U\}}$ to localize. If S,T are stopping times, one uses the multiplier $I_{]S,T]}$...

2) Differential geometric. The first main property, of course, is the fact that it is intrinsic (this is another expression of the principle of Schwartz : incidently, the principle of Schwartz itself might be recalled by a notation like $\int_0^t K_s \langle d^2X_s, \Theta \rangle$). More generally, let $F : M \to N$ be a map, on let Θ be a form of order 2 on N. Let also Z be the semimartingale $F\circ X$ with values in N. Then we have

$$\int_{Z_0^t} \Theta = \int_{X_0^t} F^*(\Theta) \qquad (F^*(\Theta) \text{ is the pull-back of } \Theta \text{ on } M).$$

Also note the following simple formulas

$$(22) \qquad \int_{X_0^t} d^2f = f(X_t)-f(X_0)$$

$$\int_{X_0^t} df.dg = \frac{1}{2}\langle f(X),g(X) \rangle_t$$

DEFINITION. Let ω be a (C^∞) form of order 1 on M. Then we define its (Stratonovich) integral along the chain $K\circ X$ as

$$(23) \qquad \int_{K\circ X_0^t} \omega = \int_{K\circ X_0^t} d\omega$$

Assume a linear connection Γ is given on M. Then the Ito integral of ω along the chain $K\circ X$ is

$$(24) \qquad (\Gamma)\int_{K\circ X_0^t} \omega = \int_{K\circ X_0^t} \Gamma\omega \ .$$

Let us pause for a discussion, since these definitions are the main point in this report : here, Ito and Stratonovich integrals are given the same status, both can be used with arbitrary predictable multipliers. (So the usual statement that S-integrals are less general than I-integrals is no longer valid here : this is due to the fact that we are working with C^∞ forms. For quite general forms the d operation would require more regularity than the Γ operation). The Ito integral requires more structure than the S-integral, which is the main geometric object, as was discovered by Ito himself, and confirmed by all the subsequent work on the subject. Finally, we remark that no approximation procedure, no smoothing of the path, has been used to define the S-integral.

SOME PROPERTIES. a) The main property of the Stratonovich integral, as noted by many authors (personnally I learnt it in Yor [1]) is the following : if ω is a closed form , then $\int_{X_0^t} \omega$ is just the integral of ω, in the differential geometric sense, along the continuous path $X_0^t(\omega)$. For an exact form $\omega=df$, this reduces to (22), the general case requiring a localization.

b) Let $F : M \to N$ be a map, and ω be a form on N, Z_t be $F(X_t)$. Then we have $\int_{Z_0^t} \omega = \int_{X_0^t} F^*(\omega)$. This corresponds to the second order formula just before (22), and the obvious property that $F^*(d\omega)=dF^*(\omega)$. The corresponding relation for Ito integrals is a rare event. Indeed, denoting by the same letter Γ two connections on M and N, the property that $\Gamma(F^*(\omega))=F^*(\Gamma\omega)$ is extremely restrictive. For reference below, note the formula

(25) If $\omega=a_\alpha dx^\alpha$ on N , $F^*(\Gamma\omega) = \Gamma(F^*(\omega)) + a_\alpha((d-\Gamma)dF^\alpha + \Gamma^\alpha_{\beta\gamma}dF^\beta . dF^\gamma)$

where the "greek" coordinates x^α refer to N .

c) The main property of Ito integrals is their relation to martingales. X is a martingale with values in M (relative to Γ) if and only if $Y^\omega_t = (\Gamma)\int_{X_0^t} \omega$ is a real valued local martingale for any form ω of order 1. In this context, the trivial identity

(26) $$\int_{X_0^t} \omega = (\Gamma)\int_{X_0^t} \omega + \int_{X_0^t} (d-\Gamma)\omega$$

appears as the true expression of Ito's formula in a manifold M, since 1) it reduces to it when $M=\mathbb{R}^n$ with its trivial connection, and $\omega=df$, and 2) if X is a martingale with values in M, it gives the decomposition of the left side in its local martingale and finite variation parts.

Let us give two applications of these computations to the Brownian motion X of a Riemannian manifold M. In this case, if $\Theta=a_{ij}dx^i.dx^j$ is a second order form reduced to its quadratic part, we have

(27) $$\int_{X_0^t} \Theta = \frac{1}{2}\int_0^t a_{ij}(X_s)d\langle X^i,X^j\rangle_s = \frac{1}{2}G(X_s)ds, \text{ where } G = a_{ij}g^{ij}$$

1) Applying this to formula (26), and taking formula (21) into account, we get that $G = -\delta\omega$, a nice formula due to Ikeda-Manabe [1].

2) Let us return to the situation of b), and look for the condition that $F(X_t)$ be a martingale with values in N . Looking at the right side of (25), the first term gives a martingale by integration, while the second is the purely quadratic second order form

$$a_\alpha f^\alpha_{ij}dx^i.dx^j \qquad f^\alpha_{ij} = D_{ij}F^\alpha - \Gamma^k_{ij}D_kF^\alpha + \Gamma^\alpha_{\beta\gamma}\circ F\, D_iF^\beta D_jF^\gamma$$

Applying (27), we see that the condition is the vanishing of the functions $G^\alpha = f^\alpha_{ij}g^{ij}$. This is exactly the definition of a harmonic mapping $F : M \to N$ (Hamilton [1], p.4).

d) Let us end this section with a basic property of the Stratonovich integral. Let ω be a form of order 1, and let f be a C^∞ function. If we know the real valued semimartingale $Y_t = \int_{X^t} \omega$, then we may compute the integral $\int_{X_0^t} f\omega = Z_t$ by ordinary Stratonovich integration. More generally, if K is a predictable (locally)bounded process

(28) $$\int_{X_0^t} f\omega = \int_0^t K_s f(X_s)*dY_s = \int_0^t K_s f(X_s)dY_s + \frac{1}{2}\int_0^t K_s d\langle f(X),Y\rangle_s$$

7. SEMIMARTINGALES AND PFAFF SYSTEMS

Consider a distribution of submanifolds of dimension p in M, described as usual by the space $\underline{\underline{I}}$ (stable under multiplication by C^∞ functions) of all C^∞ forms which vanish on the distribution. Locally we may describe the distribution by the vanishing of forms ω^α $(\alpha=1,\dots,n-p)$. Since we want to avoid localization difficulties, we assume that the ω^α are independent at each point and describe the distribution globally. A differentiable curve h(t) then is an integral curve of the distribution if and only if

$$\int_{h_0^t} \omega^\alpha = 0 \quad \text{for } \alpha = 1,2,\dots,n-p \ .$$

It is entirely natural to say that X is an <u>integral semimartingale</u> for the distribution if we have

$$(29) \qquad Y_t^\alpha = \int_{X_0^t} \omega^\alpha = 0 \text{ for } \alpha = 1,2,\dots,n-p \ .$$

This property doesn't depend on the choice of the basis ω^α . Indeed, let ω be any other form that vanishes on the distribution, and let $Y_t = \int_{X_0^t} \omega$; writing $\omega = g_\alpha \omega^\alpha$ we have $Y_t = \int_0^t g_\alpha(X_s) * dY_s^\alpha = 0$. On the other hand, the <u>geometric</u> meaning of (29) isn't at all obvious, except in the trivial case of a completely integrable system : then we may assume that $\omega^\alpha = dF^\alpha$ are (locally) exact forms, and (29) simply means that the semimartingale moves in some integral manifold $F^\alpha =$ constant ($\alpha = 1,\dots,n-p$).

We write (29) in its explicit second order expression :

$$(30) \qquad \int_{X_0^t} d\omega = 0 \text{ for } \omega \in \underline{\underline{I}}$$

and remark that $\omega \in \underline{\underline{I}} \Rightarrow f\omega \in \underline{\underline{I}}$, and so $\int_{X_0^t} df.\omega = \int_{X_0^t} d(f\omega) - fd\omega = 0$. Using the Schwartz principle in the reverse direction, we may say that a <u>second order</u> tangent vector L belongs to the distribution if $< L, d\omega > = 0$ for all $\omega \in \underline{\underline{I}}$ (therefore, $< L, \omega.\Theta > = 0$ for any form Θ). It turns out that any distribution has some non trivial second order integral fields : namely, if A and B are first order integral fields, then AB+BA is a second order integral field, thanks to the formulas

$$(31) \qquad < AB, d\omega > = A< B,\omega > - \tfrac{1}{2}< A\wedge B, \partial\omega > \quad (\partial, \text{ exterior differential })$$

$$(32) \qquad < AB+BA, d\omega > = A< B,\omega > + B< A,\omega > \ .$$

So AB and BA are second order integral fields if and only if AB-BA is a (first order) integral field.

8. THE LIFTING OF A SEMIMARTINGALE THROUGH A CONNECTION

The results on this section were explained to me by Schwartz. They are generalizations to general connections and general semimartingales of the classical "stochastic parallel displacement" theory, due to Ito and Dynkin.

The extension to general connections can be found also in Malliavin [1], for brownian semimartingales. The geometric "second order language" can possibly bring some additional clarity to the subject.

We shall use the "horizontal subspace" point of view for connections. For simplicity, instead of considering a fiber space, we consider just a product $W=U\times M$, with global coordinates (x^i) on the "base" M ($1\leqq i\leqq n$) and (x^α) on the "fiber" U ($1\leqq\alpha\leqq p$). As usual, π denotes the projection on the base, but we mention it as little as we can : if g is a function on M, we also denote by g the function $g\circ\pi$ on W . This concerns in particular the coordinates x^i, and D_i has a double meaning, as $\partial/\partial x^i$ on M and $\partial/\partial x^i$ on W.

A <u>connection</u> Γ is a distribution of subspaces $H_{x,u}\subset T_{x,u}(W)$, called <u>horizontal subspaces</u>, such that

$$\forall (x,u)\in W\ ,\ \pi_* \text{ \underline{is an isomorphism of} } H_{x,u} \text{ \underline{onto} } T_x(M)$$

(so $H_{x,u}$ is supplementary to $V_{x,u}=\mathrm{Ker}(\pi_*|_{x,u})$, the <u>vertical</u> subspace of $T_{x,u}(W)$) . Then any tangent vector $t\in T_x(M)$ has an unique <u>horizontal lift</u> $H(t)$ at $(x,u)\in W$. To compute $H(t)$ it is sufficient to know

$$(33)\qquad H(D_i) = \tilde{D}_i = D_i - \Gamma_i^\alpha(x,u)D_\alpha \qquad (\text{ often denoted by } \nabla_i\ )$$

Going back to the preceding section, we see that the distribution of horizontal subspaces is associated to the forms

$$(34)\qquad \theta^\alpha = du^\alpha + \Gamma_i^\alpha dx^i \quad .$$

Then any semimartingale (X_t) on the "base" M has a unique lift $\tilde{X}_t = (X_t^i,U_t^\alpha)$, with prescribed initial values U_0^α , which satisfies the Stratonovich differential equations $\int_{\tilde{X}_0^t} \theta^\alpha = 0$, that is

$$(35)\qquad dU_t^\alpha + \Gamma_i^\alpha(X_t,U_t)*dX_t^i = 0$$

of course, even if Γ is C^∞ , one must be careful about the possibility of an explosion in (35), and $\tilde{X}_t$ may have a finite lifetime.

We want to compute the second order tangent vector to the lifted semimartingale $\tilde{X}$, that is

$$d^2\tilde{X}_t = dX_t^iD_i + dU_t^\alpha D_\alpha + \tfrac{1}{2}d\langle X^i,X^j\rangle_t D_{ij} + d\langle X^i,U^\alpha\rangle_t D_{i\alpha} + \tfrac{1}{2}d\langle U^\alpha,U^\beta\rangle_t D_{\alpha\beta} \quad .$$

and of course, the natural way consists in transforming (35) into an Ito equation

$$(36)\qquad dU_t^\alpha + \Gamma_i^\alpha(\tilde{X}_t)dX_t^i + \tfrac{1}{2}(D_j\Gamma_i^\alpha - \Gamma_j^\beta D_\beta\Gamma_j^\alpha)(\tilde{X}_t)\ d\langle X^i,X^j\rangle_t = 0$$

and to compute from (36) the brackets $\langle U^\alpha,X^i\rangle$ and $\langle U^\alpha,U^\beta\rangle$. However, there is a nicer way to state things. In the preceding section, we have seen that the Schwartz principle can be used in the reverse direction, to express that a <u>second order tangent</u> vector satisfies to a Pfaff system. So here we should be able to define the horizontal lift $H(\lambda)$ of a <u>second order</u> tangent vector

λ at x, as a differential geometric operation[1], and then to have simply

(37) $$d^2\tilde{X}_t = H(d^2X_t) \quad (\text{ lifted from } X_t \text{ to } \tilde{X}_t \in W\ )$$

and the same result for the local characteristics vector . Now this is an entirely trivial matter : one first checks that there is a unique lift of λ (at x) to (x,u) over x, satisfying the second order system. Then one must only know $H(D_i)=\tilde{D}_i$ and $H(D_{ij})$. On the other hand, we know from (32) that $\tilde{D}_i\tilde{D}_j+\tilde{D}_j\tilde{D}_i$ satisfies the second order system, and for functions depending only on the x^i it reduces to $D_iD_j+D_jD_i=2D_{ij}$, so $H(D_{ij})=\frac{1}{2}(\tilde{D}_i\tilde{D}_j+\tilde{D}_j\tilde{D}_i)$. Finally

(38) $$H(\lambda^iD_i+\lambda^{ij}D_{ij}) \quad = \quad \lambda^i\tilde{D}_i + \lambda^{ij}\tilde{D}_i\tilde{D}_j$$

a formula due to Dynkin, and extended by Malliavin. From the formula

$$H(XY+YX) = H(X)H(Y)+H(Y)H(X)$$

we deduce at once

(39) $$H(XY) = H(X)H(Y) - \tfrac{1}{2}\bar{R}(X,Y)$$

where $\bar{R}(X,Y)$ is a vertical tangent vector , equal to $H(X)H(Y)-H(Y)H(X) - H([X,Y])$ (we write $\bar{R}$, not R, because we are operating on functions, not on vector fields, so the components $\bar{R}(X,Y)=\xi^i\eta^j\bar{R}^\alpha_{ij}D_\alpha$ will be opposite to those of the usual curvature tensor).

Let us now give two examples .

1) Linear connections . Here U will be a linear space, with its linear coordinates u^α, and $\Gamma(x,u)$ will be linear in u :

(40) $$\Gamma^\alpha_i(x,u) = u^\varepsilon\Gamma^\alpha_{\varepsilon i}(x)$$

This covers the case of T(M), on a coordinate patch (on T(M), of course, we have more structure, since the latin and greek indexes are in equal numbers and we know how to express the dx^i as a linear combination of the u^α. Often the u^α are simply the dx^i themselves). This case is well known, and the theory of the lifting of semimartingales reduces, on T(M), to the classical "stochastic parallel displacement".

2) Stochastic differential equations. Here we consider a manifold U with coordinates (u^α), and at each point of U a system of n independent vector fields $A_i=a^\alpha_iD_\alpha$ (we assume this is possible). Then we take $M=\mathbb{R}^n$ with its linear coordinates x^i, $\Gamma^\alpha_i(x,u)=a^\alpha_i(u)$, and lifting a semimartingale X with values in $\mathbb{R}^n$ amounts to solving the Stratonovich differential equation

(41) $$dU^\alpha_t = a^\alpha_i(U_t)*dX^i_t \ .$$

1. This idea is due to Schwartz. He also remarks that, if h is a path in M through x, and $\tilde{h}$ is the lifted horizontal path through (x,u), then the acceleration of h lifts to that of $\tilde{h}$ (and this property characterizes the lifting of second order tangent vectors).

REFERENCES

1) On probability theory in manifolds

The main reference will certainly be the important book of N. Ikeda and S. Watanabe (to appear in 1981). Among the basic papers of Malliavin and his school, I have quoted only [1], whose bibliography is rather complete.

BISMUT (J.M.) [1]. Principes de mécanique Aléatoire. To appear.

DUNCAN (T.E.) [1]. Optimal control in a Riemannian tangent bundle. Measure theory, Oberwolfach 1979 (D. Kölzow ed.). Lecture Notes in M. 794, Springer 1980.

DYNKIN (E.B.). [1]. Diffusion of tensors. Soviet Math. Dokl. 9, 1968, p. 532-535.

ELWORTHY (D.) [1]. Stochastic dynamical systems and their flows. Stochastic analysis, Academic Press, 1978, p. 79-96.

IKEDA (N.) and MANABE (S.) [1]. Integral of differential forms along the path of diffusion processes. Publ. RIMS, Kyoto Univ., 15, 1979, p.827-52.

ITO (K.) [1]. Stochastic differentials. Appl. M. and optimization 1, 1975, p. 374-381.

--- [2]. Stochastic parallel displacement. Proc. Victoria Conf. on probabilistic methods in Diff. Eqs, p.1-7. LN 451, Springer 1975.

MALLIAVIN (P.) [1]. Géométrie différentielle stochastique. Presses de l' Université de Montréal, 1978.

SCHWARTZ (L.) [1]. Semimartingales sur des variétés, et martingales conformes sur des variétés analytiques complexes. LN 787, Springer 1980.

--- [2]. Equations différentielles stochastiques sur des variétés, relèvement d'équations différentielles stochastiques et de semimartingales par des connexions sur des espaces fibrés. Preprint.

YOR (M.) [1]. Formule de Cauchy relative à certains lacets browniens. Bull. Soc. M. France 105, 1977, p. 3-31.

--- [2]. Sur quelques approximations d'intégrales stochastiques. Sém. Prob. XI, LN. 581, Springer 1977, p. 518-528.

2) On second order differential geometry

There is an extensive literature on second order geometry, but most of it seemed to me desperately abstract, and impossible to use for probabilists (this isn't a statement about its intrinsic value !). No paper at all mentions second order differentials. So let me quote only three papers, the first of which is a classic.

AMBROSE (W.), PALAIS (R.S.) and SINGER (I.M.). Sprays. Anais Acad. Bras. Ciencias, 32, 1960, p. 163-178.

DOMBROWSKI (P.). On the geometry of the tangent bundle. J. Reine Angew. Math. 210, 1962, p. 73-88.

GRIFONE (J.). Structure presque tangente et connexions. Ann. Inst. Fourier 22.1, 1972, p. 287-334.

3) Other

HAMILTON (R.S.) [1]. Harmonic maps of manifolds with boundary. LN 471, 1975.

NELSON (E.) [1]. Dynamical theories of brownian motion. Princeton Univ. Press, 1967.

Additions : I have just received a preprint by R.M. Dowell (Warwick University), not directly connected to the topics in this talk, but mentioning earlier work of the author on Langevin's equation in manifolds. Let me mention also that I have written an account of n-th order differential forms (not to be published) which may be sent on request to fans of this subject.

P.A. Meyer
IRMA[1]
7 rue René Descartes
67084- Strasbourg Cedex, France

1. Laboratoire associé au CNRS.

Homogenization and Stochastic Parallel Displacement

Mark A. Pinsky

Northwestern University

1. Introduction. The purpose of this paper is to reveal a hitherto unobserved connection between Brownian motion on a Riemannian manifold on the one hand and the currently popular "homogenization" techniques on the other.

For our purposes, homogenization refers to the following situation. We are given a two-component Markov process $X_\varepsilon(t) = (\xi_\varepsilon(t), Y_\varepsilon(t))$ depending upon a parameter $\varepsilon>0$. Neither component is Markovian by itself. When $\varepsilon\to 0$ the component processes "de-couple"; the Y process becomes Markovian (in fact a diffusion) where the drift and diffusion coefficients are expressible as averages over the now-extinct process $\xi(t)$. Limit theorems of this type have been studied by many authors (e.g. Bensoussan-Lions-Papanicolaou [2]) and are implicit in other works, e.g. the lecture of S. Watanabe in this conference.

Stochastic differential equations on differentiable manifolds were initiated by Ito [6] and further developed by many authors [3,4, 5,7,8,9,10,13,14]. The lectures of P. A. Meyer in this conference give the most recent viewpoint. In our approach we work on the bundle of frames (not necessarily orthonormal) of arbitrary degree. In order to use the homogenization technique, we first consider the isotropic transport process, a discontinuous Markov process on a frame bundle of one higher degree. This process is piecewise geodesic, with a Poisson-type jump mechanism; in particular the infinitesimal generator is readily determined. We insert a small parameter, which corresponds to

Supported by NSF Grant MCS 78-02144

an accelerated version of the original process. When the parameter tends to zero, we obtain the horizontal Brownian motion on the original frame bundle.

This construction has the following two advantages over previous approaches to the subject: 1) It provides a geometrically natural approximation to both Brownian motion and stochastic parallel displacement on a Riemannian manifold. 2) It provides a succinct, coordinate-free formula for the infinitesimal generator of the limiting processes, viz.

$$A_k = PZ_k^2$$

Here Z_k is the canonical horizontal vector field on $T_{k+1}(M)$, the frame bundle of one higher degree; P is a projection operator, which maps functions from $T_{k+1}(M)$ to functions on $T_k(M)$, by forming the average with respect to the tangent vector ξ.

In case $k = 0$, A_k is proportional to the Laplace-Beltrami operator of the Riemannian manifold M. In case $k = d$ (the dimension of M), A_k is proportional to the horizontal Laplacian of Bochner, when restricted to orthonormal frames. For intermediate values of k, the operator has also been studied, e.g. by Dynkin [4].

2. Connections, Geodesics and Horizontal Vector Fields

In this section we give a brief review of the necessary concepts from differential geometry.

M is a Riemannian manifold, which means the following:

(R_1) M is an infinitely differentiable manifold of dimension d.

(R_2) Each tangent space M_x carries an inner product $\langle\ ,\ \rangle$.

(R_3) For any $X,Y\in\mathfrak{X}(M)$ (infinitely differentiable vector fields) $x\to\langle X,Y\rangle$ is an infinitely differentiable function.

A Riemannian connection is a map

$$\mathfrak{X}(M)\times\mathfrak{X}(M)\to\mathfrak{X}(M)$$
$$(X,Y)\to\nabla_X Y$$

satisfying the following conditions:

(C_0) $X_1\langle X_2,X_3\rangle = \langle\nabla_{X_1}X_2,X_3\rangle+\langle X_2,\nabla_{X_1}X_3\rangle$ $\qquad (X_1,X_2,X_3\in\mathfrak{X}(M))$

(C_1) $\nabla_{X_1+X_2}X_3 = \nabla_{X_1}X_3+\nabla_{X_2}X_3$ $\qquad$ (" " " ")

(C_2) $\nabla_{X_3}(X_1+X_2) = \nabla_{X_3}X_1+\nabla_{X_3}X_2$ $\qquad$ (" " " ")

(C_3) $\nabla_{fX}Y = f\nabla_X Y$ $\qquad (f\in C^\infty(M),X,Y\in\mathfrak{X}(M))$

(C_4) $\nabla_X(fY) = f\nabla_X Y+(Xf)Y$ $\qquad$ (" " ")

A fundamental proposition of Riemanian geometry states the following:

Proposition 2.1 There exists at least one Riemannian connection. It ia unique under the further symmetry condition $\nabla_X Y-\nabla_Y X-[X,Y] = 0$, $[X,Y] = XY-YX$.

If $\partial_1,\dots,\partial_n$ are coordinate vector fields, we write $\nabla_{\partial_i}\partial_j = \sum_{k=1}^{d}\Gamma_{ij}^{\ell}\partial_\ell$, $1\le i,j\le d$ which defines the coefficients Γ_{ij}^{ℓ}.

Parallel displacement on a Riemannian manifold is defined as follows. Let $(\gamma(t), t\geq 0)$ be a smooth curve in $M, \eta_0 \varepsilon M_{\gamma(0)}$. Then there exists a unique field of vectors $(\bar{\eta}(t), t\geq 0)$ such that

$$(D_1) \qquad \bar{\eta}(t) \in M_{\gamma(t)}$$

$$(D_2) \qquad \bar{\eta}(0) = \eta_0$$

$$(D_3) \quad \nabla_{\gamma(t)}\bar{\eta}(t) = 0$$

$\bar{\eta}(t)$, which exists by the fundamental theorem on ordinary differential equations is called the parallel displacement of η_0 along $(\gamma(t), t\geq 0)$. Properties (C_0) and (D_3) immediately imply that if $\eta_0, \eta_0' \varepsilon M_{\gamma(0)}$ then

$$(D_4) \quad \langle\bar{\eta}(t), \bar{\eta}'(t)\rangle = \langle\eta_0, \eta_0'\rangle$$

Therefore parallel displacement preserves length and angle.

A smooth curve $(\gamma(t)\ t\geq 0)$ is called a geodesic if

$$\nabla_{\dot{\gamma}(t)}\dot{\gamma}(t) = 0$$

(the tangent vector $\dot{\gamma}$ is not a vector field on M, but this causes no difficulty in the theory). Given $x\varepsilon M$, $\xi\varepsilon M_x$, the fundamental theorem on ordinary differential equations implies that there exists a geodesic $(\gamma(t), 0\leq t\leq\delta)$ such that $\gamma(0) = x$, $\dot{\gamma}(0) = \xi$. We shall require in addition that M is complete, i.e. that we may take $\delta = +\infty$, $\forall x\varepsilon M$, $\xi\varepsilon M_x$.

The framebundle of a Riemannian manifold is defined by

$$T^{(k+1)}(M) \doteq \{(x,\xi,\eta_1,\ldots,\eta_k): x\varepsilon M, \xi\varepsilon M_x, \eta_1\varepsilon M_x,\ldots,\eta_k\varepsilon M_x, |\xi| = 1\}$$

(k is a fixed but unspecified integer). This has the structure of an infinitely differentiable manifold. The canonical horizontal vector field is defined by

$$Z_k f(x,\xi,\eta_1,\cdots\cdots\cdots,\eta_k) = d/dt\ f(\gamma(t),\xi(t),\overline{\eta}_1(t),\cdots\cdots,\overline{\eta}_k(t))|_{t=0}$$

where $\overline{\eta}_j(t) = \overline{\eta}(t;\eta_j)$ is the parallel displacement of η_j along the geodesic $\gamma(t)$, $\gamma(0) = x$, $\dot{\gamma}(0) = \xi$. In local coordinates we have the formula (with summation convention)

$$Z_k f(x,\xi,\cdots,\eta_k) = \xi^i \frac{\partial f}{\partial x^i} - \Gamma^{\ell}_{ij}\xi^i\xi^j \frac{\partial f}{\partial \xi^{\ell}} - \Gamma^{\ell}_{ij}\xi^i\eta_1^j \frac{\partial f}{\partial \eta_1^{\ell}} - \cdots - \Gamma^{\ell}_{ij}\xi^i\eta_k^j \frac{\partial f}{\partial \eta_k^{\ell}}$$

Proposition 2.2 Let $u(t,x,\xi,\eta_1,\ldots,\eta_k) = f(\gamma(t),\dot{\gamma}(t),\overline{\eta}_1(t),\ldots,\overline{\eta}_k(t))$

Then
$$\frac{du}{dt} = Z_k u$$

Proof. Uniqueness and smooth dependence of systems of ordinary differential equations. We shall need later an infinitesimal form of parallel displacement. For this purpose, define the following system of real-valued functions on $T^{(k+1)}(M)$

$$N_{ij}(x,\xi,\eta_1,\ldots,\ldots,\eta_k) = \langle \eta_i,\eta_j \rangle \qquad (1 \leq i,j \leq k)$$

Corollary 2.3 $Z_k N_{ij} = 0$, $1 \leq i,j \leq k$.

Proof. From (D_4), N_{ij} is constant along integral curves of Z_k, hence $d/dt\ N_{ij} = 0$. From Proposition 2.2, $Z_k N_{ij} = 0$.

3. Isotropic Transport Process

Let (Ω,B,P) be a probability triple with the following random variables:

(3.1) A sequence $(e_n)_{n\geq 1}$ of independent random variables with the common distribution $P\{e_n \varepsilon dt\} = e^{-t}dt$

(3.2) A sequence $(x^{(n)},\xi^{(n)},\eta_1^{(n)},\ldots\ldots,\eta_k^{(n)})\varepsilon T^{(k+1)}(M)$ defined inductively as follows

$$x^{(0)} = x,\ \xi^{(0)} = \xi,\ \eta_1^{(0)} = \eta_1,\ldots,\eta_k^{0} = \eta_k\ .$$

$$x^{(n+1)} = \gamma_{x^{(n)},\xi^{(n)}}(e_{n+1}) \qquad (n=0,1,\ldots)$$

$$\eta_j^{(n+1)} = \overline{\eta}(e_{n+1};\eta_j^{(n)}) \qquad (1\leqslant j\leqslant k, n=0,1,\ldots\ )$$

$$P\{\xi^{(n+1)}\varepsilon d\xi|x^{(0)},\ldots,\xi_k^{(n)}\} = \mu_{x^{(n+1)}}(d\xi) \qquad (n=0,1,\ldots)$$

where $\mu_x(d\xi)$ is the uniform law on $\{\xi\varepsilon M_x : |\xi|=1\}$ i.e. the unique probability law which is invariant under the orthogonal group defined by means of the inner product $< \ >$.

The <u>isotropic transport process</u> is defined by

$$x(t) = \gamma_{(x^{(n)},\xi^{(n)})}(t-\tau_n) \qquad (\tau_n\leqslant t<\tau_{n+1})$$

$$\xi(t) = \dot{\gamma}_{(x^{(n)},\xi^{(n)})}(t-\tau_n) \qquad (\text{" \ " \ "})$$

$$\eta_j(t) = \overline{\eta}(t-\tau_n;\eta_j^{(n)}) \qquad (\text{" \ " \ "})$$

where $\tau_n = e_1+\ldots+e_n$. It can be shown that this defines a Markov jump process on $T^{(k+1)}(M)$.

Let $\mathcal{C}$ be the space of differentiable functions on M which vanish at ∞. In order to compute the infinitesimal generator of the isotropic transport process, we introduce the following operators:

$$Pf(x,\eta_1,\ldots,\eta_k) \triangleq \int_{M_x} f(x,\xi,\eta_1,\ldots\eta_k)\mu_x(d\xi) \qquad (f\varepsilon\mathcal{C})$$

$$T_t^0 f(x,\xi,\eta_1,\ldots,\eta_k) \triangleq f(\gamma(t),\dot{\gamma}(t),\overline{\eta}_1(t),\ldots,\overline{\eta}_k(t)) \qquad (f\varepsilon\mathcal{C}, t>0)$$

$$R^0_\lambda f(x,\xi,\eta_1,\ldots,\eta_k) \doteq \int_0^\infty e^{-\lambda t} f(\gamma(t),\dot\gamma(t),\bar\eta_1(t),\ldots,\bar\eta_k(t))dt \quad (f\varepsilon C,\ \lambda>0)$$

$$T_t f(x,\xi,\eta_1,\ldots,\eta_k) \doteq E\{f(x(t),\xi(t),\eta_1(t),\ldots,\eta_k(t))\} \quad (f\varepsilon C,\ t>0)$$

$$R_\lambda f(x,\xi,\eta_1,\ldots,\eta_k) \doteq E\{\int_0^\infty e^{-\lambda t} f(x(t),\xi(t),\eta_1(t),\ldots,\eta_k(t))dt\} \quad (f\varepsilon C,\ t>0)$$

It will be clear from what follows that these operators map C into C.

__THEOREM__. $T_t f - f = \int_0^t [Z_k+P-I]T_s f\ ds = \int_0^t T_s(Z_k+P-I)f\ ds, \quad f\varepsilon C$

To prove this, it suffices to obtain the corresponding result for Laplace transforms. To prove this, we first obtain

__Lemma 3.1__. R^0_λ maps C into C and $(\lambda-Z_k)\,R^0_\lambda f = f = R^0_\lambda(\lambda-Z_k)f$.

__Proof__. From the smooth dependence on initial conditions, T^0_t maps C into C. The result now follows from Laplace transformation of Proposition 2.2.

__Lemma 3.2__. $R_\lambda f = R^0_{1+\lambda}f + R^0_{1+\lambda}PR_\lambda f$.

__Proof__.

$$R_\lambda f = E\{\int_0^{\tau_1} + \int_{\tau_1}^\infty\}e^{-\lambda t}f(x(t),\xi(t),\eta_1(t),\ldots\ldots,\eta_k(t))\ dt$$

The first integral is

$$E\{\int_0^\infty I_{(t<\tau_1)}e^{-\lambda t}f(x(t),\xi(t),\bar\eta_1(t),\ldots\ldots,\bar\eta_k(t))dt\}$$

$$= \int_0^\infty e^{-t}e^{-\lambda t}f(\gamma(t),\dot\gamma(t),\bar\eta_1(t),\ldots\ldots,\bar\eta_k(t))dt$$

$$= R^0_{1+\lambda}f$$

The second integral is

$$E\{\int_{\tau_1}^{\infty} e^{-\lambda t} f(x(t),\xi(t),\eta_1(t),\ldots\ldots,\eta_k(t))dt\}$$

$$= E\{\int_0^{\infty} e^{-\lambda(\tau_1+s)} f(x(\tau_1+s),\xi(\tau_1+s),\eta_1(\tau_1+s),\ldots,\eta_k(\tau_1+s))ds\}$$

$$= E\{e^{-\lambda\tau_1} E\int_0^{\infty} e^{-\lambda s} f(x(\tau_1+s),\xi(\tau_1+s),\eta_1(\tau_1+s),\ldots,\eta_k(\tau_1+s))ds\}$$

$$= E\{e^{-\lambda\tau_1} E\int_0^{\infty} e^{-\lambda s} f(x(s;x^{(1)}),\xi(s;\xi^{(1)}),\eta_1(s;\eta_1^{(1)}),\ldots,\eta_k(s;\eta_k^{(1)})ds\}$$

$$= E\{e^{-\lambda\tau_1}(R_\lambda f)(x^{(1)},\xi^{(1)},\eta_1^{(1)},\ldots,\eta_k^{(1)})\}$$

$$= E\{e^{-\lambda\tau_1}(PR_\lambda f)(x^{(1)},\xi^{(0)},\eta_1^{(1)},\ldots,\eta_k^{(1)})\}$$

$$= \int_0^{\infty} e^{-\lambda t}(PR_\lambda f)(\gamma(t),\dot{\gamma}(t),\bar{\eta}_1(t),\ldots,\bar{\eta}_k(t))e^{-t}dt$$

$$= R^0_{1+\lambda} PR_\lambda f$$

Lemma 3.2 immediately implies the series representation

$$R_\lambda f = \sum_{n=0}^{\infty} (R^0_{1+\lambda}P)^n R^0_{1+\lambda} f$$

From this it may be shown that R_λ maps C into C, in particular $R_\lambda f$ is in the domain of Z_k.

<u>Lemma 3.3</u>. $(\lambda - Z_k - P + I)R_\lambda f = f$

<u>Proof</u>. Apply $(I+\lambda-Z_k)$ to Lemma 3.2 and use Lemma 3.1.

Thus we have proved that $R_\lambda f - \lambda^{-1} f = \lambda^{-1}(Z_k+P-I)R_\lambda f$. By inversion of the Laplace transform the first part of the theorem is proved. To prove the second part it suffices to show that $\phi \doteq R_\lambda(\lambda - Z_k - P + I)f - f = 0$, for any $f \in C$. Clearly $\phi \in C$ and from Lemma 3.3, $(\lambda - Z_k - P + I)\phi = 0$. If we multiply this equation by ϕ and integrate over $T^{(k+1)}(M)$, the Z_k term disappears (integration by parts) and we are left with $\lambda\int\phi^2 = \int\phi P\phi - \int\phi^2 \leq 0$ a contradiction unless $\phi = 0$.

4. Convergence to a diffusion process

Let L_m^k be the Banach space of m-times continuously differentiable functions on $T^{(k)}(M)$ which vanish at infinity. The infinitesimal operator of stochastic parallel displacement is defined by

$$A_k f \doteq PZ_k^2 f \qquad f \in L_2^k$$

We make the following technical assumption.

Hypothesis 4.1 There exists a dense set $\mathcal{D} \subset L_0$ such that $(\lambda - A_k)\mathcal{D}$ is dense in L_0. This is satisfied in case M is compact or if M has Ricci curvature bounded from below (Yau [15]).

Let ε be a small parameter. Replace Z_k by εZ_k in the construction of the previous section, obtaining a process

$$({}^{\varepsilon}x(t), {}^{\varepsilon}\xi(t), {}^{\varepsilon}\eta_1(t), \ldots, {}^{\varepsilon}\eta_k(t))$$

Define

$${}^{\varepsilon}T_t f(x,\xi,\eta_1,\ldots,\eta_k) = E\, f({}^{\varepsilon}x(t/\varepsilon^2), {}^{\varepsilon}\xi(t/\varepsilon^2), {}^{\varepsilon}\eta_1(t/\varepsilon^2), \ldots, {}^{\varepsilon}\eta_k(t/\varepsilon^2))$$

$${}^{\varepsilon}R_\lambda f(x,\xi,\eta_1,\ldots,\eta_k) = E\{\int_0^\infty e^{-\lambda t} f({}^{\varepsilon}x(t/\varepsilon^2), {}^{\varepsilon}\xi(t/\varepsilon^2), {}^{\varepsilon}\eta_1(t/\varepsilon^2), \ldots, {}^{\varepsilon}\eta_k(t/\varepsilon^2))\}$$

$${}^{0}R_\lambda f(x,\eta_1,\ldots,\eta_k) = (\lambda - A_k)^{-1} f$$

Theorem If $f \in L_0^k$, then $\lim_{\varepsilon\to 0} {}^{\varepsilon}R_\lambda f = {}^{0}R_\lambda f$

For the proof, we follow the analytical method of Papanicolaou [11].

For this purpose, let $f_0 \varepsilon L_3^k$ and set

$$F_\varepsilon = f_0+\varepsilon f_1+\varepsilon^2 f_2$$
$$f_1 = Z_k f_0$$
$$f_2 = Z_k f_1 - PZ_k f_1$$

Lemma 4.1 $[\varepsilon^{-1}Z_k+\varepsilon^{-2}(P-I)]F_\varepsilon = PZ_k^2 f_0+\varepsilon Z_k f_2.$

Proof. Expand, collect like powers of ε and use the facts that $Pf = f$ $(f\varepsilon L_0^k)$ and $PZ_k f = 0$ $(f\varepsilon L_1^k)$.

Proof of the theorem. From section 3, we have for any $F\varepsilon C$

$$^{\varepsilon}R_\lambda F-\lambda^{-1}F = \lambda^{-1}\, {}^{\varepsilon}R_\lambda(\varepsilon^{-1}Z_k+\varepsilon^{-2}(P-I))F$$

Choosing $F = F_\varepsilon$ above and simplifying, we have

$$^{\varepsilon}R_\lambda(\lambda f_0-PZ_k^2 f_0) = f_0+\varepsilon F_1+\varepsilon^2 F_2$$

where

$$F_1 = f_1-{}^{\varepsilon}R_\lambda f_1-{}^{\varepsilon}R_\lambda Z_k f_2$$

$$F_2 = f_2-\lambda\, {}^{\varepsilon}R_\lambda f_2$$

By hypothesis, there exists a dense set $\tilde{\mathcal{D}} \subseteq L_0^k$ such that given $f\varepsilon\tilde{\mathcal{D}}$, the equation $(\lambda-A_k)f_0 = f$ has a unique solution $f_0 = {}^0R_\lambda f$. Thus $\lim_{\varepsilon\to 0} {}^{\varepsilon}R_\lambda f = {}^0R_\lambda f$, which proves the theorem for $f\varepsilon\tilde{\mathcal{D}}$. To obtain a general $f\varepsilon L_0^k$, use the fact that both ${}^{\varepsilon}R_\lambda$ and ${}^0R_\lambda$ are contraction operators.

5. Coordinate Formulae

It remains to identify the infinitesimal operator with the familiar forms of the Laplacian. We consider three cases.

Case 1 k=0. In this case, let $f \in L_2^0$. Then

$$Z_0 f = \xi^\alpha f_{x^\alpha}$$

$$Z_0^2 f = (\xi^i \frac{\partial}{\partial x^i} - \Gamma_{ij}^k \xi^i \xi^j \frac{\partial}{\partial \xi^k}) \xi^\alpha \frac{\partial f}{\partial x^\alpha}$$

We now use the fact [12] that $P(\xi^i \xi^j) = d^{-1} g^{ij}$. Thus

$$PZ_0^2 f = d^{-1} g^{i\alpha}(f_{x^i x^\alpha} - \Gamma_{i\alpha}^\gamma f_{x^\gamma})$$

which is the covariant form of the Laplace-Beltrami operator.

Case 2 k=1. Again, for $f \in L_2^1$, we have

$$Z_1 f = \xi^\alpha \frac{\partial f}{\partial x^\alpha} - \xi^\alpha \eta^\beta \Gamma_{\alpha\beta}^k \frac{\partial f}{\partial \eta^k}$$

$$Z_1^2 f = (\xi^i \frac{\partial}{\partial x^i} - \Gamma_{ij}^k \xi^i \xi^j \frac{\partial}{\partial \xi^k} - \Gamma_{ij}^k \xi^i \eta^j \frac{\partial}{\partial \eta^k})(\xi^\alpha f_{x^\alpha} - \Gamma_{\alpha\beta}^\gamma \xi^\alpha \eta^\beta f_{\eta^\gamma})$$

When we apply P and simplify, there results the expression

$$PZ_1^2 f = d^{-1} g^{i\alpha}[f_{x^i x^\alpha} - 2\Gamma_{\alpha\beta}^k \eta^\beta f_{x^i \eta^k} + \Gamma_{\alpha\beta}^\gamma \Gamma_{ij}^k \eta^\beta \eta^j f_{\eta^\gamma \eta^k} - \Gamma_{i\alpha}^k f_{x^k} + B_{i\alpha}^k f_{\eta^k}]$$

$$B_{i\alpha}^k \doteq \Gamma_{i\alpha}^j \Gamma_{jm}^k \eta^m + \Gamma_{ij}^m \Gamma_{\alpha m}^k \eta^j - \Gamma_{\alpha m,i}^k \eta^m$$

which corresponds to the formulae obtained by Dynkin [4] and Airault [1].

Case 3 k=d. In this case we propose to show that for $f \in L_2^d$,

$$PZ_d^2 \big|_{O(M)} f = d^{-1} \Delta f$$

where Δ is the horizontal Laplacian on O(M), the bundle of orthonormal frames over M. For this purpose, it suffices to compute in a normal coordinate chart centered at x. Thus $g^{ij}(x) = \delta^{ij}$, $\Gamma^k_{ij}(x) = 0$. From the above examples, the left hand side is

$$d^{-1}[f_{x^i x^i} - \Gamma^k_{ij,i}\eta^j_\alpha \frac{\partial f}{\partial(\eta^k_\alpha)}]$$

where $(\eta^k_1, \cdot\cdot, \eta^k_\alpha)$ are the vectors of an orthonormal frame. On the other hand, the horizontal Laplacian is

$$\Delta f = \sum_{\beta=1}^{d} E^2_\beta f$$

where

$$E_\beta f = \eta^i_\beta \frac{\partial f}{\partial x^i} - \eta^i_\beta \eta^j_\alpha \Gamma^k_{ij} \frac{\partial f}{\partial(\eta^k_\alpha)} \quad .$$

It is now a routine matter to make the stated identification.

6. Stochastic Parallel Displacement

The operator PZ^2_k is the infinitesimal generator of a diffusion process X(t) on $T^k(M)$. This process may be constructed by solving stochastic differential equations [5].

It is our purpose here to justify the term "stochastic parallel displacement". For this purpose, let

$$N_{ij}(x, \eta_1, \ldots, \eta_k) \doteq \langle \eta_i, \eta_j \rangle \qquad (1 \leq i, j \leq k)$$

a real-valued function on $T^k(M)$.

Proposition $N_{ij}(X(t)) = N_{ij}(X(0))$.

Proof. First of all Corollary 2.3 implies that $N_{ij}(X(t))$ is a local martingale. Indeed

$$PZ_k^2 N_{ij} = P(Z_k N_{ij}) = 0$$

Secondly, the associated process of quadratic variation is

$$\langle N_{ij}\rangle = \int_0^t a_{ij} ds$$

$$a_{ij} = PZ_k^2(N_{ij}^2) - 2N_{ij}(PZ_k^2)N_{ij}$$

$$= PZ_k\{2N_{ij}Z_kN_{ij}\} - 2N_{ij}PZ_k\{Z_kN_{ij}\}$$

$$= 0$$

again by Corollary 2.3. Thus the local martingale is constant, which was to be proved.

References

1. H. Airault, Subordination du processus dans le fibre tangent et formes harmoniques, Comptes Rendus Academie des Sciences Paris, Serie A 282 (1976), 1311-1314.

2. A. Bensoussan, J. Lions, G. Papanicolaou, Boundary layers and homogenization of transport processes. Publications of the Research Institute for Mathematical Sciences, Kyoto University vol 15 (1979), 53-158.

3. P. Malliavin, Géometrie Differentielle Stochastique, Univ. Montreal Press, 1978.

4. E. B. Dynkin, Diffusion of Tensors, Dokl. Akad. Nauk, SSSR, Tom 179 (1968), No. 6, p. 532-535.

5. N. Ikeda and S. Watanabe, *Stochastic Differential Equations*, forthcoming book.

6. K. Itô, On stochastic differential equations in a differentiable manifold, Nagoya Math J. 1, 35-47 (1950).

7. K. Itô, The Brownian motion and tensor fields on Riemannian manifolds, International Congress of Mathematicians, Stockholm 1962 p. 536-539.

8. K. Itô, Stochastic parallel displacement, in Springer Verlag Lecture Notes in Mathematics, vol. 451, pp. 1-7.

9. P. Malliavin, Formules de la moyenne, calcul de perturbations et théormes d'annulation pour les formes harmoniques, Journal of Functional Analysis 17 (1974), 274-291.

10. H. P. McKean, Stochastic Integrals, Academic Press, 1969.

11. G. Papanicolaou, Probabilistic Problems and Methods in Singular Perturbations, Rocky Mountain Journal of Mathematics 6 (1976), 653-673.

12. M. Pinsky, Isotropic transport process on a Riemannian manifold, Transactions of the A.M.S. 218 (1976), 353-360.

13. M. Pinsky, Stochastic Riemannian Geometry, in Probabilistic Analysis and Related Topics, Academic Press, 1978.

14. P. H. Roberts and H. D. Ursell, Random walk on a sphere and on a differentiable manifold, Philosophical Transactions Royal Society 252 (1962) 317-356.

15. S. T. Yau, The heat kernel of a complete Riemannian manifold, Journal des Mathématiques Pures et Appliquées, 1978. vol. 57, 191-201

BESSEL PROCESSES AND INFINITELY DIVISIBLE LAWS

by

Jim PITMAN (*) and Marc YOR (**)

1. INTRODUCTION.

In recent years there has been a renewed interest in the topic of infinitely divisible laws on the line, something which may perhaps be attributed to the belated realisation that the Lévy-Khintchine description of such laws does not always provide a practical criterion for deciding whether a given completely monotone function is in fact the Laplace transform of an infinitely divisible distribution.
Bondesson's paper [2] provides an excellent account of recent developments, most of which have been analytic in nature.

One aim of our paper is to bring out the relevance of probabilistic methods in this subject. It seems that many of the most important infinitely divisible laws discovered recently can be associated with a diffusion process in such a way that decompositions of the law derive from first entrance or last exit decompositions for the diffusion.

We shall focus attention on results which show ratios of Bessel functions to be completely monotone ; in the mean time, we are led to new results which concern the two-parameter family of (generalized) Bessel diffusions, and which bring out the significant rôle played by these processes in the description of Bessel bridges and other processes closely connected with Brownian motion (BM) in R^d.

(*) University of California
Department of Statistics
BERKELEY

Calif. 94720 - U.S.A.

Research of this author was supported in part by NSF Grant No MCS 78-25301

(**) Université P. et M. Curie
Laboratoire de Calcul des Probabilités
4 place Jussieu
F- 75230 PARIS CEDEX 05

Consider now the usual modified Bessel functions $I_\nu(r)$ and $K_\nu(r)$. P. Hartman [21] showed by purely analytic arguments that, for any $r > 0$, the functions

$$(1.a)\qquad \lambda \to \frac{I_{\sqrt{2\lambda}}(r)}{I_o(r)} \quad \text{and} \quad (1.b)\qquad \lambda \to \frac{K_o(r)}{K_{\sqrt{2\lambda}}(r)}$$

are the Laplace transforms of infinitely divisible distributions on $\mathbb{R}_+$, which we shall call the first and second Hartman laws with parameter r. In fact, Hartman showed more generally that the same result holds for

$$(1.c)\qquad \lambda \to \frac{I_{\sqrt{2\lambda}}(r)}{I_o(r)} \times \frac{I_o(R)}{I_{\sqrt{2\lambda}}(R)}$$

and $\qquad (0 < r < R < \infty)$

$$(1.d)\qquad \lambda \to \frac{K_o(r)}{K_{\sqrt{2\lambda}}(r)} \times \frac{K_{\sqrt{2\lambda}}(R)}{K_o(R)} .$$

The Laplace transform *(1.a)* appears in Ito-Mc Kean [28] on p. 271, although in an integrated form, and it also appears in Edwards [7] and again Hartman and Watson [20]. Probabilistic interpretations of *(1.a)*, *(1.c)* and *(1.d)* were given in Yor [51], and will be reviewed in section 2. In particular, if $(R_t, t \geq 0)$ is the radial part of a two dimensional Brownian motion, then, for any $\nu \geq 0$:

$$(1.e)\qquad E\left[\exp(-\frac{1}{2}\nu^2 \int_0^t ds/R_s^2) \mid R_o = a, R_t = y\right] = I_\nu(ay/t) \;/\; I_o(ay/t)$$

That is to say, the first Hartman law with parameter ay/t is the conditional law of $\int_0^t ds/R_s^2$ given $R_o = a$ and $R_t = y$.

The key to *(1.e)*, and apparently to most results concerning ratios of Bessel functions as completely monotone (c.m.) functions of $\sqrt{\nu}$, is a suitable Cameron - Martin - Girsanov type result which relates one Bessel distribution to another. One advantage of this approach is that many ratios of Bessel functions $I_\nu(r)$ and $K_\nu(r)$, known from the literature to be c.m. in $\sqrt{r}$ and in $\sqrt{\nu}$ separately, turn out to be jointly c.m. in these variables.

The representation *(1.e)* of the first Hartman law is disappointing in several respects. It involves a conditioned process (or <u>Bessel bridge</u>, which is an inhomogeneous diffusion), and as a result the connection with the other Hartman laws *(1.c)* is so obscured that it is not clear why the law is infinitely divisible.
Also, the second Hartman law is still more deeply hidden in the Bessel bridge.
These matters are rectified with the help of the two parameter family of Bessel diffusions $BES(\mu,\delta)$ of Watanabe [46], with index $\mu \geq 0$ and drift $\delta \geq 0$. Our justification for the term "drift" is the result of Rogers and Pitman [40] that if X is a BM in R^d with a drift vector of magnitude δ, started at the origin (which is important if $\delta > 0$!) then the radial part of X is a $BES(\mu,\delta)$ with index $\mu = (d-2)/2$.
In section 4 we present $BES(\mu,\delta)$ as a $BES(\mu)$ process "conditioned to reach a high level before an independent exponentially distributed random time with rate $\frac{1}{2}\delta^2$", a notion made precise in Section 3.

We then obtain a new and unified presentation of all the Hartman laws in terms of a single $BES(0,\delta)$ process $(\rho_t, t \geq 0)$ started at zero. To be precise, take

$$(1.f) \qquad R_t = \left[(X_t+\delta t)^2 + Y_t^2\right]^{1/2}$$

where X and Y are independent real valued BMs under P with $X_o = Y_o = 0$. For $x > 0$ let

$$(1.g) \qquad T_x = \inf\{t : R_t = x\}.$$

be the first time that the radial motion hits x. We show that

$$(1.h) \qquad E \exp\left[-\frac{1}{2}\nu^2 \int_{T_x}^{\infty} ds/R_s^2\right] = I_\nu(\delta x) \,/\, I_o(\delta x).$$

Now, the strong Markov property of $(R_t, t \geq 0)$ makes the infinite divisibility of the first Hartman law plain, by decomposition of the integral at T_y for $y > x$, and the laws with the Laplace transforms *(1.c)* appear as factors. Next, time reversal reveals a "dual" representation for the second Hartman law, namely

$$E \exp\left[-\frac{1}{2} \nu^2 \int_{L_x}^{\infty} ds/R_s^2\right] = K_o(\delta x) \;/\; K_\nu(\delta x)$$

where $L_x = \sup\{t : R_t = x\}$ is the last time at x.

In view of the last exit decomposition of Pittenger and Shih [39] (see also Getoor and Sharpe [16], Williams [49]), the infinite divisibility of the second Hartman law and associated factors with transforms *(1.d)* is now obvious.

In section 5 we shall prove the following theorem :

Theorem (1.1) : Let (X_t) be a BM in $\mathbb{R}^d$, $d \geq 2$, started at 0, with a constant drift vector $v \neq 0$, and let $R_t = |X_t|$. Then

$$(1.j) \qquad X_t = R_t \,\Theta\left(\int_t^{\infty} ds/R_s^2\right)$$

where $(\Theta(t), t \geq 0)$ is a BM in the unit sphere S^{d-1} of $\mathbb{R}^d$, starting at $\Theta_o = v/|v|$, and independent of the BES$((d-2)/2, |v|)$ process $(R_t, t \geq 0)$.

The reader is warned that it is critically important in the above theorem that X starts at 0.

This result is to be compared with the classical skew-product for $\mathbb{R}^d$-valued BM. (Ito - Mc Kean [28], § 7.15), which applies when $v = 0$. Amongst other things, the skew-product representation *(1.j)* explains the result of Reuter, mentioned in the discussion of Kendall's paper [30], that for BM with drift, with T_x as in *(1.g)*, the hitting angle $\Theta(T_x)$ is independent of the hitting time T_x. Indeed, it is plain from *(1.j)* that $\Theta(T_x)$ is independent of the whole radial motion prior to T_x, which improves the result of Wendel [48], and the same holds true for L_x instead of T_x.

Inspection of *(1.h)* and *(1.j)* reveals that we have a new representation of the result of Hartman and Watson [20] that for $d = 2$ the distribution of $\Theta(T_x)$ on the circle (which is von Mises with parameter δx) is a mixture of wrapped normal distributions, as well as the corresponding results for $d > 2$. By the skew-product representation of complex BM with no drift, another such representation of the von Mises distribution on the circle, with parameter ay/t, is provided by *(1.e)*.

We note here the remarkable fact that in all these representations arising naturally from BM the mixing law is the same. Remarkable, because, as will be shown in Section 10 to settle a question raised by Hartman and Watson themselves, the mixing law is not unique.

Also in Section 5, we show how the invariance of the $BES(\mu,\delta)$ family under time inversion, discovered by Watanabe [46], enables one to give a simple representation of the Bessel bridges in terms of this family. Consequently, the second Hartman law and the factors whose transforms appear in *(1.c)* and *(1.d)* may be reinterpreted in the context of Bessel bridges, but the actual translations are left to the reader.

It also turns out that the integrals of ds/R_s^2 appearing in the formulae above transform very simply in this representation, which helps explain the dependence of the formula *(1.e)* on ay/t alone, and the resulting ubiquity of the first Hartman law.

In section 6 we leave Bessel processes for a while, to develop a simple general formula for the density of the (infinitely divisible) law of the last time L_y that a transient diffusion on the line hits a point y, but we return to apply this result to Bessel processes in Section 7. In particular, if $(R_t, t \geq 0 \; ; \; P_x^\nu)$ is a $BES(\nu,0)$ started at x, we recover the result of Getoor [13] that

$$(1.k) \qquad P_0^\nu(L_y \in dt) = (y^2/2)^\nu [\Gamma(\nu) t^{\nu+1}]^{-1} \exp(- y^2/2t) dt,$$

with

$$(1.l) \qquad E_0^\nu \exp(- \frac{\alpha^2}{2} L_y) = 2\Gamma(\nu)^{-1} (\alpha y/2)^\nu K_\nu(\alpha y)$$

There are equally explicit formulae for $BES(\nu,\delta)$.

Let κ_ν^y be the probability on $(0,\infty)$ defined by the right hand side of *(1.k)*. (Note that $2y^2$ is just as scale parameter, and that $\kappa_\nu^{1/\sqrt{2}}$ will be denoted κ_ν in Section 9). The laws κ_ν^y seem to have been first encountered by Hammersley [19], who showed that

$$\text{(1.m)} \qquad \kappa_\nu^y \ \underline{\text{is the}}\ W_x \ \underline{\text{distribution of}} \int_0^{T_0} X_t^{\rho-2}dt \text{ for } x = (y/2\nu)^{2\nu}, \rho = 1/\nu$$

where W_x governs the real valued BM $(X_t, t \geq 0)$ starting at x, and T_a is the hitting time of a for this BM. These laws appeared again in Ismail and Kelker [25], where it was pointed out that if $(B(t), t \geq 0)$ is a BM on the line starting at 0, independent of R with distribution $\kappa_\nu^{\sqrt{\nu}}$, then $\sqrt{2}\,B(R)$ has the Student t-distribution with $2\nu^2$ degrees of freedom, and that consequently the infinite divisibility of the Student t-distribution is implied by the infinite divisibility of κ_ν^y. Grosswald [18] established this infinite divisibility of κ_ν^y by an analytic argument, subsequently simplified by Ismail [23], but these authors seem to have been unaware of Hammersley's result *(1.m)*, from which the infinite divisibility of κ_ν^y is obvious by decomposition of the integral at T_a for $a < x$. Similarly, the infinite divisibility of κ_ν^y is plain in *(1.k)* by a decomposition at L_b for $b < y$. The connection between the two representations of κ_ν^y in *(1.k)* and *(1.m)* is provided by representation of Bessel processes used in Getoor - Sharpe [15] and a time reversal. Indeed, Getoor and Sharpe rediscovered *(1.m)* by remarking that if the process $(2\rho^{-1} X_t^{\rho/2}, 0 \leq t \leq T_0; W_x)$ in *(1.m)* is time changed by the additive functional $A_t = \int_0^t X_s^{\rho-2}\,ds$, the result is a Bessel process with index $-\nu$ started at y, where $\nu = \rho^{-1}$ and $y = 2\rho^{-1} x^{\rho/2}$, while the time reversal result of Williams [49] or Sharpe [41] shows that this latter process with index $-\nu$ when reversed from the time $A(T_0)$ that it hits zero is a $BES(\nu)$ started at 0 and killed at the time $L_y = A(T_0)$ that it last hits y (see *Theorem (3.3)* and *Remark (4.2)* (ii) below).

In section 8 we apply Watanabe's time inversion theorem to obtain the distributions of

$$\tau_y = \inf\{t : R_t = ty\} \quad \text{and} \quad \sigma_r = \sup\{t : R_t = ty\}$$

for a $BES(\mu,\delta)$ process R. In particular, the fact, noted by both Hammersley [19] and Ismail and Kelker [25], that κ_ν^y is the distribution of the inverse of a gamma variable with index ν and scale parameter $y^2/2$, implies that τ_y for a $BES(\mu,0)$ has this gamma distribution. We also encounter an infinitely divisible distribution discovered in a different context by Feller [9].

Section 9 is devoted to the probabilistic interpretation of certain ratios of Bessel functions studied by Ismail [24] and Ismail and Kelker [26] ; we give new proofs and extensions of many of their results.
In particular, we consider the functions of $\alpha,\nu > 0$

$$(1.n) \quad (i) \ \frac{1}{\sqrt{\alpha}} \frac{I_\nu(\sqrt{\alpha})}{I_{\nu-1}(\sqrt{\alpha})} \quad \text{and} \quad (ii) \ \frac{1}{\sqrt{\alpha}} \frac{K_{\nu+1}(\sqrt{\alpha})}{K_\nu(\sqrt{\alpha})}.$$

View these functions first for fixed $\nu > 0$.
According to Cieselski - Taylor [5] and Getoor - Sharpe [15], 2ν times the function *(1.n)* (i) is the Laplace transform in α of the total time spent below $y = 1/\sqrt{2}$ by $BES(\nu)$ started at zero. Of course, the total time below y is the time below y before L_y. In our turn, we show that 2ν times the function *(1.n)* (ii) is the Laplace transform in α of the time $BES(\nu)$ spends <u>above</u> y before L_y, and we give a formula for the joint transform of the times spent above and below y before L_y. Now for fixed α, Ismail showed in [24] that the function *(1.n)* (ii) is c.m. as a function of ν^2. We show that the same is true of *(1.n)* (i), and that in fact both functions in *(1.n)* are jointly c.m. in (α,ν^2).

Many of the results of Ismail and Kelker mentioned above have been recently extended by Hartman [22], from Bessel functions to solutions of more general disconjugate second order differential equation depending on a a parameter μ, and in particular to solutions of Whittaker's form of the confluent hypergeometric equation. While we still do not know how to interpret Hartman's most general results probabilistically, in the concluding remarks of Section 12 we mention a Bessel process interpretation of some of this results for Whittaker functions, which is related to the pole-seeking Bessel processes of Kendall [30]. Finally, classical formulae for Bessel functions, used throughout the paper, are displayed in Section 13 for the reader's convenience.

TABLE OF CONTENTS.

2. BESSEL PROCESSES (Without drift).

A Bessel process with index ν, (and no drift), to be denoted BES(ν), is the diffusion process with values in $\mathbb{R}_+ = [0,\infty)$ whose generator is

$$G_\nu = \frac{1}{2}\frac{d^2}{dx^2} + \frac{2\nu+1}{2x}\frac{d}{dx}.$$

We shall for the most part only be interested in BES(ν) for $\nu \geq 0$, when 0 is an entrance boundary point for the process, which is never visited again. For $\nu < 0$ one must specify boundary conditions at 0 to describe the process completely, but we shall not go into this, as on the only occasion when a negative index arises we shall be stopping the process when it first hits zero.

Often $d = 2\nu + 2$ is called the "dimension" of BES(ν), since BES($\frac{1}{2}d - 1$) is the radial part of Brownian motion in R^d for $d = 1,2,\ldots$ The index $\nu = 0$, corresponding to $d = 2$, will play a fundamental rôle in what follows.

On the space $\Omega = C(\mathbb{R}_+,\mathbb{R}_+)$, let $R_t(\omega) = \omega(t)$, $\underline{\underline{F}} = \sigma\{R_s,\ s \geq 0\}$, and $\underline{\underline{F}}_t = \sigma\{R_s,\ s \leq t\}$, $t \geq 0$. For $x \in \mathbb{R}_+$, denote the law on $(\Omega, \underline{\underline{F}})$ of a BES(ν) started at x by P^ν_x. We recall now a number of basic results from [51]. The key result for comparison of BES(ν) processes with different indices ν is the following Proposition, which is a slight refinement of *Lemme (4.5)* of [51].

Proposition (2.1) : Let $a > 0$, $\nu > 0$ and let T be an $(\underline{\underline{F}}_{t+})$ stopping time such that

(2.a) $\quad E^0_a\, T^{\nu/2} < \infty.$

Then

(2.b) $\quad P^\nu_a(T < \infty) = 1$, and

(2.c) on $\underline{\underline{F}}_{T+}$, $\quad \dfrac{dP^\nu_a}{dP^0_a} = \left(\dfrac{R_T}{a}\right)^\nu \exp\left(-\frac{1}{2}\nu^2\int_0^T ds/R_s^2\right).$

Remarks (2.2) :

1) The probabilities P_0^μ and P_0^ν are mutually singular on $\underline{\underline{F}}_{0+}$ for $\mu,\nu \geq 0$ with $\mu \neq \nu$, because *theorem (3.3)* (i) of Shiga and Watanabe [42] shows that the BES(μ) and BES(ν) processes escape from zero at different rates.

2) Similarly, for $a,\mu,\nu \geq 0$ with $\mu \neq \nu$ the laws P_a^μ and P_a^ν are mutually singular on $\underline{\underline{F}}_{(t,\infty)} = \sigma(R_u, u \geq t)$. This follows from the previous remark, by time inversion (see *theorem (5.5)* below).

Proof of the proposition : For T bounded this is just a restatement of *Lemme (4.5)* in [51]. To pass to T satisfying *(2.a)*, consider the P_a^0 martingale L_t^ν defined by

$$L_t^\nu = (R_t/a)^\nu \exp(-\frac{1}{2}\nu^2 \int_0^t ds/R_s^2),$$

and observe that $(L_{t\wedge T}^\nu)$ is uniformly integrable. Indeed :

$$\sup_{t\leq T} L_t^\nu \leq \sup_{t\leq T} (R_t/a)^\nu,$$

where the last random variable is integrable by Burkholder's inequality and the assumption *(2.a)*. □

As a consequence of *Proposition (2.1)*, random variables of the form

$$\textit{(2.d)} \qquad \int_S^T ds/R_s^2$$

will appear throughout the paper.

In an attempt to reduce the acreage of formulae, we now introduce the notation $C(S,T)$ for the integral *(2.d)*, writing simply C_T or $C(T)$ for $C(0,T)$.

"C" stands for "clock", to remind the reader that when $d = 2\nu + 2$ is an integer these integrals provide the stochastic clock (or random time substitution) which reduces the angular part of Brownian motion in R^d to a Brownian motion on the sphere S^{d-1}. (Ito - Mc Kean [28], p. 270). We shall also use later the fact that $BES(\nu)$ started at $x > 0$, and time-changed with the inverse of (C_t) may be expressed as : $x\exp\{\beta_t + \nu t\}$ where β_t is a real-valued Brownian motion starting at 0 (this was already noticed and used by D. Williams [49] for $d = 3$).

Suppose now that T is an $(\underline{\underline{F}}_{t+})$ stopping time with

$$(2.e) \qquad E_a^o T^\alpha < \infty \text{ for all } \alpha > 0.$$

According to *Proposition (2.1)*, for $\lambda,\mu \geq 0$

$$(2.f) \text{ on } \underline{\underline{F}}_{T+}, \quad \frac{dP_a^\lambda}{dP_a^\mu} = \left(\frac{R_T}{a}\right)^{\lambda-\mu} \exp(-\frac{1}{2}\nu^2 C_T),$$

where $\qquad \nu^2 = \mu^2 - \lambda^2$.

Conditioning on R_T now shows that the distributions on $\mathbb{R}_+$ of R_T under P_a^λ and P_a^μ are mutually absolutely continuous with density

$$(2.g) \qquad \frac{P_a^\lambda(R_T \in dr)}{P_a^\mu(R_T \in dr)} = \left(\frac{r}{a}\right)^{\lambda-\mu} E_a\left[\exp(-\frac{1}{2}\nu^2 C_T) \mid R_T = r\right].$$

On the other hand, from Molchanov [37] (see also Kent [31]) we know that the $BES(\nu)$ process has transition density

$$(2.h) \qquad p_t^\nu(a,r) = \frac{r}{t}\left(\frac{r}{a}\right)^\nu \exp(-\frac{a^2+r^2}{2t})\, I_\nu(\frac{ar}{t})$$

By comparison of the two formulae above, one immediately obtains the result of *théorème (4.7)* in [51], namely for $\mu, \nu \geq 0$

$$(2.i) \qquad E_a\left[\exp(-\tfrac{1}{2}\nu^2 C_t) \mid R_t = r\right] = I_\lambda(\tfrac{ar}{t}) \,/\, I_\mu(\tfrac{ar}{t}),$$

where $\lambda = (\mu^2 + \nu^2)^{1/2}$.

By a similar application of *Proposition (2.1)*, the formula of Kent [31] for the Laplace transform of the hitting time

$$T_b = \inf\{t : R_t = b\}$$

of $BES(\nu)$ develops into the following joint transform of T_b and $C(T_b)$, which was given as *théorème (4.10)* in [51] :

Proposition (2.3) : For $\mu, \nu \geq 0$, $\beta, b > 0$,

$$(2.j) \qquad E_a^\mu \exp\left[-\tfrac{1}{2}\nu^2 C(T_b) - \tfrac{1}{2}\beta^2 T_b\right] = \left(\frac{b}{a}\right)^\mu \frac{\mathcal{L}_\lambda(a\beta)}{\mathcal{L}_\lambda(b\beta)}$$

where $\lambda = (\mu^2 + \nu^2)^{1/2}$, and $\mathcal{L} = K$ for $b < a$, $\mathcal{L} = I$ for $a < b$. In particular

$$(2.k) \qquad E_a^\mu\{\exp\left[-\tfrac{1}{2}\nu^2 C(T_b)\right] ; T_b < \infty\} = \left(\frac{b}{a}\right)^{\mu + \varepsilon\lambda}$$

where $\varepsilon = +1$ if $b < a$, $\varepsilon = -1$ for $a < b$.

From the formula *(2.j)* it is easy to see that the Hartman laws with transforms *(1.a)* and *(1.d)* can be described as the distributions of $C(T_b)$ for a $BES(0)$ process started at a and conditioned to hit another level b before an independent exponentially distributed random time. We return to this point in Section 4 after first considering this kind of conditioning operation in a slightly more general setting.

However, we feel that, since our Cameron-Martin-Girsanov type *Proposition (2.1)* is one of the keys to our results, we ought to give another simple application of it, before passing to (perhaps) less standard manipulations later (see section 4 for instance).
Indeed, we now show how Wendel's formulae ([47]) may be deduced from *Proposition (2.1)*.

Let $(B_t, t \geq 0)$ be a BM in $\mathbb{R}^d$, where $d \geq 2$, starting at $x \neq 0$. Put $r = |x|$, $\theta = x/r$, $R_t = |B_t|$, $\Theta_t = B_t/R_t$, and consider for example the problem of calculating

$$(2.l) \qquad E(S_n^\ell(\Theta_T) \exp(-\tfrac{1}{2} s^2 T) \; ; \; R_T = a)$$

where $T = T_a \wedge T_b$ is the first time R hits a or b, and S_n^ℓ is a spherical harmonic of degree n. By the skew-product decomposition of X (see Itô - Mc Kean [28], p. 270), one has, for $t \geq 0$:

$$(2.m) \qquad E(S_n^\ell(\Theta_t) \,|\, R_u, u \geq 0) = S_n^\ell(\theta) \exp\left(-\tfrac{1}{2} n(n+d-2) C_t\right)$$

where C_t is the clock defined in *(2.d)*.

On the other hand, from *Proposition (2.1)* we have that for $\mu, \nu \geq 0$

$$(2.n) \qquad E_r^\mu[\exp(-\tfrac{1}{2} \nu^2 C_T - \tfrac{1}{2} s^2 T) \; ; \; R_T = a]$$

$$= (a/r)^{\mu-\lambda} E_r^\lambda[\exp(-\tfrac{1}{2} s^2 T) \; ; \; R_T = a]$$

where $\lambda = (\mu^2 + \nu^2)^{1/2}$, and we hope the reader will forgive us for using the same notation for the co-ordinate process R as for $R = |B|$. Thus, taking $\mu = (d-2)/2$ and $\nu^2 = n^2 + n(d-2)$ in *(2.m)*, it emerges that the expectation *(2.l)* is identical to

$$(2.o) \qquad S_n^\ell(\theta) \, (a/r)^{-n} \, E_r^\lambda \, [\exp(-\tfrac{1}{2} s^2 T) \; ; \; R_T = a]$$

where $\lambda = (d-2)/2 + n$.

Finally, this last P_r^λ expectation can be calculated from the known Laplace transforms of T_a and T_b under P_r^λ by a routine application of the strong Markov property (see Itô - Mc Kean [28] p. 30), and it is found to be

$$(2.p) \qquad (\frac{a}{r})^\lambda \frac{I_\lambda(bs)\, K_\lambda(rs) - I_\lambda(rs)\, K_\lambda(bs)}{I_\lambda(bs)\, K_\lambda(as) - I_\lambda(as)\, K_\lambda(bs)}.$$

Substituting *(2.p)* for the P_r^λ expectation in *(3.o)* now yields the formula (9) of Wendel [47], and we leave it to the reader to check that the other formulae of Wendel can be obtained in exactly the same way.

3. CONDITIONED DIFFUSIONS.

In this section let

$$\{X_t, 0 \le t < \zeta \le \infty\ ;\ P_x, x \in (A,B)\}$$

be a regular diffusion on a sub-interval (A,B) of $[-\infty,\infty]$. To avoid unnecessary complications, we assume that

$$\zeta = \inf\{t > 0 : X_{t-} = A \text{ or } B\}, \tag{3.a}$$

so the process is killed when it reaches either boundary. Given $\alpha \ge 0$, we wish to record some basic results concerning the diffusions $\{P_x^{\alpha\uparrow}, x \in (A,B)\}$ and $\{P_x^{\alpha\downarrow}, x \in (A,B)\}$ obtained by first killing the original diffusion at a constant rate α, then conditioning this killed process to hit B in the $\uparrow$ case, and to hit A in the $\downarrow$ case. Since the original process may never hit these boundaries, as for example in the application to Bessel processes which we have in mind, this conditioning is to be understood in the sense of Doob [6] and Williams [49]. Following Williams's description of this operation with no killing (i.e. $\alpha = 0$) in section 2 of [49], we take $P_x^{\alpha\uparrow}$ to be defined by the requirement that for each $x < b < B$, the process X run up to the time T_b has the same law under $P_x^{\alpha\uparrow}$ as it does under P_x conditional on $(T_b < U_\alpha)$, where U_α is an exponentially distributed killing time with rate α independent of X. Putting $\underline{\underline{F}}_t = \sigma(X_s, 0 \le s \le t)$, this is just to say that for $x < b < B$

$$\frac{dP_x^{\alpha\uparrow}}{dP_x} = e^{-\alpha T_b} \Big/ \phi_\alpha(x,b) \quad \text{on } \underline{\underline{F}}_{T_b^+}, \tag{3.b}$$

where

$$\phi_\alpha(x,y) = E_x\, e^{-\alpha T_y}. \tag{3.c}$$

For $\alpha = 0$, $e^{-\alpha T_y}$ should be interpreted as the indicator of the event $(T_y < \infty)$. Note that $\phi_\alpha(x,y) > 0$ for all $\alpha \ge 0$, $x,y \in (A,B)$ by the assumption that X is regular.

Before going further, we recall some well known facts concerning $\phi_\alpha(x,y)$, which may be found for example in either Itô and Mc Kean [28] or Breiman [3]. Take a point $x_o \in I$ and define

$$\text{(3.d)} \qquad \phi_{\alpha\uparrow}(y) = \phi_\alpha(y,x_o), \qquad y \leq x_o$$

$$= 1/\phi_\alpha(x_o,y), \quad y > x_o.$$

Since the identity

$$\text{(3.e)} \qquad \phi_\alpha(x,z) = \phi_\alpha(x,y)\ \phi_\alpha(y,z),$$

is valid whenever $x \leq y \leq z$, one gets :

$$\text{(3.f)} \qquad \phi_\alpha(x,y) = \phi_{\alpha\uparrow}(x) \,/\, \phi_{\alpha\uparrow}(y), \quad x < y,$$

which shows that the choice of reference point x_o affects $\phi_{\alpha\uparrow}$ only by a constant factor. Similarly there is a function $\phi_{\alpha\downarrow}$ which gives the analog of *(3.f)* for $x > y$. These functions $\phi_{\alpha\uparrow}$ and $\phi_{\alpha\downarrow}$ may be determined as solutions subject to appropriate boundary conditions of the equation

$$\text{(3.g)} \qquad (G-\alpha)\phi = 0,$$

where G is the generator of the diffusion.

Proposition (3.1) : Let T be an $(\underline{\underline{F}}_t)$ stopping time, $x \in I$

$$\frac{dP_x^{\alpha\uparrow}}{dP_x} = e^{-\alpha T}\ \phi_{\alpha\uparrow}(X_T) \,/\, \phi_{\alpha\uparrow}(x) \quad \text{on} \quad \underline{\underline{F}}_{T+} \cap (T < \zeta)$$

Proof : For $T \leq T_b$, this follows from *(3.b)* after conditioning on $\underline{\underline{F}}_{T+}$, using *(3.f)* and the strong Markov property. For general T consider $T \wedge T_b$, let b tend to B and use *(3.a)*. □

It follows easily from the above proposition that the probabilities $\{P_x, x \in (A,B)\}$ define a new diffusion process which is transient with

$$(3.h) \qquad P_x^{\alpha\uparrow}(X_{\zeta-} = B) = 1,$$

except if $\alpha = 0$ and the original diffusion $\{P_x\}$ is recurrent, when $P_x^{o\uparrow} = P_x$.

Clearly, the probability $P_x^{\alpha\uparrow}(\zeta < \infty)$ is either 1 for all x or 0 for all x.

Since

$$(3.i) \qquad E_x^{\alpha\uparrow} e^{-\beta\zeta} = \lim_{b \to B} E_x^{\alpha\uparrow} e^{-\beta T_b},$$

the $\alpha\uparrow$ motion hits B in finite time a.s. iff the limit $\lim_{b \to B} \phi_{\alpha\uparrow}(b) \big/ \phi_{\alpha+\beta,\uparrow}(b)$, is strictly positive for some (or equivalently all) $\beta > 0$.

Let $P_t(x,dy)$ be the transition function of the original diffusion. Then from *Proposition (3.1)*, it is plain that the $\alpha\uparrow$ diffusion has transition function

$$(3.j) \qquad P_t^{\alpha\uparrow}(x,dy) = P_t(x,dy)\, e^{-\alpha t}\, \phi_{\alpha\uparrow}(y) \big/ \phi_{\alpha\uparrow}(x).$$

A formal calculation based on *(3.j)* shows that the generator $G^{\alpha\uparrow}$ must be

$$(3.k) \qquad G^{\alpha\uparrow} = \phi_{\alpha\uparrow}^{-1}(G-\alpha)\phi_{\alpha\uparrow},$$

and in particular if

$$G = a(x)\frac{d^2}{dx^2} + b(x)\frac{d}{dx},$$

a further calculation using *(3.g)* reveals that

$$(3.l) \qquad G^{\alpha\uparrow} = G + 2\, a(x)\frac{\phi'_{\alpha\uparrow}(x)}{\phi_{\alpha\uparrow}(x)}\frac{d}{dx},$$

where $\phi'_{\alpha\uparrow} = \frac{d}{dx}\phi_{\alpha\uparrow}$. As we shall not make any use of these formulae for generators in what follows, we shall not attempt a careful justification, but rather refer the reader to Kunita [33] and Meyer [34], where such matters are deftly handled in a much more general context.

Of course, after some obvious substitutions such as $\phi_{\alpha\downarrow}$ for $\phi_{\alpha\uparrow}$, everything above applies equally well to the $\alpha\downarrow$ process obtained by conditioning X killed at rate α to hit the lower boundary point A instead of the upper boundary point B. As the reader can easily verify, we have

Proposition (3.2) : If either the $\alpha\uparrow$ or $\alpha\downarrow$ process is conditioned $\beta\uparrow$, the result is the $(\alpha + \beta)\uparrow$ process.

In particular, taking $\beta = 0$, we see that the $\alpha\uparrow$ and $\alpha\downarrow$ processes are dual in the sense of section 2.5 of Williams [49]. As a consequence, either process can be presented as a time reversal of the other. To be precise, for $y \in (A,B)$, let

$$(3.m) \qquad L_y = \sup\{t : X_t = y\}.$$

Then we have

Theorem (3.3) (Williams [49], *theorem (2.5)*) :

Fix $\alpha \geq 0$. Suppose that the $\alpha\downarrow$ process hits A in finite time with probability one.
Then A is an entrance point for the $\alpha\uparrow$ process, and for each $y \in (A,B)$ the processes

$$\{X(\zeta - t),\ 0 < t < \zeta;\ P_y^{\alpha\downarrow}\} \text{ and } \{X(t), 0 < t < L_y\ ;\ P_A^{\alpha\uparrow}\}$$

are identical in law.

Remark : Williams states this theorem in the case $\alpha = 0$, starting from a process satisfying hypotheses which make it identical to its own $0\uparrow$ process, and with the rôles of A and B reversed.

However the apparent extension above to a general $\alpha \geq 0$ is only a superficial one, by *Proposition (3.2)* with $\beta = 0$. Williams proved his theorem by first establishing a special case and then arguing that the result could be transferred to the general case by the method of time substitution.

The result can also be deduced from the time reversal theorem of Nagasawa [38], via the work of Sharpe [41]. The connection with Sharpe's work is easily made after noting that

(3.n) $$s(x) = -\,\phi_{\alpha\downarrow}(x) \Big/ \phi_{\alpha\uparrow}(x)$$

serves as a scale function for the $\alpha\uparrow$ process, with $s(B-) = 0$ and $s(A+) = \infty$.

The reader should be well prepared by now for the conditioned Bessel processes of the next section.

But lest our change of hitting rate from α to $\frac{1}{2}\delta^2$, and our use of the term"drift" for δ in that section seem mysterious, we recommend the following trivial exercise :

Exercise (3.4) :

Show that, for Brownian motion (BM) on the line with zero drift, $\delta > 0$,

i) $\phi_{\frac{1}{2}\delta^2\uparrow}(x) = e^{+\delta x}$;

ii) the $\frac{1}{2}\delta^2\uparrow$ process is BM with drift $+\,\delta$;

iii) changing $\uparrow$ to $\downarrow$ above changes $+$ to $-$;

iv) for BM with drift γ, the $\frac{1}{2}\delta^2\uparrow$ process is BM with drift $+\sqrt{\gamma^2+\delta^2}$;

v) the recipe iii) applies to iv) too.

(Hint : use *(3.2)*).

To conclude this section, we record the following result, which will not be required until the end of section 4.

Given a random time L, define $\underline{\underline{F}}_{L-} = \sigma\{F_t(t < L),\ F_t \in \underline{\underline{F}}_t, t \geq 0\}$, and let L_y be as in *(3.m)*.

Theorem (3.5) : <u>Suppose that the regular diffusion</u> $\{P_x\}$ <u>is transient, meaning that for</u> $x,y \in (A,B)$, $P_x(L_y < \infty) = 1$. <u>Then</u>

(3.n) $$\frac{dP_x^{\alpha\uparrow}}{dP_x} = c(x,y,\alpha)e^{-\alpha L_y} \quad \underline{\text{on}}\ \underline{\underline{F}}_{(L_y)-} \cap (L_y > 0), \quad \underline{\text{where}}$$

(3.o) $$c(x,y,\alpha) = \phi_{\alpha\uparrow}(y) \,/\, [\phi_{\alpha\uparrow}(x)\, E_y e^{-\alpha L_y}]$$

Proof : For $b > y$, let

$$L_{yb} = \sup\{t : X_t = y,\ 0 < t < T_b\}.$$

The following equalities result from the use of first the strong-Markov property of the $\alpha\uparrow$ process at time T_b, then *(3.b)*, and finally the last exit decomposition at time L_{yb} of the pre-T_b process :

$$P_x^{\alpha\uparrow}(F_t, t < L_y < T_b) = P_x^{\alpha\uparrow}(F_t, t < L_{yb})\, P_b^{\alpha\uparrow}(L_y = 0)$$

$$= P_x(F_t, t < L_{yb}, e^{-\alpha L_{yb}}\, e^{-\alpha(T_b - L_{yb})})\, P_b^{\alpha\uparrow}(L_y = 0) \,/\, \phi_\alpha(x,b)$$

$$= c(x,y,b,\alpha)\, P_x(F_t, t < L_{yb}, e^{-\alpha L_{yb}})$$

where $c(x,y,b,\alpha) = P_x(e^{-\alpha(T_b - L_{yb})} \,/\, L_{yb} > 0)\, P_b^{\alpha\uparrow}(L_y = 0) \,/\, \phi_\alpha(x,b)$.

Now let $b \uparrow B$. On the left we have $T_b \uparrow \zeta$, $P_x^{\alpha\uparrow}$ a.s., and on the right $L_{yb} \uparrow L_y$, P_x a.s., so it follows that *(3.n)* holds with

(3.p) $$c(x,y,\alpha) = \lim_{b\uparrow B} c(x,y,b,\alpha)$$

(3.q) $$= P_x^{\alpha\uparrow}(L_y > 0) \,/\, E_x\, e^{-\alpha L_y}(L_y > 0)$$

where *(3.q)* is obtained by taking $t = 0$ and $F_t = \Omega$ above and $c(x,y,\alpha) \in (0,\infty)$ by the regularity and transience of $\{P_x\}$.

Finally, to turn *(3.q)* into *(3.o)* use *(3.f)* and the obvious formulae

(3.r) $$P_x^{\alpha\uparrow}(L_y > 0) = P_x^{\alpha\uparrow}(T_y < \infty) = 1, \quad x \le y$$

$$= \phi_\alpha(x,y)\,\phi_\alpha(y,x), \quad x > y,$$

(3.s) $$E_x\, e^{-\alpha L_y}(L_y > 0) = \phi_\alpha(x,y)\, E_y\, e^{-\alpha L_y}.$$

Remark (3.6) : If A is an entrance point for the diffusion, as will be the case in our applications to Bessel processes, the Laplace transform of L_y appearing in *(3.o)* above can be computed very easily, since

(3.t) $$E_A\, e^{-\alpha L_y} = E_A\, e^{-\alpha T_y}\, E_y\, e^{-\alpha L_y}$$

while by Williams's time reversal *(3.5)*

(3.u) $$E_A\, e^{-\alpha L_y} = E_y^{0\downarrow}\, e^{-\alpha T_A},$$

whence

(3.v) $$E_y\, e^{-\alpha L_y} = \left[\lim_{a\downarrow A} \phi_\alpha(y,a) \,/\, \phi_o(y,a)\right] \,/\, \left[\lim_{a\downarrow A} \phi_\alpha(a,y)\right].$$

As will become clearer in Section 9, it is interesting to ask what can be said along the lines of *Theorem (3.5)* when the basic diffusion $\{P_x\}$ is recurrent. To focus on the most important case, fix $y > A$ and consider the probabilities $P_y^{\alpha\uparrow}$ on $\underline{\underline{F}}_{L_y -}$ for $\alpha \ge 0$.

Note that $P_y^{\alpha\uparrow}(0 < L_y < \infty)$ is 1 for $\alpha > 0$, 0 for $\alpha = 0$. Still, for $\alpha, \beta > 0$, from *(3.8)* and *(3.5)* we have

(3.w) $$\frac{dP_y^{\beta\uparrow}}{dP_y^{\alpha\uparrow}} = c(\alpha,\beta)\, e^{-(\beta-\alpha)L_y} \quad \text{on } \underline{\underline{F}}_{L_y -},$$

where

$$c(\alpha,\beta) = c^{\alpha\uparrow}(y,y,\beta) = 1/E_y^{\alpha\uparrow}\, e^{-(\beta-\alpha)L_y},$$

and, as a consequence, for $\alpha, \beta, \gamma > 0$

$$c(\alpha,\gamma) = c(\alpha,\beta)\, c(\beta,\gamma).$$

Our analogue of *Theorem (3.5)* for this recurrent case is

Theorem (3.6) : Suppose $\{P_x\}$ is recurrent.

For each $y > A$, there is a strictly positive function $\alpha \to f_y(\alpha)$ and a σ-finite measure M_y on $\underline{F}_{L_y-}$, each defined uniquely up to constant multiples, such that $M_y(L_y = \infty) = 0$ and for every $\alpha > 0$

$$\frac{dP_y^{\alpha\uparrow}}{dM_y} = f_y(\alpha)\, e^{-\alpha L_y} \quad \text{on } \underline{F}_{L_y-}.$$

Fix $\gamma > 0$. One can take $f_y(\alpha) = 1/c(\alpha,\gamma)$.

Then for $A \in \underline{F}_{L_y-}$, for every $\alpha > 0$,

$$(3.xa) \qquad M_y(A) = c(\alpha,\gamma)\, E_y^{\alpha\uparrow}(e^{\alpha L_y}\,;\, A)$$

$$(3.xb) \qquad = \lim_{\alpha\to 0} c(\alpha,\gamma)\, P_y^{\alpha\uparrow}(A).$$

Proof : The fact that *(3.xa)* defines a measure which does not depend on α is immediate from *(3.w)*. The rest of the assertions follow at once, using $M_y(A) = \lim_{\alpha\to 0} M_y(A\, e^{-\alpha L_y})$ for *(3.xb)*.

Remark (3.7) : It follows from *Proposition (3.1)* that for an arbitrary diffusion (P_x), and for any $\alpha \geq 0$, the law $BR(x,y,t)$ of the bridge obtained as the conditional distribution of $(X_s, 0 \leq s \leq t)$ given $X_0 = x$ and $X_t = y$ is the same for either the $\alpha\uparrow$, $\alpha\downarrow$, or original process. As a consequence, using the last exit decomposition of the $\alpha\uparrow$ process at time L_y, the σ-finite measure M_y may be described as follows : under M_y, L_y has σ-finite distribution

$$M_y(L_y \in dt) = f_y(\alpha)^{-1}\, e^{\alpha t}\, P_y^{\alpha\uparrow}(L_y \in dt)$$

for any $\alpha > 0$, and, conditional on $L_y = t$, the process $(X_s, 0 \leq s \leq t)$ is a $BR(y,y,t)$. For an even simpler description of M_y in terms of local time at y, see *Remark (3.9)* below.

Example (3.8) : For $\mu \geq 0$, let

$$(X_t, t \geq 0 \; ; \; W^\mu)$$

be a Brownian motion with drift μ, started at zero, and let $L = \sup\{t : X_t = 0\}$. By the method of time inversion used in Section 5, one finds easily that L has a gamma $(\frac{1}{2}, \mu^2/2)$ distribution, with

$$W^\mu(e^{-\frac{1}{2}\beta^2 L}) = (1 + \frac{\beta^2}{\mu^2})^{-\frac{1}{2}}$$

and

$$W^\mu(L \in dt) = \mu(2\pi t)^{-\frac{1}{2}} \, e^{-\mu^2 t/2}.$$

Since W^μ is the $\frac{1}{2}\mu^2 \uparrow$ process obtained from W° (see *(3.4)*), one finds from *Theorem (3.6)* that the σ-finite measure on $\underline{\underline{F}}_{L-}$ defined by

$$M(A) = \mu^{-1} \, E^\mu\left[\exp(\frac{\mu^2}{2} L) \; ; \; A\right]$$

does not depend on μ, where the nice constant is obtained by taking $\gamma = 1$ in *(3.xa)*. Thus

$$M(L \in dt) = (2\pi t)^{-\frac{1}{2}} \, dt,$$

and given $L = t$, M governs $(X_s, 0 \leq s \leq t)$ as a Brownian bridge.

Remark (3.9) : Notice that in the last example we have

$$(3.y) \qquad M_y(L_y \in dt) = p_t(y,y)\,dt,$$

where $p_t(x,y)$ is the transition density of the diffusion. In fact this formula holds quite generally, as a result of the following description of M_y, which the reader can easily verify using *(3.1)*. Under M_y the total local time ℓ_∞ at the point y has distribution which is a multiple of Lebesgue measure on $(0,\infty)$, and conditional on $\ell_\infty = u$ the process $(X_t, 0 \leq t \leq L_y)$ has the same law as $(X_t, 0 \leq t \leq \tau_u \; ; \; P_y)$, where $\tau_u = \inf\{t : \ell_\infty = u\}$. Thus, the M_y distribution of L_y is the potential measure of the subordinator, $(\tau_u, u \geq 0)$, up to a constant $c > 0$, whence,

$$M_y(0 \leq L_y \leq t) = cE_y \int_0^\infty 1(\tau_u \leq t)\,dt = cE_y \int_0^\infty 1(\ell_t > u)\,du = cE_y \, \ell_t$$

If M_y and $(\ell_t, t \geq 0)$ are appropriately normalised this leads to *(3.y)* for any diffusion with sufficiently regular transition function -see e.g. Getoor [14], and *(6.d)* below. We note that this last description of M_y makes sense with y a recurrent point for a strong Markov process with arbitrary state space.

4. CONDITIONED BESSEL PROCESSES.

For $\nu,\delta \geq 0$, consider now the $\frac{1}{2}\delta^2\uparrow$ and $\frac{1}{2}\delta^2\downarrow$ processes obtained from the $BES(\nu)$ diffusion of Section 2. From *Proposition (2.3)*, it is plain that for $\delta > 0$, $x > 0$, one can take

$$\text{(4.a}\uparrow\text{)} \qquad \phi^{\nu}_{\frac{1}{2}\delta^2\uparrow}(x) = x^{-\nu} I_\nu(\delta x),$$

$$\text{(4.a}\downarrow\text{)} \qquad \phi^{\nu}_{\frac{1}{2}\delta^2\downarrow}(x) = x^{-\nu} K_\nu(\delta x),$$

and

$$\text{(4.b}\uparrow\text{)} \qquad \phi^{\nu}_{0\uparrow}(x) = 1,$$

$$\text{(4.b}\downarrow\text{)} \qquad \phi^{\nu}_{0\downarrow}(x) = x^{-2\nu},$$

where $x > 0$. The results of the last section reveal that for $\delta \geq 0$ the $BES(\nu)$ diffusion conditioned $\frac{1}{2}\delta^2\uparrow$ is a diffusion on $[0,\infty)$ with infinite lifetime, to be referred to as $BES(\nu,\delta)\uparrow$, or simply $BES(\nu,\delta)$. Of course $BES(\nu,0)$ is just $BES(\nu)$.

From *(2.h)*, *(4.a*$\uparrow$*)* and *(3.j)* the transition density $p_t^{\nu,\delta}$ of $BES(\nu,\delta)$ is given for $\delta > 0$ by

$$\text{(4.c}\uparrow\text{)} \qquad p_t^{\nu,\delta}(x,y) = yt^{-1}\, I_\nu(\delta x)^{-1}\, I_\nu(\delta y)\, I_\nu(xy/t)\, \exp -(x^2+y^2+\delta^2t^2)/2t,$$

which shows that our $BES(\nu,\delta)$ is a process introduced by Watanabe [46], and called by him a <u>Bessel diffusion process in the wide sense with index</u> (α,c), where $\alpha = 2\nu + 2$ is the "dimension", and $c = \delta^2/2$.

<u>Remarks (4.1)</u> :

(i) Watanabe allows his α to be any strictly positive number, which corresponds to $\nu > -1$. The above definition of $BES(\nu,\delta)$ can also be extended to $\nu > -1$ if the boundary point 0 of $BES(\nu)$ is taken to be reflecting, which completes the correspondence with Watanabe, but the reader is warned that because the assumption *(3.a)* is no longer satisfied, the results of section 3 must be reinterpreted with some care to cover this case.

(ii) A further extension of the definition of $BES(\nu,\delta)$ to $\nu \le -1$ gives nothing new. Because 0 is then an exit but not entrance point for $BES(\nu)$, it follows from *(4.2)* (ii) below that $BES(\nu,\delta)$ and $BES(-\nu,\delta)$ are identical for $\nu \le -1$.

(iii) According to [40], for $d \ge 1$, the radial part of a BM in $\mathbb{R}^d$ started at 0, with drift v, is a $BES(\nu,\delta)$ for $\nu = (d-2)/2$ and $\delta = |v|$, a fact which will be refined and reproved in Section 5.

(iv) By *(4.a↑)* and *(3.1)*, the generator G_ν^δ of $BES(\nu,\delta)$ may be obtained from the generator G_ν of $BES(\nu)$ as

(4.d1) $$G_\nu^\delta = G_\nu + \psi_{\nu,\delta}(x)\,\frac{d}{dx}$$

where

(4.d2) $$\psi_{\nu,\delta}(x) = \left[x^{-\nu}\, I_\nu(\delta x)\right]^{-1} \frac{d}{dx}\left[x^{-\nu}\, I_\nu(\delta x)\right]$$

(4.d3) $$= \delta\, I_{\nu+1}(\delta x)\,/\,I_\nu(\delta x),$$

by the recurrence formula *(13.4)*. We note (see *(4.k)* below) that the extra drift term $\psi_{\nu,\delta}(x)$ is a continuous increasing function of x with

$$\psi_{\nu,\delta}(0) = 0, \quad \psi_{\nu,\delta}(\infty) = \delta.$$

Turning now to the $BES(\nu)$ process conditioned $\frac{1}{2}\delta^2\!\downarrow$, call it, $BES(\nu,\delta)\!\downarrow$, we find that except in the trivial case $\nu = \delta = 0$, for all $\nu \ge 0$ and $\delta \ge 0$ this process reaches 0 in finite time and dies there. Therefore the $BES(\nu,\delta)\!\downarrow$ process started at x can be described via Williams *theorem (3.3)* as the time reversal of a $BES(\nu,\delta)$ started at 0 and run to time L_x. For this reason, results for $BES(\nu,\delta)\!\downarrow$ can readily be reexpressed in terms of $BES(\nu,\delta)$, and it is this process which will play a dominant rôle.

Remark (4.2) :

For the sake of completeness, we record the following facts about $BES(\nu,\delta)\downarrow$.

(i) The $\downarrow$ version of formula *(4.c)* has K instead of I in the first two Bessel functions only.

(ii) In the $0\downarrow$ case of *(4.c)*, $(x/y)^{\nu}$ should be substituted for these two factors. By inspection of *(4.b↓)* and *(3.1)*, $BES(\nu,0)\downarrow$ is just $BES(-\nu)$ killed when it hits zero, a fact which is implicit in Sharpe [41].

(iii) The $\downarrow$ version of *(4.d)* has K instead of I everywhere, and a factor of -1 in *(4.d3)*. Thus from *(4.k)* below, the extra drift term in this case increases from $-\infty$ to $-\delta$ as x increases from 0 to ∞.

(iv) $BES(\frac{1}{2},\delta)\downarrow$ is BM with drift $-\delta$ killed when it hits zero, a fact which is intimately related to remarkable properties of $BES(\frac{1}{2},\delta)$ described in [40] and [49].

For $\delta > 0$, $\nu,x \geq 0$, let $P_x^{\nu,\delta}$ be the law of $BES(\nu,\delta)$ started at x on the space $C(\mathbb{R}_+,\mathbb{R}_+)$, and let $P_x^{\nu,\delta\downarrow}$ correspond to $BES(\nu,\delta)\downarrow$. We remind the reader that the "δ" refers to killing at rate $\frac{1}{2}\delta^2$. Strictly speaking we should now declare that $BES(\nu,\delta)\downarrow$ is absorbed rather than killed on reaching 0, to keep the trajectory in $C(\mathbb{R}_+,\mathbb{R}_+)$, but this won't ever be important.

By a straightforward application of *Proposition (3.1)*, we obtain the following extension of *Proposition (2.3)* to Bessel processes with drift :

Theorem (4.3) : Let $x, \delta, r > 0$, $\nu \geq 0$. Then

$$(4.e\uparrow) \qquad E_x^{\nu,\delta} \exp\left[-\tfrac{1}{2}\alpha^2 C(T_r) - \tfrac{1}{2}\beta^2 T_r\right] = \frac{I_\nu(\delta r)}{I_\nu(\delta x)} \times \frac{\mathcal{L}_\theta(\gamma x)}{\mathcal{L}_\theta(\gamma r)},$$

where $\gamma = (\beta^2 + \delta^2)^{1/2}$; $\theta = (\nu^2 + \alpha^2)^{1/2}$, and $\mathcal{L} = I$, if $x < r$; K, if $x > r$.

The corresponding formula $(4.e\downarrow)$ has K substituted for I in the first ratio of Bessel functions only. By Williams time reversal, for $r < x$, the expectation on the left side of $(4.e\downarrow)$ is identical to

$$(4.e\downarrow *) \qquad E_0^{\nu,\delta} \exp\left[-\tfrac{1}{2}\alpha^2 C(L_r, L_x) - \tfrac{1}{2}\beta^2 (L_x - L_r)\right]$$

Remark (4.4) :

We note from $(4.e\uparrow)$ the formula

$$(4.f) \qquad P_x^{\nu,\delta}(T_r < \infty) = H^{\nu,\delta}(x) \,/\, H^{\nu,\delta}(r), \; r < x,$$

where

$$(4.g) \qquad H^{\nu,\delta}(y) = K_\nu(\delta y) \,/\, I_\nu(\delta y), \quad y > 0,$$

a result which is also obvious from *(3.n)* and *(4.a)*.

Thus $-H^{\nu,\delta}(x)$ serves as a scale function for BES$(\nu,\delta)\uparrow$. The alternative expression

$$(4.h) \qquad H^{\nu,\delta}(y) = \int_y^\infty du \,/\, u(I_\nu(\delta u))^2,$$

which is the equivalent of formula *(2.5)* in Watanabe [46], is a simple consequence of the classical formula for the Wronskian : $W[K_\nu(z), I_\nu(z)] = z^{-1}$.

From well known asymptotics of Bessel functions which are displayed in Section 12 one can now obtain the asymptotics of $(4.e\uparrow)$ when $r \to \infty$ or $x \to 0$, and those of $(4.e\downarrow)$ when $r \to 0$ or $x \to \infty$.

In particular, one obtains the following formulae, the first two of which imply the interpretations *(1.h)* and *(1.i)* of the Hartman laws, and the third of which is equivalent to a result of Kent ([31], *theorem (4.1)*), by virtue of *Corollary (5.6)* below.

Corollary (4.5) :

(4.i) $$E_0^{\nu,\delta} \exp\left[-\tfrac{1}{2}\alpha^2 C(T_y,\infty)\right] = I_\theta(\delta y) / I_\nu(\delta y),$$

(4.i*) $$E_0^{\nu,\delta} \exp\left[-\tfrac{1}{2}\alpha^2 C(L_y,\infty)\right] = K_\nu(\delta y) / K_\theta(\delta y),$$

where $\theta = (\nu^2 + \alpha^2)^{1/2}$;

(4.j) $$E_0^{\nu,\delta} \exp\left[-\tfrac{1}{2}\beta^2 T_y\right] = \frac{I_\nu(\delta y)}{I_\nu(\gamma y)} \left(1 + \frac{\beta^2}{\delta^2}\right)^{\nu/2},$$

(4.j*) $$E_0^{\nu,\delta} \exp\left[-\tfrac{1}{2}\beta^2 L_y\right] = \frac{K_\nu(\gamma y)}{K_\nu(\delta y)} \left(1 + \frac{\beta^2}{\delta^2}\right)^{\nu/2},$$

where $\gamma = (\delta^2 + \nu^2)^{1/2}$.

Proof : Proceed thus from *Theorem (4.3)*.

(i) In *(e↑)* put $x = y$, $\beta = 0$ and let $r \to \infty$.

(i*) In *(e↓*)* put $r = y$, $\beta = 0$ and let $x \to \infty$.

(j) In *(e↑)* put $r = y$, $\alpha = 0$ and let $x \to 0$.

(j*) In *(e↓*)* put $x = y$, $\alpha = 0$ and let $r \to 0$.

For $\mu,\nu \geq 0$, define

(4.k) $$I_{\mu:\nu}(x) = I_\mu(x) / I_\nu(x) \ ; \ K_{\mu:\nu}(x) = K_\mu(x) / K_\nu(x), \ x > 0.$$

We note that on putting $\theta=\mu$ the formulae *(4.i)* and *(4.i*)* above make obvious the result of Hartman and Watson [20] , *Proposition 7.1*, that for $0\leq\nu<\mu$, both $I_{\mu:\nu}$ and $K_{\nu:\mu}$ are continuous distribution functions on $(0,\infty)$.

It now emerges that for fixed $\delta > 0$ and $x > 0$, the BES(μ,δ) laws $P_x^{\mu,\delta}$ for $\mu \geq 0$ are mutually absolutely continuous.

To be precise, we have

Theorem (4.6) : Let $\delta, x > 0$, $\mu, \nu \geq 0$.

$$\frac{dP_x^{\mu,\delta}}{dP_x^{\nu,\delta}} = I_{\nu:\mu}(\delta x) \exp\left[-\tfrac{1}{2}(\mu^2-\nu^2)C_\infty\right] \quad \text{on } \underline{\underline{F}}_\infty, \tag{4.l}$$

and for every $(\underline{\underline{F}}_{t+})$ stopping time T,

$$\frac{dP_x^{\mu,\delta}}{dP_x^{\nu,\delta}} = I_{\mu:\nu}(\delta R_T)\, I_{\nu:\mu}(\delta x) \exp\left[-\tfrac{1}{2}(\mu^2-\nu^2)C_T\right] \quad \text{on } \underline{\underline{F}}_{T+} \tag{4.m}$$

Note : On $(T = \infty)$, $I_{\mu:\nu}(\delta R_T) = 1$ by convention.

Proof : It is enough to concider the case $\mu > \nu$ and prove *(4.m)*. For bounded stopping times T the result follows at once from *Propositions (2.1)* and *(3.1)*. To extend to unbounded T let M_T denote the right hand expression above. Since $(M_t, t \geq 0)$ is an $(\underline{\underline{F}}_t)$-martingale under $P_x^{\nu,\delta}$, it only remains to show that this martingale is uniformly integrable, or, what is the same, that its almost sure limit as $t \to \infty$ has $P_x^{\nu,\delta}$ expectation equal to 1. But this is immediate from *(4.i)* and the fact mentioned above that $I_{\mu:\nu}(z)$ increases to 1 as $z \to \infty$.

Corollary (4.7) :

Let $Z \geq 0$ be an $\underline{\underline{F}}_\infty$-measurable random variable. For fixed $x, \delta > 0$, the function

$$\nu \to E_x^{\nu,\delta} Z, \; \nu \geq 0$$

is right continuous, and continuous except possibly for a jump down from ∞ at $\hat{\nu} = \inf\{\nu : E_x^{\nu,\delta} Z < \infty\}$.

Proof : Use *(4.6)*, the continuity of $\mu \to I_\mu(y)$, and the monotone and dominated convergence theorems.

Note : The above result is clearly false if either $\delta = 0$ or $x = 0$.

Corollary (4.8) : For $\delta, x, y > 0$,

$$\frac{dP_x^{\mu,\delta}}{dP_x^{\nu,\delta}} = K_{\nu:\mu}(\delta y)\, I_{\nu:\mu}(\delta x) \exp\left[-\tfrac{1}{2}(\mu^2-\nu^2)C_{L_y}\right] \quad \text{on } \underline{F}_{L_y -} \cap (L_y > 0).$$

Proof : Starting from *(4.6)* for $T = \infty$, condition on $\underline{F}_{L_y -}$, and use the last exit decomposition at time L_y together with *(4.i*)*.

Corollary (4.9) : For $x, \mu, \nu > 0$,

$$\frac{dP_x^{\mu}}{dP_x^{\nu}} = \left(\frac{\mu}{\nu}\right)\left(\frac{y}{x}\right)^{\mu-\nu} \exp\left[-\tfrac{1}{2}(\mu^2-\nu^2)C_{L_y}\right] \quad \text{on } \underline{F}_{L_y -} \cap (L_y > 0).$$

Proof : This results from *(4.8)* on letting $\delta \downarrow 0$, using *(13.ℓ)*, the passage to the limit being justified by *(3.5)*.

5. THE RADIAL AND ANGULAR PARTS OF BROWNIAN MOTION WITH DRIFT.

This section offers two different approaches to BM with drift in R^d and its decomposition into radial and angular parts, using firstly the Cameron - Martin formula, and secondly time inversion, to transform to the more familiar case with no drift.

D. Williams seems to have initiated the use of the Cameron - Martin (CM) formula to calculate distributions associated with the radial and angular parts of BM with drift (cf : the end of Kent's paper [31]). The method is also used more or less explicitly in a number of recent papers ([17], [32], [48]) but in none of these papers is the argument developed to its fullest extent.

Fix an integer $d \leq 1$, and for $\delta \geq 0$, let $\{\Omega, \underline{\underline{F}}_t, B_t, t \geq 0 ; P^{\vec{\delta}}\}$ be the canonical realisation of Brownian motion in $\mathbb{R}^d$ started at the origin with a constant drift $\vec{\delta}$ of magnitude $\delta = |\vec{\delta}|$ in the direction $\vec{u} = \vec{\delta}/\delta \in S^{d-1}$, the unit sphere in $\mathbb{R}^d$. Thus $\Omega = C(\mathbb{R}^+, \mathbb{R}^d)$, $B_t(\omega) = \omega(t)$, $\underline{\underline{F}}_t = \sigma(B_s, 0 \leq s \leq t)$, $\{B_t, t \geq 0 ; P^o\}$ is standard BM in $\mathbb{R}^d$ starting at the origin, and $P^{\vec{\delta}}$ is the P^o distribution of $(B_t + t\vec{\delta}, t \geq 0)$. According to the CM formula, for any $(\underline{\underline{F}}_t)$ stopping time T,

$$(5.a) \qquad \frac{dP^{\vec{\delta}}}{dP^o} = \exp\left[(B_T, \vec{\delta}) - \tfrac{1}{2}\delta^2 T\right] \qquad \text{on } \underline{\underline{F}}_T \cap (T < \infty),$$

where $(\, , \,)$ is the inner product in R^d. (See e.g. McKean [29] p.97 or Freedman [11], §1.11).

The applications of the CM formula below hinge largely on the product form of the Radon - Nikodym derivative, which can be exploited by virtue of the following general (and trivial) Lemma.

Lemma (5.1) : Let P and Q be probabilities on a measurable space $(\Omega,\underline{\underline{F}})$, with

$$\frac{dQ}{dP} = GH$$

where $G \geq 0$ is $\mathcal{G}$-measurable, $H \geq 0$ is $\mathcal{H}$-measurable, and $\mathcal{G},\mathcal{H} \subset \underline{\underline{F}}$.

(i) For $\underline{\underline{F}}$-measurable $Z \geq 0$,

$$Q(Z|\mathcal{G}) = P(ZH|\mathcal{G}) \;/\; P(H|\mathcal{G}), \quad Q \text{ a.s.},$$

where $0/0 = 0$.

(ii) If $\mathcal{G}$ and $\mathcal{H}$ are P-independent, they are also Q-independent, and :

$$P(G)\;P(H) = 1,$$

$$\frac{dQ}{dP} = G/PG \text{ on } \mathcal{G}, \quad H/PH \text{ on } \mathcal{H}.$$

For $k \geq 0$, let $vM(k)$ be the von Mises distribution on S^{d-1} centered at $\vec{u}$, with concentration parameter k. That is,

(5.b) $$vM(k,d\theta) = U(d\theta)\; C_d(k)^{-1} \exp k(\theta,\vec{u}),$$

where $U = vM(0)$ is the uniform probability on S^{d-1}, and $C_d(k)$ is the normalising constant

(5.c1) $$C_d(k) = \int U(d\theta) \exp k(\theta,\vec{u})$$

(5.c2) $$= \Gamma(\nu+1)\;(k/2)^{-\nu}\; I_\nu(k),$$

where $\nu = (d-2)/2$, and the formula *(5.c2)* will be later derived in *(5.4)* (iv).

Starting from the easy case $\delta = 0$, using part (i) of the Lemma above and the CM formula *(5.a)*, one easily obtains the following proposition, which is implicit in [40].

Proposition (5.2) : Let $R_t = |B_t|$, $\mathcal{R}_t = \sigma(R_s, 0 \leq s \leq t)$, $\theta_t = B_t/R_t$, and let T be an $(\mathcal{R}_t)$ stopping time. Then, the $P^{\vec{\delta}}$ conditional law of θ_T given $\mathcal{R}_T$ on $(T < \infty)$ is $vM(\delta R_T)$.

Now for $T_r = \inf\{t : R_t = r\}$, it is obvious by symmetry that under P^o the uniformly distributed angle Θ_{T_r} is independent of the stopped radial process $(R_t, 0 \leq t \leq T_r)$ generating $\mathcal{R}_{T_r}$. Moreover, $P^{\vec{\delta}}(T_r < \infty) = 1$ for all $\delta \geq 0$, $r > 0$.
Thus part (ii) of the Lemma and the CM formula *(5.a)* imply

Proposition (5.3) : Let $\vec{\delta} \in \mathbb{R}^d$, $r > 0$, and let $T_r = \inf\{t : R_t = r\}$, where $\{B_t, t \geq 0 ; P^{\vec{\delta}}\}$ is a BM with drift $\vec{\delta}$ in $\mathbb{R}^d$ starting at the origin, $R_t = |B_t|$, $\Theta_t = B_t/R_t$.

(i) The hitting angle Θ_{T_r} and the radial process up to time T_r, $(R_t, 0 \leq t < T_r)$, are $P^{\vec{\delta}}$ independent.

(ii) The $P^{\vec{\delta}}$ distribution of Θ_{T_r} is $vM(\delta r)$.

(iii) The distribution of $(R_t, 0 \leq t < T_r)$ under $P^{\vec{\delta}}$ is identical to the distribution of the same process under P^o conditional on $(T_r < U_\delta)$, where U_δ is an exponentially distributed time with rate $\frac{1}{2}\delta^2$, which is P_o independent of this process.

(iv) $E_o\left\{\exp\left[- r(\Theta_{T_r}, \vec{\delta})\right]\right\} = 1/E_o\left\{\exp\left[- \frac{1}{2}\delta^2 T_r\right]\right\}$.

Remarks (5.4) : The first four remarks refer to the correspondingly numbered assertions above.

(i) This extends independence results to be found in Kent [31], Stern [43], and Wendel [48]. But, see (v) below for a further extension.

(ii) This may be found in Kent [31], and in Gordan and Hudson [17] for $d = 2$. The joint distribution of Θ_{T_r} and T_r in this case was first obtained by Reuter (see [30]).

(iii) An immediate consequence of this is the result of [40] that $(R_t, 0 \le t < \infty ; P^{\vec{\delta}})$ is a $\mathrm{BES}(\nu,\delta)$ as defined in section 4 for $\nu = (d-2)/2$.

(iv) Since $(R_t, 0 \le t < \infty ; P^0)$ is a $\mathrm{BES}(\nu)$, from *(4.a+)*, *(3.f)* and *(12.2)* one has the well known formula

$$1/E_0\left\{\exp\left[-\tfrac{1}{2}\delta^2 T_r\right]\right\} = \Gamma(\nu+1)\ (\delta r/2)^{-\nu}\ I_\nu(\delta r),$$

and the evaluation *(5.c2)* of the constant in the $vM(k)$ density is a consequence.

(v) There is another way of identifying the $P^{\vec{\delta}}$ distribution of $(R_t, 0 \le t < \infty)$ which is quite instructive. Assuming $d \ge 2$, let $X_t = B_t - t\vec{\delta}$, so $(X_t, t \ge 0)$ is a BM with no drift, and use Itô's formula to obtain

$$R_t = |B_t| = \sum_{i=1}^{d} \int_0^t \frac{B_s^i\, dB_s^i}{R_s} + \frac{d-1}{2}\int_0^t \frac{ds}{R_s}$$

$$= \beta_t + \int_0^t (\theta_s, \vec{\delta})\, ds + \frac{d-1}{2}\int_0^t \frac{ds}{R_s},$$

where $\beta_t = \sum_{i=1}^{d} \int_0^t \frac{B_s^i\, dX_s^i}{R_s}$ is a real valued BM without drift. It now develops that

$$R_t - \int_0^t E[(\theta_s, \vec{\delta}) \mid \mathcal{R}_s]\, ds - \frac{d-1}{2}\int_0^t \frac{ds}{R_s}$$

is an $(\mathcal{R}_t)$-martingale, whose increasing process is t, which shows that it is a BM. But from *(5.2)* we have $E[(\theta_s, \vec{\delta}) \mid \mathcal{R}_s] = \psi_\delta(R_s)$, where

(5.d) $$\psi_\delta(r) = \int_{S^{d-1}} (\theta, \vec{\delta})\ vM(\delta r, d\theta),$$

whence $(R_t, t \ge 0)$ is a diffusion with generator

$$\frac{1}{2}\frac{d^2}{dr^2} + \left(\psi_\delta(r) + \frac{d-1}{2r}\right)\frac{d}{dr}.$$

To complete the calculation of ψ_δ, calculate the integral *(5.d)* by differentiating the identity *(5.c)* with respect to k, with the result that ψ_δ is identical to the extra drift term $\psi_{\nu,\delta}$ in the generator *(4.d1)* of BES(ν,δ), for $\nu = (d-2)/2$.

(vi) Parts (i) and (iii) of *Proposition (5.3)* admit a further extension, which is now explained in the case $d = 2$ with $\mathbb{R}^2$ viewed as the complex plane. Fix $r > 0$ and define the <u>quotient process</u> $Q = (Q_t, 0 \le t < T_r)$ by

$$Q_t = \overline{\Theta}_{T_r} B_t, \quad 0 \le t < T_r,$$

where $\overline{\theta}$ is the complex conjugate of θ.
Thus the path of Q is obtained from that of B by killing at time T_r and then rotating the killed trajectory so that it dies at r on the real axis instead of at B_{T_r} on the circle of radius r.
Then, the same argument shows that (i) and (iii) hold with the quotient process Q instead of R, hence too with R replaced by S or (R,S), where $S_t = |B_t - B_{T_r}|$ is the distance to the hitting point. This is as far as one can go with the independence result : Since the whole process up to time T_r can be recovered from Q and Θ_{T_r}, the σ-field generated by Q is maximal among sub-σ-fields of $\underline{\underline{F}}_{T_r}$ which are $P^{\vec{\delta}}$ independent of Θ_{T_r}. If d is not 1, 2 or 4, the lack of a group structure on S^{d-1} makes any definition of Q somewhat arbitrary, but a similar result can still be formulated. (See Kent [32], section 9)

(vii) By a further application of *(5.1)* (i) to *(5.a)*, for $d \ge 1$, $r > 0$,

(5.e1) <u>the</u> $P^{\vec{\delta}}$ <u>conditional law of</u> $(B_t, 0 \le t \le T_r) | \Theta_{T_r} = \theta$
<u>is identical to</u>

(5.e2) <u>the</u> P^o <u>conditional law of</u> $(B_t, 0 \le t \le T_r) | \Theta_{T_r} = \theta,\ T_r < U_\delta$,
<u>where</u> U_δ <u>is exponential with rate</u> $\frac{1}{2}\delta^2$ <u>independent of</u>
$(B_t, 0 \le t \le T_r)$ <u>under</u> P^o.

On the other hand, for $d \ge 2$ one can start from the classical skew-product decomposition of BM in $\mathbb{R}^d$ with no drift, use the reversibility of BM on S^{d-1} as in the discussion of "spinning" in Itô - McKean [28], section 7.17 to argue that under P^o for each $r > 0$,

(5.f) $$B_t = \Phi_r\left(\int_t^{T_r} ds/R_s^2\right),\quad 0 < t \le T_r,$$

where Φ_r is a BM on S^{d-1} started with the uniform distribution and independent of the BES(ν) process $(R_t, t \ge 0)$.

Thus the law in *(5.e2)* can be described as the skew-product *(5.f)* with Φ_r started at the fixed point θ and T_r conditioned to be less than U_δ, which transforms $(R_t, 0 \le t \le T_r)$ into BES(ν,δ). The identity *(5.e)* now reveals that the skew product *(5.f)* holds under $P^{\vec{\delta}}$ with $(R_t, t \ge 0)$ a BES(ν,δ) and Φ_r a BM on S^{d-1} started with the vM(δr) distribution of *(5.3)* (ii). Since for $\delta > 0$, the strong law of large numbers implies $\Theta_{T_r} \to \vec{u}$, $P^{\vec{\delta}}$ almost surely as $r \to \infty$, it can now be argued that for $\delta > 0$ one can let $r \to \infty$ in *(5.f)* to obtain the skew-product description of the whole drifting motion $(B_t, 0 \le t < \infty\ ;\ P^{\vec{\delta}})$ described in *Theorem (1.1)*. This was how we first discovered that result, but we shall not attempt to give a detailed justification of the above argument. Rather, we turn now to the alternative method of time inversion, which gives a very crisp proof of *Theorem (1.1)*.

Proof of *Theorem (1.1)* by time inversion.

Let $(B(t), t \ge 0)$ be a BM in $\mathbb{R}^d$ starting at 0, with drift vector v, where $d \ge 2$, and $v \ne 0$.

Define

$$B_0(t) = B(t) - vt,\quad t \ge 0,$$

So, B_0 is a standard BM in $\mathbb{R}^d$ starting at 0.

By the well known invariance of B_0 under time inversion, the process

(5.g) $$(s\,B(1/s), s > 0) = (v + s\,B_0(1/s), s > 0)$$

is a BM in $\mathbb{R}^d$ starting at v.

Thus, according to the classical skew-product formula for this process (see e.g. Itô - Mc Kean [28], p. 270), there is a BM $(\Theta(u), u \geq 0)$ in S^{d-1} independent of $(|s\,B(1/s)|, s > 0)$, and hence of $(|B(s)|, s > 0)$, such that

$$(5.h) \qquad s\,B(1/s) = |s\,B(1/s)|\ \Theta\left(\int_0^s du\ /\ |u\,B(1/u)|^2\right).$$

The substitutions : $s \to 1/_s$ and : $u \to 1/_u$ now reveal that

$$(5.i) \qquad B(s) = |B(s)|\ \Theta\left(\int_s^\infty du\ /\ |B(u)|^2\right),$$

and the theorem is proved.

The above time inversion provides yet another means of identifying the radial process $(|B(t)|, t \geq 0)$ as $BES_0(\nu,\delta)$: indeed, $(s\,B(1/_s), s > 0)$ is $BES_{|v|}(\nu)$ (a subscript is now being used to indicate the starting position of a process). Then, the identification is a consequence of the following remarkable result, which will be the key to several further developments.

Theorem (5.5) : (Watanabe [46], *theorem (2.1)*).

For all $\nu > -1$, $\gamma,\delta \geq 0$, a process $(U(t), t > 0)$ is a $BES_\gamma(\nu,\delta)$ if and only if $(t\,U(1/_t), t > 0)$ is a $BES_\delta(\nu,\gamma)$.

Watanabe's time inversion *(5.5)* and the Pythagorean property of the $BES(\nu)$ family, discovered by Shiga and Watanabe [42], imply the following

Corollary (5.6) : Let $(X_t, t \geq 0)$ be a $BES_0(\lambda,\alpha)$, and let $(Y_t, t \geq 0)$ be an independent $BES_0(\mu,\beta)$, where $\lambda,\mu > -1$, $\alpha,\beta \geq 0$. Then the process

$$(5.j) \qquad ((X_t^2 + Y_t^2)^{1/2},\ t \geq 0)$$

is a $BES_0\ (\lambda + \mu + 1,\ (\alpha^2 + \beta^2)^{1/2})$.

In particular, for $\delta \geq 0$, $\nu > -{}^1/_2$, if $X_t = |B_t + \delta t|$ where $\delta \geq 0$ and $(B_t, t \geq 0)$ is a BM_o on $\mathbb{R}_1$ and $(Y_t, t \geq 0)$ is a $BES_o(\nu - \frac{1}{2})$, then the process in *(5.j)* is a $BES_o(\nu, \delta)$.

This last presentation of $BES_o(\nu, \delta)$ underlies the work of Kent [31].

Proof : Let $A_t = t\,X({}^1/_t)$, $B_t = t\,Y({}^1/_t)$.

By Watanabe's inversion, A and B are independent $BES_\alpha(\lambda)$ and $BES_\beta(\mu)$ processes. Now, by the Pythagorean property of [42], $(A^2 + B^2)^{1/2}$ is a $BES_\gamma(\nu)$ with starting place $\gamma = (\alpha^2 + \beta^2)^{1/2}$ and index $\nu = \lambda + \mu + 1$ (which corresponds to adding the dimensions).

Inverting once more yields the desired conclusion. We now indicate how Watanabe's time inversion can be used to obtain a very simple description of Bessel bridges. Somewhat more generally, consider a family of diffusions on a subset S of $\mathbb{R}^d$ indexed by a parameter $\gamma \in S$ in such a way that if P^γ_δ governs the co-ordinate process $(X_t, t \geq 0)$ as the γ-diffusion starting at δ, then the laws $\{P^\gamma_\delta, \gamma, \delta \in S\}$ satisfy the

Inversion Hypothesis *(5.7)*. For $\gamma, \delta \in S$, the P^γ_δ distribution of $(sX({}^1/_s), s > 0)$ is P^δ_γ. We have two examples in mind :

Example 1 : $S = \mathbb{R}^d$ and P^γ_δ governs BM with drift vector γ started at δ. Then *(5.7)* is a variant of the familar time inversion property of BM.

Example 2 : $S = \mathbb{R}_+$ and P^γ_δ governs $BES_\delta(\nu, \gamma)$, where $\nu > -1$ is fixed. Then *(5.7)* amounts to Watanabe's time inversion.

For such a family of diffusions there is an extremely simple description of the bridges obtained by conditioning the two ends of the sample path over a fixed time interval :

Theorem (5.8) : Let $\{P^\gamma_\delta\}$ be a family of diffusions on S satisfying the inversion hypothesis *(5.7)*.
Let $t > 0$, $\gamma, \delta \in S$. Then, the processes

(5.k1) $\{X(u),\ 0 < u < t\ ;\ P^\gamma_\delta | X_t = \rho\}$,

(5.k2) $\{u\, X(\frac{1}{u} - \frac{1}{t}),\ 0 < u < t\ ;\ P^\delta_{\rho/t}\}$, and

(5.k3) $\{(\frac{t-u}{t})\, X(\frac{tu}{t-u}),\ 0 < u < t\ ;\ P^{\rho/t}_\delta\}$

are identical in law.

Proof : The first identity comes from the application of the Markov property to $(s\, X(\frac{1}{s}), s > 0)$, under P^γ_δ, at time $s = 1/t$. The second identity is obvious from *(5.7)*.

Remarks (5.9) :

(i) The law of the bridge *(5.k1)* does not depend on γ.

(ii) For BM the conclusion of *Theorem (5.8)* is a well known method for obtaining distributional properties of the Brownian bridge - see e.g. Breiman [3], p. 290. In such applications, a key property of the transformations of space-time involved in *(5.k)* is that they preserve straight lines.

(iii) The transformations of *(5.k)* also act quite simply on integrals. For example, for a positive function $V(u,x)$, one finds that for $t > 0$

(5.l1) the P^γ_δ distribution of $\int_0^t V(u, X_u)\,du \,|\, X_t = \rho$ is identical to

(5.l2) the $P^\delta_{\rho s}$ distribution of $\int_0^\infty V(\frac{1}{s+u}, \frac{X_u}{s+u})\, \frac{du}{(s+u)^2}$,

where $s = 1/t$. In particular, if $V(u, X_u) = 1/X_u^2$, the integrand in *(5.l2)* simplifies to $1/X_u^2$. As a result, either of the formulae *(1.e)* and *(4.i)* can be seen immediately to be a consequence of the other.

Some other consequences of these time-inversion tricks are featured in section 8.

6. THE DENSITY OF THE LAST EXIT TIME FOR A DIFFUSION.

Getoor [13] gave the explicit formula *(1.k)* for the density of the last time L_y that a $BES_0(\nu)$ process hits a level $y > 0$. We observe here that Getoor's formula admits an extension to a large class of transient diffusions on the line ; this enables us to write down the analogues of Getoor's formula for a $BES_x(\nu,\delta)$ (see section 7). Afficionados of the general theory of last exits could certainly deduce our result from Getoor - Sharpe [16], but we find it simpler to work it out directly using Tanaka's local time formula.

We consider the canonical realisation on $C(\mathbb{R}_+,\mathbb{R}_+)$ of a regular diffusion $(R_t, t \geq 0 \,;\, P_x, x \in R_+)$, with infinite lifetime, and suppose for simplicity that

(6.a) $\quad P_x(T_0 < \infty) = 0, \quad x > 0,$ and

(6.b) $\quad P_x(\lim_{t\to\infty} R_t = \infty) = 1, \quad x > 0.$

As a consequence of *(6.a)* and *(6.b)*, a scale function s for this diffusion satisfies $s(0+) = -\infty$ and $s(\infty) < \infty$. One can therefore suppose $s(\infty) = 0$, and then for $x,y > 0$, the function

(6.c1) $\quad u_y(x) = P_x(T_y < \infty) = P_x(L_y > 0)$

is given in terms of s by

(6.c2) $\quad u_y(x) = [s(x) \,/\, s(y)] \wedge 1.$

Let Γ be the infinitesimal generator of the diffusion, and take the speed measure m to be such that $\Gamma = \frac{1}{2}\frac{d}{dm}\frac{d}{ds}$.

According to Itô - Mc Kean ([28], p. 149), there exists a continuous function :

$$\overset{\bullet}{p} : (0,\infty)^3 \ni (t,x,y) \to \overset{\bullet}{p}_t(x,y)$$

which is strictly positive, and such that the semigroup (P_t) of the diffusion is given by

(6.d) $\quad P_t(x,dy) = \overset{\bullet}{p}_t(x,y)\, m(dy).$

We can now state the following result :

Theorem (6.1) : Let (R_t, P_x) be a regular diffusion on $\mathbb{R}_+$ satisfying the hypotheses *(6.a)* and *(6.b)*. Then

(i) For all $x,y > 0$

$$P_x(L_y \in dt) = \frac{-1}{2\, s(y)}\, \dot{p}_t(x,y)dt \tag{6.e}$$

(ii) For $x \leq y$ the formula *(6.e)* defines an infinitely divisible probability distribution on $(0,\infty)$.

Notes : (i) For $x > y$, the distribution defined by *(6.e)* on $(0,\infty)$ is an infinitely divisible sub-probability with total mass given by *(6.c)*.

(ii) If 0 is an entrance point for the diffusion, simple formulae for the Laplace transform of the law *(6.e)* can be obtained from *(3.v)*.

(iii) In practice the generator Γ of the diffusion will coïncide on $C^2(0,\infty)$ with

$$\gamma = \frac{1}{2}\, a(x)\, \frac{d^2}{dx^2} + b(x)\, \frac{d}{dx}.$$

Suppose simply that $a,b \in C^\infty(0,\infty)$, with $a > 0$. Then there exists, by the hypo-ellipticity of γ, a function $p : (0,\infty)^3 \ni (t,x,y) \to p_t(x,y)$, of class C^∞ on $(0,\infty)^3$ such that

$$P_t(x,dy) = p_t(x,y)dy.$$

In this case, one has

$$\frac{dm}{dy} = (s'a)^{-1}\,(y),$$

hence

$$\dot{p}_t(x,y) = p_t(x,y)\,(s'a)\,(y),$$

and the formula *(6.e)* becomes

$$P_x(L_y \in dt) = -\frac{1}{2}\, p_t(x,y)\, \left(\frac{s'a}{s}\right)(y)dt,\quad t > 0. \tag{6.e'}$$

Proof : For $x > 0$, $y > 0$, $t > 0$,

$$P_x(L_y \geq t | R_t) = u_y(R_t), \quad \text{whence}$$

$$P_x(L_y < t) = E_x[1 - u_y(R_t)].$$

Put $M_t = s(R_t)$, and recall that for each $x > 0$, M is a P_x-local martingale with continuous paths.

One can thus apply the generalised Itô (or Tanaka) formula to the process

$$u_y(R_t) = (\frac{M_t}{s(y)}) \wedge 1.$$

According to this formula, if $(\Lambda_t^z, t \geq 0)$ denotes the local time of M at z (see Meyer [35] Chapter VI ; II), the process

$$(u_y(R_t) - \frac{1}{2\, s(y)} \Lambda_t^{s(y)}, \ t \geq 0)$$

is a P_x-martingale starting at $u_y(x)$ which is square-integrable (it even belongs to BMO - see [35] théorème 4, p. 334). It follows that

$$P_x(0 < L_y < t) = \frac{(-1)}{s(y)} \ E_x \ \Lambda_t^{s(y)}.$$

On the other hand, Itô - Mc Kean ([28] p. 175) show that for all

$$(t,x,y) \in (0,\infty)^3,$$

$$\frac{\partial}{\partial t} E_x \ \Lambda_t^{s(y)} = \dot{p}_t(x,y),$$

whence the formula *(6.e)*. The *proof* of *Theorem (6.1)* is completed by observing that the infinite divisibility of L_y is the special case $A(t) = t$ of the following more general fact which will be used later :

Lemma (6.2) : Let $(A(t), t \geq 0)$ be an additive functional of a regular diffusion process $(R_t, t \geq 0 \ ; \ P_x)$.

Then for all x and y such that $P_x(0 < L_y < \infty) > 0$ the law of $A(L_y)$ conditional on $0 < L_y < \infty$ is infinitely divisible.

Proof : Conditioning on $(T_y < \infty)$ and decomposing the conditional diffusion at time T_z for z between x and y shows that $A(T_y)$ is infinitely divisible, and it suffices therefore to treat the case: $x = y$. Let $(\tau_s, s \geq 0)$ be the right continuous inverse of a local time process Λ for the point y. Then

$$A(L_y) = A(\tau_\sigma)$$

where $\sigma = \Lambda(L_y)$. But by Itô's excursion theory (see e.g. Meyer [36]) the process

$$(A(\tau_s), 0 \leq s \leq \sigma \; ; \; P_y)$$

has the same law as a process

$$(Y_s, 0 \leq s \leq \sigma)$$

where Y is a subordinator and σ is an exponentially distributed random time independent of Y. Since the exponential law is infinitely divisible, the conclusion is immediate.

7. DISTRIBUTION OF LAST EXIT TIMES FOR BES(ν,δ).

In this section we record explicit formulae for the distribution of

$$L_y = \sup\{t : R_t = y\}$$

when $(R_t, t \geq 0 ; P_x^{\nu,\delta})$ is a $BES_x(\nu,\delta)$, $y > 0$, and consider also the joint distribution of L_y and the clock $C(L_y)$.

From *(6.e')* and the formulae *(4.g)* and *(4.h)* for the scale function of BES(ν,δ), for all $\nu,x,\delta \geq 0$,

$$(7.a) \qquad P_x^{\nu,\delta}(L_y \in dt) = p_t^{\nu,\delta}(x,y)\ [G(\delta,y)]^{-1}\ dt, t > 0$$

where

$$G(\delta,y) = 2y\ I_\nu(\delta y)\ K_\nu(\delta y), \quad \delta > 0$$

$$= y/\nu, \qquad \delta = 0.$$

The formulae below for densities follow immediately from *(7.a)* and the formulae *(2.h)* and *(4.c)* for the transition density of BES(ν,δ). The corresponding formulae for the Laplace transforms can either be derived from the density expressions using the well known formula

$$(7.b) \qquad \int_0^\infty \exp(-\tfrac{1}{2}\alpha^2 t)\ p_t^\nu(x,y)dt = 2y\ (y/x)^\nu\ I_\nu[\alpha(x\wedge y)]\ K_\nu[\alpha(x\vee y)],$$

or they can be obtained using *(3.v)*. Notice that the total mass of the law of L_y on $(0,\infty)$ can in each case be obtained by setting $\alpha = 0$ in the formula for the Laplace transform. Formula *(7.$\tilde{d}$)* was obtained earlier as *(4.j*)* and is included only for the sake of completeness.

Case $\delta > 0$, $x > 0$, $\nu \geq 0$:

$$(7.c) \qquad P_x^{\nu,\delta}(L_y \in dt) = \frac{dt\ I_\nu(xy/t)\ \exp[-(x^2+y^2+\delta^2t^2)/2t]}{2t\ I_\nu(\delta x)\ K_\nu(\delta y)}$$

$$(7.\tilde{c}) \qquad E_x^{\nu,\delta}[\exp(-\tfrac{1}{2}\alpha^2 L_y) ; L_y > 0] = \frac{I_\nu[\beta(x\wedge y)]\ K_\nu[\beta(x\vee y)]}{I_\nu(\delta x)\ I_\nu(\delta y)}$$

where $\beta = (\alpha^2 + \delta^2)^{1/2}$.

<u>Case</u> $\delta > 0,\ x = 0,\ \nu \geq 0$:

$$(7.d)\quad P_0^{\nu,\delta}(L_y \in dt) = \frac{dt\ y^{\nu}\ \exp[-(y^2+\delta^2 t^2)/2t]}{2t\ (\delta t)^{\nu}\ K_{\nu}(\delta y)}\ ;$$

$$(7.\tilde{d})\quad E_0^{\nu,\delta}\exp(-\tfrac{1}{2}\alpha^2 L_y) = \frac{\beta^{\nu} K_{\nu}(\beta y)}{\delta^{\nu} K_{\nu}(\delta y)}$$

where $\beta = (\alpha^2 + \delta^2)^{1/2}$.

<u>Case</u> $\delta = 0,\ x > 0,\ \nu > 0$:

$$(7.e)\quad P_x^{\nu}(L_y \in dt) = dt(\nu/t)\ (y/x)^{\nu}\ I_{\nu}(xy/t)\ \exp[-(x^2+y^2)/2t]$$

$$(7.\tilde{e})\quad E_x^{\nu}[\exp(-\tfrac{1}{2}\alpha^2 L_y)\ ;\ L_y > 0] = 2\nu(y/x)^{\nu}\ I_{\nu}[\alpha(x\wedge y)]\ K_{\nu}[\alpha(x\vee y)].$$

Finally, the case $\delta = 0,\ x = 0,\ \nu > 0$ can be obtained from either the last case by letting x tend to 0, or the previous case by letting δ tend to zero. The result is the formulae *(1.k)* and *(1.l)* of Getoor.

From the Laplace transforms *(7.$\tilde{c}$)* and *(7.$\tilde{e}$)* and the formulae of *Corollaries (4.8)* and *(4.9)* for change of law on $\underline{F}_{L_y^-}$, we now obtain analogues of the formulae of *Proposition (2.3)* and *Theorem (4.3)* with L_y instead of T_y, valid for all $\nu \geq 0,\ x > 0$:

$$E_x^{\nu,\delta}\{\exp[-\tfrac{1}{2}\alpha^2 C(L_y) - \tfrac{1}{2}\beta^2 L_y]\ ;\ L_y > 0\}$$

$$= \frac{I_{\theta}[\gamma(x\wedge y)]\ K_{\theta}[\gamma(x\vee y)]}{I_{\nu}(\delta x)\ K_{\nu}(\delta y)}, \qquad \text{if} \quad \delta > 0$$

$$= 2\nu\ I_{\theta}[\beta(x\wedge y)]\ K_{\theta}[\beta(x\vee y)]\left(\frac{y}{x}\right)^{\nu} \qquad \text{if} \quad \delta = 0.$$

where $\gamma = (\beta^2 + \delta^2)^{1/2}$, $\theta = (\nu^2 + \alpha^2)^{1/2}$;

In particular

$$(7.g) \qquad E_x^{\nu}\{\exp[-\tfrac{1}{2}\alpha^2 C(L_y)] \; ; L_y > 0\} = \frac{\nu}{\theta}\left(\frac{y}{x}\right)^{\nu+\varepsilon\theta},$$

where $\varepsilon = +1$ if $x > y$, -1 otherwise.

In view of the formula *(2.k)*, the only really new information in *(7.g)* comes in the case $x = y$, when *(7.g)* becomes

$$(7.h) \qquad E_y^{\nu} \exp[-\tfrac{1}{2}\alpha^2 C(L_y)] = (1+\alpha^2/\nu^2)^{-\frac{1}{2}},$$

which is to say that under P_y^{ν} the variable $C(L_y)$ has the gamma $(\tfrac{1}{2}, \tfrac{1}{2}\nu^2)$ distribution.

Remark : For another means of obtaining distributions associated with the clock $(C_t, t \geq 0)$ of a $BES_y(\nu)$ process R_t, consider the process $\hat{R}_s = R(A_s)$ where A is an inverse of the clock C. The process $\hat{R}$ is a diffusion with infinitesimal generator

$$\tfrac{1}{2} x^2 \frac{d^2}{dx^2} + (2\nu+1)\, x\frac{d}{dx},$$

which can be written as

$$\hat{R}_s = y \exp(B_s + \nu s), s \geq 0,$$

where B is a BM_0 on the line. For example, to obtain *(7.h)* , observe that

$$C(L_y) = \sup\{s : \hat{R}_s = y\}$$

$$= \sup\{s : B_s + \nu s = 0\},$$

and see example *(3.8)*.

8. THE FIRST AND LAST TIMES THAT $BES(\mu,\delta)$ HITS A LINE.

Given a $BES_x(\nu,\delta)$ process, for $y > 0$ consider the random times

$$\tau_y = \inf\{t : R_t = yt\},$$

and

$$\sigma_y = \sup\{t : R_t = yt\}$$

which are the first and last times that the trajectory hits a line in space-time of slope y which passes through the origin. As an immediate consequence of Watanabe's time inversion *theorem (5.5)* and our formulae for the distribution of T_y and L_y, we are able to write down formulae for the distributions of τ_y and σ_y.

It would also be interesting to obtain more general formulae for the case of a line with intercept $a \geq 0$, and in particular to calculate the probability that $BES_x(\mu,\delta)$ ever hit such a line. This would lead to the distribution of the maxima and minima of a Bessel bridge by *theorem (5.8)*, but we don't know how to manage the case $a > 0$.

As a start, we observe from *theorem (5.5)* and the nice way the clock transforms with time inversion, that the law of the triple

(8.a) $\qquad (\tau_y^{-1},\ C(0,\tau_y),\ C(\tau_y,\infty)) \qquad$ under $\quad P_\delta^{\nu,x}$

is identical to the law of the triple

(8.$\hat{a}$) $\qquad (L_y,\ C(L_y,\infty),\ C(0,L_y)) \qquad$ under $\quad P_x^{\nu,\delta}$.

Thus the Laplace transforms of the two triples are identical, and that of the latter could be written down using *(7.f)*, *(4.i)*, *(4.i*)* and the independence of the outer components and the inner one given $(L_y > 0)$. In particular one finds that under $P_\delta^{\nu,x}$ given $(\tau_y < \infty)$, the clock $C(\tau_y)$ is independent of τ_y^{-1}, hence independent also of $R(\tau_y) = y\,\tau_y$, and even of the whole future process $(R(\tau_y + s), s \geq 0)$.

And for $\nu = 0$, the law of $C(\tau_y)$ given $(\tau_y < \infty)$ turns out to be the second Hartman law of *(1.b)* with parameter δy, where δ is the starting place here, regardless of the drift $x \geq 0$.
Moreover, similar remarks apply to σ_y, since the identity of *(8.a)* and *(8.$\hat{a}$)* holds equally well with σ_y instead of τ_y, and T_y instead of L_y. There is independence of $C(\sigma_y)$ and $(R(\sigma_y + s), s \geq 0)$ given $(\sigma_y > 0)$, and for $\nu = 0$ the law of $C(\sigma_y)$ given $(\sigma_y > 0)$ is just the first Hartman law with parameter δy. Obviously when $d = 2\nu + 2$ is an integer one can go further and express these results in terms of the radial and angular parts of a BM in $\mathbb{R}^d$, but we leave this to the reader.

From the results of the previous section, we obtain explicit formulae for both the density and the Laplace transform of τ_y. From *(7.a)* and the identity in law of τ_y^{-1} and L_y, we have for all $\nu, x, \delta \geq 0$,

$$(8.b) \qquad P_\delta^{\nu,x}(\tau_y \in dt) = \frac{dt}{t^2} p_{1/t}^{\nu,\delta}(x,y) \; [G(\delta,y)]^{-1}, \qquad t > 0,$$

where here again, and in the formulae below, δ is the starting point and x the drift. From *(8.b)* and *(4.c)*, we obtain, in the

<u>Case</u> $\delta > 0,\ x > 0$

$$(8.c) \qquad P_\delta^{\nu,x}(\tau_y \in dt) = \frac{dt\ I_\nu(xyt) \exp\left(-\frac{1}{2}\left[(x^2+y^2)t + \delta^2/t\right]\right)}{2t\ I_\nu(\delta x)\ K_\nu(\delta y)},$$

$$(8.\tilde{c}) \qquad E_\delta^{\nu,x} \exp(-\tfrac{1}{2}\lambda^2 \tau_y) = \frac{I_\nu(\delta a_\lambda)\ K_\nu(\delta b_\lambda)}{I_\nu(\delta x)\ K_\nu(\delta y)}$$

where $a = a_\lambda(x,y)$ and $b_\lambda = b_\lambda(x,y)$ are defined by the requirements

$$0 < a_\lambda \leq b_\lambda < \infty,$$

$$a_\lambda^2 + b_\lambda^2 = x^2 + y^2 + \lambda^2,$$

$$a_\lambda b_\lambda = xy,$$

or, to be more explicit

$$a_\lambda = \tfrac{1}{2}\,[(x+y)^2 + \lambda^2]^{1/2} - \tfrac{1}{2}\,[(x-y)^2 + \lambda^2]^{1/2},$$

$$b_\lambda = \tfrac{1}{2}\,[(x+y)^2 + \lambda^2]^{1/2} - \tfrac{1}{2}\,[(x-y)^2 + \lambda^2]^{1/2}.$$

To derive *(8.$\tilde{c}$)* from *(8.c)* one uses the following identity, valid for $0 < a \leq b < \infty$, which results from *(7.b)* after substituting $x = a,\ y = b,\ \varepsilon = \delta$, using *(2.h)*, and making the change of variable $u = 1/t$:

$$\int_0^\infty \frac{du}{u}\, I_\nu(abu)\, \exp\Big(-\tfrac{1}{2}\,[(a^2+b^2)u + \delta^2/u]\Big) = 2\, I_\nu(\delta a)\, K_\nu(\delta b).$$

The derivation of the corresponding formulae in the remaining cases is straightforward. One obtains

<u>Case</u> $\delta > 0,\ x = 0$:

(8.d) $$P_\delta^\nu(\tau_y \in dt) = \frac{dt\ (yt)^\nu \exp\left[-\tfrac{1}{2}\,(y^2 t + \delta^2/t)\right]}{2t\ \delta^\nu\ K_\nu(\delta y)}\ ;$$

(8.$\tilde{d}$) $$E_\delta^\nu \exp(-\tfrac{1}{2}\,\lambda^2\,\tau_y) = (1 + \lambda^2/y^2)^{-\nu/2}\, K_\nu[\delta(y^2+\lambda^2)^{1/2}]/K_\nu(\delta y).$$

<u>Case</u> $\delta = 0,\ \nu, x > 0$:

(8.e) $$P_0^{\nu,x}(\tau_y \in dt) = dt\ \nu t^{-1}\ (y/x)^\nu\ I_\nu(xyt)\ \exp\left[-\tfrac{1}{2}\,(x^2+y^2)t\right],$$

(8.$\tilde{e}$) $$E_0^{\nu,x} \exp(-\alpha\tau_y) = [\psi(x,y,\alpha)]^\nu,$$

where

$$\psi(x,y,\alpha) = [\alpha + \tfrac{1}{2}\,x^2 + \tfrac{1}{2}\,y^2 - ((\alpha + \tfrac{1}{2}\,x^2 + \tfrac{1}{2}\,y^2)^2 - x^2y^2)^{1/2}]/x^2$$

<u>Case</u> $\delta = 0,\ x = 0,\ \nu > 0$:

$$(8.f) \qquad P_o^\nu(\tau_y \in dt) = dt\ \Gamma(\nu)^{-1}\ (\tfrac{1}{2}y^2)^\nu\ t^{\nu-1}\ \exp - \tfrac{1}{2}y^2 t$$

$$(8.\mathfrak{f}) \qquad E_o^\nu \exp(-\tfrac{1}{2}\lambda^2\ \tau_y) = (1 + \lambda^2/y^2)^{-\nu},$$

which is to say that in this case τ_y has a gamma distribution with index ν and scale parameter $y^2/2$.

Remarks : (i) Some very curious facts emerge from the above formulae. From the Laplace transforms it is clear that if either $\delta = 0$ or $x = 0$, the law of τ_y for a $BES_\delta(\nu,x)$ process is infinitely divisible, but we do not have any probabilistic explanation to offer. For $\delta > 0,\ x = 0$ the infinite divisibility of τ_y is seen using *(8.$\mathfrak{d}$)*, *(4.j*)* and *(8.$\mathfrak{f}$)*, which exhibit the remarkable fact that the P_δ^ν distribution of τ_y is the convolution of the P_o^ν distribution of τ_y and the $P_o^{\nu,y}$ distribution of L_δ.

(ii) The infinitely divisible law *(8.e)* was encountered by Feller [9] (see also [10]) in the study of first passage times for a continuous time random walk : for positive integers ν the law *(8.e)* is the distribution of the first passage time to ν of a compound Poisson process starting at zero with jumps of + 1 at rate $2x^2/(x^2+y^2)^2$ and jumps of − 1 at rate $2y^2/(x^2+y^2)^2$.

(iii) We do not know if the distribution of τ_y is infinitely divisible for $\delta,\ x > 0$.

9. COMPLETELY MONOTONE FUNCTIONS ASSOCIATED WITH LAST EXIT TIMES OF BES(ν)

Our aim in this section is to explain how the complete monotonicity of certain ratios and products of Bessel functions, many of which were studied by Ismail and Kelker [26], can be related to behaviour of Bessel diffusions prior to last exit times.

Ismail and Kelker showed by purely analytic arguments that for $\nu > -1$, $\theta > 0$, there is an infinitely divisible probability distribution $\iota^{\nu,\theta}$ on $(0,\infty)$ with Laplace transform in $\alpha > 0$

$$\text{(9.a1)} \qquad \frac{\Gamma(\nu+\theta+1)\ 2^{\theta}\ I_{\nu+\theta}(\sqrt{\alpha})}{\Gamma(\nu+1)\ (\sqrt{\alpha})^{\theta}\ I_{\nu}(\sqrt{\alpha})} = \int_0^{\infty} e^{-\alpha x}\ \iota^{\nu,\theta}(dx),$$

and that as $\theta \to \infty$, $\iota^{\nu,\theta}$ converges to the infinitely divisible law ι^{ν} with Laplace transform

$$\text{(9.a2)} \qquad (\sqrt{\alpha})^{\nu}/\Gamma(\nu+1)\ 2^{\nu}\ I_{\nu}(\sqrt{\alpha}) = \int_0^{\infty} e^{-\alpha x}\ \iota^{\nu}(dx) = E_0^{\nu}\ \exp(-\alpha T_y/2y^2),$$

where the identification of ι^{ν} as the P_0^{ν} distribution of $(T_y/2y^2)$ is due to Kent [31].
Here T_y is the hitting time of y for a $\text{BES}_0(\nu)$ process, and the usual reflecting boundary condition at 0 must be stipulated for $-1 < \nu < 0$. Ismail and Kelker also proved that for $\nu > 0$, $\theta > 0$, $y > 0$, there is an infinitely divisible probability distribution $\kappa^{\nu,\theta}$ on $(0,\infty)$ with Laplace transform in $\alpha > 0$

$$\text{(9.b1)} \qquad \frac{\Gamma(\nu+\theta)\ 2^{\theta}\ K_{\nu}(\sqrt{\alpha})}{\Gamma(\nu)\ (\sqrt{\alpha})^{\theta}\ K_{\nu+\theta}(\sqrt{\alpha})} = \int_0^{\infty} e^{-\alpha x}\ \kappa^{\nu,\theta}(dx),$$

and that as $\theta \to \infty$, $\kappa^{\nu,\theta}$ converges to the infinitely divisible law κ^{ν} with Laplace transform

$$(9.b2)\qquad \frac{2(\sqrt{\alpha})^{\nu}}{\Gamma(\nu)2^{\nu}}\, K_{\nu}(\sqrt{\alpha}) = \int_0^{\infty} e^{-\alpha x}\, \kappa^{\nu}(dx) = E_o^{\nu} \exp(-\alpha L_y/2y^2),$$

where the idenfication of κ^{ν} as the P_o^{ν} distribution of $L_y/2y^2$ is made by Getoor's formula *(1.1)*. In view of the obvious Laplace transform identities corresponding to the convolution identity

$$(9.c)\qquad \iota^{\nu} = \iota^{\nu,\theta} * \iota^{\nu+\theta},\quad \nu > -1,\ \theta > 0,$$

and its companion with κ instead of ι, valid for $\nu > 0$, these results of Ismail and Kelker are equivalent to the assertion that for each $y > 0$ there exist on some probability space two processes with independent increments

$$(9.d)\qquad (T_y^{\nu},\ \nu > -1) \text{ and } (L_y^{\nu},\ \nu > 0),$$

each with decreasing trajectories coming down from ∞ to 0, such that for each ν in the appropriate range, $T_y^{\nu}/2y^2$ has law ι^{ν}, and $L_y^{\nu}/2y^2$ has law κ^{ν}. Then, taking simply $y = 1/\sqrt{2}$, the increment $T_y^{\nu} - T_y^{\nu+\theta}$ would have distribution $\iota^{\nu,\theta}$.

Note that because the trajectories have a finite limit at infinity but not at their start, this increment would be independent of $T_y^{\nu+\theta}$ but not of T_y^{ν}, in contrast to the usual case of a process with independent increments starting at zero. It would be interesting to find a presentation of $BES_0(\nu)$ for varying ν in which such processes were embedded, (perhaps even for all y with independent increments in (y,ν)), as the results of Ismail and Kelker would then follow immediately from those of Getoor and Kent. We do not know of any such representation, but in the course of our investigations, we shall provide probabilistic proofs of the existence of $\iota^{\nu,\theta}$ for all $\nu > -1$, $\theta > 0$, and of $\kappa^{\nu,\theta}$ for all $\nu \geq 1$, $\theta > 0$. The gap in our argument for the second case if $0 < \nu < 1$ is curious, but stems from the fact, obvious on differentiating *(9.a2)* and *(9.b2)* at $\alpha = 0$, that

$$(9.e1)\qquad E_o^{\nu}\, T_y = y^2/2(\nu+1) < \infty,\quad \nu > -1$$

while

$$(9.a2) \qquad E_o^\nu L_y = y^2/2(\nu-1) < \infty, \ \nu > 1$$

$$= \infty, \quad 0 < \nu \leq 1.$$

Let λ^ν denote the Lévy measure of the infinitely divisible law ι^ν, $\nu > -1$, so the Laplace transform of ι^ν in *(9.a2)* is

$$(9.f) \qquad \exp\left[-\int_0^\infty (1-e^{-\alpha x})\ \lambda^\nu(dx)\right], \quad \alpha > 0.$$

It is clear that the existence of laws $\iota^{\nu,\theta}$ satisfying *(9.a1)* is equivalent to

$$(9.g) \qquad \nu \to \lambda_y^\nu \ \text{ is decreasing}, \ \nu > -1,$$

(where 'decreasing' means "decreasing when evaluated on any Borel set") since $\iota^{\nu,\theta}$ then appears as the infinitely divisible law with Lévy measure $\lambda^\nu - \lambda^{\nu+\theta}$. But we can determine λ^ν by taking the negative logarithm and differentiating with respect to α in *(9.f)* and *(9.a2)*. After using the recurrence formula *(13.4)* for the derivative of I_ν, the result is

$$(9.h) \qquad \int_0^\infty x\ e^{-\alpha x}\ \lambda^\nu(dx) = \frac{I_{\nu+1}(\sqrt{\alpha})}{2\sqrt{\alpha}\ I_\nu(\sqrt{\alpha})}, \ \alpha > 0$$

which proves that *(9.a1)* holds for $\nu > -1$, $\theta = 1$, with the measure

$$\iota^{\nu,1}(dx) = 4(\nu+1)x\ \lambda^\nu(dx)$$

This is a probability measure, in keeping with *(9.e1)* for $y = 1/\sqrt{2}$, but it is certainly not obvious at this stage that $\iota^{\nu,1}$ is infinitely divisible. Still the problem of showing that λ^ν decreases is now reduced to showing that

$$\nu \to \iota^{\nu,1}\ /\ (\nu+1) \ \text{ is decreasing}, \quad \nu > -1.$$

The argument is completed by appealing to the result of Getoor and Sharpe [15], that

$$(9.i) \qquad \iota^{\nu,1} \ \text{ is the } P_y^{\nu+1} \text{ distribution of } A(L_y)/2y^2, \ \nu > -1,$$

where $(A(t), t \geq 0)$ is the additive functional of the Bessel process $(R_t, t \geq 0)$ defined by

(9.j) $$A(t) = \int_0^t 1(R_s \le y)\,ds,$$

the final touch being the fact that for $y > 0$,

(9.k1) $\mu \to P_y^\mu/\mu$ is decreasing on $\underline{F}_{L_y-}$, $\mu > 0$,

which is obvious after using *(4.9)* to write

(9.k2) $$\mu^{-1} P_y^\mu(A) = \nu^{-1} E_y^\nu[\exp\{-\tfrac{1}{2}(\mu^2-\nu^2)\,C(L_y)\}\,;\,A], \quad A \in \underline{F}_{L_y-},$$

for an arbitrary fixed $\nu > 0$.

Turning now to the production of $\kappa^{\nu,\theta}$, let η^ν be the Lévy measure of κ^ν, the P_o^ν distribution of $L_y/2y^2$. We want to show that

(9.l) $\nu \to \eta^\nu$ is decreasing, $\nu > 0$.

The steps used to obtain *(9.i)* give this time

(9.m1) $$\int_0^\infty x\,e^{-\alpha x}\,\eta^\nu(dx) = \frac{K_{\nu-1}(\sqrt{\alpha})}{2\sqrt{\alpha}\,K_\nu(\sqrt{\alpha})}, \quad \nu > 0.$$

The substitution $\nu = \mu + 1$ now reveals that

(9.m2) $$\kappa^{\mu,1}(dx) = 4\mu x\,\eta^{\mu+1}(dx), \quad \mu > 0,$$

but the appearance of $\mu + 1$ on the right is most frustrating. Indeed, we shall establish below an analogue of *(9.i)*, namely

(9.n) $\kappa^{\mu,1}$ is the P_y^μ distribution of $B(L_y)/2y^2$, $\mu > 0$,

where $(B(t), t > 0)$ is the additive functional

(9.o) $$B(t) = \int_0^t 1(R_s > y)\,ds,$$

and in view of *(9.m2)* and *(9.k)*, it follows that *(9.l)* holds for $\nu > 1$.

This gives the existence of $\kappa^{\nu,\theta}$ for $\nu > 1$, $\theta > 0$, and $\kappa^{1,\theta}$ can be obtained by letting $\nu \to 1$, but we are cheated of the result for $0 < \nu < 1$.

Remarks (9.1) :

(i) That the trajectories of the processes *(9.d)* tend to 0 as $\nu \to \infty$ is obvious from the formulae *(9.e)* and *(9.f)*. This gives a new proof of Ismail and Kelker's result that $\iota^{\nu,\theta}$ and $\kappa^{\nu,\theta}$ converge to ι^{ν} and κ^{ν} as $\theta \to \infty$, and a glance at the Laplace transforms of these laws reveals that we have given a probabilistic proof of the asymptotic formulae

$$I_{\lambda}(x) \sim (x/2)^{\lambda}/\Gamma(\lambda+1),\ K_{\lambda}(x) \sim \tfrac{1}{2}\,\Gamma(\lambda)\ (x/2)^{-\lambda},\ \lambda \to \infty.$$

(ii) That the trajectories of $(T_y^{\nu}, \nu > -1)$ and $(L_y^{\nu}, \nu > 0)$ come down from ∞ at the start is obvious from the formulae *(9.a2)* and *(9.b2)*. This corresponds to the fact that state 0 is a trap for BES(-1), so $P_o^{-1}(T_y = \infty) = 1$, while state y is recurrent for BES(0), so $P_o^{0}(L_y = \infty) = 1$.

(iii) In *Theorem (1.8)* of Ismail and Kelker [26] it is stated that the representation *(9.b1)* obtains for an infinitely divisible probability $\iota^{\nu,\theta}$ for all $\nu > -1$, but the result is only proved for $\nu > 0$. The result is in fact false for $-1 < \nu \leq 0$, because the asymptotic formulae for the behaviour of the Bessel functions at zero show that the function on the left is unbounded as $\alpha \to 0$. Still, from *(9.m1)* we recover the result of Grosswald [18] that $K_{\nu-1}(\sqrt{\alpha})\ /\ \sqrt{\alpha}\ K_{\nu}(\sqrt{\alpha})$ is completely monotone in α for all $\nu > 0$, with a representation of this function as the Laplace transform of $2x\ \eta^{\nu}(dx)$, a measure which has infinite mass for $0 < \nu \leq 1$ in view of *(9.e2)*. Ismail [24] showed that $K_{\nu-\theta}(\sqrt{\alpha})\ /\ (\sqrt{\alpha})^{\theta}\ K_{\nu}(\sqrt{\alpha})$ is actually c.m. for all real ν and $\theta > 0$, but we do not have any explanation of this for $\nu \leq \theta + 1$, except for the case $\theta = 1$ which appears in *(9.8)* **(v)** below.

(iv) In combination with the strong Markov property, the identities *(9.i)* and *(9.a)* for $\theta = 1$ reveal that for $\mu > 0$ the total time $A(L_y)$ spent by $BES_0(\mu)$ below y has the same law as T_y for $BES_0(\mu-1)$, a result noted by Getoor and Sharpe [15], and due originally to Cieselski and Taylor [5] in the case $\mu = (d-2)/2$, $d = 3,4,\ldots$. Curiously, the companion identities *(9.n)* and *(9.a)* for κ with $\theta = 1$ do not seem to combine to yield such an attractive result.

(v) Ismail and Kelker [26] give formulae for the densities of $\iota^{\nu,\theta}$ and $k^{\nu,\theta}$ in terms of the Bessel functions J_μ and Y_μ.

We come now to the proof of *(9.n)*. We shall establish the following result, which encompasses *(9.i)*, *(9.n)* and *(7.$\tilde{e}$)* for $x = y$ by giving the joint Laplace transform of the times spent below and above y by $BES_y(\nu)$ before L_y. We use the notation of *(9.j)* and *(9.o)*.

Proposition (9.2) : For $\nu > 0$

$$E_y^\nu \exp\{-(\alpha A(L_y)+\beta B(L_y))/2y^2\} = \frac{2\nu\, I_\nu(\sqrt{\alpha})\; K_\nu(\sqrt{\beta})}{\sqrt{\alpha}\; I_{\nu-1}(\sqrt{\alpha})\, K_\nu(\sqrt{\beta})+\sqrt{\beta}K_{\nu-1}(\sqrt{\beta})\, I_\nu(\sqrt{\alpha})}.$$

Note : To recover *(9.i)* and *(9.n)* put either α or β equal to zero and use the recurrence formulae *(13.4)*. To recover *(7.$\tilde{e}$)* for $x = y$, put $\alpha = \beta$ and use the formula *(13.8)* for the Wronskian.

Proof : Fix $\nu > 0$ and put $f(x) = E_x^\nu \exp\{-(\alpha A(L_y) + \beta B(L_y))/2y^2\}$. Since $f(0) = E_0 \exp(-\alpha T_y/2y^2) f(y)$, it suffices in view of *(9.a2)* to calculate $f(0)$. But by *remark (4.2)* (ii) and Williams time reversal *theorem (3.3)*, or by the result of Sharpe [41], $f(0) = g(y)$, where

$$g(x) = E_x^{-\nu} \exp\{-(\alpha A(T_0) + \beta B(T_0))/2y^2\}$$

where $P_x^{-\nu}$ governs a $BES_x(-\nu)$ process up to the time T_0 when it hits zero. Now g is a solution of

$$G_{-\nu} g = \alpha g \text{ on } (0,y) \; ; \; G_{-\nu} g = \beta g \text{ on } (y,\infty),$$

where $G_{-\nu}$ is the generator of $BES(-\nu)$.

The determination of g is now completed in the manner of [15] Section 8, the constants in the general solution being determined by the boundary conditions

$$g(0) = 1 \ ; \ g(y-) = g(y+) \ ; \ g'(y-) = g'(y+).$$

It is also possible to derive *(9.2)* from the special cases *(9.i)* and *(7.$\tilde{e}$)* mentioned earlier, using the independence of the excursions above and below y - see the end of this section.

We address now the question of complete monotonicity of the ratios *(1.n)* in the Introduction. Our description of the positive measures on $[0,\infty)^2$ which have these ratios as Laplace transforms is based on the limit as $\mu \downarrow 0$ of the measures P^{μ}_{y}/μ on $\underline{\underline{F}}_{L_y -}$, which, as we have already noted, increase as μ decreases.

Proposition (9.3) : Fix $y > 0$. Then for $\mu > 0$, and $Z \geq 0$, $\underline{\underline{F}}_{L_y -}$ measurable

$$\text{(i)} \qquad \frac{1}{\mu} E^{\mu}_{y} Z \exp\left[\tfrac{1}{2}\mu^2 C(L_y)\right] \stackrel{\text{def}}{=} M_y(Z)$$

does not depend on μ. This M_y is a σ-finite measure on $\underline{\underline{F}}_{L_y -}$ with infinite total mass. Moreover, for each $\delta > 0$

$$\text{(ii)} \qquad M_y(Z) = 2\, I_o(\delta y)\, K_o(\delta y)\, E^{o,\delta}_{y}\, Z \exp(\tfrac{1}{2}\delta^2 L_y),$$

and

$$\text{(iii)} \qquad M_y(Z) = \lim_{\mu \downarrow 0} \frac{1}{\mu} E^{\mu}_{y} Z = \lim_{\delta \downarrow 0} \operatorname{Log}(\delta^{-2})\, E^{o,\delta}_{y} Z.$$

Proof : The identities (i) and (ii) follow immediately from the formulae of *(4.8)* and *(4.9)* for change of law on $\underline{\underline{F}}_{L_y -}$, and (iii) follows like *(3.xb)*, using *(13.2)*.

Remark (9.4) : In view of $(7.\tilde{c})$, the identity (ii) shows that M_y here is identical to the M_y associated with $BES_y(0)$ in *(3.6)*. Thus the description of M_y in *(3.7)* applies, where from *(7.c)*, and *(2.h)* we find

$$M_y(L_y \in dt) = dt\; p_t^o(y,y)/y,$$

in keeping with *(3.y)* .

The measure M_y can also be described as the image of the measure M associated with BM in *(3.8)* after the space transformation and random time change described at the end of section 7.

Proposition (9.4) : Let $Z \geq 0$ be $\underline{F}_{L_y}$-measurable. The function

(i) $$\nu \to (\sqrt{\nu})^{-1}\, E_y^{\sqrt{\nu}}\,(Z)$$

is the Laplace transform of a positive measure on $(0,\infty)$. Let

$$f(\alpha,\nu) = \nu^{-1}\, E_y^{\nu}\, \exp(-\alpha Z).$$

Then the function

(ii) $$(\alpha,\nu) \to f(\alpha,\sqrt{\nu})$$

is the Laplace transform of a positive measure on $[0,\infty)^2$.

If $Z = A(L_y)$ for an additive functional A, then, for each $\nu > 0$, the function

(iii) $$\alpha \to \nu f(\alpha,\nu) \quad (= E_y^{\nu}\,(\exp -\alpha Z))$$

is the Laplace transform of an infinitely divisible law on $[0,\infty)$, and the same is true of the function

(iv) $$\nu \to f(\alpha,\sqrt{\nu}) \;/\; f(\alpha,0+),$$

for each $\alpha > 0$ such that $f(\alpha,0+) < \infty$.

Note : It will be seen below in *(9.10)* (i) that in fact $f(\alpha,0+) < \infty$ for all $\alpha > 0$ except in the trivial case when $A_t = 0$ for all t.

Proof : From *(9.3)* (i) we have

$$\nu^{-1} E_y^\nu Z = M_y Z \exp(-\tfrac{1}{2}\nu^2 C(L_y)),$$

which shows that (i) is the Laplace transform in ν of the image under the map $\omega \to \frac{1}{2} C(L_y)(\omega)$ of the measure $Z(\omega) M_y(d\omega)$ on $\underline{\underline{F}}_{L_y -}$.

The proof for (ii) is almost the same, and (iii) is a case of *Lemma (6.2)*. Turning to (iv), we have as in *Lemma (6.2)* that, for each $\nu > 0$, the law of the pair $(A(L_y), C(L_y))$ under P_y^ν is infinitely divisible.

Thus, for each $\alpha > 0$, $C(L_y)$ has an infinitely divisible law under the probability

$$Q_y^\nu(F) = E_y^\nu[\exp(-\alpha Z) ; F] / E_y^\nu[\exp(-\alpha Z)], \quad F \in \underline{\underline{F}}_{L_y -},$$

$$= \frac{M_y[\exp(-\alpha Z - \frac{1}{2}\nu^2 C(L_y)) ; F]}{M_y[\exp(-\alpha Z - \frac{1}{2}\nu^2 C(L_y))]}$$

$$\xrightarrow[(\nu\to 0)]{} M_y[\exp(-\alpha Z) ; F] / M_y \exp(-\alpha Z),$$

and (iv) follows because the collection of infinitely divisible laws on R_+ is closed under weak convergence

Remarks (9.5) :

(i) Suppose that for each ν, Z has a right-continuous density $g^\nu(z)$, $z > 0$. Then by approximating $g^\nu(z)$ by $P^\nu(z \leq Z \leq z + \varepsilon)/\varepsilon$, and using *(9.4)* (i) with the indicator of the event $(z \leq Z \leq z + \varepsilon)$ instead of Z, we see that for each $z > 0$, the function

$$(9.p1) \qquad \nu \to (\sqrt{\nu})^{-1} g^{\sqrt{\nu}}(z)$$

is the pointwise limit of c.m. functions, hence itself c.m..

(ii) The proof of (ii) shows that

$$f(\alpha,\nu) = \nu^{-1} E_y^{\nu} \exp(-\alpha Z) = \int_0^{\infty} \exp(-\tfrac{1}{2}\nu^2 t)\, h_\alpha(t)\,dt,$$

where $h_\alpha(\cdot)$ is a density for the $M_y(d\omega)e^{-\alpha Z(\omega)}$ distribution of $C(L_y)$. Assuming that for each α this density is right-continuous in t, the same sort of argument used in the last remark shows that for each $t > 0$, the function

(9.p2) $\qquad \alpha \to h_\alpha(t)$

<u>is c.m.</u>.

<u>*Examples (9.6)*</u> :

(i) Take $Z = L_y$ in *(9.4)*. From *(7.$\tilde{e}$)* we have

$$f(\alpha,\nu) = 2I_\nu(\sqrt{2\alpha}\, y)\, K_\nu(\sqrt{2\alpha}\, y),$$

hence from *(9.4)* (ii), for each $r > 0$ the function

(9.q1) $\qquad (\alpha,\nu) \to I_{\sqrt{\nu}}(\sqrt{\alpha}\, r)\, K_{\sqrt{\nu}}(\sqrt{\alpha}\, r)$ <u>is c.m.</u>,

a result which we already know from *(7.f)*.
This complements the fact, noted in [51], that as a consequence of *(2.j)* each of Hartman's functions *(1.a)*, *(1.b)*, *(1.c)* and *(1.d)* becomes completely monotone in (α,ν) after the substitution $\lambda = \nu/2$ and the introduction of an extra factor of $\sqrt{\alpha}$ in the argument of each Bessel function. The infinitely divisible laws on the line arising from *(9.4)* (iii) and (iv) have already appeared in *(7.$\tilde{e}$)* and *(7.f)*. From *(9.p1)* we recover the c.m. of the first Hartman function *(1.a)* while the conclusion of *(9.p2)* seems very complicated.

(ii) Take $Z = A(L_y)$, the total time spent below y. From *(9.i)* the P_y^{ν} distribution of $Z/2y^2$ is $\iota^{\nu-1,1}$, and *(9.4)* (ii) and *(9.a2)* yield the c.m. property of the first of the two functions in *(1.n)*.
Taking $y = 1/\sqrt{2}$, Ismail and Kelker ([26] ; *theorem* 1.9 and formula *(4.15)*) show that Z has density

$$g^{\nu}(z) = 4\nu\, \Sigma_{\nu-1}(z), \quad z > 0$$

where for $\mu \geq -1$,

(9.q2) $$\Sigma_\mu(z) = \sum_{n=1}^{\infty} \exp(-j_{\mu,n}^2 z), \qquad (j_{\mu,n} : n = 1,2,\dots)$$

being the increasing sequence of positive zeros of the Bessel function of the first kind J_μ. From *(9.p1)* we learn that for each $z > 0$ the function

(9.q3) $\nu \to \Sigma_{\sqrt{\nu-1}}(z)$ is c.m.,

hence also (see *(9.7)* below)

(9.q3') $\nu \to \Sigma_{\sqrt{\nu}}(z)$ is c.m..

(iii) Take $Z = B(L_y)$, the time spent above y before L_y. From *(9.n)* and *(9.4)* (ii) we obtain the c.m. property of the second of the functions in *(1.n)*. On the other hand, following Grosswald [18] and Ismail [24], we have, for $\nu \geq -1$,

(9.q4) $$\frac{K_\nu(\sqrt{\alpha})}{\sqrt{\alpha}\, K_{\nu+1}(\sqrt{\alpha})} = \int_0^\infty e^{-\alpha z}\, k_\nu(z)\,dz,$$

where

$$k_\nu(z) = \frac{2}{\pi^2}\int_0^\infty dt\; t^{-1} e^{-tz} \left(J_{\nu+1}^2 + Y_{\nu+1}^2\right)^{-1}(\sqrt{t}).$$

Taking $y = 1/\sqrt{2}$, the P_y^ν density of Z at z is therefore $2\nu k_\nu(z)$, and from *(9.p1)* we find that for each $z > 0$ the function

(9.q5) $\nu \to k_{\sqrt{\nu}}(z)$ is c.m..

Ismail also showed that

$$K_\nu(x) \,/\, K_{\nu+1}(x) = \int_0^\infty \exp(-\tfrac{1}{2}\nu^2 t)\, m_x(t)\,dt,$$

where

$$m_x(t) = \frac{1}{2\pi}\int_0^\infty ds\; e^{-st/2}\, \frac{\mathrm{Im}\{K_{-i\sqrt{s}}(x)\, K_{1+i\sqrt{s}}(x)\}}{|K_{1+i\sqrt{s}}(x)|^2}$$

which identifies the function $h_\alpha(t)$ of *(9.p2)* in this case as $2m_{\sqrt{\alpha}}(t) / \sqrt{\alpha}$. Thus for each $t > 0$, the function

(9.q6) $\alpha \to m_{\sqrt{\alpha}}(t)/\sqrt{\alpha}$ is c.m..

Note *(9.7)* : According to *Theorem 5.4* of Ismail [24], for each $x > 0$, $n = 1,2,\ldots$, $\theta \geq 0$, the function

$$\nu \to K_{\theta+\sqrt{\nu}}(x) \; / \; K_{\theta+\sqrt{\nu}+n}(x) \qquad \underline{\text{is c.m}}$$

As noted by Ismail, it suffices to prove this when $n = 1$. We note here that by the Criterion 2 of Section XIII of Feller [10] for the composition of two functions to be c.m., it also suffices to consider the case $\theta = 0$ since : $\nu \to (\sqrt{\nu} + \theta)^2$ is c.m.. By the same kind of bootstrap argument, one learns from *(1.p)* that also for each $\alpha > 0$, $n = 1,2,\ldots$, $\theta \geq 0$,

$$(9.r1) \qquad (\alpha,\nu) \to \frac{K_{\theta+\sqrt{\nu}}(\sqrt{\alpha})}{(\sqrt{\alpha})^n \, K_{\theta+\sqrt{\nu}+n}(\sqrt{\alpha})} \qquad \underline{\text{is c.m.}},$$

and for each $\alpha > 0$, $n = 1,2,\ldots$, $\theta \geq -1$,

$$(9.r2) \qquad (\alpha,\nu) \to \frac{I_{\theta+\sqrt{\nu}+n}(\sqrt{\alpha})}{(\sqrt{\alpha})^n \, I_{\theta+\sqrt{\nu}}(\sqrt{\alpha})} \qquad \underline{\text{is c.m.}}.$$

To conclude this section, we show how many of the results above admit interesting interpretations in terms of local time and excursion theory. Consider in this paragraph two arbitrary additive functionals $(A(t), t \geq 0)$ and $(B(t), t \geq 0)$, and write $A_u = A(\tau_u)$, $B_u = B(\tau_u)$, where $(\tau_u, 0 \leq u \leq \ell_\infty)$ is the right continuous inverse of the local time process $(\ell_t, t \geq 0)$ at a fixed point y. We assume that ℓ is normalised as an occupation density, so that for $\nu > 0$ we have from *(7.b)*

$$(9.s1) \qquad E_y^\nu \, \ell(L_y) = E_y^\nu \, \ell_\infty = \int_0^\infty p_t^\nu(y,y)\,dt = y/\nu.$$

Then by applying the Itô excursion theory as in the proof of *(6.2)*, for $\nu > 0$ we find that under P_y^ν the total local time ℓ_∞ is exponentially distributed with rate ν/y, and conditional on $\ell_\infty = u$ the law of (A_u, B_u) is infinitely divisible with

$$(9.s2) \qquad E_y^\nu[\exp(-\alpha A(L_y) - \beta B(L_y)) \,|\, \ell_\infty = u] =$$

$$(9.s3) \qquad E_y^\nu[\exp(-\alpha A_u - \beta B_u) \,|\, \ell_\infty \geq u] = \exp - u\,\psi_y^\nu(\alpha,\beta), \; \bullet$$

where the exponent $\psi = \psi_y^\nu(\alpha,\beta)$ and the Laplace transform

$$(9.s4) \qquad \phi = \phi_y^\nu(\alpha,\beta) = E_y^\nu \exp(-\alpha A(L_y)-\beta B(L_y))$$

are related by the reciprocal formulae

$$(9.s5) \qquad \phi = \nu/(\nu+\psi y), \quad \psi = \nu(1-\phi)/\phi y,$$

the first of these identities being obtained by integrating *(9.s2)* and *(9.s3)* with respect to the exponential law of ℓ_∞ with rate ν/y.

For the A and B in *Proposition (9.2)*, the jumps of the killed subordinators $(A_u, 0 \le u \le \ell_\infty)$ and $(B_u, 0 \le u \le \ell_\infty)$ come from disjoint sets of excursions, hence they are independent conditional on ℓ_∞ (see Itô [27] or Meyer [36]). Thus we have in this instance

$$(9.s6) \qquad \psi_y^\nu(\alpha,\beta) = \psi_y^\nu(\alpha,0) + \psi_y^\nu(0,\beta).$$

This observation leads to a new proof of *Proposition (9.2)*, since we know $\phi_y^\nu(\alpha,0)$ and $\phi_y^\nu(\alpha,\alpha)$ from *(9.i)* and *(7.ẽ)*. Thus it is a simple matter to compute $\psi_y^\nu(\alpha,0)$ and $\psi_y^\nu(\alpha,\alpha)$ using *(9.s5)*, then $\psi_y^\nu(0,\alpha)$ from *(9.s6)*. Making use along the way of the Bessel recurrences *(13.4)* and the Wronskian formula *(13.3)* and putting $a = y\sqrt{2\alpha}$ the results are

$$(9.s7) \qquad \begin{cases} \psi_y^\nu(\alpha,\alpha) = \dfrac{1}{2y}\left[(I_\nu(a)\,K_\nu(a))^{-1} - 2\nu\right]; \\[2ex] \psi_y^\nu(\alpha,0) = \dfrac{a}{2y}\dfrac{I_{\nu+1}(a)}{I_\nu(a)}\,; \quad \psi_y^\nu(0,\alpha) = \dfrac{a}{2y}\dfrac{K_{\nu-1}(a)}{K_\nu(a)}, \end{cases}$$

and *(9.2)* follows after another application of *(9.s6)* and *(9.s5)*.

Remarks (9.8) :

(i) After some slight modifications the formulae *(9.s7)* remain valid for $\nu \leq 0$, with $\psi_y^\nu(\alpha,\beta)$ defined by *(9.s3)*. They hold without alteration for $\nu = 0$, since, for arbitrary additive functionals A and B, $y > 0$, the function

(9.s8) $(\alpha,\beta,\nu) \to \psi_y^\nu(\alpha,\beta)$ is continuous for $\alpha,\beta,\nu \geq 0$, as a consequence of the following formula : for $\underline{\underline{F}}_{\tau_u}$-measurable $Z \geq 0$,

(9.s9) $E_y \, Z(\tau_u < \infty) = E_y^o \, Z \exp\{- \frac{1}{2} \nu^2 \, C(\tau_u)\}$.

To obtain *(9.s9)* apply *(2.c)* to $T = t \wedge \tau_u$ and let $t \to \infty$.
A time reversal as in the first proof of *(9.2)* can be used to obtain the formulae for $\nu < 0$. For $-1 < \nu < 0$ with the usual reflecting boundary condition at zero, the result is that the formula for $\psi_y^\nu(\alpha,0)$, the exponent for time below y, is still the same, while a term of ν/y must be added to the formulae for $\psi_y^\nu(0,\alpha)$ and $\psi_y^\nu(\alpha,\alpha)$. And for $\nu \leq -1$ with zero absorbing one has simply $\psi_y^\nu = \psi_y^{-\nu}$ in all cases.

(ii) Inspection of the formulae of Kent [31] embedded in *(2.3)* for $\nu \geq 0$, and of similar formulae of Kent [31] and Getoor and Sharpe [15] for $\nu < 0$, reveals that for all real $\nu, y > 0, \alpha > 0$, $\psi_y^\nu(\alpha,0)$ and $\psi_y^\nu(0,\alpha)$ are respectively equal to the values at y of one half the left derivative and minus one half the right derivative of the function

$$x \to E_x^\nu[\exp(-\alpha T_y) \,|\, T_y < \infty].$$

As shown in Section 6.2 of Ito-Kean [28], this can be proved directly using excursion theory for any diffusion on the line which near y looks enough like Brownian motion with drift, thereby providing a fresh derivation of *(9.s7)*, *(9.i)*, *(9.n)* and *(9.2)*, together with extensions of these results to more general diffusions

(iii) From the formula for $\psi_y^\nu(\alpha,\alpha)$, *(7.b)*, and the fact that the inverse local time process $(\tau_u, u \geq 0)$ jumps to ∞ after an exponential time with rate ν/y, one finds that for $\nu > 0$

$$E_y^\nu[\exp(-\alpha\tau_u) \; ; \; \tau_u < \infty] = \exp[-u/g_\alpha^\nu(y,y)]$$

where

$$g_\alpha^\nu(y,y) = 2y \; I_\nu(y\sqrt{2\alpha}) \; K_\nu(y\sqrt{2\alpha}) = \int_0^\infty e^{-\alpha t} \; p_t^\nu(y,y)dt$$

This is a special case of a well known formula which holds for any Markov process with nice enough transition function - see e.g. Getoor [14], formulae *(7.9)* and *(7.15)*.

(iv) If $\exp(-\psi(\alpha))$ is the Laplace transform in α of an infinitely divisible probability law on $[0,\infty)$, then $\psi(0) = 0$, and $\psi(\alpha)$ has a c.m. dérivative

$$\text{(9.t1)} \qquad \frac{d}{d\alpha}\psi(\alpha) = c + \int_0^\infty e^{-\alpha t} \; t\lambda(dt),$$

where c is a positive constant which will always be zero in our applications, and λ is the usual Lévy measure satisfying $\lambda(t,\infty) < \infty$ for all $t > 0$. It follows by the Criterion 2 of Feller [10], XIII.4, that

(9.t2) if f is c.m. then so is $f(\psi)$,

and an integration by parts shows also that $\psi(\alpha)/\alpha$ is c.m. with

$$\text{(9.t3)} \qquad \psi(\alpha)/\alpha = c + \int_0^\infty e^{-\alpha t} \; \lambda(t,\infty)dt.$$

Applying these observations to the ψ's in *(9.s7)*, we learn from *(9.t1)* that for $\nu > -1$ the function of α (with the substitution $a = y\sqrt{2\alpha}$)

$$\text{(9.t4)} \qquad 2y^{-1}\frac{d}{d\alpha}\psi_y^\nu(\alpha,0) = 1 - I_{\nu-1}(a)\; I_{\nu+1}(a)/I_\nu^2(a) \quad \underline{\text{is c.m.}},$$

and similarly for $\nu > 0$

$$\text{(9.t4')} \qquad 2y^{-1}\frac{d}{d\alpha}\psi_y^\nu(0,\alpha) = -1 + K_{\nu-1}(a)\; K_{\nu+1}(a)/K_\nu^2(a) \quad \underline{\text{is c.m.}}.$$

We note that *(9.t4)* implies the inequality of Thiruvenkatchar and Nanjundiah [45],

$$(9.t5) \qquad 0 \leq I_\nu^2(a) - I_{\nu-1}(a)\, I_{\nu+1}(a) \leq I_\nu^2(a)/(\nu+1), \quad a > 0, \ \nu > -1,$$

which is similar to the result of Szácz [44] for J_ν. From *(9.t4')* we obtain a companion inequality for K_ν which we have not seen in the literature :

$$(9.t5') \qquad 0 \leq K_{\nu-1}(a)\, K_{\nu+1}(a) - K_\nu^2(a) \leq K_\nu^2(a)/(\nu-1), \quad a > 0, \ \nu > 1.$$

The Laplace transforms *(9.t4)* and *(9.t4')* determine the corresponding Lévy measures by *(9.t1)*, but the Lévy measures in question are specified much more simply by the alternative formula *(9.t3)*. Indeed after dividing by α in *(9.s7)* and taking $y = 1/\sqrt{2}$ we recognise a multiple of the Laplace transform of $\iota^{\nu,1}$ in *(9.a1)* and a multiple of the function in *(9.q4)* with $\nu-1$ instead of ν. Thus the Lévy measure λ_y^ν for the distribution of the time spent by $BES_y(\nu)$ below y before the time τ_1 that local time at y first reaches 1 (given $\tau_1 < \infty$) is given by

$$(9.t6) \qquad \lambda_y^\nu(t,\infty) = y^{-1}\, \Sigma_\nu(t/2y^2), \quad y > 0, \ \nu \geq -1,$$

where Σ_ν is defined in *(9.q2)*, and the corresponding Lévy measure for time spent above y is

$$(9.t7) \qquad \eta_y^\nu(t,\infty) = (2y)^{-1}\, k_{|\nu|-1}\, (t/2y^2), \quad y > 0, \ \nu \geq -1,$$

where k_ν is defined below *(9.q4)*. Adding the two measures gives the Lévy measure for τ_1.

Recalling that the inverse local time process $(\tau_u, u \geq 0)$ jumps to ∞ at an exponentially distributed time with rate ν/y, for $\nu > 0$ these results can be re-expressed as follows, without reference to local time.

For fixed $y > 0$ and $t > 0$, let N^+ be the number of excursions of the process $(R_s, s \geq 0)$ above y of duration at least t,

$$N^+ = \#\{s : R_s = y,\ R_{s+u} > y \quad \text{for} \quad 0 < u < t\},$$

let N^- be defined similarly as the number of excursions below y of duration at least t, and let $N^* = N^- + N^+$ be the total number of excursions away from y of duration at least t.
Then according to the Itô excursion theory, each of the random variables $N = N^-$, $N^+ - 1$ and $N^* - 1$ has a geometric distribution on $\{0,1,2,\ldots\}$ with

(9.t8) $\quad P_y^\nu(N = n) = \theta^n\nu/(\theta+\nu)^{n+1}, \quad n = 0,1,2,\ldots$ and

(9.t8') $\quad E_y^\nu N = \theta/\nu,$

where θ is found from *(9.t6)* and *(9.t7)* to be given by

(9.t8")
$$\begin{aligned} \theta &= \Sigma_\nu(t/2y^2) && \text{for } N = N^- \\ &= \tfrac{1}{2} k_{\nu-1}(t/2y^2) && \text{for } N = N^+ - 1 \\ &= \Sigma_\nu(t/2y^2) + \tfrac{1}{2} k_{\nu-1}(t/2y^2) && \text{for } N = N^* - 1. \end{aligned}$$

From *(9.t8')* and *(9.4)* (i) we obtain a complement to *(9.q5)* which gives some information about $k_\nu(z)$ as a function of ν for $-1 \leq \nu \leq 0$, namely that for each $z > 0$

(9.t9) $\quad \nu \to \frac{1}{\nu} k_{\sqrt{\nu}-1}(z)$ <u>is c.m.</u>.

<u>*Remark (9.8)* (v) :</u>

A miscellany of c.m. functions can be read from *(9.s7)* and *(9.t2)*. In particular, taking $f(x) = x^{-1}$ in *(9.t2)* shows that for $\nu \geq 0$ the following functions of α are c.m. :

(9.u1)
$$\frac{1}{\sqrt{\alpha}} \frac{I_\nu(\sqrt{\alpha})}{I_{\nu+1}(\sqrt{\alpha})}, \quad \frac{1}{\sqrt{\alpha}} \frac{K_\nu(\sqrt{\alpha})}{K_{\nu-1}(\sqrt{\alpha})} -$$

Each of these functions $\phi(\alpha)$ is unbounded, so the associated measure has infinite total mass. But still this measure is in all cases infinitely divisible, meaning that for $0 < \theta \le 1$ the function $[\phi(\alpha)]^\theta$ is c.m., as is seen by taking $f(x) = x^{-\theta}$ instead of x^{-1}. The c.m. property of the second function in *(9.u1)* was proved by Ismail [24] using an integral representation similar to *(9.q4)*, and he also showed in [24] that the function $K_{\nu+\beta}(\sqrt{\alpha})/K_\nu(\sqrt{\alpha})$ is c.m. in α for $\nu \ge 0$, $\beta \ge 0$. Since we know from *(4.k)* that for such ν,β, $I_\nu(\sqrt{\alpha})/I_{\nu+\beta}(\sqrt{\alpha})$ is decreasing in α, it is natural to conjecture that this function too may be c.m..

Remarks (9.8) (vi) :

We note that the special case $Z = 1$ of *(9.r1)* and the fact that τ_u is exponential with rate ν/y under P_y^ν can be rewritten as

$$(9.v) \qquad E_y^o \exp(-\tfrac{1}{2}\nu^2 C(\tau_u)) = \exp(-u\nu/y),$$

an identity which could also be derived by the time change described at the end of section 7.

In particular, *(9.v)* shows that the processes

$$(C(\tau_u), u \ge 0 \; ; \; P_y^o) \quad \text{and} \quad (T_{u/y}, u \ge 0 \; ; \; W_o)$$

are identical in law, where T_x is the hitting time of x for a BM_o on the line. It is a short step from *(9.v)* to deduce that the total angle process $(\phi(t), t \ge 0)$ swept out by a two dimensional Brownian motion started at radius $y > 0$ forms a Cauchy process when watched only when the radial part is at y and indexed by local time.

A closely related fact, plain from *(2.k)* and the skew-product, is that the process $(\phi(T_{y\exp(s)}), s \ge 0)$ is a Cauchy process, where T_r is the hitting time of the circle of radius r. As shown by David Williams in an unpublished manuscript, this leads quickly to Spitzer's law that $2\Phi(t)/\log t$ has a limiting Cauchy distribution (cf. Ito-Mc Kean [28], p. 270).

Proposition (9.9) : Let $(\tau_u, u \geq 0)$ be the inverse local time at $y > 0$, and as in *(9.4)*, let

$$f(\alpha,\nu) = \nu^{-1} E_y^{\nu} \exp\{-\alpha A(L_y)\}, \quad \alpha \geq 0, \ \nu > 0,$$

where $(A(t), t \geq 0)$ is an additive functional. Then for $\mu \geq 0$, $\alpha > 0$, $\nu > 0$

$$(9.w) \qquad E_y^{\mu}[\exp\{-\alpha A(\tau_u) - \tfrac{1}{2}\nu^2 C(\tau_u)\} \; ; \; \tau_u < \infty] = \exp\{-u/yf(\alpha,\lambda)\}$$

where $\lambda = (\mu^2 + \nu^2)^{1/2}$.

Proof : For $\mu > 0$, let

$$\phi^{\mu}(\alpha,\nu) = E_y^{\mu} \exp\{-\alpha A(L_y) - \tfrac{1}{2}\nu^2 C(L_y)\}$$

$$= \mu f(\alpha,\lambda) \qquad \text{by } (4.8).$$

The corresponding exponent $\psi^{\mu}(\alpha,\nu)$ is given by *(9.s5)* as

$$\psi^{\mu}(\alpha,\nu) = (f(\alpha,\lambda)^{-1} - \mu)/y,$$

and this implies *(9.w)* for $\mu > 0$. The extension to the case $\mu = 0$ is justified by *(9.s8)*.

Remarks (9.10) :

(i) The formula *(9.w)* shows that, provided A_t is not identically zero, the definition of $f(\alpha,\nu)$ may be extended continuously to all $\alpha \geq 0$, $\nu \geq 0$, with $f(\alpha,\nu) = \infty$ iff $\alpha = \nu = 0$ (see note below *(9.4)*). Then *(9.w)* holds for all $\mu,\alpha,\nu \geq 0$.

(ii) By remark *(3.9)*, formula *(9.w)* for $\mu = 0$ shows that for $Z = A(L_y)$, the measure in *(9.4)* (ii) is y^{-1} times the potential measure of the two-dimensional subordinator

$$\{(A(\tau_u), \tfrac{1}{2} C(\tau_u), u \geq 0 \; ; \; P_y^0\}.$$

10. THE von MISES DISTRIBUTION IS NOT A UNIQUE MIXTURE OF WRAPPED NORMALS.

Consider the von-Mises distribution $vM(k)$ on the circle S^1, with parameter $k > 0$, centered at $\theta = 0$, specified as in *(5.b)* by

$$\text{(10.a)} \qquad vM(k,d\theta) = \left[2\pi\, I_0(k)\right]^{-1} \exp(k \cos\theta), \; -\pi < \theta \leq \pi.$$

According to Hartman and Watson [20], for each $k > 0$ the $vM(k)$ distribution can be presented as a mixture of the wrapped normal distributions $wN(v)$, $v > 0$,

$$\text{(10.b)} \qquad vM(k) = \int_0^\infty wN(v)\, \eta_k(dv),$$

where $wN(v)$ is the distribution modulo 2π of a normally distributed random variable with mean 0 and variance v, and η_r is the distribution on $(0,\infty)$ with Laplace transform *(1.a)*. As noted below *Theorem (1.1)*, this result admits a direct probabilistic expression in terms of a Brownian motion in R^2 with drift $\delta = k$ started at the origin. For $vM(k)$ is then the distribution of Θ_{T_1} (see *(5.2)*), while $wN(v)$ is the conditional distribution of Θ_{T_1} given $C(T_x,\infty) = v$ by *Theorem (1.1)*, and η_k is the distribution of $C(T_x,\infty)$ by *(4.i)*. Similar interpretation of *(10.b)* but in terms of conditioned processes obtained from BM with no drift can be read from *(1.e)*.

Hartman and Watson raised, the question of whether the mixing measure η_k in *(10.b)* is unique. We answer this question here in the negative : for each k, η_k is not even an extreme point of the convex set of possible mixing measures.
To see this, observe that $vM(k)$ is determined by its sequence of Fourier coefficients, the n th of which can be expressed using *(10.b)* as

$$\int_0^\infty \eta_k(dv)\, e^{-\frac{1}{2} n^2 v} = \int_0^1 F_k(dx)x^{n^2},$$

where F_k is the law on $[0,1]$ obtained from η_k on $[0,\infty)$ by the map $v \to e^{-\frac{1}{2}v}$. Thus η_k is unique if and only if there exists no other distribution on $[0,1]$ with the same n^2-moments as F_k.

According to a famous theorem of Müntz, for a sequence of non-negative integers $0 \leq n(0) < n(1) < \ldots$, linear combinations of the functions $x^{n(i)}$ are dense in $C[0,1]$ iff $\Sigma 1/_{n(i)} = \infty$.
Since $\Sigma 1/_{n^2} < \infty$, it follows that there do exist distributions F on $[0,1]$ which are not determined by their n^2-moments. That F_k is such a distribution for each k is implied by the following Lemma, whose formulation and proof were suggested by Jim Reeds :

Lemma (10.1) : Let F be a probability on $[0,1]$, with density f with respect to Lebesgue measure, such that for some $\varepsilon > 0$,

$$f(x) \geq \varepsilon \quad \text{if} \quad x \leq \varepsilon.$$

If $\Sigma 1/_{m(i)} < \infty$ then F is not an extreme point of the convex set of probability measures G on $[0,1]$ with the same $m(i)$-moments as F.

Proof : According to a variant of Müntz's theorem, the functions $x^{n(0)}, x^{n(1)} \ldots$ are complete in $L^2[0,1]$ iff $\Sigma 1/_{n(i)} = \infty$.
By a change of variable, the same is true in $L^2[0,\varepsilon]$. Thus, by taking $n(0) = 1$, $n(i) = m(i-1)+1$, there exists ϕ in $L^2[0,\varepsilon]$ which is orthogonal to both 1 and $x^{m(i)+1}$ for all i. Extend ϕ to $[0,1]$ by letting it be zero on $(\varepsilon,1]$, and put $g(x) = \int_0^x \phi(y)dy$. By an integration by parts

$$(m(i)+1) \int_0^1 x^{m(i)} g(x)dx = - \int_0^1 x^{m(i)+1} \phi(x)dx = 0.$$

Thus for any $\delta > 0$ with $\delta \leq \varepsilon/\sup(-a,b)$, where $a = \inf g$, $b = \sup g$, both $(f + \delta g)(x)dx$ and $(f - \delta g)(x)dx$ are probability measures on $[0,1]$ with the same $m(i)$-moments as f. Since the average of these two probabilities is F, the conclusion of the Lemma is evident.

To see that the Lemma applies to $F = F_k$ for each k, we argue that F_k has a continuous density f_k such that $f_k(0+) = \infty$.
This is an immediate consequence of the formula *(5.7)* (ii) of [51] for the continuous density h_k of η_k, which shows that

$$h_k(u) \sim c(k)\, u^{-3/2} \quad \text{as} \quad u \to \infty$$

for some constant $c(k)$.

11. OTHER WRAPPINGS

Let L_x be the last time at x for the radial part $(R_t, t \geq 0)$ of a BM in R^2 starting at zero with drift $\delta > 0$ in the direction $\theta = 0$. According to *Theorem (1.1)* and formula *(1.i)*, the distribution of the angle $\Theta(L_x)$ is a mixture of wrapped normal distributions $wN(v)$ with mixing measure the second Hartman law with Laplace transform *(1.b)* for parameter $r = \delta x$. The n th Fourier coefficient of this distribution is therefore $K_o(r)/K_n(r)$, and since these coefficients have a finite sum by *(9.1)*(i), we learn that $\Theta(L_x)$ has a bounded continuous density

$$\text{(11.a)} \qquad \frac{1}{2\pi} \sum_{n \in Z} e^{in\theta} K_o(r)/K_n(r), \quad -\pi < \theta \leq \pi.$$

But we do not know of any more explicit formula for this density analogous to the formula for the von-Mises density of $\Theta(T_x)$ obtained in the same way from the first Hartman law :

$$\text{(11.b)} \qquad \frac{1}{2\pi} \sum_{n \in Z} e^{in\theta} I_n(r)/I_o(r) = e^{r \cos\theta}/2\pi\, I_o(r).$$

We note that, for $x \leq y$, $\theta(T_x)$ is the sum modulo 2π of $\Theta(L_y)$ and $\Theta(T_x) - \Theta(L_y)$; these two random variables are independent by *Theorem (1.1)*, and there is an explicit formula for the distribution of $\Theta(T_x) - \Theta(L_y)$. Indeed, by *Theorem (1.1)*, the law of this random variable is a mixture of wrapped normal distributions with mixing measure the distribution of the clock $C(L_y)$ for a $BES_x(0)$, which has Laplace transform

$$\alpha \to I_{\sqrt{2\alpha}}(\delta x)\, K_{\sqrt{2\alpha}}(\delta y) \;/\; I_o(\delta x)\, K_o(\delta y)$$

by *(7.f)*. Assuming for simplicity that $\delta = 1$, it follows that the n th Fourier coefficient for the law of $\Theta(T_x) - \Theta(L_y)$ is $I_n(x)\, K_n(y)/I_o(x)\, K_o(y)$, whence $\Theta(T_x) - \Theta(L_y)$ has bounded continuous density given by the left hand side of the following identity :

$$\text{(11.c)} \qquad \frac{\sum_{n \in Z} e^{in\theta} I_n(x)\, K_n(y)}{2\pi\, I_o(x)\, K_o(y)} = \frac{K_o(\sqrt{x^2+y^2-2xy\cos\theta})}{2\pi\, I_o(x)\, K_o(y)}$$

The identity of *(11.c)* is equivalent to the classical identity (see Erdelyi [8], 7.6.1 (3))

$$\text{(11.d)} \qquad K_0(\sqrt{x^2+y^2-2xy\cos\theta}) = I_0(x)\,K_0(y) + 2\sum_{n=1}^{\infty} I_n(x)\,K_n(y)\cos n\theta,$$

which contains as the special case $\theta = -\pi$ the addition formula

$$\text{(11.e)} \qquad K_0(x+y) = I_0(x)\,K_0(y) + 2\sum_{n=1}^{\infty} (-1)^n I_n(x)\,K_n(y).$$

We indicate now a probabilistic proof of the identity *(11.c)*, hence of *(11.d)* and *(11.e)*, which serves as a complement to Feller's proof in [9] of the addition formula corresponding to *(11.e)* for I_0.

By conditioning on L_y, the n th Fourier coefficient of the distribution of $\Theta(T_x) - \Theta(L_y)$ is

$$\text{(11.f)} \qquad \int_0^{\infty} P_x^{0,1}(L_y \in dt)\; E_x^0\left[\exp\left[-\tfrac{1}{2} n^2 C_t\right] \middle| R_t = y\right].$$

But from *(1.e)* and *(11.b)*, the second factor is

$$I_n(xy/t)/I_0(xy/t) = \int_{-\pi}^{\pi} e^{in\theta}\, e^{(xy/t)\cos\theta}\, d\theta/2\pi\, I_0(xy/t),$$

while the first factor is given by *(7.c)* with $\nu = 0$, $\delta = 1$.
After these substitutions in *(11.f)* and a change in the order of integration, the right side of *(11.c)* emerges as the density by virtue of the integral representation

$$K_0(r) = \int_0^{\infty} \exp(-\,t/2 - r^2/2t)\,dt/2t, \qquad r > 0,$$

(Erdelyi [8], 7.12 (23)).

Remarks (11.1) :

(i) Let $q_t^{\delta}(z,z')$ be the transition density of BM in the complex plane with drift δ in the real direction. It is easy to verify that

$$\int_0^{\infty} q_t^{\delta}(z,z')dt = \pi^{-1} \; K_o(|z-z'|) \exp \delta \mathrm{Re}(z-z')$$

and, put together with *(11.a)*, this yields a formula for the density on the circle of radius x of the equilibrium (or capacitary) measure of this circle for BM with drift δ (see Getoor-Sharpe [16] or Chung [4]). The theory of last exit distributions as developed on these papers can now be applied quite routinely to obtain formulae for the joint density of the time and place of the last exit from the circle of radius x for BM with drift δ started at an arbitrary initial position z rather than $z = 0$. Though quite complicated these formulae are as explicit as *(11.a)*. Integrating these formulae gives a means of calculating the hitting probability for the circle which seems simpler than a direct attack based on the CM formula and the formulae of Wendel [47].

(ii) Everything above can be extended to integer dimensions $d \geq 2$, and perhaps even to real dimensions in the manner of Kent [31], but the results do not appear to deserve space.

12. CONCLUDING REMARKS.

We mention first a characterization of certain processes which admit a skew-product decomposition similar to that of BM with no drift in $\mathbb{R}^d$. Let us work on the canonical space $\Omega = C(\mathbb{R}_+,\mathbb{R}^d)$, $d \geq 2$. We consider probabilities P on Ω satisfying

Hypothesis (12.1) : There exists $b : \mathbb{R}_+ \times \Omega \to \mathbb{R}^d$, uniformly bounded and $(\underline{\underline{F}}_t)$-predictable such that

$$X_t(\omega) = x + B_t(\omega) + \int_0^t b(s,\omega)ds,$$

where $x \in \mathbb{R}^d \setminus \{0\}$, and $(B_t, t \geq 0)$ is a BM in $\mathbb{R}^d$ starting at 0.

Here of course $\underline{\underline{F}}_t = \sigma(X_s, 0 \leq s \leq t)$. We put $X_t = R_t\,\Theta_t$, where $R_t > 0$, $\Theta_t \in S^{d-1}$, and let $\underline{\underline{R}}_t = \sigma(R_s, 0 \leq s \leq t)$.

Proposition (12.2) : Under Hypothesis *(12.1)*, (X_t) admits a skew-product representation

(12.a1) $\qquad X_t = R_t\,\Phi(A_t)$

where $(\Phi(s), s \geq 0)$ is a BM in S^{d-1} independent of (R_t) and (A_t) is an increasing process adapted to $(\underline{\underline{R}}_t)$ if and only if both

(12.a2) $\qquad A_t = \int_0^t ds/R_s^2,$

and there exists $r : \mathbb{R}_+ \times \Omega \to \mathbb{R}$, a bounded predictable scalar valued process, such that

(12.b) $\qquad b(t,\omega) = r(t,\omega)\,\Theta(t,\omega) \quad$ a.e. $\quad dt\,dP$.

Proof : As is easily verified, if $0 = s_0^n < s_1^n < \ldots < s_m^n(s) = s$ is a sequence of refining partitions of $[0,s]$ with $\lim_n \sup_k |s_{k+1}^n - s_k^n| = 0$, and Φ is a BM in S^{d-1}, then, denoting the Euclidean norm in $\mathbb{R}^d$ by $|\cdot|$,

$$\sum_k |\Theta(s_{k+1}^n) - \Theta(s_k^n)|^2 \to (d-1)s$$

in probability as $n \to \infty$. Using the fact that under *(12.a1)* the laws P and W_x (Wiener measure starting at x) are equivalent on $\underline{\underline{F}}_t$ for each $t > 0$, one deduces from this that the representation *(12.a1)* holds iff the clock is of the familiar Brownian form *(12.a2)*. Let (L_t) now be a (continuous) version of the density of P with respect to W_x. Now *(12.a1)* holds iff

$$P(F|\underline{\underline{R}}_t) = W^x(F|\underline{\underline{R}}_t), \quad F \in \underline{\underline{F}}_t, t > 0,$$

and it is immediate that this is in turn equivalent to

L_t <u>is $\underline{\underline{R}}_t$-measurable,</u>

and again to

(12.d) L_t <u>is an $\underline{\underline{R}}_t$-martingale</u>.

Now such a martingale is represented as

(12.e) $$L_t = \exp\{\int_0^t r(s,\omega)\,d\beta_s - \tfrac{1}{2}\int_0^t r^2(s,\omega)\,ds\}$$

where r is a real valued predictable process with

$$\int_0^t r^2(s,\omega)\,ds < \infty, \quad \beta_t = \int_0^t (\sum_{i=1}^n X_s^i\, dX_s^i)/R_s.$$

According to Girsanov's theorem, *(12.e)* obtains iff, under P,

$$X_t(\omega) - \int_0^t r(s,\omega)\,\Theta_s(\omega)\,ds \quad \underline{\text{is a martingale,}}$$

which, finally, is equivalent to *(12.b)*.

Remarks :

(i) The Proposition above may be compared to the deeper result of Galmarino [12] in a Markovian setting, where it is shown that a skew-product representation characterises all isotropic diffusions.

(ii) It does not seem to be so easy to describe a nice class of processes admitting the kind of reverse skew-product in *Theorem (1.1)*.

An important example of a process satisfying *(12.a1)* has been studied extensively by Kendall [30]. This process admits the skew-product *(12.a1)*, as noted by Williams in the discussion after Kendall's paper (p. 414 of [30]).

This is the special case where $r(t,\omega) = c$ is a real valued constant, and in Kendall's application $d = 2$ and $c = -\delta \leq 0$.

The corresponding $\mathbb{R}^d$-valued diffusion is the Markov process with infinitesimal generator

(12.f) $$\frac{1}{2}\Delta + c\frac{x}{|x|}\cdot\nabla .$$

From *(12.e)* its radial part is the diffusion on $(0,\infty)$ with infinitesimal generator

(12.g) $$G_\nu + c\frac{d}{dx}$$

where G_ν is the generator of $BES(\nu)$.

Let $({}^cP_x^\nu, x \geq 0)$ be the distributions of this process, and $({}^cp_t^\nu(x,y))$ the semigroup of its densities. Again from *(12.e)* we have after a transfer to the canonical space $C(\mathbb{R}_+,\mathbb{R}_+)$

(12.h) $$\frac{d({}^cP_x^\nu)}{dP_x^\nu} = \exp(c\beta_T - \tfrac{1}{2}c^2T) \quad \text{on} \quad \underline{F}_{T+} \cap (T < \infty) ,$$

for any $(\underline{\underline{F}}_t)$ stopping time T, where $(\underline{\underline{F}}_t)$ is now as in Section 2 the filtration generated by the co-ordinate process $(R_t, t \geq 0)$ on $C(\mathbb{R}_+, \mathbb{R}_+)$, and

$$\beta_t = R_t - x - \frac{2\nu+1}{2} \int_0^t ds/R_s \quad \underline{\text{is a}} \quad P_x^\nu \quad \underline{\text{BM}}.$$

It follows that

$$(12.i) \qquad \frac{{}^c p_t^\nu(x,y)}{p_t^\nu(x,y)} \exp(\tfrac{1}{2} c^2 t) = E_x^\nu(\exp(c\beta_t)/R_t = y)$$

$$= \exp c(y-x) \, E_x^\nu\left[\exp\{-\tfrac{1}{2}(2\nu+1)\, c \int_0^t ds/R_s\} \middle| R_t = y\right].$$

It would be interesting to obtain an explicit knowledge of ${}^c p_t^\nu(x,y)$, as this would lead us to the conditional distribution of β_t given R_t, and that of $\int_0^t ds/R_s$ given R_t, a nice complement to *(2.i)*. A step in this direction is made by the following proposition, from which one can obtain a formula for the Laplace transform in t of ${}^c p_t^\nu(x,y)$: this has been derived jointly with P. Mc Gill. Unfortunately, this transform seems quite difficult to invert.

Let $W_{\kappa\mu}(t)$ and $M_{\kappa\mu}(t)$ be the usual principal solutions at $t = \infty$, resp : 0, of Whittaker's form of the confluent hypergeometric equation

$$(12.j) \qquad y'' - [1/4 + (\mu^2 - 1/4)/t^2 - \kappa/t]\, y = 0.$$

Proposition (12.2) :

$$(12.k) \qquad E_a^0\{\exp - \tfrac{1}{2}\mu^2 \int_0^{T_b} ds/\rho_s^2 - \lambda \int_0^{T_b} ds/\rho_s - \tfrac{1}{2}\theta^2 \, T_b\} = \left(\frac{b}{a}\right)^{1/2} \frac{\mathcal{L}_{\kappa\mu}(2\theta a)}{\mathcal{L}_{\kappa\mu}(2\theta b)}$$

<u>where</u> $\kappa = -\lambda/\theta$, <u>and</u> $\mathcal{L}_{\kappa\mu} = W_{\kappa\mu}$ <u>if</u> $b < a$

$= M_{\kappa\mu}$ <u>if</u> $a > b$.

Proof : Fix b,μ,λ,θ and let $W(a)$ be the expectation in *(12.k)*. Then on $(0,b)$ and on (b,∞), W solves

$$G_o W = \left[\frac{1}{2}\frac{\mu^2}{a^2} + \frac{\lambda}{a} + \frac{1}{2}\theta^2\right]W,$$

where

$$G_o = \frac{1}{2}\frac{d^2}{da^2} + \frac{1}{2a}\frac{d}{da}$$

is the BES(0) generator. A change of variables now reveals that

$W(a) = (2\theta a)^{-1/2} y(2\theta a)$ where y is a solution of Whittaker's equation *(12.j)*, and *(12.k)* results after the usual consideration of boundary conditions.

The above proposition provides probabilistic interpretations and extensions of some of the results of Hartman [22] (see in particular *(4.13)* and *(4.14)* of [22]). Also, if P governs the pole-seeking Brownian motion of Kendall [30] with generator *(12.f)*, in $d = 2$ dimensions with polar drift c starting at a point $x \in \mathbb{R}^2$ with $|x| = a$ and angle $\theta = 0$, after using first the skew-product *(12.1)*, then the change of law formula *(12.h)*, and finally *(12.k)*, we obtain

$$\text{(12.l)} \qquad E\exp(i\mu\Theta(T_b) - \alpha T_b) = e^{c(b-a)}\left(\frac{b}{a}\right)^{1/2} \frac{\mathcal{L}_{\kappa\mu}(2\gamma a)}{\mathcal{L}_{\kappa\mu}(2\gamma b)},$$

where the $\mathcal{L}_{\kappa\mu}$ are as in *(12.k)*, and now $\gamma = (c^2 + 2\alpha)^{1/2}$, $\kappa = -c/2\gamma$. In the case $c \leq 0$ and $a > b$, this is the formula (32) of Kendall [30].

13. APPENDIX. FORMULAE FOR BESSEL FUNCTIONS

We record here for the reader's convenience those formulae for the Bessel functions I_ν and K_ν which have been used in the paper. These formulae and definitions of I_ν and K_ν may be found in Abramowitz and Stegun [1] or Erdelyi [8].

(13.1) Negative indices. For integer n and real ν

$$I_{-n}(z) = I_n(z) \; ; \quad K_{-\nu}(z) = K_\nu(z).$$

(13.2) Asymptotics. As $z \to 0$, ν fixed

$$I_\nu(z) \sim (\tfrac{1}{2} z)^\nu / \Gamma(\nu+1), \quad \nu \neq 1,-2,\ldots$$

$$K_0(z) \sim - \operatorname{Log} z$$

$$K(z) \sim \tfrac{1}{2}\Gamma(\nu)\,(\tfrac{1}{2} z)^{-\nu}, \quad \nu > 0.$$

(13.3) Wronskian

$$W\{K_\nu(z), I_\nu(z)\} = I_\nu(z)\, K_{\nu+1}(z) + I_{\nu+1}(z)\, K_\nu(z) = 1/z$$

(13.4) Derivatives and recurrences For $L_\nu = I_\nu$ or $e^{\nu\pi i} K_\nu$,

$$\frac{d}{dz} L_\nu(z) = L_{\nu-1}(z) - \nu L_\nu(z)/z = L_{\nu+1}(z) + \nu L_\nu(z)/z.$$

$$\frac{d}{dz}\left[z^{\sigma\nu} L_\nu(z)\right] = z^{\sigma\nu} L_{\nu-\sigma}(z), \quad \sigma = \pm 1.$$

$$L_{\nu-1}(z) - L_{\nu+1}(z) = 2\nu\, L_\nu(z)/z.$$

REFERENCES

[1] M. ABRAMOWITZ, I. STEGUN : Handbook of Mathematical Functions. New York - Dover - 1970.

[2] L. BONDESSON : A general result on infinite divisibility, The Annals of Probability, n° 6, 7 (1979), 965-979.

[3] L. BREIMAN : Probability. Addison - Wesley, Reading, Mass. 1968.

[4] K.L. CHUNG : Probabilistic approach in potential theory to the equilibrium problem. Ann. Inst. Fourier, Grenoble, 23, n° 3, 313-322, 1973.

[5] Z. CIESIELSKI, S.J. TAYLOR: First passage times and sojourn times for Brownian motion in space and the exact Hausdorff measure of the sample path. Trans. Amer. Math. Soc. 103, 434-450, 1962.

[6] J.L. DOOB : Conditional Brownian motion and the boundary limits of harmonic functions. Bull. Soc. Math. France. 85 (1957), 431-458.

[7] S.F. EDWARDS : Statistical mechanics with topological constraints : I. Proc. Phys. Soc., 91, 513-519 (1967).

[8] A. ERDELYI, and al : Tables of Integral Transforms, vol. 1, Mc Graw - Hill, New York, 1953.

[9] W. FELLER : Infinitely divisible distributions and Bessel functions associated with random walks, J. Siam Appl. Math., vol 14, 4 (1966), 864-875.

[10] W. FELLER : An Introduction to Probability Theory and its Applications, vol II, Wiley, New York, 1966.

[11] D. FREEDMAN : Brownian motion and diffusion. San Francisco - Cambridge - London ; Holden-Day (1971).

[12] A.R. GALMARINO : Representation of an isotropic diffusion as a skew-product, Zeitschrift für Wahr., 1 (1963), 359-378.

[13] R.K. GETOOR : The brownian escape process, The Annals of Probability, n° 5, 7 (1979), 864-867.

[14] R.K. GETOOR : Excursions of a Markov process. Annals of Probability, 7, n° 2, 244-266 (1979).

[15] R.K. GETOOR, M.J. SHARPE : Excursions of brownian motion and Bessel processes, Zeitschrift für Wahr., 47 (1979), 83-106.

[16] R.K. GETOOR, M.J. SHARPE : Last exit times and additive functionals, The Annals of Probability, vol. 1, 4 (1973), 550-569.

[17] L. GORDON, M. HUDSON : A characterization of the von-Mises distribution. Ann. Stat. 5, 813-814, 1977.

[18] E. GROSSWALD : The Student t-distribution of any degree of freedom is infinitely divisible, Zeitschrift für Wahr., 36 (1976), 103-109.

[19] J.M. HAMMERSLEY : On the statistical loss of long - period comets from the solar system II. Proceedings of the 4th Berkeley Symp. on Math. Stat. and Probability. (1960) Volume III : Astronomy and Physics.

[20] P. HARTMAN, G.S. WATSON : "Normal" distribution functions on spheres and the modified Bessel functions, Ann. Probability, 2 (1974), 593-607.

[21] P. HARTMAN : Completely monotone families of solutions of n th order linear differential equations and infinitely divisible distributions, Ann. Scuola Norm. Sup. Pisa, IV, vol III (1976), 267-287.

[22] P. HARTMAN : Uniqueness of principal values, Complete monotonicity of Logarithmic Derivatives of principal solutions, and Oscillation theorems. Math. Annalen 241, 257-281, 1979.

[23] M.E. ISMAIL : Bessel Functions and the infinite divisibility of the Student t-distribution. Annals of Proba, 5, n° 4, 582-585, 1977.

[24] M.E. ISMAIL : Integral representations and Complete monotonicity of various quotients of Bessel functions. Canadian J. Maths. XXIX, 1198-1207, 1977.

[25] M.E. ISMAIL, D.H. KELKER : The Bessel Polynomials and the Student t-distribution. Siam J. Math. Anal. 7, n° 1, 82-91, 1976.

[26] M.E. ISMAIL, D.H. KELKER : Special Functions, Stieltjes Transforms, and Infinite divisibility. Siam J. Math. Anal. vol 10, n° 5, 884-901, 1979.

[27] K. ITO : Poisson point processes attached to Markov processes. Proc. 6th Berkeley Symp. on Math. Stat. and Probability, vol III (1970/71).

[28] K. ITO, H.P. Mc KEAN : Diffusion processes and their sample paths, Springer-Verlag, 1965.

[29] H.P. Mc KEAN : Stochastic Integrals. Academic Press (1969).

[30] D.G. KENDALL : Pole - seeking Brownian motion and bird navigation (with discussion). J. Royal Statist. Soc. B 36 365-417 (1974).

[31] J. KENT : Some probabilistic properties of Bessel functions, Annals of Probability, n° 5, 6 (1978), 760-770.

[32] J. KENT : The infinite divisibility of the von-Mises Fisher distribution for all values of the parameter in all dimensions. Proc. London Math. Soc. 3, 35, 359-384 (1977).

[33] H. KUNITA : Absolute continuity of Markov processes and their extended generators. Nagoya Math. J. 36, 1-26, 1969.

[34] P.A. MEYER : Démonstration probabiliste de certaines inégalités de Littlewood - Paley ; Exposé II : L'opérateur carré du champ. Sém. Probas Strasbourg X. Lect. Notes in Maths n° 511 - Springer (1976).

[35] P.A. MEYER : Un cours sur les intégrales stochastiques, Lect. Notes in Math. 511, Springer, 1976.

[36] P.A. MEYER : Processus de Poisson ponctuels, d'après Ito. Sém. Probas. Strasbourg Vn°191 Springer(1971).

[37] S.A. MOLCHANOV : Martin boundaries for invariant Markov processes on a solvable group, Theo. Proba. Appl., 12 (1967), 310-314.

[38] M. NAGASAWA : Time reversions of Markov processes. Nagoya Math. J. 24, 177-204, 1964.

[39] A.O. PITTENGER, C.T. SHIH: Coterminal families and the strong Markov property. Trans. Amer. Math. Soc. 183, 1-42, 1973.

[40] J. PITMAN, L. ROGERS : Markov functions of Markov processes. To appear in Annals of Proba.

[41] M.J. SHARPE : Some transformations of diffusions by time reversal, Preprint.

[42] T. SHIGA, S. WATANABE : Bessel diffusions as a one-parameter family of diffusions processes, Zeitschrift für Wahr., 27 (1973), 37-46.

[43] F. STERN : An independence in Brownian motion with constant drift. Ann. Prob. 5, 571-2, 1977.

[44] O. SZACZ : Inequalities concerning ultraspherical polynomials and Bessel functions. Proc. Amer. Math. Soc. 1, n° 2, 256-267, 1950.

[45] V.R. THIRUVENKATACHAR, T.S. NANJUNDIAH : Inequalities concerning Bessel functions and orthogonal polynomials. Proc. Indian Ac. Sci. 33 A, p. 373-384 (1951).

[46] S. WATANABE : On Time Inversion of One-Dimensional Diffusion processes, Zeitschrift für Wahr., 31 (1975), 115-124.

[47] J.W. WENDEL : Hitting spheres with brownian motion, The Annals of Probability, vol. 8, 1 (1980), 164-169.

[48] J.G. WENDEL : An independence property of Brownian motion with drift. Ann. Prob. 8, n° 3, 600-601, 1980.

[49] D. WILLIAMS : Path decomposition and continuity of local time for one-dimensional diffusions, I. Proc. London Math. Soc., Ser 3, 28, 738-68, 1974.

[50] D. WILLIAMS : Diffusions, Markov Processes, and Martingales. Vol. 1 : Foundations. J. Wiley (1979).

[51] M. YOR : Loi de l'indice du lacet brownien, et distribution de Hartman-Watson, Zeitschrift für Wahr., 53, 71-95 (1980).

Euclidean Quantum Mechanics and Stochastic Integrals

R.F. Streater
Bedford College
Regent's Park
London NW1 4NS

Contents

§1. Quantum mechanics and probability

In the quantum mechanics of a system of m degrees of freedom, the space of wave-functions is taken to be the complex Hilbert space $\mathcal{H} = L^2(\mathbb{R}^m)$. A function $\in \mathcal{H}$ is usually written $\psi(q)$ or $\psi(q_1, \ldots q_m)$. We note that ψ and $e^{i\alpha}\psi$ represent the same state for any $\alpha \in \mathbb{R}$. States are normalized by the condition

$$||\psi||^2 \equiv \langle\psi, \psi\rangle = 1.$$

Observables of the system are represented by self-adjoint operators of which the most important are:

the momentum operator for the jth degree of freedom, $p_j = -i\frac{\partial}{\partial q_j}$, $1 \leq j \leq m$;

the position operator for the jth degree of freedom, q_j, co-ordinate multiplication, $1 \leq j \leq m$;

the energy operator, $H = \frac{\underline{p}^2}{2} + V(q)$.

Here, V is a real function of $q = (q_1, \ldots q_m)$, acting as a multiplication operator, called the potential energy; the operator $\underline{p}^2/2$ means $-\frac{1}{2}\sum_{j=1}^{n}\frac{\partial^2}{\partial q_j^2}$. To be useful, H must be self-adjoint rather than merely symmetric. For then, by the spectral theorem, H generates a one-parameter unitary group, $U(t) = \exp(-iHt)$; we interpret U(t) as the time-evolution operator: in the Heisenberg picture, the "configuration" at time t is described by the m self-adjoint operators

$$q_j(t) = U(t)q_jU^{-1}(t), \quad 1 \leq j \leq m \tag{1}$$

Clearly, $q_j(t)$ is the global operator solution to Heisenberg's equation of motion

$$\frac{i\,dq_k(t)}{dt} = [H, q_k(t)]$$

subject to the boundary condition

$$q_k(0) = q_k \qquad 1 \leq k \leq m$$

We note that $q_k(t)$ does not in general commute with $q_j(s)$ if $s \neq t$.

Interesting theories are obtained if V is such that H is a non-negative operator, and has a simple eigen-value at zero, the lowest point of the spectrum. The corresponding eigen-function, ψ_0 say, is called the ground state or vacuum state. Usually, ψ_0 may be chosen to be positive as a function of $q = (q_1, \ldots q_m)$. This exemplifies the principle that the fundamental vibration of a physical system has no nodes. It is then convenient to replace $\mathcal{H} = L^2(\mathbb{R}^m, dq)$ by the unitarily equivalent Hilbert space $\mathcal{H}' = L^2(\mathbb{R}^m, |\psi_0(q)|^2 dq)$, the unitary map $W: \mathcal{H} \to \mathcal{H}'$ being given by

$$(W\psi)(q) = \psi(q)/\psi_0(q).$$

In the new realization, the ground state $W\psi_0$ is just the function 1, and an observable, represented by an operator A on $\mathcal{H}$, is taken to be represented by $A' = WAW^{-1}$. As the operators $q_1, \ldots q_m$ commute with W, they also represent the configuration in the new realization on $\mathcal{H}'$.

The probabilistic interpretation of the theory is given by relating each state ψ and observable A to a probability measure $\rho_{A,\psi}$ on $\mathbb{R}$, constructed as follows. Define the function, $\mathbb{R} \to \mathbb{C}$:

$$F_{A,\psi}(s) = \langle\psi, e^{isA}\psi\rangle_{L^2}, \quad s \in \mathbb{R}.$$

Then $F_{A,\psi}$ is continuous in s, is equal to 1 at $s = 0$, and is of positive type. Hence, by Bochner's theorem, $F_{A,\psi}(s)$ is the

Fourier transform of a probability measure, $\rho_{A,\psi}$. We interpret $\rho_{A,\psi}$ as the probability density of a random variable representing the observable A, the system being in the state ψ. It naturally turns out that we get the same interpretation whether we use $\mathcal{H}$ with vectors ψ and observables {A}, or $\mathcal{H}'$ with vectors ψ' and observables A': $\rho_{A,\psi} = \rho_{A',\psi'}$, where $F_{A',\psi'}(s) = \langle\psi', e^{iA's}\psi'\rangle_{\mathcal{H}'}$.

We can do this construction of probability measures simultaneously for observables $A_1, A_2, \ldots A_n$ that <u>commute</u> in the strong sense that the unitary groups $e^{is_k A_k}$, $k = 1,2,\ldots N$ commute. Then,

$$\langle\psi, \exp(i \sum_{j=1}^{N} s_j A_j)\psi\rangle \quad = F_{A_1,\ldots A_N,\psi}(s_1,\ldots s_N)$$

is the simultaneous characteristic function of N random variables on $\mathbb{R}^N$ with some probability measure $\rho_{A_1,\ldots A_n,\psi}$. For example, we may choose N = m, and $A_j = q_j$, $\psi = \psi_0$. Then the random variables become the functions q_j on $\mathbb{R}^m$ and the probability measure turns out to be $|\psi_0(q)|^2 dq$, as in the formulation $\mathcal{H}'$. Alternatively, we may choose $A_j = p_j$ to obtain the "momentum space realization" of quantum mechanics. But p_j and q_j do not commute, and it is impossible to construct a simultaneous probability measure on $\mathbb{R}^{2m}$, whose marginal densities coincide with $\rho_{p_j,\psi}$ and $\rho_{q_j,\psi}$. Thus, quantum mechanics is a <u>different</u> model of probability from the classical one. It has sometimes been said that quantum mechanics contains classical probability as a special case, so it is a more general theory rather than a different theory. But this is only true in a formal sense: any set of (classical) random variables on a probability space (Ω, P) can be regarded as (mutually commuting) self-adjoint operators on $L^2(\Omega, P)$, and these operators have an interpretation as a very special quantum theory; moreover, the quantum-defined interpretation coincides with the original probabilistic meaning as random variables. But the quantum mechanics of a realistic theory contains non-commuting operators; it contains $p_1, \dots p_m$ as well as $q_1, \dots q_m$, and it contains H as well as $q_1(t), q_2(t), \dots q_m(t)$ for all t; it is not true that this contains classical mechanics as a special case.

I will not attempt to relate classical mechanics to a quantum theory; rather the opposite - I shall embed quantum mechanics in a classical probability theory. This is done by enlarging the probability space $(\mathbb{R}^m, \rho_{q_1,\dots q_m,\psi_0})$ arising from $q_1, \dots q_m$ in the ground state, to a larger probability theory that contains all the information about the quantum dynamics. But, how to recover this information is rather subtle.

§2. A Wightman-like formulation

Certain properties of a quantum theory are mentioned in §1; they are desirable and often hold (that is, they hold for a wide class of potential functions V(q)). We can formalise a framework for a quantum theory with one degree of freedom (m = 1) by four axioms, in analogy with Wightman's axioms for quantum field theory [1]:

A quantum theory with one degree of freedom is a quadruplet $(\mathcal{H}, U(t), q, \psi_0)$ where

1. U(t) is a continuous unitary representation of the group $\mathbb{R}$ (time-evolution) on a Hilbert space $\mathcal{H}$, and the generator H (where $U(t) = e^{-itH}$) is a positive operator: $H \geq 0$.
2. ψ_0 is invariant under U(t) and is unique up to a phase.

3. q is a symmetric operator on a dense linear domain $\mathcal{D} \subseteq \mathcal{H}$; $\mathcal{D}$ is invariant under q and U(t), $t \in \mathbb{R}$; and $\psi_0 \in \mathcal{D}$.

4. The vacuum ψ_0 is a cyclic vector for the set of operators $\{q(t): t \in \mathbb{R}\}$, where $q(t) = U(t)qU^{-1}(t)$.

(Cyclicity means that the set of vectors $\{q(t_1)\ldots q(t_n)\psi_0 : t = (t_1,\ldots t_n) \in \mathbb{R}^n, n = 0,1,\ldots\}$ spans $\mathcal{H}$.)

In these circumstances, the famous Wightman functions $W_n(t_1,\ldots t_n)$ can be defined:

$$W_n(t_1,\ldots t_n) = \langle\psi_0, q(t_1)\ldots q(t_n)\psi_0\rangle$$

The set of functions $\{W_n : n = 0,1,\ldots\}$ uniquely determines the dynamics, in the sense that two theories with the same W-functions are unitarily equivalent. The W-functions are the quantum analogues of the joint moments of a stochastic process. I say quantum analogues because $q(t_1)q(t_2) \neq q(t_2)q(t_1)$ in general. This implies that the W-functions are not symmetric in $(t_1,\ldots t_n)$, unlike the moments of a process.

The axioms imply a list of properties for W_n. Thus:

(a) $W_n(t_1,\ldots t_n)$ is continuous and is the boundary value of a function holomorphic in the domain

$$\mathrm{Im}(t_1 - t_2) > 0,\ \mathrm{Im}(t_2 - t_3) > 0,\ldots\ \mathrm{Im}(t_{n-1} - t_n) > 0.$$

(the energy is positive)

(b) W_n is translation invariant:

$$W_n(t_1,\ldots t_n) = W_n(t_1 + s, t_2 + s,\ldots t_n + s),\ s \in \mathbb{R}$$

(ψ_0 is invariant)

(c) $W_n(t_1,\ldots t_n) = \overline{W_n(t_n,\ldots t_1)},\ t \in \mathbb{R}^n$

(q is a symmetric operator)

(d) Let $\mathcal{A}$ be the non-abelian associative algebra over $\mathbb{C}$ generated by the symbols $t_1, t_2, \ldots$, furnished with the conjugation * defined by $(t_1 \ldots t_n)^* = t_n \ldots t_1$. Let $W: \mathcal{A} \to \mathbb{C}$ be the linear map obtained by linear extension of $W: t_1 \ldots t_n \mapsto W_n(t_1, \ldots t_n)$. Then W is a positive linear form. (metric of $\mathcal{H}$ is positive definite).

e) For each $k \in (1, 2, \ldots n)$, $1 < k < n$, we have

$$W_n(t_1, \ldots t_k, t_{k+1} + \lambda, \ldots t_n + \lambda) \to$$
$$W_k(t_1, \ldots t_k) W_{n-k}(t_{k+1}, \ldots t_n) \text{ as } \lambda \to \infty$$

(uniqueness of the vacuum).

The axioms 1)-4) and the properties a)-e) are equivalent. This is our version of the Wightman reconstruction theorem, namely: given any set of functions $\{W_n\}$ with properties a)-e), there exists a Wightman quantum mechanics ($\mathcal{H}$, U(t), q, ψ_0) of which they are the W-functions. I have already mentioned the uniqueness.

The theory so defined will obey axioms 1)-4), but the Hamiltonian will not necessarily be of the form $H = \frac{p^2}{2} + V$. Apart from technicalities arising from conditions on the domain $\mathcal{D}$, the Wightman formulation is a considerable generalization of the conventional quantum theory described in §1.

The reconstruction theorem is a non-commutative version of the moment problem - constructing a stochastic process from its moments. It is also similar to the Gelfand-Segal theorem [2], the algebra $\mathcal{A}$ taking the rôle of the c*-algebra and the functional W taking the rôle of the state.

A stochastic process is not always determined by its moments;

this ambiguity arises in this theory in that the operator $q(t)$ obtained from the construction is merely symmetric on $\mathcal{D}$, and might not determine a unique self-adjoint extension.

§3. Schwinger functions

Suppose we have the Wightman functions $\{W_n(t_1, \ldots t_n), n = 1, 2, \ldots\}$ of a theory obeying axioms 1)-4). According to property (a), $W_n(t_1, \ldots t_n)$ has an analytic continuation to purely imaginary time $t_j = is_j$, $j = 1, \ldots n$ where s_j are real numbers obeying

$$s_j - s_{j+1} > 0, \qquad j = 1, \ldots n-1. \tag{2}$$

Define a function

$$S_n(s_1, \ldots s_n) = W_n(is_1, \ldots, is_n)$$

in the region (2). We define S_n in the other permuted regions of $\mathbb{R}^n$ by requiring S_n to be symmetric under permutations of $s_1, \ldots s_n$. If two arguments are equal, we define S_n as the corresponding boundary value of the analytic function.

Obviously, $\{S_n, n = 1, 2, \ldots\}$ determines $\{W_n, n = 1, 2, \ldots\}$ and so $\{S_n\}$ determines $(\mathcal{H}, U(t), q, \psi_0)$. $\{S_n\}$ are called the Schwinger functions of the theory. One might attempt to find a list of conditions on $\{S_n\}$ analogous to a)-e), that are necessary and sufficient for $\{S_n\}$ to be the Schwinger functions of some theory obeying axioms 1)-4). In field theory such conditions have been found [3] and are known as the Osterwalder-Schrader axioms.

The Schwinger functions are sometimes called the Euclidean Greens functions. This is because in a relativistic field theory, they are invariant under the Euclidean group, and for a field obeying a linear equation, the two-point function W_2 is a Green's function for the equation.

After this lengthy account of the concepts involved, here comes the punch-line: the Schwinger functions of a quantum theory (as described in §1) are the moments of a classical stochastic process X:

$$S_n(s_1, \ldots s_n) = \mathbb{E}(X(s_1) \ldots X(s_n)).$$

Sufficient conditions for this to hold (besides the axioms 1)-4)) are that $H = p^2/2 + V$ where V is smooth and of short range, and that $\psi_0(q) > 0$ for all q. This and improvements are proved in Barry Simon's stimulating book [4].

A quantum theory in the more general sense of §2 (obeying only 1)-4)) does not in general lead to a stochastic process X. Conversely, the moments of a general process are not always the Schwinger functions of some quantum theory. This question is resolved in the papers of Fröhlich [5] and Klein and Landau [6].

Example. Harmonic oscillator.

The Hamiltonian is

$$H_0 = \frac{1}{2}(p^2 + q^2 - 1) \geq 0;$$

The ground state is

$$\psi_0(q) = \pi^{-\frac{1}{4}}\exp(-q^2/2) > 0,$$

and the eigenvalue 0 of H is simple. We find

$$\langle\psi_0, q(t_1)q(t_2)\psi_0\rangle = \frac{1}{2}e^{i(t_1-t_2)}.$$

This leads by analytic continuation to

$$S_2(s_1, s_2) = \frac{1}{2} e^{-|s_1-s_2|} = \mathbb{E}(X(s_1)X(s_2))$$

where X is the Ornstein-Uhlenbeck process (called the oscillator process in [4]).

The Wiener process arises in a similar discussion of the free Schrödinger particle, with V = O and Hamiltonian $p^2/2$ [4]. In this case there is no ground state in $\mathcal{H}$, as the invariant function is $\psi_0(q) = 1$ for all q, and this is not in $L^2(\mathbb{R}, dq)$. So this example does not quite fit into this general framework.

For a quantum system at a non-zero temperature, the Schwinger functions are the moments of a periodic stochastic process ([4], [5], [6]).

§4. The Gell-Mann-Low formula

In 1954, M. Gell-Mann and F. Low derived an explicit formula for the Green's functions of a quantum theory. They used the idea of a Feynman path integral. Formally, their formula may be analytically continued to imaginary time to give the Schwinger functions in terms of a Feynman-Kac integral.

To aid our brief derivation, let us take $H_0 = \frac{1}{2}(p^2 + q^2 - 1)$ as the free Hamiltonian, which leads to a functional integral with the oscillator measure, rather than $H_0 = p^2/2$, which leads to the Wiener measure. The two corresponding functional integrals are related by a Cameron-Martin transformation.

The ground state of H_0 on $L^2(\mathbb{R}, dq)$ is $\psi_0(q) = \pi^{-\frac{1}{4}}e^{-q^2/2}$. From now on we shall use the equivalent formulation on $\mathcal{H}' = L^2(\mathbb{R}, |\psi_0|^2 dq)$, as explained in §1, related to $\mathcal{H}$ by the unitary operator W (multiplication by $\pi^{\frac{1}{4}}e^{q^2/2}$). On $\mathcal{H}'$, the energy operator is $WH_0W^{-1} = H_0'$ and this turns out to be $-\frac{1}{2}(D^2 - qD)$ where $D = \frac{\partial}{\partial q}$. This is familiar as the generator of Brownian motion with linear drift. In $\mathfrak{H}'$, the ground state is the function $\psi_0'(q) = 1$.

Suppose now $V(q)$ is a potential such that $H = H_0' + V \geq 0$, and H has a simple eigenvalue 0 with eigenfunction Ψ_0 that may be chosen positive. Then $\langle\psi_0', \Psi_0\rangle = \langle 1, \Psi_0\rangle = \int\Psi_0(q)\,|\psi_0(q)|^2 dq > 0$. Moreover, in the sense of strong convergence in $\mathfrak{H}'$, we have

$$\Psi_0 = \underset{T\to\infty}{\text{s-lim}} \frac{e^{-TH}1}{\|e^{-TH}1\|} . \qquad (3)$$

The group $U(t)$ can be analytically continued to imaginary $t = -is$, $s > 0$, to give the contraction semigroup e^{-sH}, $s \geq 0$. The Schwinger functions are the analytic continuations, to $t_k = is_k$, of

$$W_n(t_1,\dots t_n) = \langle\Psi_0, e^{-it_1H} q\, e^{i(t_1-t_2)H} q \dots q e^{it_nH}\Psi_0\rangle$$

and may therefore be expressed as

$$S_n(s_1,\dots s_n) = \langle\Psi_0, e^{s_1H} q\, e^{-(s_1-s_2)H} q \dots q e^{-s_nH}\Psi_0\rangle$$

This makes sense if $s_j - s_{j+1} > 0$, $j = 1,\dots n-1$, since $e^{s_1H}\Psi_0 = \Psi_0$, $e^{-s_nH}\Psi_0 = \Psi_0$. (We take it as given that the semigroup e^{-sH} maps $\mathfrak{D}$ (of axiom 3) into itself).

This formula for S_n has a neat expression in terms of a functional integral. Let Ω be the set of continuous paths $Q: \mathbb{R} \to \mathbb{R}$, and furnish Ω with the Ornstein-Uhlenbeck measure μ_0. Let $\mathfrak{F}_s$ be the σ-ring generated by $Q(s)$ (for each s). Let $\mathfrak{F}_{\leq s}$ be the σ-ring generated by $\{\mathfrak{F}_u: u \leq s\}$, and let $\mathfrak{F}_\infty$ be the σ-ring generated by $\{Q(s),\ s \in \mathbb{R}\}$. Let $\mathfrak{F}$ be the filtration

$\{\mathfrak{F}_{\leq u}\}_{u \in \mathbb{R}}$. Let $\mathbb{E}_0$ be the conditional expectation, given $\mathfrak{F}_0$. Then $\mathbb{E}_0$ is the orthogonal projection from $L^2(\Omega, \mu, \mathfrak{F}_\infty)$ onto $L^2(\Omega, \mu, \mathfrak{F}_0)$. It is known that the latter may be identified with $\mathfrak{H}' = L^2(\mathbb{R}, |\psi_0(q)^2|dq)$ in such a way that $Q(0)$ may be identified with q and the function 1 with the function $1 = \psi_0'(q)$. This follows from the fact that the probability density of $Q(0)$ is $\pi^{-\frac{1}{2}}e^{-q^2}$, equal to $|\psi_0(q)|^2$, the density of q. This, then, is the desired embedding of the quantum Hilbert space, $\mathfrak{H}'$, into a larger space.

Let $W(t)$: $L^2(\Omega, \mu_0, \mathfrak{F}_\infty) \to L^2(\Omega, \mu_0, \mathfrak{F}_\infty)$ be the time-displacement of the process: $W(t)$ is the unique unitary operator such that for each $\{s_1, \ldots s_n\}$,

$$W(t)[Q(s_1)Q(s_2)\ldots Q(s_n)] = Q(s_1 + t)\ldots Q(s_n + t).$$

Then the Feynman-Kac formula [4] may be expressed as

$$e^{-Ht} = \mathbb{E}_0 e^{-\int_0^t V(Q(s))ds} W(t)\mathbb{E}_0, \tag{4}$$

where H denotes the operator $H_0' + V$ on $\mathfrak{H}'$ identified with its image on $L^2(\Omega, \mu_0, \mathfrak{F}_0)$ under the identification, and given the same symbol for convenience.

Let $\mathbb{E}_s$ denote the conditional expectation, given $\mathfrak{F}_s$. Then $\mathbb{E}_s$ is the orthogonal projection from $L^2(\Omega, \mu_0, \mathfrak{F}_\infty)$ onto $L^2(\Omega, \mu_0, \mathfrak{F}_s)$, and so may be given the symbol E_s (meaning the projection as an operator on $L^2(\Omega, \mu_0, \mathfrak{F}_\infty)$). The well-known Markov property of the Ornstein-Uhlenbeck process may then be expressed as

$$E_{\leq s} E_t E_{\geq u} = E_{\leq s} E_{\geq u} \quad \text{if } s \leq t \leq u \tag{5}$$

where $E_{\leqslant s}$ and $E_{\geqslant s}$ are similarly defined as projections onto $L^2(\Omega, \mu_0, \mathfrak{F}_{\leqslant s})$ and the corresponding sub-space generated by the process $\{Q(u), u \geqslant s\}$, $L^2(\Omega, \mu_0, \mathfrak{F}_{\geqslant s})$.

The Markov property, (5), has the consequence that we can remove E_t from a scalar product, $\langle F, E_t G\rangle$, if F is in $L^2(\Omega, \mu_0, \mathfrak{F}_{\leqslant t})$ and G is in $L^2(\Omega, \mu_0, \mathfrak{F}_{\geqslant t})$:

$$\langle F, E_t G\rangle = \langle E_{\leqslant t} F, E_t E_{\geqslant t} G\rangle = \langle F, E_{\leqslant t} E_t E_{\geqslant t} G\rangle$$

$$= \langle F, E_{\leqslant t} E_{\geqslant t} G\rangle \quad \text{by (4)}$$

$$= \langle F, G\rangle$$

We also note that $W(t)E_s W^{-1}(t) = E_{s+t}$.

We can now evaluate the Schwinger functions by using the limit (3), the Feynman-Kac formula (4) and the Markov property, (5). From (3), we have

$$S_n(s_1,\dots s_n) =$$

$$\lim_{T\to\infty} \frac{\langle 1, e^{-TH} e^{s_1 H} Q(0) e^{-(s_1-s_2)H} Q(0)\dots e^{-(s_{n-1}-s_n)H} Q(0) e^{-(T+s_n)H} 1\rangle}{\langle 1, e^{-2TH} 1\rangle}, \tag{6}$$

This is valid provided $T > s_1 > s_2 > \dots > s_n > -T$. From (4), we insert

$$E_0 \exp\left[-\int_0^{s_j-s_{j+1}} V(Q(s))ds\right] W(s_j - s_{j+1})E_0$$

for the semigroup $e^{-(s_j-s_{j+1})H}$, $j = 1,\dots n-1$, and a factor $E_0 \exp\left[-\int_0^{T+s_n} V(Q(s))ds\right] W(T+s_n)E_0$ for $e^{-(T+s_n)H}$, acting on 1. The terms $W(t+s_n)E_0$ can be ignored, as $E_0 1 = 1$, $W(s)1 = 1$. The remaining expression $\exp\left[-\int_0^{T+s_n} V(Q(s))ds\right]$ lies in the future, as $T+s_n > 0$. Also $E_0 Q(0) E_0 = E_0 Q(0)$. Putting these remarks together, the last few terms in the numerator in (6) become

$$e^{-(s_{n-1}-s_n)H} Q(0) e^{-(T+s_n)H} 1 =$$

$$= E_0 \exp\left[-\int_0^{s_{n-1}-s_n} V(Q(s))ds\right] W(s_{n-1} - s_n) E_0 Q(0) \times$$
$$\times \exp\left[-\int_0^{T+s_n} V(Q(s))ds\right] 1.$$

Now take $W(s_{n-1} - s_n)$ to the right, and use the invariance of 1. The term becomes

$$E_0 \exp\left[-\int_0^{s_{n-1}-s_n} V(Q(s))ds\right] E_{s_{n-1}-s_n} Q(s_{n-1} - s_n) \times$$
$$\times \exp\left[-\int_{s_{n-1}-s_n}^{T+s_{n-1}} V(Q(s))ds\right] 1.$$

The terms on the left of $E_{s_{n-1}-s_n}$ are in the past of $s_n - s_{n-1}$, those on the right are in the future of $s_n - s_{n-1}$. Hence, by (5), we may remove the conditional expectation $E_{s_n-s_{n-1}}$. Collecting up terms, (which all commute), this part of the numerator comes to

$$E_0 \exp\left[-\int_0^{T+s_{n-1}} V(Q(s))ds\right] Q(s_{n-1} - s_n) 1 \tag{7}$$

Proceeding in the same way with the next factor in the numerator, namely

$$E_0 \exp\left[-\int_0^{s_{n-2}-s_{n-1}} V(Q(s))ds\right] W(s_{n-2} - s_{n-1}) E_0 Q(0),$$

and combining with (7), we get

$$E_0 \exp\left[-\int_0^{T+s_{n-2}} V(Q(s))ds\right] Q(s_{n-2} - s_{n-1}) Q(s_{n-2} - s_n) 1.$$

We may proceed to the end, and use translation invariance to obtain

$$S_n(s_1,\dots s_n) = \lim_{T\to\infty} \frac{\langle 1,\ E_0 \exp\left[-\int_{-T}^{T} V(Q(s))ds\right] Q(-s_1)\dots Q(-s_n)1\rangle_{\mathcal{H}'}}{\langle 1,\ E_0 \exp\left[-\int_{-T}^{T} V(Q(s))ds\right] 1\rangle_{\mathcal{H}'}} .$$

Now, $\int dq|\psi_0(q)|^2 E_0(\dots) = \mathbf{E}(\dots)$, the expectation with respect to μ_0, giving finally

$$S_n(s_1,\dots s_n) = \mathbf{E}(G\ Q(-s_1)\ \dots\ Q(-s_n)) \tag{8}$$

where
$$G = \lim_{T\to\infty} \frac{e^{-\int_{-T}^{T} V(Q(s))ds}}{\mathbf{E}\left(e^{-\int_{-T}^{T} V(Q(s))ds}\right)} \tag{9}$$

when this limit exists. Thus, the Schwinger function is indeed the nth moment of a process, namely of $Q(-s)$, relative to the measure $d\mu = G d\mu_0$ on Ω; G has the form of a Gibbs factor $e^{-\beta H_1}/\mathbf{E}e^{-\beta H_1}$ in a classical statistical mechanics of "paths" $Q(s)$ with energy $\int_{-\infty}^{\infty} V(Q(s))ds$. The limit, $T \to \infty$, is called the "thermodynamic limit" and it was indeed by using the methods of correlation inequalities, borrowed from classical statistical mechanics, that Albeverio and Hoegh-Krohn obtained the first proof of the existence of Schwinger functions [7] (for a model relativistic field theory, not for a system with one degree of freedom). Historically, the Gell-Mann-Low formula, (8), was derived without much rigour; later it was used, after proving the limits involve exist, to show the existence of Wightman theories in two dimensions. It is now used to prove (see [4]) analytical estimates in quantum theory that are hard and unobvious without the use of functional integrals.

§5. Magnetic Fields

Let us now briefly summarise so far. To the quantum mechanics with Hamiltonian $H = H_0 + V$, $H_0 = \frac{1}{2}(p^2 + q^2 - 1)$, there corresponds a stationary stochastic process $(\Omega, d\mu, \mathfrak{F}_{\leq t})$, where $d\mu$ is the Gibbs measure $G d\mu_0$, μ_0 being the measure of the Ornstein-Uhlenbeck process. The Schwinger functions of the quantum mechanics are the moments of the process $Q(-s)$ relative to $d\mu$. The physical Hilbert space is the subspace $\mathfrak{H}' = L^2(\Omega, d\mu, \mathfrak{F}_0)$ $= E_0 L^2(\Omega, d\mu, \mathfrak{F}_\infty)$, the space of time-zero. The quantum Hamiltonian H is the generator of the semi-group $E_0 W(t) E_0$, where $W(t)$ is the time-translation of the process.

A similar theory can be set up for $m > 1$ degrees of freedom: in that case, Ω consists of paths in $\mathbb{R}^m$.

Suppose now $m = 3$, and let us interpret (q_1, q_2, q_3) as the coordinates of a particle in real space. We wish to couple the system with Hamiltonian $p^2/2 + V$ to a magnetic field $\underline{B}$ described by a given magnetic potential, $\underline{A}$: thus $\underline{B} = \text{curl}\, \underline{A}$, $A \in C_0^\infty(\mathbb{R}^3)$. The "minimum coupling" rule is: replace the vector operator $\underline{p} = -i\underline{\partial}$, in the Hamiltonian, by the "covariant derivative" $\underline{p} - \underline{A}$. For example the energy operator for an otherwise free particle in a magnetic potential $\underline{A}(\underline{q})$ is

$$H_0(\underline{A}) = \frac{1}{2}\left\{\left(-i\frac{\partial}{\partial q_1} - A_1(\underline{q})\right)^2 + \left(-i\frac{\partial}{\partial q_2} - A_2(\underline{q})\right)^2 + \left(-i\frac{\partial}{\partial q_3} - A_3(\underline{q})\right)^2\right\} \tag{10}$$

acting on $L^2(\mathbb{R}^3)$. This is not a real operator (it does not preserve the reality of a wave-function). Nevertheless, probabilistic methods can be used to represent the Schwinger functions, namely, the Feynman-Kac-Ito formula.

Theorem

Let $(\Omega, \mu_0, \mathcal{F}_\infty)$ be the m-dimensional Ornstein-Uhlenbeck process space. Suppose $A_j, V \in C_0^\infty(\mathbb{R}^m)$, $= j=1,\dots m$, and let

$$H = \frac{1}{2}(-i\underline{\partial} - \underline{A})^2 + \frac{1}{2}(\underline{q}^2 - 1) + V(\underline{q})$$

Then $$e^{-tH} = E_0\, e^{F(\underline{Q},t)} W(t) E_0$$

where W is time-displacement of the O-U process, and

$$F(\underline{Q}, t) = -i\int_0^t \underline{A}(\underline{Q}(s)).d\underline{Q} - \frac{i}{2}\int_0^t \operatorname{div} \underline{A}\, ds - \int_0^t V(\underline{Q}(s))ds$$

so

$$F(\underline{Q}, t) = -i\int_0^t \underline{A}(\underline{Q}(s))\circ d\underline{Q} - \int_0^t V(\underline{Q}(s))ds.$$

(the Feynman-Kac-Ito formula).

Simon [4] gives two proofs of a similar theorem involving the Wiener measure, from which this result can be obtained by a Cameron-Martin transformation.

Rather than repeat Simon's proof here* I give an argument which may clarify why the phase of F is the Stratanovich integral $\int_0^t \underline{A}\circ d\underline{Q}$, rather than the Ito integral (Simon has given an argument showing that the Ito integral is not gauge invariant).

For our argument, we must work on $\mathcal{B}$, the space of paths beginning at zero, furnished with the measure db of Brownian motion; we shall arrive at a F-K-I formula for the Hamiltonian (10).

Let $P_1, \dots P_m$ be self-adjoint operators on a Hilbert space $\mathcal{H}$. Then if $t > s$,

*I thank M. Pinsky for correcting a mistake in my lecture at this point.

$$e^{-\frac{1}{2}(P_1^2+\ldots+P_m^2)(t-s)} = \mathbb{E}_{db}(e^{i\,\underline{P}.(\underline{b}(t)-\underline{b}(s))}) \tag{11}$$

as can be seen by expanding, keeping the order of the non-commuting operators P ,... P_m in mind.

For any N we may divide (0, t) into N intervals (t_k, t_{k+1}), $k = 0,\ldots N-1$, where $t_k = \frac{kt}{N}$. The corresponding steps in Brownian motion are independent, so by (11) we may write

$$e^{-\frac{1}{2}(\Sigma P_j^2)t} = \prod_{k=0}^{N-1} (\mathbb{E}\, e^{i\,\underline{P}.[\underline{b}(t_{k+1})-\underline{b}(t_k)]})$$

$$= \mathbb{E}\left[e^{i\,\underline{P}.(b(t_1)-\underline{b}(t_0))} \ldots e^{i\,\underline{P}.(\underline{b}(t_N)-\underline{b}(t_{N-1}))}\right]$$

We now take $\underline{P}$ to be the covariant momentum: $P_j = p_j - A_j$ $= -i\frac{\partial}{\partial q_j} - A_j$, $j = 1,\ldots m$.

We note that A_j is a function of q, so that, as operators on $L^2(\mathbb{R}^m)$, p_j and A_j do not commute. Choose N large; the lesson learnt from the Ito calculus is that we get the right answer if we keep all infinitesimals up to second order. Thus, if X and Y are infinitesimal operators, we can use

$$e^{X+Y} = e^{\frac{X}{2}} e^{Y} e^{\frac{X}{2}} \tag{12}$$

as this is correct up to X^2, XY, YX and Y^2.

Put $X_k = -i\underline{A}.(\underline{b}(t_k) - \underline{b}(t_{k-1}))$,

$Y_k = -\underline{\partial}.(\underline{b}(t_k) - \underline{b}(t_{k-1}))$

and use (12). Thus

$$e^{-H_0'(\underline{A})t} = \lim_{N\to\infty} \mathbb{E}_{db}(e^{\frac{1}{2}X_1} e^{Y_1} e^{\frac{1}{2}X_1} e^{\frac{1}{2}X_2} e^{Y_2} e^{\frac{1}{2}Y_2} \ldots e^{\frac{1}{2}X_N} e^{Y_N} e^{\frac{1}{2}X_n}) \tag{13}$$

Now move all the translation operators e^{Y_k} to the right; they build up to $e^{-\underline{\partial}.\underline{b}(t)}$, changing the magnetic potential factors

$e^{\frac{1}{2}\underline{A}(q)\cdot\Delta\underline{b}_k}$ to $e^{\frac{1}{2}\underline{A}(\underline{q}+\Delta\underline{b}_k)\cdot\Delta\underline{b}_k}$ as it passes. We may then collect up all the exponents (as they commute), obtaining for the phase, a stochastic integral $\sum\frac{1}{2}[\underline{A}(\underline{q} + \underline{b}_k) + \underline{A}(\underline{q} + \underline{b}_{k+1})]\cdot(\underline{b}_{k+1} - \underline{b}_k)$, $\underline{b}_k = \underline{b}(t_k)$. This gives rise to the Stratanovich integral in the limit:

$$e^{-H_0(\underline{A})t} = \mathbb{E}_{db}\left(e^{i\int \underline{A}\bullet d\underline{b}}\, e^{-\underline{\partial}\cdot db}\right).$$

This concludes our discussion of the theorem.

The Feynman-Kac-Ito formula immediately gives us the following inequality: for all $\underline{A}$,

$$|\langle f, e^{-tH(A,V)}g\rangle| \leq \langle |f|, e^{-tH(0,V)}|g|\rangle$$

since the phase $\exp(-i\int \underline{A}\bullet d\underline{b})$ has unit modulus. This was suggested by Nelson, and it gives a quick proof of the "diamagnetic inequalities" first proved by Simon using other methods. See [4].

The F-K-I formula suggests a general way to quantize systems in an electromagnetic field: we just multiply the Gibbs factor G in the functional integral by $\exp(-i\int_\omega \underline{A}\bullet d\underline{\omega})$. This has been used by physicists to quantize a system on any Riemanifold $\mathcal{M}$; if $\mathcal{M}$ has a boundary $\partial\mathcal{M}$, we restrict the functional integral to paths $\underline{\omega}(t)$ that do not hit $\partial\mathcal{M}$ in the time interval $[0, t]$. We would then need to show, post hoc, that the quantum theory thus obtained (by projection to the time-zero space) was the one we sought, e.g. that the Hamiltonian is the covariant Laplacian on $\mathcal{M}$ with Dirichlet boundary conditions on $\partial\mathcal{M}$. This step is usually omitted.

§6. Non-abelian gauge fields

The famous factor $e^{-i\int_{\underline{\omega}} \underline{A}\bullet d\underline{\omega}}$ is a (1-dimensional) unitary representation of a random element (i.e. an element depending on $\underline{\omega}$) of the "gauge group" U(1). In this form it has a natural generalization to other, non-abelian, Lie groups.

Let G, the gauge group, be a given Lie group, and let V be a continuous unitary representation of G on a complex Hilbert space L, with dim L = ℓ. We say that a gauge field is given, if we are given the following. To each pair x, y in $\mathbb{R}^m$, and each continuous path $\underline{\omega}$ from x to y, is given a group element $g(x, y; \underline{\omega})$ obeying

1. $g(x, x', 0) = 1_G$, the identity of G.
2. $g(x, y; \underline{\omega}) = [g(y, x; -\underline{\omega})]^{-1}$.
3. $g(x, y; \underline{\omega})g(y, z; \underline{\omega}') = g(x, z; \underline{\omega} \cup \underline{\omega}')$.

We require the map g to be jointly continuous in x, y and measurable in $\underline{\omega}$.

The space of wave-functions for a "multiplet" of ℓ particles related by the symmetry group G is taken to be the Hilbert space $K = L \otimes L^2(\mathbb{R}^m)$; a wave-function in K is, equivalently, described by a column vector $\{\psi_i\}_{1\leq i\leq \ell}$ of functions in $L^2(\mathbb{R}^m)$.

The "connection" between x and y via the path $\underline{\omega}$ is the unitary operator $V(g(x, y; \underline{\omega})): L \to L$; we write this as $V(x, y; \underline{\omega})$. Thus, in the special case of §5, G = U(1) and L = $\mathbb{C}$, $V(x, y; \underline{\omega}) = e^{-i\int_x^y A\bullet d\underline{\omega}}$, so each magnetic potential determines a connection.

The m components of the "gauge potential" are the m self-adjoint operators $A_j(x)$ defined on a dense domain in L, by the limit

$$-iA_j(x) = \lim_{\lambda\to 0} \lambda^{-1}\{V(x, x + \lambda\hat{j}, \omega) - 1_L\},\ 1 \leq j \leq \ell$$

Here, $\hat{j}$ is the unit vector along x_j, and ω is the straight line between x and $x + \lambda\hat{j}$. The covariant derivative is

$$\nabla_j\psi = \lim_{\lambda\to 0} \lambda^{-1}\{V(x, x + \lambda\hat{j};\ \omega)\psi(x + \lambda\hat{j}) - \psi(x)\}$$

$$= (\frac{\partial}{\partial x_j} - iA_j(x))\psi(x).$$

According to the theory of path-ordered integrals [8] we can recover V from $\underline{A}$ by

$$V(x, y;\ \omega)\psi(y) = Pe^{i\int_x^y \underline{\nabla}.d\omega}\psi(x)$$

for each smooth path $\underline{\omega}$. We hope to make sense of this relation for all paths in Ω, and would expect the correct form to involve an ordered Stratanovich integral.

The energy operator for a Schrödinger particle in a given gauge potential $\underline{A}$ is related to the covariant Laplacian $\Delta(\underline{A})$ thus:

$$H(\underline{A}) = -\frac{1}{2}\Delta(\underline{A}) = -\frac{1}{2}\sum_{j=1}^{m}\nabla_j.\nabla_j$$

We can now pursue the Euclidean method, as in §5, to obtain

$$e^{-H(\underline{A})[t-s]} = \mathbb{E}_{db}\left(e^{\underline{\nabla}.[\underline{\omega}(t)-\underline{\omega}(s)]}\right)$$

This shows that the semigroup is the expectation of a random unitary operator. Again, let us use (12)

$$e^{X+Y} = e^{\frac{X}{2}}e^{Y}e^{\frac{X}{2}}$$

for infinitesimal random operators. Then

$$e^{-H(\underline{A})t} = \left(e^{-H(\underline{A})\frac{t}{N}}\right)^N = \prod_{k=1}^{N} \mathbf{E} e^{\underline{\nabla}.\delta\underline{\omega}_k}$$

where $\delta\underline{\omega}_k = \underline{\omega}(t_k) - \underline{\omega}(t_{k-1})$, $t_k = \frac{kt}{N}$.

Thus

$$e^{-H(\underline{A})t} = \mathbf{E}\left(e^{\underline{\nabla}.\delta\underline{\omega}_1}.e^{\underline{\nabla}.\delta\underline{\omega}_2} \ldots e^{\underline{\nabla}.\delta\underline{\omega}_N}\right)$$

$$= \mathbf{E} \lim_{N\to\infty}\left(e^{-\frac{i}{2}\underline{A}.\delta\underline{\omega}_1} e^{\underline{\partial}.\delta\underline{\omega}_1} e^{-\frac{i}{2}\underline{A}.\delta\underline{\omega}_1}\right) \ldots$$

$$\ldots\left(e^{-\frac{i}{2}\underline{A}.\delta\underline{\omega}_N} e^{\underline{\partial}.\delta\underline{\omega}_N} e^{-\frac{i}{2}\underline{A}.\delta\underline{\omega}_N}\right)\Bigg)$$

if the limit exists.

Let us move all the translation operators to the right. They build up to $e^{\underline{\partial}.\underline{\omega}}$. We are left with a path ordered product of unitaries,

$$\prod_{k=1}^{N}\left(e^{-i\frac{1}{2}\underline{A}(\underline{x}+\underline{\omega}_k).\delta\underline{\omega}_k} e^{-i\frac{1}{2}A(\underline{x}+\underline{\omega}_{k+1}).\delta\underline{\omega}_k}\right)e^{\underline{\partial}.\underline{\omega}}$$

which has the Stratanovich character. Progress towards proof of convergence of products like this is reported by Parthasarthy and Sinha [9].

I would like to thank Miss K. Anderson for her careful typing of the manuscript.

References

[1] R.F. Streater and A.S. Wightman, PCT, Spin and Statistics and All That. Benjamin/Cummings 2nd Ed. N.Y. 1978.

[2] G.G. Emch, Algebraic Methods in Statistical Mechanics and Quantum Field Theory. Wiley-Interscience, 1972.

[3] K. Osterwalder and R. Schrader, Axioms for Euclidean Greens functions. Commun. Math. Phys. 31, 83 (1973); and 42, 281 (1975).

[4] B. Simon, Functional Integration and Quantum Physics. (Academic Press, 1979).

[5] J. Fröhlich, The reconstruction of quantum fields from Euclidean Green's functions at arbitrary temperatures; Helv. Phys. Acta 48, 355-363 (1975).

[6] A. Klein and L.J. Landau, Stochastic Processes associated with KMS states. Preprint, University of California, Irvine.

[7] S. Albeverio and R. Hoegh-Krohn, Uniqueness of the Physical Vacuum and the Wightman Functions in the Infinite Volume Limit for Some non Polynomial Interactions. Commun. Math. Phys. 30, 171-200 (1973).

[8] J. Dollard and C. Friedman, On strong product integration. Jour. Funct. Anal. 28, 309 (1978).

[9] K. Parthasarathy and B. Sinha, a Random Kato-Trotter product formula. Preprint, India Statistical Institute, New Delhi.

The Malliavin Calculus and its Applications

Daniel W. Stroock*

*This work was partially supported by N.S.F. Grant MCS 77-14881.

Lecture #1

0. Statement of the General Problem:

Let $(E,\mathcal{F},P)$ be a probability space. Given an $\mathcal{F}$-measurable function $\Phi : E \to R^1$, let $\mu^\Phi = P \circ \Phi^{-1}$ be the distribution of Φ under P. The central problem addressed in these lectures will be the development of techniques to answer the question: when is μ^Φ absolutely continuous and, if it is, what can be said about its density? Actually, we will be studying this question only when $(E,\mathcal{F},P)$ is Wiener space; but in the hope that the underlying ideas will be clearer in a more general setting, we will begin with an abstract treatment.

To begin with, consider how an analyst would attack this problem. An analyst confronted with the problem of showing that a measure μ on R^1 has a density would attempt to obtain estimates of the form:

$$(0.1) \qquad \left|\int \varphi^{(n)} d\mu\right| \le C_n \|\varphi\|_u \quad , \quad \varphi \in C_b^{(n)}(R^1) \quad ,$$

where $\varphi^{(n)} = D^n\varphi$ is the $n^{\underline{th}}$ derivative of φ and $\|\cdot\|_u$ denotes the uniform norm. Indeed, suppose that (0.1) holds for $0 \le n \le N$ and $\varphi \in C_b^{(N)}(R^1)$. Taking $\varphi_\xi(x) = e^{i\xi x}$, we would then have:

$$|(i\xi)^n \hat{\mu}(\xi)| = \left|\int \varphi_\xi^{(n)} d\mu\right| \le C_n$$

for all $0 \le n \le N$ and $\xi \in R^1$, where

$$\hat{\mu}(\xi) \equiv \int e^{i\xi x} \mu(dx) \quad .$$

In particular, if $N \ge 2$, then $\hat{\mu} \in L^1(R^1)$ and so, by standard Fourier theory, μ has a density $f \in C_b(R^1)$. In fact, so long as $0 \le m \le N-2$:

$$f^{(m)}(x) = \frac{1}{2\pi}\int e^{-ix\xi}(i\xi)^m \hat{\mu}(\xi) d\xi \quad ;$$

and so $f \in C_b^{(N-2)}(R^1)$ with $\|f\|_{C_b^{(N-2)}(R^1)}$ depending only on the $\{C_n : 0 \le n \le N\}$.

The preceding indicates that we would be well-advised to seek estimates of the form

$$|E^P[\varphi^{(n)} \circ \Phi]| \le C_n \|\varphi\|_u \quad , \quad \varphi \in C_b^{(n)}(R^1) \quad . \tag{0.2}$$

The problem is that the usual technique for obtaining such estimates is integration by parts and in the present setting it is not at all clear what we can use to replace this time-honored technique. What Malliavin has done is produce just such a replacement.

1. Abstract Formulation of Malliavin's Technique:

The basic idea underlying Malliavin's technique is to introduce an elliptic self-adjoint operator on $L^2(P)$. The elliplicity of the operator is used to test smoothness, the self-adjointness is used to integrate by parts.

(1.1) Definition: A densely defined linear operator $\mathcal{L}$ on $L^2(P)$ with domain $\mathcal{D}(\mathcal{L})$ is a symmetric diffusion operator if:

i) $\mathcal{L}$ on $\mathcal{D}(\mathcal{L})$ is self-adjoint;

ii) $1 \in \mathcal{D}(\mathcal{L})$ and $\mathcal{L}1 = 0$;

iii) there is a linear subspace $\mathcal{D} \subseteq \{\Phi \in \mathcal{D}(\mathcal{L}) \cap L^4(P) : \mathcal{L}\Phi \in L^4(P)$ and $\Phi^2 \in \mathcal{D}(\mathcal{L})\}$ such that $\text{graph}(\mathcal{L}|_{\mathcal{D}})$ is dense in $\text{graph}(\mathcal{L})$;

iv) if $\langle\Phi,\Psi\rangle \equiv \mathcal{L}(\Phi,\Psi) - \Phi\mathcal{L}\Psi - \Psi\mathcal{L}\Phi$ for $\Phi,\Psi \in \mathcal{D}$, then $\langle\cdot,\cdot\rangle : \mathcal{D} \times \mathcal{D} \to L^2(P)$ is a non-negative bilinear form;

v) if $\Phi = (\Phi_1,\ldots,\Phi_n) \in \mathcal{D}^n$ and $F \in C_b^2(R^n)$, then $F \circ \Phi \in \mathcal{D}(\mathcal{L})$ and

(1.2) $\mathcal{L}(F\circ\Phi) = 1/2 \sum_{i,j=1}^{n} \langle\Phi_i,\Phi_j\rangle \frac{\partial^2 F}{\partial x_i \partial x_j}\circ\Phi + \sum_{i=1}^{n} \mathcal{L}\Phi_i \frac{\partial F}{\partial x_i}\circ\Phi$;

vi) there is an extention $\mathcal{L}_1$ of $\mathcal{L}$ as a closed linear operator on $L^1(P)$ with domain $\mathcal{D}(\mathcal{L}_1)$ such that $\mathcal{D}(\mathcal{L}) = \{\Phi \in \mathcal{D}(\mathcal{L}_1) \cap L^2(P) : \mathcal{L}_1\Phi \in L^2(P)\}$.

(1.3) Example: Let $E = R^1$, $\mathcal{F} = \mathcal{B}_{R^1}$, and $P(dx) = \frac{e^{-x^2/2}}{(2\pi)^{1/2}}dx$ For $n \geq 0$ define $H_n(\cdot)$ to be the $n^{\underline{th}}$ Hermite polynomial:

$$H_n(x) = \frac{(-1)^n}{(n!)^{1/2}} e^{x^2/2} D^n e^{-x^2/2} .$$

As is well-known, $\text{span}\{H_n : 0 \leq n \leq N\} = \text{span}\{x^n : 0 \leq n \leq N\}$ for all $N \geq 0$ and $\{H_n : n \geq 0\}$ is an orthonormal basis in $L^2(P)$. Hence, if $\mathcal{D}(\mathcal{L}) \equiv \{\Phi \in L^2(P) : \sum_0^\infty n^2(\Phi,H_n)^2 < \infty\}$ (where $(\Phi,\Psi) = E^P[\Phi\cdot\Psi]$) and

$$\mathcal{L}\Phi \equiv \sum_0^\infty \frac{-n}{2}(\Phi,H_n)H_n$$

for $\Phi \in \mathcal{D}(\mathcal{L})$, then $\mathcal{L}$ obviously satisfies i) and ii) of (1.1) . Moreover, if $\mathcal{D} \equiv \{\Phi \in L^2(P) : (\Phi,H_n) = 0$ for all by a finite number of n's$\}$, then it is clear that $\mathcal{D}$ satisfies iii) .

Checking iv) , v) , and vi) requires us to give another interpretation to $\mathcal{L}$. To this end, note that

$$1/2\,(D^2 - xD)H_n = -\frac{n}{2}H_n \quad , \quad n \geq 0 .$$

Thus for $\Phi \in \mathcal{D}$,

(1.4) $$\mathcal{L}\Phi = 1/2(D^2 - xD)\Phi .$$

Because $\mathcal{L}$ is closed, it is clear that if $\Phi \in L^2(P)$ and $\{(\Phi,H_n)\}_1^\infty$ is rapidly decreasing, then $\Phi \in \mathcal{D}(\mathcal{L})$ and (1.4) continues to hold.

This means that $\mathcal{S}(R^1) \subseteq \mathcal{D}(\mathcal{L})$ ($\mathcal{S}(R^n)$ is the space of Schwartz functions on R^n) and that $\mathcal{L}\Phi$ satisfies (1.4) for $\Phi \in \mathcal{S}(R^1)$. Finally, we can use an easy approximation argument to conclude that if $C_\uparrow^2(R^1) = \{\Phi \in C^2(R^1) : \Phi, \Phi', \text{ and } \Phi'' \text{ are slowly increasing}\}$, then $C_\uparrow^2(R^1) \subseteq \mathcal{D}(\mathcal{L})$ and (1.4) holds for $\Phi \in C_\uparrow^2(R^1)$. In particular, if $F \in C_\uparrow^2(R^1)$ and $\Phi = (\Phi_1,\ldots,\Phi_n) \in \mathcal{D}^n$, then $F \circ \Phi \in C_\uparrow^2(R^1)$ and from (1.4):

$$\mathcal{L}(F \circ \Phi) = 1/2\,(D^2 - xD)(F \circ \Phi)$$

$$= 1/2 \sum_{i,j=1}^{n} \Phi_i' \Phi_j' \frac{\partial^2 F}{\partial x_i \partial x_j} \circ \Phi + \sum_{i=1}^{n} (1/2\, \Phi_i'' - 1/2\, x\Phi_i') \frac{\partial F}{\partial x_i} \circ \Phi$$

$$= 1/2 \sum_{i,j=1}^{n} \Phi_i' \Phi_j' \frac{\partial^2 F}{\partial x_i \partial x_j} \circ \Phi + \sum_{i=1}^{n} \mathcal{L}\Phi_i \frac{\partial F}{\partial x_i} \circ \Phi \quad .$$

Taking $F(x,y) = x \cdot y$, we now see that $\langle \Phi, \Psi \rangle = \Phi' \Psi'$; and so both iv) and v) are now proved.

To prove vi) , first observe that, by (1.4) , for $F \in C_\uparrow^2(R^1)$:

$$\mathcal{L}F \overset{L^2(P)}{=\!=\!=} \lim_{t \downarrow 0} \frac{T_t F - F}{t}$$

where

$$T_t \Phi(x) = \int \gamma(1 - e^{-t}, y - xe^{-t/2}) \Phi(y) dy$$

and $\gamma(t,x) = (2\pi t)^{-1/2} \exp(-x^2/2t)$. Clearly $\{T_t : t \ge 0\}$ is a Feller continuous conservative Markov semi-group on $C_b(R^1)$. Moreover, since

$$(2\pi)^{1/2} \gamma(1 - e^t, y - xe^{-t/2}) e^{y^2/2}$$

$$= \frac{1}{(1 - e^{-t})^{1/2}} \exp\left(- \frac{x^2 e^{-t} - 2xye^{-t/2} + y^2 e^{-t}}{2(1 - e^{-t})}\right)$$

is symmetric in x and y, .

(1.5) $$E^P[\Phi T_t\Psi] = E^P[\Psi T_t\Phi] \quad , \quad \Phi,\Psi \in C_b(R^1) \quad .$$

Taking $\Psi = 1$ in (1.5), it follows that $P = P \circ T_t^{-1}$ and so for and $1 \le p < \infty$:

$$E^P[|T_t\Phi|^p]^{1/p} \le E^P[T_t|\Phi|^p]^{1/p} = E^P[|\Phi|^p]^{1/p} \quad .$$

From this it is easy to see that for each $1 \le p < \infty$, $\{T_t : t \ge 0\}$ has a unique extention as a strongly continuous contraction semi-group $\{T_t^{(p)} : t > 0\}$ on $L^p(P)$. Moreover, because of (1.5), $\{T_t^{(2)} : t > 0\}$ is a semi-group of self-adjoint contractions; and because $\frac{T_t^{(2)}F - F}{t} \to \mathcal{L}F$ for $F \in C_\uparrow^2(R^1)$ and $\text{graph}(\mathcal{L}|_{C_\uparrow^2(R^1)})$ is dense in $\text{graph}(\mathcal{L})$, $\mathcal{L}$ is contained in and therefore equal to the $L^2(P)$-generator of $\{T_t^{(2)} : t > 0\}$. But this means that if $\mathcal{L}_1$ is the $L^1(P)$-generator of $\{T_t^{(1)} : t > 0\}$, then $\mathcal{L}_1$ is a closed extention of $\mathcal{L}$. Finally if $\Phi \in \mathcal{D}(\mathcal{L}_1) \cap L^2(P)$ and $\mathcal{L}_1\Phi \in L^2(P)$, then

$$T_t^{(2)}\Phi - \Phi = T_t^{(1)}\Phi - \Phi = \int_0^t T_s^{(1)}\mathcal{L}_1\Phi ds = \int_0^t T_s^{(2)}\mathcal{L}_1\Phi ds$$

and so $\Phi \in \mathcal{D}(\mathcal{L})$. That is, vi) is satisfied and we see that $\mathcal{L}$ is a symmetric diffusion operator.

In case the reader is wondering why I have adopted such a circuitous route to arrive at such mundane results about the Ornstein-Uhlenbeck generator, I beg his patience and ask him to wait until the next section before he gets an explanation.

Before we can see how Malliavin puts diffusion operators to work, we need a couple of preliminary results.

Theorem (1.7): Let $\mathcal{L}$ be a symmetric diffusion operator on $L^2(P)$. Then $\langle\cdot,\cdot\rangle$ admits a unique extention to $\mathcal{D}(\mathcal{L}) \times \mathcal{D}(\mathcal{L})$ so that $(\Phi,\Psi) \in (\mathcal{D}(\mathcal{L}))^2 \to \langle\Phi,\Psi\rangle \in L^1(P)$ is a non-negative bilinear symmetric map which is continuous with respect to the graph$(\mathcal{L})$-norm. Moreover,

$$E^P[\Phi\mathcal{L}\Psi] = -1/2\, E^P[\langle\Phi,\Psi\rangle] \quad , \quad \Phi,\Psi \in \mathcal{D}(\mathcal{L}) \quad ; \tag{1.8}$$

and for any $\Phi = (\Phi_1,\ldots,\Phi_n) \in \mathcal{D}(\mathcal{L})^n$, $\Psi \in \mathcal{D}(\mathcal{L})$, and $F \in C_b^2(R^n)$: $F\circ\Phi \in \mathcal{D}(\mathcal{L}_1)$, $\mathcal{L}_1(F\circ\Phi)$ is given by the right hand side of (1.2), and if $\mathcal{L}_1(F\circ\Phi) \in L^2(P)$, then for $\Psi \in \mathcal{D}(\mathcal{L})$:

$$\langle F\circ\Phi,\Psi\rangle = \sum_1^n \frac{\partial F}{\partial x_i}\circ\Phi\langle\Phi_i,\Psi\rangle \quad . \tag{1.9}$$

Finally, if $\Phi,\Psi \in \mathcal{D}(\mathcal{L})$, then $\Phi\cdot\Psi \in \mathcal{D}(\mathcal{L}_1)$ and

$$\mathcal{L}_1(\Phi\cdot\Psi) = \Phi\mathcal{L}\Psi + \Psi\mathcal{L}\Phi + \langle\Phi,\Psi\rangle \quad . \tag{1.10}$$

Proof: We first show that (1.8) holds for $\Phi,\Psi \in \mathcal{D}$. But

$$\mathcal{L}(\Phi\cdot\Psi) = \Phi\mathcal{L}\Psi + \Psi\mathcal{L}\Phi + \langle\Phi,\Psi\rangle$$

and for any $\Xi \in \mathcal{D}(\mathcal{L})$:

$$E^P[\mathcal{L}(\Xi)] = E^P[1\mathcal{L}(\Xi)] = E^P[\Xi\mathcal{L}(1)] = 0 \quad .$$

Thus

$$2E^P[\Phi\mathcal{L}\Psi] + E^P[\langle\Phi,\Psi\rangle] = 0$$

and so (1.7) holds for $\Phi,\Psi \in \mathcal{D}$.

To prove that the desired extention of $\langle\cdot,\cdot\rangle$ exists, it suffices to show that if $\{\Phi_n\}_1^\infty \subset \mathcal{D}$ and $(\Phi_n,\mathcal{L}\Phi_n) \to (\Phi,\mathcal{L}\Phi)$ in $(L^2(P))^2$, then $\{\langle\Phi_n,\Phi_n\rangle\}_1^\infty$ converges in $L^1(P)$. But by (1.8):

$$E^P[\langle\Phi_n-\Phi_m,\Phi_n-\Phi_m\rangle] = -2E^P[(\Phi_n-\Phi_m)\mathcal{L}(\Phi_n-\Phi_m)] \to 0$$

as $m,n \to \infty$. Moreover, if $\langle\Phi\rangle \equiv \langle\Phi,\Phi\rangle^{1/2}$, then by the usual reasoning for non-negative bilinear forms, $\langle\cdot\rangle$ satisfies Minkowski's inequality. In particular:

$$|\langle\Phi_n,\Phi_n\rangle - \langle\Phi_m,\Phi_m\rangle| = |\langle\Phi_n\rangle - \langle\Phi_m\rangle|(\langle\Phi_n\rangle + \langle\Phi_m\rangle)$$

$$\le \langle\Phi_n - \Phi_m\rangle(\langle\Phi_n\rangle + \langle\Phi_m\rangle)$$

and so

$$\|\langle\Phi_n,\Phi_n\rangle - \langle\Phi_m,\Phi_m\rangle\|_{L^1(P)} \le E^P[\langle\Phi_n - \Phi_m\rangle^2]^{1/2} E^P[(\langle\Phi_n\rangle + \langle\Phi_m\rangle)^2]^{1/2}$$

$$\le 2 \sup_k E^P[\langle\Phi_k,\Phi_k\rangle]^{1/2} E^P[\langle\Phi_n - \Phi_m, \Phi_n - \Phi_m\rangle]^{1/2}$$

$$\to 0$$

as $m,n \to \infty$.

The proof of (1.10) is now easy. In fact, given $\Phi \in \mathcal{D}(\mathcal{L})$, choose $\{\Phi_n\}_1^\infty \subseteq \mathcal{D}$ so that $(\Phi_n,\mathcal{L}\Phi_n) \to (\Phi,\mathcal{L}\Phi)$ in $(L^2(P))^2$. Then $\Phi_n^2 \to \Phi^2$ in $L^1(P)$ and

$$\mathcal{L}_1\Phi_n^2 = \mathcal{L}\Phi_n^2 = 2\Phi_n\mathcal{L}\Phi_n + \langle\Phi_n,\Phi_n\rangle \to 2\Phi\mathcal{L}\Phi + \langle\Phi,\Phi\rangle$$

in $L^1(P)$. Thus, since $\mathcal{L}_1$ is closed, $\Phi^2 \in \mathcal{D}(\mathcal{L}_1)$ and (1.9) holds with $\Phi = \Psi$. When $\Phi \neq \Psi$, (1.10) now follows by polarization.

If $\Phi = (\Phi_1,\ldots,\Phi_n) \in (\mathcal{D}(\mathcal{L}))^n$ and $F \in C_b^2(R^n)$, then it is easy to check that $F\circ\Phi \in \mathcal{D}(\mathcal{L}_1)$ and that $\mathcal{L}_1(F\circ\Phi)$ is given by the right hand side of (1.2) . One simply uses the fact that $\mathcal{L}_1$ is closed.

In view of the continuity result about $\langle\cdot,\cdot\rangle$, we need only prove (1.9) for $\Phi = (\Phi_1,\ldots,\Phi_n) \in \mathcal{D}^n$ and $\Psi \in \mathcal{D}$. Choose $\rho \in C_0^\infty(R^1)$, so that $\rho(x) = x$ for $x \in [-1,1]$ and define $\rho_k(x) = k\rho(x/k)$. By iii) in (1.1) :

$$\mathcal{L}((F\circ\Phi)\cdot(\rho_k\circ\Psi)) = (F\circ\Phi)\mathcal{L}(\rho_k\circ\Psi) + (\rho_k\circ\Psi)\mathcal{L}(F\circ\Phi) + \sum_{i=1}^n (\rho_k'\circ\Psi)(\frac{\partial F}{\partial x_i}\circ\Phi)\langle\Phi_i,\Psi\rangle$$

Hence, by (1.10) :

$$\langle F\circ\Phi,\rho_k\circ\Psi\rangle = \sum_{i=1}^{n}(\rho_k'\circ\Psi)(\frac{\partial F}{\partial x_i}\circ\Phi)\langle\Phi_i,\Psi\rangle$$

But $(\rho_k\circ\Psi,\mathcal{L}(\rho_k\circ\Psi))\to(\Psi,\mathcal{L}\Psi)$ in $(L^2(P))^2$ and so $\langle F\circ\Phi,\rho_k\circ\Psi\rangle\to$ $\langle F\circ\Phi,\Psi\rangle$. Since $\rho_k'\circ\Psi\to 1$, (1.9) follows immediately.

(1.11) Definition: For $1\le p<\infty$, define $\mathcal{D}_p = \{\Phi\in\mathcal{D}(\mathcal{L}_1)\cap L^p(P)$ $\mathcal{L}\Phi\in L^p(P)\}$; and for $2\le p<\infty$, define $\mathcal{K}_p = \{\Phi\in\mathcal{D}_p : \langle\Phi,\Phi\rangle^{1/2}\in$ $L^p(P)\}$. Finally, define $\mathcal{K} = \bigcap_{2\le p<\infty}\mathcal{K}_p$.

(1.12) Remark: By Theorem (1.7) , $\mathcal{K}_2 = \mathcal{D}_2 = \mathcal{D}(\mathcal{L})$. Also, each $\mathcal{K}_p$ can be made into a Banach space with norm

$$|||\Phi|||_{\mathcal{K}_p} = (\|\Phi\|^p_{L^p(P)} + \|\mathcal{L}\Phi\|^p_{L^p(P)} + \|\langle\Phi,\Phi\rangle^{1/2}\|^p_{L^p(P)})^{1/p} .$$

(By Theorem (1.7) , $|||\cdot|||_{\mathcal{K}_2}$ on $\mathcal{K}_2$ determines a norm which is equivalent to the graph($\mathcal{L}$) - norm.) Finally, $\mathcal{K}$ can be made into a countably normed Frechét space in which convergence corresponds to convergence in $|||\cdot|||_{\mathcal{K}_p}$ for each $2\le p<\infty$.

Corollary (1.13): Let $2\le\alpha\le p<\infty$ and suppose $F\in C^2(R^n)$ satisfies: $|F(x)|\le C_0(1+|x|^2)^{\alpha/2}$, $\max_{1\le i\le n}|\frac{\partial F}{\partial x_i}(x)|\le C_1(1+|x|^2)^{(\alpha-1)/2}$, and $\max_{1\le i,j\le n}|\frac{\partial^2 F}{\partial x_i\partial x_j}(x)|\le C_2(1+|x|^2)^{(\alpha-2)/2}$. If $\Phi = (\Phi_1,\ldots,\Phi_n)\in\mathcal{K}_p^n$, then $F\circ\Phi\in\mathcal{D}_{p/\alpha}$ and $\mathcal{L}_1(F\circ\Phi)$ is given by the right hand side of (1.2) . In particular, if $2\le p<\infty$ and $\Phi,\Psi\in\mathcal{K}_p$, then $\Phi\cdot\Psi\in\mathcal{D}_{p/2}$. Finally, if $2<p<\infty$, $1\le q<p/2$, and $\Phi\in\mathcal{K}_p$ is a non-negative function satisfying $1/\Phi\in L^{3p/p-2q}(P)$, then $1/\Phi\in\mathcal{D}_q$.

Proof: The first part is proved by approximating F with functions from $C_b^2(R^n)$ and observing that $\|$right hand side of (1.2)$\|_{L^{p/\alpha}(P)}$ is dominated by

$$1/2 \sum_{i,j=1}^{n} (\|\langle \Phi_i, \Phi_j \rangle\|_{L^{p/2}(P)} \|\frac{\partial^2 F}{\partial x_i \partial x_j} \circ \Phi\|_{L^{p/\alpha-2}(P)})^{p/\alpha}$$
$$+ \sum_{i=1}^{n} (\|\mathcal{L}\Phi_i\|_{L^p(P)} \|\frac{\partial F}{\partial x_i} \circ \Phi\|_{L^{p/\alpha-1}(P)})^{p/\alpha} \quad .$$

The second assertion is just a special case of the first. Finally, the last part can be easily proved by considering $F_\varepsilon \circ \Phi$ where $F_\varepsilon(x) = 1/(x^2+\varepsilon^2)^{1/2}$. The desired result follows upon letting $\varepsilon \downarrow 0$.

Q.E.D.

We are at last ready to see how all of this machinery can be put to work on the problem which we posed at the beginning.

Suppose that $\Phi \in \mathcal{K}$ has the properties that $\langle \Phi, \Phi \rangle \in \mathcal{K}$ and that $1/\langle \Phi, \Phi \rangle \in \bigcap_{1 \le p < \infty} L^p(P)$. Given $\varphi \in C_b^2(R^1)$, we then have:

$$E^P[\varphi' \circ \Phi] = E^P[\frac{\langle \varphi \circ \Phi, \Phi \rangle}{\langle \Phi, \Phi \rangle}]$$
$$= E^P[\frac{\mathcal{L}((\varphi \circ \Phi)\cdot\Phi) - (\varphi \circ \Phi)\cdot\mathcal{L}\Phi - \Phi\cdot\mathcal{L}(\varphi \circ \Phi)}{\langle \Phi, \Phi \rangle}]$$
$$= E^P[\varphi \circ \Phi(\Phi\cdot\mathcal{L}(1/\langle \Phi, \Phi \rangle)) - \frac{\mathcal{L}\Phi}{\langle \Phi, \Phi \rangle} - \mathcal{L}(\frac{\Phi}{\langle \Phi, \Phi \rangle}))]$$
$$= -E^P[\varphi \circ \Phi(\langle \Phi, 1/\langle \Phi, \Phi \rangle \rangle + \frac{2\mathcal{L}\Phi}{\langle \Phi, \Phi \rangle})] \quad .$$

Hence

$$|E^P[\varphi' \circ \Phi]| \le C\|\varphi\|_u$$

where

$$C = \|\langle \Phi, 1/\langle \Phi, \Phi \rangle \rangle\|_{L^1(P)} + 2\|\frac{\mathcal{L}\Phi}{\langle \Phi, \Phi \rangle}\|_{L^1(P)}$$
$$\le \|\langle \Phi, \Phi \rangle\|_{L^1(P)}^{1/2} \|\langle 1/\langle \Phi, \Phi \rangle \, , \, 1/\langle \Phi, \Phi \rangle \rangle\|_{L^1(P)}^{1/2}$$
$$+ 2\|\mathcal{L}\Phi\|_{L^2(P)} \|1/\langle \Phi, \Phi \rangle\|_{L^2(P)}$$
$$\le 3|||\Phi|||_{\mathcal{K}_2} |||1/\langle \Phi, \Phi \rangle|||_{\mathcal{K}_2} \quad .$$

In other words, if $\Phi \in \mathcal{D}(\mathcal{L})$ and $1/\langle\Phi,\Phi\rangle \in \mathcal{D}(\mathcal{L})$, then

(1.14) $$\left|\int \varphi' d\mu^{\Phi}\right| \leq C\|\varphi\|_u \quad .$$

A refinement of the technique used in section (0.1) shows that (1.14) by itself already implies that μ^{Φ} is absolutely continuous and that $\frac{d\mu^{\Phi}}{dx} \in L^p(R^1)$ for $1 \leq p < \infty$.

We conclude this lecture with the reformulation of the preceding in the form which will be most useful to us in the sequel.

(<u>1.15</u>) <u>Definition</u>: Given $M \subseteq \mathcal{D}(\mathcal{L})$, define $\mathcal{S}(M) = \{\Phi, \mathcal{L}\Phi, \langle\Phi,\Psi\rangle : \Phi,\Psi \in M\}$. Now use induction on $n \geq 1$ to define $\mathcal{K}^{(n)}$ and $\mathcal{S}^{(n)}(\Phi)$ for $\Phi \in \mathcal{K}^{(n)}$ so that: i) $\mathcal{K}^{(1)} \equiv \mathcal{K}$ and $\mathcal{S}^{(1)}(\Phi) \equiv \mathcal{S}(\{\Phi\})$; and ii) if $\mathcal{K}^{(n)}$ and $\mathcal{S}^{(n)}(\Phi)$ for $\Phi \in \mathcal{K}^{(n)}$ have been defined, then $\mathcal{K}^{(n+1)} \equiv \{\Phi \in \mathcal{K}^{(n)} : \mathcal{S}^{(n)}(\Phi) \subseteq \mathcal{D}(\mathcal{L})\}$ and $\mathcal{S}^{(n+1)}(\Phi) \equiv \mathcal{S}(\mathcal{S}^{(n)}(\Phi))$ for $\Phi \in \mathcal{K}^{(n+1)}$. Finally, let $\mathcal{K}^{(\infty)} = \bigcap_1^\infty \mathcal{K}^{(n)}$.

<u>Lemma</u> (<u>1.16</u>): If $\Phi = (\Phi_1,\ldots,\Phi_n) \in (\mathcal{K}^{(\infty)})^n$ and $F \in C_\uparrow^\infty(R^n)$ (the space of $C^\infty(R^n)$ function which together with all their derivatives are slowly increasing), then $F \circ \Phi \in \mathcal{K}^{(\infty)}$. Moreover, if $\Phi \in \mathcal{K}^{(\infty)}$ and $1/\Phi \in \bigcap_{1 \leq p < \infty} L^p(P)$, then $1/\Phi \in \mathcal{K}^{(\infty)}$.

<u>Proof</u>: In both cases, the proof is a straight forward inductive argument which turns on Corollary (1.13).

Q. E. D.

<u>Theorem</u> (<u>1.17</u>): Let $\Phi = (\Phi_1,\ldots,\Phi_n) \in (\mathcal{K}^{(\infty)})^n$ and set $A = ((\langle\Phi_i,\Phi_j\rangle))_{1\leq i,j\leq n}$ and $\Delta = \det A$. For $1 \leq i \leq n$ and $\Psi \in \mathcal{K}^{(\infty)}$, define

$$\mathcal{K}_i\Psi = -\sum_{j=1}^n [A^{(ij)}(2\Psi\mathcal{L}\Phi_j + \langle\Psi,\Phi_j\rangle) + \Psi\langle\Phi_j,A^{(ij)}\rangle] \quad ,$$

where $((A^{(ij)}))$ is the cofactor matrix of A.

Then for $\varphi \in C_b^2(R^n)$:

$$(1.18) \qquad E^P[(\frac{\partial\varphi}{\partial x_i} \circ \Phi)(\Delta\Psi)] = E^P[(\varphi \circ \Phi)\mathcal{K}_i\Psi] \quad , \quad \Psi \in \mathcal{K}^{(\infty)} \quad .$$

In particular, if $1/\Delta \in \bigcap_{1 \le p < \infty} L^p(P)$, then

$$(1.19) \qquad E^P[(\frac{\partial\varphi}{\partial x_i} \circ \Phi)\Psi] = E^P[(\varphi \circ \Phi)\bar{\mathcal{K}}_i\Psi] \quad , \quad \Psi \in \mathcal{K}^{(\infty)} \quad ,$$

where

$$\bar{\mathcal{K}}_i\Psi = \mathcal{K}_i(\Psi/\Delta) \quad .$$

Finally, assume $1/\Delta \in \bigcap_{1 \le p < \infty} L^p(P)$ and for $\alpha = (\alpha_1, \ldots, \alpha_n) \in \eta^n$ define $\bar{\mathcal{K}}^\alpha = \bar{\mathcal{K}}_1^{\alpha_1} \circ \cdots \circ \bar{\mathcal{K}}_n^{\alpha_n}$. Then for $\varphi \in C_b^\infty(R^n)$:

$$(1.20) \qquad E^P[((D^\alpha\varphi) \circ \Phi)\Psi] = E^P[(\varphi \circ \Phi)\bar{\mathcal{K}}^\alpha\Psi] \quad , \quad \Psi \in \mathcal{K}^{(\infty)} \quad .$$

<u>Proof</u>: The proof of (1.18) is precisely like the argument used to arrive at (1.14) . Once (1.18) has been proved, (1.19) and (1.20) follow easily.

Q. E. D.

<u>Corollary (1.21)</u>: Let everything be as in Theorem (1.17) . If $1/\Delta \in \bigcap_{1 \le p < \infty} L^p(P)$, then the distribution μ^Φ of Φ under P is absolutely continuous with respect to Lebesgue measure on R^n and $\frac{d\mu^\Phi}{dx} \in C_b^\infty(R^n)$.

<u>Proof</u>: Once one has (1.20) , one obtains

$$|E^P[(D^\alpha\varphi) \circ \Phi]| \le C_\alpha \|\varphi\|_u \quad , \quad \varphi \in C_b^\infty(R^n) \quad ,$$

simply by taking $\Psi = 1$. Now the argument proceeds in exactly the same way as in the introduction.

Q. E. D.

Lecture #2

2. A Symmetric Diffusion Operator for Wiener Space:

Let $\Theta = \Theta_d$ be the space of continuous maps $\theta : [0,\infty) \to R^d$ such that $\theta(0) = 0$. For $t \geq 0$, define $\mathcal{B}_t = \sigma(\theta(s) : 0 \leq s \leq t)$ and set $\mathcal{B} = \sigma(\theta(s) : s \geq 0)$. We will use $\mathcal{W} = \mathcal{W}_d$ to denote the standard Wiener measure on $(\Theta, \mathcal{B})$. It is our goal in this section to describe a symmetric diffusion operator $\mathcal{L}$ for $(\Theta, \mathcal{B}, \mathcal{W})$. In order not to encumber the ideas and the notation too much, throughout the remainder of the present section we will be assuming that $d = 1$. (Nonetheless, without any substantial changes, everything that we do here can be done when $d > 1$.) It may be helpful to the reader to occasionally refer back to example (1.3) as we proceed. The analogy between that example and what we are going to do here provides a good touchstone.

In order to describe the operator $\mathcal{L}$ which we have in mind, it is necessary for us to review a few facts about Wiener's theory of "homogenious chaos." To begin with, we must define iterated stochastic integrals. To this end, let $\Delta_n = \{(t_1,\ldots,t_n) \in R^n : 0 \leq t_1 < \cdots < t_n < \infty\}$ If $f \in L^2(\Delta_n)$ and $= f(t_1,\ldots,t_n) = f_1(t_1)\cdots f_n(t_n)$ where $\{f_i\}_1^n \subseteq L^2(\Delta_1)$, then $\int_{\Delta^{(n)}} f d\theta^{(n)}$ is defined inductively so that

$$\int_{\Delta_1} f_1 d\theta^{(1)} = \int_0^\infty f_1(t_1) d\theta(t_1)$$

and

$$\int_{\Delta_k} f_1 \cdots f_k d\theta^{(k)} = \int_0^\infty f_k(t_k) d\theta(t_k) \int_{\Delta_{k-1}(t_k)} f_1 \cdots f_{k-1} d\theta^{(k-1)}$$

where $d\theta(t)$-integrals are taken in the sense of Itô and $\Delta_\ell(t) = \{(t_1,\ldots,t_\ell) : 0 \leq t_1 \leq \cdots \leq t_\ell \leq t\}$. Still restricting ourselves to f's which are products, it is easy to check that

$$(2.1) \qquad E^{\mathcal{W}}\left[\int_{\Delta_m} f d\theta^{(m)} \int_{\Delta_n} g d\theta^{(n)}\right] = \begin{cases} 0 & \text{if } m \neq n \\ \int_{\Delta_m} fg d\vec{t} & \text{if } m = n \end{cases} .$$

In particular, for fixed $n \geq 1$, (2.1) allows us to establish a unique linear isometry $f \to \int_{\Delta_n} f d\theta^{(n)}$ from $L^2(\Delta_n)$ into $L^2(\mathcal{W})$ so that $\int_{\Delta_n} f d\theta^{(n)}$ is defined as above when f is a product. Clearly (2.1) continues to hold for all $f \in L^2(\Delta_m)$ and $g \in L^2(\Delta_n)$. Thus if $Z_n \equiv \{\int_{\Delta_n} f d\theta^{(n)} : f \in L^2(\Delta_n)\}$, then $Z_m \perp Z_n$ in $L^2(\mathcal{W})$ for unequal $m,n \geq 1$. Moreover, since $E^{\mathcal{W}}[\int_{\Delta_n} f d\theta^{(n)}] = 0$, if $Z_0 \equiv R^1$, then $Z_0 \perp Z_n$ for $n \geq 1$. In other words $\{Z_n\}_0^\infty$ are mutually orthogonal closed linear subspaces of $L^2(\mathcal{W})$. The remarkable theorem which Wiener proved is that

$$(2.2) \qquad L^2(\mathcal{W}) = \bigoplus_0^\infty Z_n .$$

There are by now many ways to establish (2.2). For our purposes the most useful is Wiener's original proof as interpreted (and made comprehensible) by Itô. The idea is as follows. Let Γ on $(R^{Z^+}, \mathcal{B}_{R^{Z^+}})$ be the measure μ^{Z^+}, where μ is the standard normal measure on R^1. The space $L^2(\Gamma)$, known as Fock space, has a natural basis coming from the Hermite basis $\{H_n\}_0^\infty$ in $L^2(\mu)$ (cf. example (1.3)). Namely, if $A \equiv \{\alpha \in \eta^{Z^+} : |\alpha| < \infty\}$ (here $\eta = \{0,1,\ldots,n,\ldots\}$), define $H_\alpha : R^{Z^+} \to R^1$ so that

$$H_\alpha(x) = H_{\alpha_{k_1}}(x_{k_1}) \cdots H_{\alpha_{k_N}}(x_{k_N})$$

where $k_1 < \cdots < k_N$ are the elements of $[\alpha] \equiv \{k : \alpha_k \neq 0\}$. By an elementary argument, it is easy to check that $\{H_\alpha : \alpha \in A\}$ is an orthonormal basis in $L^2(\Gamma)$. In particular, if $\mathcal{H}_n = \overline{\text{span}\{\mathcal{H}_\alpha : |\alpha| = n\}}$, $n \geq 0$, then the $\mathcal{H}_n$'s are mutually orthogonal closed subspaces of $L^2(\Gamma)$ and $L^2(\Gamma) = \bigoplus_0^\infty \mathcal{H}_n$. Thus if we can construct an isomorphism Λ from $L^2(\Gamma)$ onto $L^2(\mathcal{W})$ such that $\Lambda(\mathcal{H}_n) \subseteq Z_n$, $n \geq 0$, then (2.2) will have been proved. To this end, let $\{f_k\}_1^\infty$ be an orthonormal basis in $L^2(\Delta_1)$. Then

$$(2.3) \qquad \theta \to \theta(\vec{f}) \equiv \left(\int_0^\infty f_1 d\theta, \ldots, \int_0^\infty f_k d\theta, \ldots\right)$$

is a measure preserving map of $(\Theta, \mathcal{B}, \mathcal{W})$ into $(R^{Z^+}, \mathcal{B}_{R^{Z^+}}, \Gamma)$. Hence the map $\Lambda_{\vec{f}}$ defined on $L^2(\Gamma)$ by

$$\Lambda_{\vec{f}}(F) = F \circ \theta(\vec{f})$$

is an isometric embedding of $L^2(\Gamma)$ into $L^2(\mathcal{W})$. Not immediately clear is the fact that $\Lambda_{\vec{f}}$ is onto. The way to see this is to first observe that $\text{range}(\Lambda_{\vec{f}}) = \text{range}(\Lambda_{\vec{g}})$ where $\{g_k\}_1^\infty$ is any other orthonormal basis in $L^2(\Delta_1)$. Indeed, if $\mathcal{U}(\vec{g},\vec{f}) = (((g_i,f_j)_{L^2(\Delta_1)}))_{1\leq i,j<\infty}$, then $x \to \mathcal{U}(\vec{g},\vec{f})x \equiv (\sum_j \mathcal{U}(\vec{g},\vec{f})_{1j}x_j, \ldots, \sum_j \mathcal{U}(\vec{g},\vec{f})_{kj}x_j, \ldots)$ determines a measure preserving transformation of $(R^{Z^+}, \mathcal{B}_{R^{Z^+}}, \Gamma)$ into itself such that $\mathcal{U}(\vec{g},\vec{f}) \circ \mathcal{U}(\vec{f},\vec{g})x = x$ (a.s., Γ). Hence $T_{(\vec{g},\vec{f})} : L^2(\Gamma) \to L^2(\Gamma)$ given by $T_{(\vec{g},\vec{f})}F = F \circ \mathcal{U}_{(\vec{g},\vec{f})}$ is a unitary map on $L^2(\Gamma)$; and clearly $\Lambda_{\vec{g}} = \Lambda_{\vec{f}} \circ T_{(\vec{g},\vec{f})}$. Having shown that $\text{range}(\Lambda_{\vec{f}})$ is independent of the choice of $\{f_k\}_1^\infty$, it is a simple matter to show that for any $n \geq 1$, $F \in C_b(R^n)$, and $0 = t_0 < t_1 < \cdots < t_n$, $F\left(\frac{1}{t_1^{1/2}}\theta(t_1), \frac{1}{(t_2 - t_1)^{1/2}}(\theta(t_2) - \theta(t_1)), \ldots, \frac{1}{(t_n - t_{n-1})^{1/2}}(\theta(t_n) - \theta(t_{n-1}))\right) \in \text{range}(\Lambda_{\vec{f}})$; simply let

$\{g_k\}_1^\infty$ be any orthonormal basis in $L^2(\Delta_1)$ such that
$g_k = \frac{1}{(t_k - t_{k-1})^{1/2}} \chi_{[t_k, t_{k-1})}$ for $1 \le k \le n$. From this it is obvious that $\Lambda_{\vec{f}}$ maps $L^2(\Gamma)$ onto $L^2(\mathfrak{W})$.

To complete the proof of (2.2) it remains to show that $\Lambda_{\vec{f}} H_\alpha \in Z_{|\alpha|}$. There are two ingredients to this step. First, one needs to know that if $f \in L^2(\Delta_1)$, $\|f\|_{L^2(\Delta_1)} = 1$, and $f^{(n)} \in L^2(\Delta_1)$ is given by $f^{(n)}(t_1,\ldots,t_n) = f(t_1)\cdots f(t_n)$, then

(2.4) $$\int_{\Delta_n} f^{(n)} d\theta^{(n)} = \frac{1}{(n!)^{1/2}} H_n\left(\int_0^\infty f(t)d\theta(t)\right) .$$

The proof of (2.4) comes from the identity

$$\sum_0^\infty \frac{\lambda^n}{(n!)^{1/2}} H_n\left(\int_0^\infty f(t)d\theta(t)\right) = \exp\left(\lambda\int_0^\infty f(t)d\theta(t) - \lambda^2/2\right)$$
$$= 1 + \sum_1^\infty \lambda^n \int_{\Delta_n} f^{(n)} d\theta^{(n)} , \quad \lambda \in R^1 ;$$

which in turn comes, on the one hand, from the generating function for $\{H_n\}_0^\infty$ and, on the other hand, Itô's theory of the "stochastic exponential" (cf. exercise 4.6.14 in [3] or McKean [2]). The second ingredient is the classical fact that if $F \subseteq Z^+$ is a finite set then for $\lambda = \{\lambda_k : k \in F\} \in R^{|F|}$ and $n \ge 0$:

(2.5) $$\|\lambda\|^n_{\ell^2(F)} (n!)^{1/2} H_n\left(\frac{1}{\|\lambda\|_{\ell^2(F)}} \sum_{k\in F} \lambda_k x_k\right)$$
$$= \sum_{\{\alpha:[\alpha]=F \text{ and } |\alpha|=n\}} \binom{n}{\alpha} (\alpha!)^{1/2} \lambda^\alpha H_\alpha(x) .$$

Since, by (2.5) ,

$$H_n\left(\frac{1}{\|\lambda\|_{\ell^2(F)}} \sum_{k\in F} \lambda_k \int_0^\infty f_k(t)d\theta(t)\right) \in Z_n$$

for all $\lambda \neq 0$, it follows from (2.5) that $H_\alpha(\theta(\vec{f})) \in Z_{|\alpha|}$

for all $\alpha \in A$. We have therefore shown that $\Lambda_{\vec{f}}\mathcal{H}_n \subseteq Z_n$, $n \geq 0$. Of course, since $\Lambda_{\vec{f}}$ is onto, this not only implies (2.2) , it also implies that

(2.6) $$\Lambda_{\vec{f}}\mathcal{H}_n = Z_n \quad , \quad n \geq 0 \quad .$$

With these preliminaries we are at last ready to introduce $\mathcal{L}$. For $n \geq 0$, let $\Pi_n : L^2(\mathcal{W}) \to Z_n$ be the orthogonal projection operator onto Z_n . Define $\mathcal{D}(\mathcal{L}) = \{\Phi \in L^2(\mathcal{W}) : \sum_1^\infty n^2\|\Pi_n\Phi\|^2 < \infty\}$ and let $\mathcal{L}$ be the self-adjoint operator with domain $\mathcal{D}(\mathcal{L})$ such that

$$\mathcal{L}\Phi = \sum_0^\infty -\frac{n}{2}\Pi_n\Phi \quad .$$

We will show that $\mathcal{L}$ is a symmetric diffusion operator on $(\Theta, \mathcal{B}, \mathcal{W})$. We already know that $\mathcal{L}$ satisfies i) and ii) of definition (1.1) .

To prove that $\mathcal{L}$ satisfies iii) , iv) , and v) , we must first introduce the space $\mathcal{D}$. To this end, let $\{f_k\}_1^\infty$ be an orthonormal basis in $L^2(\Delta_1)$ and take $\mathcal{D} = \text{span}\{\mathcal{H}_\alpha(\theta(\vec{f})) : \alpha \in A\}$. By (2.4) , $\mathcal{D} \subseteq \text{span}(\bigcup_0^\infty Z_n)$ and so $\mathcal{D} \subseteq \mathcal{D}(\mathcal{L})$. In fact, $\mathcal{L}\mathcal{D} \subseteq \mathcal{D}$. Moreover, again by (2.4) , $(\Phi, \mathcal{L}\Phi) \in \overline{\text{graph}(\mathcal{L}|_{\mathcal{D}})}$ for all $\Phi \in \bigcup_0^\infty Z_n$ and so graph $(\mathcal{L}|_{\mathcal{D}})$ is dense in graph$(\mathcal{L})$. Next, because Γ is the distribution of $\theta(\vec{f})$ under $\mathcal{W}$, $H_\alpha(\theta(\vec{f})) \in \bigcap_{1 \leq p < \infty} L^p(\mathcal{W})$ for each $\alpha \in A$, and so $\mathcal{D} \subseteq \bigcap_{1 \leq p < \infty} L^p(\mathcal{W})$. Finally, since in R^1 we know that $\text{span}\{H_m(x) : 0 \leq m \leq n\} = \text{span}\{x^m : 0 \leq m \leq n\}$ for all $n \geq 0$, it is clear that $\mathcal{D}$ forms an algebra. Combining these remarks, we conclude that $\mathcal{D}$ satisfies iii) .

To check iv) and v) we will first need the following lemma.

<u>Lemma (2.7)</u>: If $n \geq 1$ and $F \in C_\uparrow^2(R^n)$, then $F(\theta(f_1), \ldots, \theta(f_n))$

$\in \mathcal{D}(\mathcal{L})$ and:

(2.8) $$\mathcal{L}(F(\theta(f_1),\dots,\theta(f_n))$$

$$= 1/2\Big(\sum_{i=1}^{n}\frac{\partial^2 F}{\partial x_i^2}(\theta(f_1),\dots,\theta(f_n)) - \sum_{i=1}^{n}\theta(f_i)\frac{\partial F}{\partial x_i}(\theta(f_1),\dots,\theta(f_n))\Big) .$$

Proof: Because $\{\theta(f_i)\}_1^n$ are independent $N(0,1)$ random variables under $\mathbb{W}$, the integrability requirements on $F(\theta(f_1),\dots,\theta(f_n))$ and the right hand side of (2.8) pose no problems.

Next set $A_n = \{\alpha \in A : \alpha_k = 0 \text{ for } k \geq n+1\}$. Then

$$1/2\Big(\sum_1^n \frac{\partial^2 H_\alpha}{\partial x_i^2}(x) - \sum_1^n x_i \frac{\partial H_\alpha}{\partial x_i}(x)\Big) = -\frac{|\alpha|}{2} H_\alpha(x)$$

for all $\alpha \in A_n$. Since $H_\alpha(\theta(\vec{f})) \in Z_{|\alpha|}$, this proves (2.8) for $F \in \operatorname{span}\{H_\alpha : \alpha \in A_n\}$. To complete the proof, we proceed, as in example (1.3) , to first check (2.8) for $F \in \mathcal{S}(R^n)$ and then for $F \in C_\uparrow^2(R^n)$.

Q. E. D.

Starting from (2.8) , it is now clear that

$$\langle H_\alpha(\theta(\vec{f})), H_\beta(\theta(\vec{f}))\rangle$$

$$= \sum_{i \in [\alpha] \cup [\beta]} \frac{\partial H_\alpha}{\partial x_i}(\theta(\vec{f}))\frac{\partial H_\beta}{\partial x_i}(\theta(\vec{f}))$$

and so $\langle \Phi, \Phi\rangle \geq 0$ for all $\Phi \in \mathcal{D}$. Thus iv) is satisfied. Also, if $\alpha^{(1)},\dots,\alpha^{(n)} \in A$ and $F \in C_b^2(R^n)$, then by Lemma (2.7) , $F(H_{\alpha^{(1)}}(\theta(\vec{f})),\dots,H_{\alpha^{(n)}}(\theta(\vec{f}))) \in \mathcal{D}(\mathcal{L})$ and by (2.8) :

$$\mathcal{L}(F(H_{\alpha^{(1)}}(\theta(\vec{f})),\dots,H_{\alpha^{(n)}}(\theta(\vec{f})))$$

$$= 1/2 \sum_{\ell,\ell'=1}^{n} \langle H_{\alpha^{(\ell)}}(\theta(\vec{f})), H_{\alpha^{(\ell')}}(\theta(\vec{f}))\rangle \frac{\partial^2 F}{\partial x_\ell \partial x_{\ell'}}(H_{\alpha^{(1)}}(\theta(\vec{f})),\dots,H_{\alpha^{(n)}}(\theta(\vec{f})))$$

$$+ \sum_{\ell=1}^{n} \mathcal{L}(H_{\alpha^{(\ell)}}(\theta(\vec{f})))\frac{\partial F}{\partial x_\ell}(H_{\alpha^{(1)}}(\theta(\vec{f})),\dots,H_{\alpha^{(n)}}(\theta(\vec{f}))) .$$

From this it is obvious that v) holds.

We must now show that $\mathcal{L}$ admits an extention $\mathcal{L}_1$ of the sort required by vi) . First observe that since $\theta \to \theta(\vec{f})$ is measure preserving from $(\Theta,\mathcal{B},\mathbb{W})$ to $(R^{Z^+},\mathcal{B}_{R^{Z^+}},\Gamma)$, we can define an isometry $\Lambda^{(p)}_{\vec{f}}: L^p(\Gamma) \to L^p(\mathbb{W})$ for each $1 \leq p < \infty$ so that $\Lambda^{(p)}_{\vec{f}} F = F \circ \theta(\vec{f})$. Moreover, since $\Lambda^{(2)}_{\vec{f}} = \Lambda_{\vec{f}}$ is onto, $\Lambda^{(p)}_{\vec{f}}$ is onto for each $1 \leq p < \infty$. Next, let $\mu_\tau(\xi,d\eta)$ be the measure on R^1 given by $\mu_\tau(\xi,d\eta) = \gamma(1-e^{-\tau},\eta-\xi e^{-\tau/2})d\eta$ and define $P(\tau,x,\cdot)$ on $(R^{Z^+},\mathcal{B}_{R^{Z^+}})$ by

$$P(\tau,x,\cdot) = \prod_{k\in Z^+} \mu_\tau(x_k,\cdot) \quad .$$

Then $P(\tau,x,\cdot)$ is a Feller continuous, conservative Markov transition function. Moreover,

$$\text{(2.9)} \qquad \begin{aligned} &\int F(x)\Gamma(dx)\int G(y)P(\tau,x,dy) \\ &= \int G(x)\Gamma(dx)\int F(y)P(\tau,x,dy) \end{aligned}$$

for all $F,G \in C_b(R^{Z^+})$. Indeed, one first checks (2.9) for F which depend only on a finite number of coordinates (this is done in the same manner as we used in (1.3)) and then one passes to limits. Just as in (1.3) , one concludes from (2.9) that if $\{S_\tau : \tau > 0\}$ is the Feller semi-group on $C_b(R^{Z^+})$ given by

$$(S_\tau F)(x) = \int F(y)P(\tau,x,dy) \quad ,$$

then for each $1 \leq p < \infty$ there is a unique strongly continuous contraction semi-group $\{S^{(p)}_\tau : \tau > 0\}$ on $L^p(\Gamma)$ such that $S^{(p)}_\tau F = S_\tau F$ for $F \in C_b(R^{Z^+})$. We now define $T^{(p)}_\tau = \Lambda^{(p)}_{\vec{f}} \circ S^{(p)}_\tau \circ (\Lambda^{(p)}_{\vec{f}})^{-1}$ Then, for each $1 \leq p < \infty$, $\{T^{(p)}_\tau : \tau > 0\}$ is a strongly continuous contraction semi-group on $L^p(\mathcal{W})$; and for $1 \le q \le p < \infty$, $T_\tau^{(q)}$ on $L^p(\mathcal{W})$ coincides with $T_\tau^{(p)}$. Thus, if we identify the $L^2(\mathcal{W})$-generator of $\{T_\tau^{(2)} : \tau > 0\}$ as being $\mathcal{L}$, then, just as in (1.3) , the $L^1(\mathcal{W})$-generator $\mathcal{L}_1$ of $\{T_\tau^{(1)} : \tau > 0\}$ will be the extention of $\mathcal{L}$ required by vi) . But, because of Lemma (2.7) , this identification boils to checking that

$$\frac{S_\tau H_\alpha - H_\alpha}{\tau} \xrightarrow{L^2(\Gamma)} 1/2 \sum_{k \in [\alpha]} \left(\frac{\partial^2 H_\alpha}{\partial x_k^2} - x_k \frac{\partial H_\alpha}{\partial x_k}\right)$$

for all $\alpha \in A$; a fact which is easy to derive.

We have now shown that $\mathcal{L}$ is a symmetric diffusion operator on $(\Theta, \mathcal{B}, \mathcal{W})$. Of course, we did this under the assumption that $d = 1$; but as we said at the beginning of this present section, there are no serious obstacles preventing us from doing the same thing for $d > 1$.

2.10 Example: Before going on, let us indulge in the following ridiculous exercise: to show that the distribution of $\theta(t)$ under $\mathcal{W}$ admits a C_b^∞-density for each $t > 0$. To this end, note that since $\theta(t) \in Z_1$, $\mathcal{L}\theta(t) = -1/2\ \theta(t)$. Also, by Itô's formula:

$$\theta^2(t) = 2\int_0^t \theta(s)d\theta(s) + t \quad .$$

But $\int_0^t \theta(s)d\theta(s) \in Z_2$ and $t \in Z_0$, and so

$$\mathcal{L}(\theta^2(t)) = -2\int_0^t \theta(s)d\theta(s) = t - \theta^2(t) \quad .$$

In particular, $\langle\theta(t),\theta(t)\rangle = t$. Hence, $\theta(t) \in \mathcal{K}^{(\infty)}$. Now suppose that $F \in C_\uparrow^\infty(R^1)$. Then for $\varphi \in C_b^\infty(R^1)$:

$$E^{\mathcal{W}}[\varphi'(\theta(t))F(\theta(t))] = \frac{1}{t}E^{\mathcal{W}}[\langle\varphi(\theta(t)),\theta(t)\rangle F(\theta(t))]$$

$$= \frac{1}{t}E^{\mathbb{W}}[\varphi(\theta(t))\theta(t)\mathcal{L}(F(\theta(t)))] - \frac{1}{t}E^{\mathbb{W}}[\varphi(\theta(t))\mathcal{L}(\theta(t)F(\theta(t))]$$

$$+ \frac{1}{2t}E^{\mathbb{W}}[\varphi(\theta(t))(\theta(t)F(\theta(t)))]$$

$$= \frac{1}{t}E^{\mathbb{W}}[\varphi(\theta(t))(\theta(t)F(\theta(t)))] - \frac{1}{t}E^{\mathbb{W}}[\varphi(\theta(t))\langle\theta(t),F(\theta(t))\rangle]$$

$$= \frac{1}{t}E^{\mathbb{W}}[\varphi(\theta(t))(\theta(t)F(\theta(t)))] - E^{\mathbb{W}}[\varphi(\theta(t))F'(\theta(t))] \quad .$$

That is:

$$E^{\mathbb{W}}[\varphi'(\theta(t))F(\theta(t))] = E^{\mathbb{W}}[\varphi(\theta(t))(M_t F)(\theta(t))]$$

where $M_t F(x) = \frac{x}{t}F(x) - F'(x)$, and so

$$E^{\mathbb{W}}[\varphi^{(n)}(\theta(t))] = E^{\mathbb{W}}[\varphi(\theta(t))(M_t^n 1)(\theta(t))] \quad .$$

Since it is clear that $M_t^n 1$ is an $n^{\underline{th}}$ order polynomial for each $t > 0$, we conclude that for all $n \geq 0$:

$$|E^{\mathbb{W}}[\varphi^{(n)}(\theta(t))]| \leq C_n(t)\|\varphi\|_u \quad , \quad \varphi \in C_b^\infty(R^1) \quad ,$$

and so the distribution of $\theta(t)$ under $\mathbb{W}$ has a $C_b^\infty(R^1)$ - density. Without too much trouble, it is even possible to check that $\overline{\lim_{t\downarrow 0}}\, t^{n/2}C_n(t) < \infty$ and thereby get estimates on the distribution of $\theta(t)$ as $t \downarrow 0$.

Of course it is fair to ask whether we could have possibly developed the Malliavin machinery without knowing ahead of time that $\theta(t)$ has as nice a distribution as there ever was, but the preceding exercise gives the flavor of the applications which we have in mind.

3. The Malliavin Calculus and Stochastic Integral Equations:

Until the end of this section we will again restrict ourselves to

$d = 1$. Once again this is a matter of convenience and is not done because $d > 1$ presents any essential difficulties not encountered when $d = 1$.

To begin with, suppose $\alpha, \beta \in \mathcal{D}(\mathcal{L})$ are $\mathcal{B}_t$ - measurable and let $\Phi = \alpha\cdot(\theta(t+h) - \theta(t)) + \beta h$ where $h > 0$. We want to show that $\Phi \in \mathcal{D}(\mathcal{L})$ and then compute $\mathcal{L}\Phi$. Clearly βh presents no problem and $\mathcal{L}(\beta h) = (\mathcal{L}\beta)h$. As for $\alpha(\theta(t+h) - \theta(t))$, note that if $\alpha \in Z_n$, then

$$\alpha = \int_{\Delta_n} f d\theta^{(n)}$$

for some $f \in L^2(\Delta_n(t))$ (recall that $\Delta_n(t) = \{(t_1,\dots,t_n) \in R^n : 0 \le t_1 < \cdots < t_n \le t\})$. Thus if $\tilde{f}(t_1,\dots,t_n,t_{n+1}) = f(t_1,\dots,t_n)\chi_{[t,t+h)}(t_{n+1})$, then

$$\alpha\cdot(\theta(t+h) - \theta(t)) = \int_{\Delta_n} \tilde{f} d\theta^{(n+1)} \quad .$$

Thus, if $\alpha \in Z_n$ is $\mathcal{B}_t$ - measurable, then $\alpha\cdot(\theta(t+h) - \theta(t)) \in Z_{n+1}$ and so

$$\begin{aligned}\mathcal{L}(\alpha\cdot(\theta(t+h) - \theta(t))) &= -\frac{n+1}{2}\,\alpha\cdot(\theta(t+h) - \theta(t)) \\ &= (\mathcal{L}\alpha - 1/2\,\alpha)\cdot(\theta(t+h) - \theta(t)) \quad .\end{aligned}$$

From here it is an easy matter to conclude in general that for $\mathcal{B}_t$ - measurable $\alpha, \beta \in \mathcal{D}(\mathcal{L})$ and $h > 0$, $\alpha(\theta(t+h) - \theta(t)) + \beta h \in \mathcal{D}(\mathcal{L})$ and

(3.1) $$\mathcal{L}(\alpha\cdot(\theta(t+h) - \theta(t)) + \beta h) = (\mathcal{L}\alpha - 1/2\,\alpha)\cdot(\theta(t+h) - \theta(t)) + (\mathcal{L}\beta)h \quad .$$

Starting from (3.1) , one can now show that if $\alpha : [0,\infty) \times \Theta \to R^1$ and $\beta : [0,\infty) \times \Theta \to R^1$ are progressively measurable functions such that $\alpha(t),\beta(t) \in \mathcal{D}(\mathcal{L})$ for each $t \geq 0$ and $E^{\mathbb{W}}[\int_0^T(|||\alpha(t)|||_{\mathcal{K}_2}^2 + |||\beta(t)|||_{\mathcal{K}_2}^2)dt] < \infty$, for each $T > 0$, then $\xi(t) = \int_0^t \alpha(s)d\theta(s) + \int_0^t \beta(s)ds \in \mathcal{D}(\mathcal{L})$, $t \geq 0$, and

$$\mathcal{L}\xi(t) = \int_0^t (\mathcal{L}\alpha - 1/2\,\alpha)(s)d\theta(s) + \int_0^t \mathcal{L}\beta(s)ds \quad . \tag{3.2}$$

There are a few technicalities that have to be overcome in going from (3.1) to (3.2), but they are of the sort which are well-understood by afficianados of stochastic integration theory.

Next let $\alpha,\beta : [0,\infty) \times \Theta \to R^1$ be progressively measurable functions such that $\alpha(t),\beta(t) \in \mathcal{K}_2$, $t > 0$, and $E^{\mathbb{W}}[\int_0^T(|||\alpha(t)|||_{\mathcal{K}_4}^4 + |||\beta(t)|||_{\mathcal{K}_4}^4)dt] < \infty$, $T > 0$. Set $\xi(t) = \int_0^t \alpha(s)d\theta(s) + \int_0^t \beta(s)ds$. If we know that $\xi(t) \in \mathcal{K}_2$, $t > 0$, and that $E^{\mathbb{W}}[\int_0^T |||\xi(t)|||_{\mathcal{K}_4}^4 dt] < \infty$, $T > 0$, then we could conclude from:

$$\xi^2(t) = 2\int_0^t \alpha(s)\xi(s)d\theta(s) + \int_0^t (2\beta(s)\xi(s) + \alpha^2(s))ds$$

that

$$\mathcal{L}\xi^2(t) = \int_0^t (2\mathcal{L}(\alpha(s)\xi(s)) - \alpha(s)\xi(s))d\theta(s) + \int_0^t (2\mathcal{L}(\beta(s)\xi(s)) + \mathcal{L}(\alpha^2(s)))ds \quad .$$

Since

$$\xi(t)\mathcal{L}(\xi(t)) = \int_0^t (\alpha(s)\mathcal{L}(\xi(s)) + \mathcal{L}(\alpha(s))\xi(s) - 1/2\,\alpha(s)\xi(s))d\theta(s) + \int_0^t (\beta(s)\mathcal{L}(\xi(s)) + \mathcal{L}(\beta(s))\xi(s) + \alpha(s)\mathcal{L}(\alpha(s)) - 1/2\,\alpha^2(s))ds \quad ,$$

it follows that

$$\langle\xi(t),\xi(t)\rangle = \int_0^t 2\langle\alpha(s),\xi(s)\rangle d\theta(s) + \int_0^t (2\langle\beta(s),\xi(s)\rangle + \langle\alpha(s),\alpha(s)\rangle + \alpha^2(s))ds \quad . \tag{3.3}$$

Of course we arrived at (3.3) by assuming things about $\xi(\cdot)$ which we have not yet proved. However, the assumptions that we made are trivial to verify if $\alpha(\cdot)$ and $\beta(\cdot)$ are simple (i.e. $\alpha(t) = \alpha([nt]/n)$ and $\beta(t) = \beta([nt]/n)$, $t \geq 0$, for some $n \geq 1$). Hence (3.3) is proved for simple $\alpha(\cdot)$ and $\beta(\cdot)$. But from (3.3) for simple $\alpha(\cdot)$ and $\beta(\cdot)$ plus standard approximation techniques used in stochastic integration, one can justify (3.3) for general $\alpha(\cdot)$ and $\beta(\cdot)$ which satisfy our basic assumptions.

Now let $\sigma : R^1 \to R^1$ and $b : R^1 \to R^1$ be smooth functions with bounded first derivatives and define $x(\cdot)$ by

$$(3.4) \qquad x(t) = \int_0^t \sigma(x(s))d\theta(s) + \int_0^t b(x(s))ds \quad , \quad t \geq 0 \quad .$$

If we know that $x(t) \in \mathcal{H}_2$, $t \geq 0$, and $E^{\mathcal{W}}[\int_0^T \|x(t)\|^4_{\mathcal{H}_4} dt] < \infty$, $T > 0$, then by (3.3) :

$$(3.5) \qquad \langle x(t),x(t)\rangle = \int_0^t 2\sigma'(x(s))\langle x(s),x(s)\rangle d\theta(s)$$

$$+ \int_0^t [(2b'(x(s)) + \sigma'^2(x(s)))\langle x(s),x(s)\rangle + \sigma^2(x(s))]ds \quad .$$

In order to give a rigorous derivation of (3.5) , one can use the same Picard iteration scheme that Itô used to solve (3.4) . At each stage one has to check that the iterate at that state has the required smoothness. It is then quite easy to pass to the limit and thereby justify at (3.5) . At the same time, one can derive:

$$(3.6) \; \mathcal{L}x(t) = \int_0^t [1/2\, \sigma''(x(s))\langle x(s),x(s)\rangle + \sigma'(x(s))\mathcal{L}x(s) - 1/2\, \sigma(x(s))]d\theta(s)$$

$$+ \int_0^t [1/2\, b''(x(s))\langle x(s),x(s)\rangle + b'(x(s))\mathcal{L}x(s)]ds \quad .$$

Combining (3.4) , (3.5) , and (3.6) , we see that

$\mathcal{J}(\{x(t)\}) = \{x(t), \langle x(t),x(t)\rangle, \mathcal{L}x(t)\}$ again satisfies a system of stochastic differential equations to which we can apply the same procedure. In this way, one can use induction to show that the Malliavin operations may be applied aribtrarily often to $x(t)$. In fact, by being a little careful, one can prove the following theorem.

<u>Theorem</u> (<u>3.7</u>): Let $\sigma : R^N \to R^N \otimes R^d$ and $b : R^N \to R^N$ be C^∞-functions such that $\frac{\partial \sigma}{\partial x_i}$ and $\frac{\partial b}{\partial x_i}$ are C_b^∞-functions for all $1 \leq i \leq N$. Given a $\mathcal{B}_0$-measurable $\xi_0 \in (K^{(\infty)})^N$, let $x(\cdot)$ be the progressively measurable function satisfying:

$$(3.8) \qquad x(t) = \xi_0 + \int_0^t \sigma(x(s))d\theta(s) + \int_0^t b(x(s))ds \quad .$$

Then $x(t) \in (K^{(\infty)})^N$ for all $t \geq 0$. Moreover, if $A(t) \equiv ((\langle x_i(t), x_j(t)\rangle))_{1\leq i,j\leq N}$, then

$$(3.9) \qquad A(t) = A_0 + \sum_{k=1}^d \int_0^t [S_k(x(s))A(s) + A(s)S_k(x(s))^*]d\theta_k(s)$$

$$+ \int_0^t [B(x(s))A(s) + A(s)B(x(s))^* + \sum_{k=1}^d S_k(x(s))A(s)S_k(x(s))^* + a(x(s))]ds$$

where

$$A_0 \equiv ((\langle \xi_i, \xi_j\rangle))_{1\leq i,j\leq N} \quad ;$$

$$S_k(x) = ((\frac{\partial \sigma_{ik}}{\partial x_j}(x)))_{1\leq i,j\leq N} \quad \text{for } 1 \leq k \leq d \quad ;$$

$$B(x) = ((\frac{\partial b_i}{\partial x_j}(x)))_{1\leq i,j\leq N} \quad ;$$

and

$$a(x) = \sigma\sigma^*(x) \quad .$$

Lecture #3

4. Some Preliminary Applications:

In view of Theorem (3.7) and Corollary (1.21), we know that the solution $x(t)$ to (3.8) admits a $C_b^\infty(R^N)$ - density whenever we can get estimates of the form:

$$E^{\mathbb{W}}[|1/\Delta(t)|^p] < \infty \quad , \quad 1 \leq p < \infty \tag{4.1}$$

where $\Delta(t) = \det(A(t))$ and $A(t)$ is given by (3.9). Obviously, estimates like (4.1) only can come from an analysis of equation (3.9).

To understand (3.9), observe that $A(\cdot)$ enters this equation in a linear fashion. Hence one suspects that the time-honored method of variation of parameters should be tried. After a little thought, one sees that:

$$A(t) = X(t)A_0X(t)^* + \int_0^t X(s,t)a(x(s))X(s,t)^* ds \quad , \tag{4.2}$$

where $X(t) = X(0,t)$ and $X(s,t)$, $0 \leq s \leq t$, is the solution to:

$$X(s,t) = I + \sum_{k=1}^{d} \int_s^t S_k(x(u))X(s,u)d\theta_k(u) + \int_s^t B(x(u))X(s,u)du \quad , \quad t \geq s \quad . \tag{4.3}$$

For those who are not familiar with the variation of parameter technique in this context, the best way to check (4.3) is the first note that $A(\cdot)$ is uniquely determined (path-wise) by (3.9) and then verify that the right hand side of (4.3) satisfies (3.9).

As soon as one has (4.3), it is an easy matter to recapture the classical result that $x(t)$ has a smooth density if $a(\cdot)$ is positive definite. Indeed, since $\langle\cdot,\cdot\rangle$ is a non-negative bilinear form,

$A_0 \geq 0$ and so

$$A(t) \geq \int_0^t X(s,t)a(x(s))X^*(s,t)ds \quad .$$

Thus, if $a(\cdot) \geq \varepsilon I$, where $\varepsilon > 0$, then

$$A(t) \geq \varepsilon\int_0^t X(s,t)X^*(s,t)ds \quad .$$

Hence our problem comes down to estimating

$$(4.4) \qquad Y(s,t) \equiv X(s,t)X(s,t)^* \quad , \quad 0 \leq s \leq t \quad ,$$

from below. Perhaps the easiest way to obtain such estimates is to derive an equation for $X(s,\cdot)^{-1}$, namely:

$$(4.5) \qquad X(s,t)^{-1} = I - \sum_{k=1}^{d}\int_s^t X(s,u)^{-1}S_k(x(u))d\theta_k(u) + \int_s^t X(s,u)^{-1}[-B(x(u)) + \sum_{k=1}^{d} S_k(x(u))^2]du \quad .$$

That $X(s,\cdot)^{-1}$ satisfies (4.5) can be seen by defining $X(s,\cdot)^{-1}$ by (4.5) and checking that $X(s,t)X(s,t)^{-1} \equiv I$. The argument involves elementary Itô calculus plus the pathwise uniqueness of solutions to linear stochastic integral equations. Given (4.5), one finds that:

$$(4.6) \qquad Y(s,t)^{-1} = I - \sum_{k=1}^{d}\int_s^t \{Y(s,u)^{-1},S_k(x(u))^*\}d\theta_k(u)$$
$$+ \int_s^t [\{Y(s,u)^{-1},-B(x(u))^* + \sum_{k=1}^{d}(S_k(x(u))^2)^*\} + \sum_{k=1}^{d} S_k(x(u))^*Y(s,u)^{-1}S_k(x(u))]du$$

where $\{C_1,C_2\} \equiv C_1C_2 + C_2^*C_1^*$ for $C_1,C_2 \in R^N \otimes R^N$. In particular,

$$(4.7) \qquad Tr(Y(s,t)^{-1}) = N - 2\sum_{k=1}^{d}\int_s^t \gamma_k(s,u)Tr(Y(s,u)^{-1})d\theta_k(u) + \int_s^t \beta(s,u)Tr(Y(s,u)^{-1})du \quad ,$$

where

$$\gamma_k(s,u) = \mathrm{Tr}(Y(s,u)^{-1}S_k(x(u))^*)/\mathrm{Tr}(Y(s,u)^{-1})$$

and

$$\beta(s,u) = \mathrm{Tr}(Y(s,u)^{-1}[-2B(x(u))^* + \sum_{k=1}^{d}(2(S_k(x(u))^2)^* + S_k(x(u))^*S_k(x(u)))])/\mathrm{Tr}(Y(s,u)^{-1})$$

Since $Y(s,u)^{-1}$ is non-negative definite, $|\mathrm{Tr}(Y(s,u)^{-1}C)| \le \|C\|_{op}\mathrm{Tr}(Y(s,u)^{-1})$ for any $C \in R^N \otimes R^N$, and therefore the γ_k's and the β are bounded. Moreover, (4.7) is equivalent to:

$$\mathrm{Tr}(Y(s,t)^{-1}) = N\exp[-\sum_{k=1}^{d}2\int_s^t \gamma_k(s,u)d\theta_k(u) + \int_s^t(\beta(s,u) + 2\sum_{k=1}^{d}\gamma_k^2(s,u))du] \quad .$$

Hence, by standard estimates, for $1 \le p < \infty$:

$$E^{\mathbb{W}}[(\mathrm{Tr}(Y(s,t)^{-1}))^p] \le K_p N^p \exp(B_p(t-s)) \quad . \tag{4.8}$$

Finally, since

$$A(t) \ge \varepsilon\int_0^t Y(s,t)ds \ge \varepsilon(\int_0^t \frac{1}{\mathrm{Tr}(Y(s,t)^{-1})}ds)I \quad ,$$

we now have:

$$\begin{aligned}
E^{\mathbb{W}}[(1/\Delta(t))^p]^{1/p} &\le \frac{1}{\varepsilon^N}E^{\mathbb{W}}[(\int_0^t \frac{1}{\mathrm{Tr}(Y(s,t)^{-1})}ds)^{-Np}]^{1/p} \\
&\le \frac{1}{(\varepsilon t)^N}E^{\mathbb{W}}[\frac{1}{t}\int_0^t(\mathrm{Tr}(Y(s,t)^{-1}))^{Np}ds]^{1/p} \\
&\le (\frac{N}{\varepsilon t})^N K_{Np}^{1/p}(\frac{1}{t}\int_0^t \exp(B_{N_p}(t-s))ds)^{1/p} \\
&= (\frac{N}{\varepsilon t})^N(\frac{K_{N_p}}{B_{N_p}})^{1/p}(\frac{e^{B_{Np}t}-1}{t})^{1/p} \quad .
\end{aligned}$$

We have therefore proved the following theorem.

<u>Theorem</u> (<u>4.9</u>): Let $x(t,x)$ denote the solution to (3.8) when $\xi_0 = x$. If $a(\cdot) \equiv \sigma(\cdot)\sigma^*(\cdot) \ge \varepsilon I$ for some $\varepsilon > 0$, then for each

$t > 0$ and $x \in R^N$ the distribution $P(t,x,\cdot)$ of $x(t,x)$ under $\mathbb{W}$ admits a $C_b^\infty(R^N)$-density $p(t,x,\cdot)$.

For some time, Theorem (4.9) was the cornerstone of this subject. This result was sharpened in various ways over the years, but nothing essentially new was proved until L. Hörmander wrote his ground-breaking paper [1]. To state Hörmander's result in the form most relevant to us, define

$$(4.10) \qquad V^{(k)} = \sum_{i=1}^{N} \sigma_{ik}(x)\frac{\partial}{\partial x_i} \quad , \quad 1 \le k \le d \quad .$$

The "forward generator" L^* associated with (3.8) is given by:

$$(4.11) \quad L^*f(x) = 1/2 \sum_{i,j=1}^{N} \frac{\partial^2}{\partial x_i \partial x_j}((\sigma\sigma^*)_{ij}(x)f(x)) - \sum_{i=1}^{N}\frac{\partial}{\partial x_i}(b_i(x)f(x)) \quad ,$$

and it is an elementary exercise to show that one can express the action of L^* as:

$$(4.12) \qquad L^*f = 1/2 \sum_{k=1}^{d} (V^{(k)})^2 f + Uf + cf \quad ,$$

where the coefficients $u_i(\cdot)$ of $U = \sum_{i=1}^{N} u_i(x)\frac{\partial}{\partial x_i}$ and the function $c(\cdot)$ are functionals of the $\sigma_{ik}(\cdot)$, $b_i(\cdot)$, and their derivatives. What Hörmander's theorem says is that the distribution of $x(t)$ under $\mathbb{W}$ will have a smooth density for $t > 0$ so long as Lie$\{V^{(1)},\ldots, V^{(d)},[V^{(1)}U],\ldots,[V^{(d)},U]\}$ has dimension N at each point $x \in R^N$. (Here $[X,Y]$ denotes the "Lie bracket" or commutator of the vector fields X and Y .) Of course Hörmander's theorem covers Theorem (4.9) , since $\sigma(\cdot)\sigma^*(\cdot) \ge \varepsilon I$ implies that $\dim\{V^{(1)},\ldots,V^{(d)}\} = N$ at each point. What is remarkable is how many situations not included under Theorem (4.9) it also covers. Before we get into examining some examples which show how Malliavin's technique applies to such situations, I want to mention that Malliavin himeslf outlined how his method can be used to prove Hörmander's result for the case in which $\mathrm{Lie}\{V^{(1)},\ldots,V^{(d)}\}$ has dimension N at each point. In fact, he was able to show that Hörmander's conclusions continue to hold so long as the set of x at which $\dim(\mathrm{Lie}\{V^{(1)},\ldots,V^{(d)}\}) < N$ is sufficiently "thin." Recently S. Watenabe has provided the details necessary to carry out Malliavin's program, and an excellent exposition of all this can be found in the forthcoming book on diffusions by Ikeda and Watenabe.

We will devote the rest of this lecture to some examples which are intended to indicate the sort of directions in which one can go with Malliavin's method.

Example (4.13): Let $N = 2$ and $d = 1$ and suppose that

$$\sigma(x) = \begin{pmatrix} \alpha(x_1) \\ 1 \end{pmatrix} \quad \text{and} \quad b(x) = \begin{pmatrix} \beta(x_1) \\ 0 \end{pmatrix}, \quad x = \begin{pmatrix} x_1 \\ x_2 \end{pmatrix} \in R^2$$

where $\alpha,\beta \in C_b^\infty(R^1)$ and α is uniformly positive. If $x(t)$ solves (3.8) with this choice of $\sigma(\cdot)$ and $b(\cdot)$ and with $\xi_0 = \begin{pmatrix} x_1 \\ 0 \end{pmatrix}$, then $x(t) = \begin{pmatrix} \xi(t) \\ \theta(t) \end{pmatrix}$, where

$$\xi(t) = x_1 + \int_0^t \alpha(\xi(s))d\theta(s) + \int_0^t \beta(\xi(s))ds \quad .$$

In particular, $\xi(\cdot)$ is a progressively measurable functional of $\theta(\cdot)$, and so the only chance for the distribution of $x(t)$ to have a density with respect to two dimensional Lebesgue measure is that $\xi(t)$ truly depend on $\theta(s)$ for $s < t$. It is not at all clear what conditions guarantee such dependence. We now investigate this problem using the Malliavin calculus.

Note that:

$$S(x) = \begin{pmatrix} \alpha'(x_1) & 0 \\ 0 & 0 \end{pmatrix} \quad \text{and} \quad B(x) = \begin{pmatrix} \beta'(x_1) & 0 \\ 0 & 0 \end{pmatrix} .$$

Thus if

$$\gamma(t) \equiv \exp[\int_0^t \alpha'(\xi(s))d\theta(s) + \int_0^t [b'(\xi(s)) - 1/2\, \alpha'(\xi(s))^2]ds] \quad ,$$

then

$$X(s,t) = \begin{pmatrix} \gamma(t)/\gamma(s) & 0 \\ 0 & 1 \end{pmatrix} .$$

We therefore have that

$$\begin{pmatrix} 1/\gamma(t) & 0 \\ 0 & 1 \end{pmatrix} A(t) \begin{pmatrix} 1/\gamma(t) & 0 \\ 0 & 1 \end{pmatrix}$$

$$= \int_0^t \begin{pmatrix} \alpha^2(\xi(s))/\gamma^2(s) & \alpha(\xi(s))/\gamma(s) \\ \alpha(\xi(s))/\gamma(s) & 1 \end{pmatrix} ds$$

and so

(4.14) $$\Delta(t) = \gamma^2(t)t\ \mathrm{var}_{[0,t]}(\alpha(\xi(\cdot))/\gamma(\cdot)) \quad ,$$

where $\mathrm{var}_{[a,b]}(f(\cdot)) = \frac{1}{b-a}\int_a^b |f(t) - \frac{1}{b-a}\int_a^b f(s)ds|^2 dt$ for $f \in L^2([a,b])$. By an elementary application of the mean value theorem, one can easily show that if $f \in C^1([a,b])$ then

(4.15) $$\mathrm{var}_{[a,b]}(f(\cdot)) \geq \frac{(b-a)^2}{12} \min_{[a,b]} |f'(\cdot)|$$

where equality holds if and only if $f'(\cdot)$ is constant on $[a,b]$. It would therefore be sensible for us to look at $d(\alpha(\xi(t))/\gamma(t))$. Using Itô's formula, one finds that

$$d(\alpha(\xi(t))/\gamma(t)) = \frac{1}{\gamma(t)}(-\alpha\beta' + \alpha'\beta + 1/2\, \alpha''\alpha^2)(\xi(t))dt \quad .$$

That is, $\alpha(\xi(\cdot))/\gamma(\cdot) \in C^1([0,\infty))$ and

$$(\alpha\circ\xi/\gamma)' = \frac{(-\alpha\beta'+\alpha'\beta+1/2\,\alpha''\alpha^2)\circ\xi}{\gamma} .$$

In particular, if $-\alpha\beta' + \alpha'\beta + 1/2\,\alpha''\alpha^2$ never vanishes, then from (4.14), (4.15), and standard estimates one can show that $1/\Delta(t) \in L^p(\mathbb{W})$ for all $t > 0$ and $1 \le p < \infty$. On the other hand, if $-\alpha\beta' + \alpha'\beta + 1/2\,\alpha''\alpha^2 \equiv 0$ throughout an open interval $I \ni x_1$, then since $\mathbb{W}(\xi(s) \in I,\ s \in [0,t]) > 0$ one sees that $\Delta(t) = 0$ with positive $\mathbb{W}$-probability. In order to understand what is happening here, first observe that $-\alpha\beta' + \alpha'\beta + 1/2\,\alpha''\alpha^2 = 0$ on I if and only if there is a $\lambda \in R^1$ for which $\beta = 1/2\,\alpha'\alpha + \lambda\alpha$ on I. Now define $F \in C^\infty(R^1)$ so that $F(0) = x_1$ and $F' = \alpha\circ F$. If $\eta(t) = F(\theta(t)+\lambda t)$, then $\eta(0) = x_1$ and $d\eta(t) = \alpha(\eta(t))d\theta(t) + (1/2\,\alpha'\alpha(\eta(t))+\lambda\alpha(\eta(t))dt$. Hence if $\tau = \inf\{t \ge 0 : \eta(t) \notin I\}$, then $\xi(t) = F(\theta(t)+\lambda t)$ for $t \in [0,\tau]$. We can therefore find an $\varepsilon > 0$ so that $\xi(t) = F(\theta(t)+\lambda t)$ whenever $\max_{0\le s\le t}|\theta(s)+\lambda s| \le \varepsilon$. Since $\mathbb{W}(\max_{0\le s\le t}|\theta(s)+\lambda s| \le \varepsilon) > 0$ for any $t \ge 0$ and $\varepsilon > 0$, it is now clear why the distribution of $x(t)$ fails to have a density in R^2 when $-\alpha\beta' + \alpha'\beta + 1/2\,\alpha''\alpha^2$ vanishes on an open interval containing x_1.

Example (4.16): The preceding example deals with a situation to which Hörmander's theorem could have been applied. We now want to look at a situation which does not lend itself to analysis via Hörmander's theorem.

Let $\sigma : R^d \to R^d \otimes R^d$ and $b : R^d \to R^d$ be C_b^∞-functions. Define $a = \sigma\sigma^*$. Assume that for some $1 \le N \le d$ and $\varepsilon > 0$, $a^{(N)}(\cdot) \ge \varepsilon I$, where $a^{(N)} = ((a_{ij}))_{1\le i,j\le N}$. In other words, the diffusion $x(\cdot,x)$ determined by (3.8) with this choice of σ and b and with $\xi_0 \equiv x$ is "partially elliptic": it is non-degenerate in coördinates $1 \le i \le N$

and may be degenerate in coördinates $N+1 \le i \le d$. One's probabilistic intuition would lead one to guess that the marginal distribution $P^{(N)}(t,x,\cdot)$ of $x^{(N)}(t) = (x_1(t),\ldots,x_N(t))$ under $\mathbb{W}$ should be just as nice as it would be were $a(\cdot)$ completely elliptic (i.e. $a(\cdot) \ge \varepsilon I$). On the other hand, any P.D.E.'er would say that studying $P^{(N)}(t,x,\cdot)$ is not going to be easy because, as a function of the forward variables, it satisfies no autonomous equation. It is at this point that Malliavin's method comes into its own. Indeed, the analysts's problem with $P^{(N)}(t,x,\cdot)$ is that it is a marginal (i.e. a projection); but from Malliavin's point of view, the transition function $P(t,x,\cdot)$ itself is already marginal. Indeed, $P(t,x,\cdot) = \mathbb{W}\circ x(t,x)^{-1}$. Hence, the difference between studying $P(t,x,\cdot)$ and $P^{(N)}(t,x,\cdot) = \mathbb{W}\circ x^{(N)}(t,x)^{-1}$ does not seem very great from his standpoint; and, as we are about to see, it really is not.

What we have to show is that $1/\Delta^{(N)}(t) \in \bigcap_{1\le p<\infty} L^p(\mathbb{W})$ for all $t > 0$, where

$$\Delta^{(N)}(t) = \det(A^{(N)}(t))$$

and $A^{(N)}(t) = (((\langle x_i(t),x_j(t)\rangle))_{1\le i,j\le N}$. But for any $\rho \in (0,1)$:

$$A(t) = \int_0^t X(s,t)a(x(s))X(s,t)^* ds$$

$$\ge \int_{t-\delta_\rho}^t X(s,t)a(x(s))X(s,t)^* ds$$

where

$$\delta_\rho = \max\{s \in [0,t] : \|X(u,t) - I\|_{op} \le \rho \quad , \quad t-s \le u \le t\} \quad .$$

But if $v = (v_1,\ldots,v_N,0,\ldots,0)$ and $t-\delta \le s \le t$, then

$$(v,X(s,t)a(x(s))X(s,t)^*v) \geq \varepsilon|v|^2 - \rho^2 K|v|^2$$

where $K = \max_{x\in R^d}\|a(x)\|_{op}$. Hence

$$A^{(N)}(t) \geq (\varepsilon - \rho^2 K)\delta_\rho I \quad .$$

It is therefore enough for us to show that $1/\delta_\rho \in \bigcap_{1\leq p<\infty} L^p(\mathfrak{W})$ for each $\rho \in (0,1)$. Such estimates are quite easy to obtain from the observation that

$$X(s,t) = X(t)X(s)^{-1} \quad ,$$

where $X(\cdot) \equiv X(0,\cdot)$, together with the stochastic differential equation for $X(\cdot)^{-1}$.

Actually, the line of reasoning just given can be pushed to yield the following variation on Hörmander's theorem. Let the vector fields $V^{(1)},\ldots,V^{(d)}$ be defined as in the discussion preceding (4.13) and suppose that M is a C^∞ - sub-manifold of R^N such that at each $x \in M$ the tangent space to M at x is contained in $\mathrm{Lie}\{V^{(1)},\ldots,V^{(N)}\}(x)$ Then for each $t > 0$ and $x \in M$, the marginal distribution of $x(t,x)$ on M is absolutely continuous with respect to the natural measure on M and has a $C^\infty(M)$ - density.

Example (4.17): We have just seen that Malliavin's technique can be used in situations where there is partial degeneracy. In this, our final, example, we will investigate how far one can go in this direction.

An optimist might conclude from (4.16) that if $\alpha : [0,\infty) \times \Theta \to (0,\infty)$ is a uniformly positive bounded progressively measurable function and if $x(t) = \int_0^t \sigma(s)d\theta(s)$ $(d = 1)$, then in order to show that the distribution of $\xi(t)$ has a C^∞ - density for all $t > 0$ all that one should have to check is that $x(s), \sigma(s) \in \mathcal{X}^{(\infty)}$ for all $s > 0$. His reasoning would be that $\langle x(t),x(t)\rangle = \int_0^t \langle \sigma(s),x(s)\rangle d\theta + \int_0^t [\langle \sigma(s),\sigma(s)\rangle + \sigma^2(s)]ds$ and that from this one ought to be able to obtain estimates on $\|1/\langle x(t),x(t)\rangle\|_{L^p(\mathbb{W})}$. Unfortunately, I have not been able to justify such optimism. The problem is that in general all that one can say is that $|\langle x(s),\sigma(s)\rangle|^2 \le \langle x(s),x(s)\rangle\langle \sigma(s),\sigma(s)\rangle$. Hence if $\xi(t) = \langle x(t),x(t)\rangle$, $\alpha(t) = \langle x(s),\sigma(s)\rangle/\langle x(s),x(s)\rangle^{1/2}$, and $\beta(t) = \langle \sigma(s),\sigma(s)\rangle + \sigma^2(s)$, then:

$$(4.18) \qquad \xi(t) = \int_0^t \alpha(s)|\xi(s)|^{1/2} d\theta(s) + \int_0^t \beta(s)ds ,$$

and $|\alpha(\cdot)|^2 \le \langle \sigma(\cdot),\sigma(\cdot)\rangle$ while $\beta(\cdot) \ge \sigma^2(\cdot) \ge \varepsilon$ for some $\varepsilon > 0$. From (4.18) it is possible to show that not only must $\xi(t)$ be non-negative (a fact we already know) but also $E^{\mathbb{W}}[\int_0^\infty \chi_{\{0\}}(\xi(t))dt] = 0$. However, this is a long way from $1/\xi(t) \in L^p(\mathbb{W})$; all that it allows one to conclude is that for a.e. $t > 0$ the distribution of $\xi(t)$ admits a density. Without going into more details, let it suffice to say that equations of the form (4.18) are far more delicate than ones in which $|\xi(s)|^{1/2}$ is replaced by $|\xi(t)|$. For example, if $\xi(t) = \theta(t)^2$, then $\xi(t) = 2\int_0^t |\xi(s)|^{1/2} d\theta(s) + t$ and clearly $1/\xi(t)$ is not integrable even though $\mathbb{W}(\xi(t) = 0) = 0$ for all $t > 0$. It is still conceivable that a careful analysis of equations like (4.18) might give some useful insights into these questions, but as yet the final word is not in.

In order to end on a less sour note, let us conclude with an example to which Malliavin's method can be applied with complete success. Let $\alpha : R^1 \to (0,\infty)$ be a uniformly positive C_b^∞ - function and let $\{p_n\}_0^\infty \subseteq R^1$ satisfy $|p_n| \le \lambda^n$, $n \ge 0$, for some $\lambda > 0$. Consider the process

$\xi(\cdot)$ given by

(4.19) $$\xi(t) = \int_0^t \alpha(\int_0^s p(s-u) \quad \xi(u)du)d\theta(s)$$

where $p(t) = \sum_0^\infty p_n t^n/n!$. It is not hard to show that (4.19) uniquely determines a progressively measurable $\xi(\cdot)$. In fact, if $\bar{p} \in L^1_{loc}([0,\infty))$ and

$$\bar{\xi}(t) = \int_0^t \alpha(\int_0^s \bar{p}(s-u)\bar{\xi}(u)du)d\theta(s) \quad ,$$

then

(4.20) $$E^{W}[\sup_{0\le t\le T}|\xi(t) - \bar{\xi}(t)|^2] \le AT^2\|p - \bar{p}\|^2_{L^1([0,T])} e^{B(T)T} \quad ,$$

where $A = 4\|\alpha\|^4_{C^1_b(R^1)}$ and $B(T) = 8\|\alpha\|^2_{C^1_b(R^1)}\|\bar{p}\|^2_{L^1([0,T])}$. Furthermore, $\xi(t) \in K^{(\infty)}$ for each $t \ge 0$. Thus we have a special case of the general situation described in the preceding paragraph; only in this special case will we be able to get integrability estimates on $1/\langle\xi(t),\xi(t)\rangle$ for $t > 0$.

The technique which we will use is the following. Given $N \ge 2$, define $\sigma^{(N)}(x) \in R^{N+1}$ and $b^{(N)}(x) \in R^{N+1}$ for $x = (x_0,\ldots,x_N) \in R^{N+1}$ by

$$\sigma_i^{(N)}(x) = \begin{cases} \alpha(x_1) & \text{if } i = 0 \\ 0 & \text{if } 1 \le i \le N \end{cases} ,$$

$$b^{(N)}(x) = \begin{cases} 0 & \text{if } i = 0 \\ p_{i-1}x_0 + x_{i+1} & \text{if } 1 \le i \le N-1 \\ p_{N-1}x_0 & \text{if } i = N \end{cases} ;$$

and consider the progressively measurable process $x^{(N)} : [0,\infty) \times \Theta \to R^{N+1}$ given by

$$(4.21)\quad x^{(N)}(t) = \int_0^t \sigma^{(N)}(x^{(N)}(s))d\theta(s) + \int_0^t b^{(N)}(x^{(N)}(s))ds \quad .$$

It is easy to show that if $\xi^{(N)}(\cdot) = x_0^{(N)}(\cdot)$, then

$$\xi^{(N)}(t) = \int_0^t \alpha(\int_0^s \rho^{(N)}(s-u)\xi^{(N)}(u)du)d\theta(s) \quad ,$$

where $\rho^{(N)}(t) = \sum_0^N \rho_n t^n/n!$, and therefore by (4.20) :

$$E^{\mathbb{W}}[\sup_{0\le t\le T}|\xi^{(N)}(t) - \xi(t)|^2] \to 0$$

as $N \uparrow \infty$ for each $T > 0$. In particular, if $\mu_t^{(N)}$ denotes the distribution of $\xi^{(N)}(t)$ under $\mathbb{W}$, then $\mu_t^{(N)}$ tends weakly to the distribution μ_t of $\xi(t)$ under $\mathbb{W}$. Thus if we can show that $\mu_t^{(N)}(dx) = f_t^{(N)}(x)dx$ where $f_t^{(N)} \in C_b^\infty(R^1)$ and, in addition, $\sup_{N\ge 2}\|f_t^{(N)}\|_{C_b^n(R^1)} < \infty$, then $\mu_t(dx) = f_t(x)dx$ where $f_t \in C_b^\infty(R^1)$

Notice that for each $N \ge 2$, we have the "partially elliptic" situation treated in (4.16) . Thus for each $t > 0$ and $N \ge 2$, $f_t^{(N)} \in C_b^\infty(R^1)$ exists. What is not so clear is how to get uniform estimates on $\|f_t^{(N)}\|_{C_b^n(R^1)}$. The difficult ingredient here is to show that if $X^{(N)}(s,t)$ is defined as in (4.3) for $\sigma^{(N)}(\cdot)$ and $b^{(N)}(\cdot)$, then $\mathbb{W}(\sup_{t-\delta\le s\le t}|X^{(N)}(s,t)_{00} - 1| \ge \varepsilon) \to 0$ as $\delta \downarrow 0$ at a sufficiently fast rate which is independent of $N \ge 2$. Without going into details, suffice it to say that one has to use

$$X^{(N)}(s,t) = I + \int_s^t S^{(N)}(x(s))X^{(N)}(s,u)d\theta(u)$$
$$+ \int_s^t B^{(N)}(x(s))X^{(N)}(s,u)du \quad ,$$

and observe that estimates on $\mathbb{W}(\sup_{t-\delta\le s\le t}|X^{(N)}(s,t)_{00} - 1| \ge \varepsilon)$ can be made to depend on $\sup_x\|S^{(N)}(x)\|_{op} \vee \|B^{(N)}(x)\|_{op}$ alone (i.e. independent of $N \geq 2$). One then notes that if the $\lambda > 0$ introduced in connection with $\{p_n\}_0^\infty$ is strictly smaller than one, then $\sup_x \|S^{(N)}(x)\|_{op} \vee \|B^{(N)}(x)\|_{op}$ can be bounded independent of $N \geq 2$. This gives the required result so long as $\lambda \in (0,1)$. If $\lambda \geq 1$, one can reduce to the case $\lambda < 1$ simply by replacing $\xi(t)$ with $\bar{\xi}(t) = \xi(t/2\lambda)$ and observing that the distribution of $\bar{\xi}(\cdot)$ is the same as the distribution of $\tilde{\xi}(\cdot)$ defined by

$$\tilde{\xi}(t) = \int_0^t \tilde{\alpha}\left(\int_0^s \tilde{p}(s-u)\tilde{\xi}(u)du\right)d\theta(u)$$

where $\tilde{\alpha}(x) = \frac{1}{(2\lambda)^{1/2}}\alpha(\frac{x}{2\lambda})$ and $\tilde{p}(t) = p(t/2\lambda)$.

For more details on computations of the sort outlined above, see section (6) in [4] .

References

[1] Hörmander, L., "Hypoelliptic second order differential equations," Acta Math., 119, pp. 147-171 (1967).

[2] McKean, H.P., "Geometry of differential spaces," Ann. Prob. 1, pp. 197-206 (1973).

[3] Stroock, D., and Varadhan, S.R.S., Multidimensional Diffusion Processes, Springer-Verlag (1979).

[4] Stroock, D., "The Malliavin calculus and its applications to second order parabolic differential equations," Parts I and II, to appear in vol. 13 of Math. Systems Theory.

The probability functionals (Onsager-Machlup functions) of diffusion processes

Y. Takahashi, University of Tokyo

and

S. Watanabe, Kyoto University

Introduction. For the n-dimensional Wiener measure, the functional $\exp[\ -1/2 \int_0^T |\dot{\phi}_t|^2\ dt\ ]$ is often considered as an ideal density with respect to a fictitious uniform measure on the space of all continuous paths $\phi_t: [0,T] \to R^n$. Stratonovich [10] introduced a notion of the probability functionals of diffusion processes which may be considered as such ideal densities. Also, physists call functions naturally associated with these functionals the Onsager-Machlup functions and regard them as Lagrangeans giving rise to the most probable paths [9],[2],[4],[7].

We are concerned with the following problem: given a non-singular, locally conservative diffusion process on a manifold M, to obtain an asymptotic evaluation of the probability that the paths of the diffusion lie in a small tube around a given smooth curve $\phi_t: [0,T] \to M$. Since a Riemannian structure is naturally induced by the diffusion coefficients so that the generator of the diffusion is $\frac{1}{2}\Delta + f$ (Δ : the Laplace-Beltrami operator, f: a vector field) and an intrinsic metric defining the tube should be the Riemannian distance $\rho(x,y)$, a precise formulation of the problem may be given as follows: let M be a Riemannian manifold of the dimension n , (x_t, P_x) be the diffusin process with the generator $\frac{1}{2}\Delta + f$ and $\phi_t : [0,T] \to M$ be a smooth curve. Find an asymptotic formula for the probability

$$P_{\phi_0}(\rho(x_t,\phi_t) < \varepsilon \quad \text{for all} \quad t \in [0,T])$$

as $\varepsilon \downarrow 0$.

An answer is given in the following

THEOREM

$$P_{\phi_0}(\rho(x_t,\phi_t) < \varepsilon \quad \text{for all} \quad t\in[0,T])$$

$$\sim C \exp(-\frac{\lambda_1 T}{\varepsilon^2}) \exp[-\int_0^T L(\phi_t,\dot\phi_t)dt] \quad \text{as} \quad \varepsilon \downarrow 0$$

where L is a function on the tangent bundle TM defined by

$$L(x,\dot x) = \frac{1}{2}\| f(x) - \dot x \|^2 + \frac{1}{2} \operatorname{div} f(x) - \frac{1}{12} R(x).$$

Here $\| \ \|$ is the Riemannian norm in the tangent space $T_x(M)$, $R(x)$ is the scalar curvature, $C = \psi_1(0)\int_D \psi_1(x)dx$ and $(\psi_m(x),\lambda_m)_{m=1,2,..}$ is the eigensystem for $-\frac{1}{2}\Delta_{R^n}$ (Δ_{R^n} ; the Laplacian in R^n) in the unit ball D of R^n with Dirichlet's boundary condition.

The proof will be given in the subsequent sections. Throughout this paper, the usual convention for the abbreviation of summation sign will be used.

1. A reduction of the problem by the normal coordinates along the curve.

Let the diffusion (x_t, P_x) on M and the smooth curve ϕ_t on M be given as above. On the product manifold $[0,T]\times M$, let $\tilde{\phi}=(\tilde{\phi}_t)$ be the curve $t\in[0,T]\longrightarrow(t,\phi_t)$ and introduce a coordinate system in a neiborhood U of the curve $\tilde{\phi}$ as follows. First, choose an orthonormal basis (ONB) $e=\{e_1,e_2,\ldots,e_n\}$ in the tangent space $T_{\phi_0}(M)$ and let $e^t=\{e_1^t, e_2^t,\ldots,e_n^t\}$ be the ONB in T_{ϕ_t} obtained as the parallel translate of e along the curve ϕ_t. Then there exists a neighborhood U of the curve $\tilde{\phi}$ in $[0,T]\times M$ such that the following mapping $(t,x)\in U\rightarrow(t,x^1,x^2,\ldots,x^n)\in[0,T]\times R^n$ is well-defined:

$$x=\exp_{\phi_t}(x^i e_i^t)\ . \tag{1.1}$$

Here, $\exp_{\phi_t}X$, $X\in T_{\phi_t}(M)$, stands for the exponential map, i.e. $s\rightarrow\exp_{\phi_t}(sX)$ is the geodesic $c(s)$ such that $c(0)=\phi_t$ and $\dot{c}(0)=X$. The mapping $(t,x)\in U\longrightarrow(t,x^1,x^2,\ldots,x^n)$ is a diffeomorphism of U onto some neighborhood V of the curve $t\rightarrow(t,0)$ in $[0,T]\times R^n$ and for each fixed $t\in[0,T]$, the mapping $x\rightarrow(x^1,x^2,\ldots,x^n)$ is nothing but the normal coordinate system N_t in a neighborhood of ϕ_t with respect to the frame e^t. The components of the Riemann metric tensor of M, its inverse, the Christoffel symbol and the vector field f in the coordinate system N_t: $x=(x^1,x^2,\ldots,x^n)$ for each fixed t are denoted by $g_{ij}(t,x)$, $g^{ij}(t,x)$, $\Gamma_{ij}^k(t,x)$ and $f^i(t,x)$ respectively. The following are some of the well-known properties of the normal coordinates ([1]),

(1.2) $$g^{ij}(t,x) = \delta^{ij} + \frac{1}{3}R_{imlj}(t,0)x^m x^l + O(|x|^3),$$

(1.3) $$\Gamma^i_{jk}(t,x) = \frac{1}{3}\left\{R_{jmik}(t,0)\ x^m + R_{ijkm}(t,0)\ x^m\right\} + O(|x|^2)$$

and

(1.4) $$\sum_j g^{ij}(t,x)x^j = x^i$$

where $R_{ijkl}(t,x)$ is the component of the Riemann curvature tensor in the coordinate system N_t for each t. Let the differential operator $\frac{\partial}{\partial t} + \frac{1}{2}\Delta + f$ on $U \subset [0,T] \times M$ be transformed to the differential operator $\frac{\partial}{\partial t} + \frac{1}{2}a^{ij}(t,x)\frac{\partial^2}{\partial x^i \partial x^j} + b^i(t,x)\frac{\partial}{\partial x^i}$ on $V \subset [0,T] \times R^n$ under the above diffeomorphism. Then clearly, $a^{ij}(t,x) = g^{ij}(t,x)$ and it is easy to prove that

(1.5) $$b^i(t,x) = f^i(t,x) - \dot{\phi}^i(t) - \frac{1}{2}\sum_k \Gamma^i_{kk}(t,x) + O(|x|^2),$$

where $\dot{\phi}^i(t)$ is the component of the tangent vector $\dot{\phi}_t \in T_{\phi_t}(M)$ in the coordinate system N_t for each t. By (1.3),

$$\sum_k \Gamma^i_{kk}(t,x) = \frac{2}{3}R_{ij}(t,0)x^j + O(|x|^2)$$

where $R_{ij}(t,x)$[‡] is the component of the Ricci tensor in N_t and hence

(1.6) $$b^i(t,x) = f^i(t,x) - \dot{\phi}^i(t) - \frac{1}{3}R_{ij}(t,0)x^j + O(|x|^2).$$

Let (x_t, P_{ϕ_0}) be the diffusion starting at $\phi_0 \in M$ and consider the product diffusion (t,x_t). Then, under the above diffeomorphism,

‡) $R_{ij}(t,x) = \sum_m R_{mijm}(t,x)$.

it is transformed to (t,X_t) on $[0,T]\times R^n$ and $X_t = (X_t^1,X_t^2,\ldots,X_t^n)$ is given as the solution of the following stochastic differential equation (SDE),

$$(1.7)\qquad \begin{cases} dX_t^i = \sum_{k=1}^{n}\sigma^{ik}(t,X_t)dB_t^k + b^i(t,X_t)dt \\ X_0 = 0 \end{cases}$$

where $\sigma^{ik}(t,x)$ is the square-root of $g^{ij}(t,x)$ and $B_t = (B_t^1,B_t^2,\ldots,B_t^n)$ is an n-dimensional Wiener process. By (1.2) and (1.4), we have

$$(1.8)\qquad \sigma^{ik}(t,x) = \delta^{ik} + \frac{1}{6} R_{imlk}(t,0)\, x^m x^l + O(|x|^3)$$

and

$$(1.9)\qquad \sum_{i=1}^{n} \sigma^{ik}(t,x)\, x^i = x^k .$$

Since the Riemannian distance is given by the ordinary Euclidean distance in the normal coordinates, we have

$$(1.10)\qquad P_{\phi_0}(\ \rho(x_t,\phi_t) < \varepsilon \quad \text{for all} \quad t\in[0,T]\ )$$

$$= P(\max_{t\in[0,T]} |X_t| := \|X\|_T < \varepsilon\ ).$$

Thus, our problem is reduced to the following: <u>to obtain the asymptotic evaluation of the right-hand side (RHS) of (1.10) for the solution</u> X_t <u>of (1.7)</u>.

We shall further rewrite the RHS of (1.10) by the Girsanov transformation. Let

$$\gamma^i(t,x) = -\frac{x^i}{2|x|^2}\Big(\sum_j g^{jj}(t,x) - n\Big), \quad i = 1,2,\dots,n, \tag{1.11}$$

and consider the following SDE for $Y_t = (Y_t^1, Y_t^2, \dots, Y_t^n)$

$$\begin{cases} dY_t^i = \sum_{k=1}^{n} \sigma^{ik}(t,Y_t)dB_t^k + \gamma^i(t,Y_t)dt \\ Y_0 = 0 \end{cases} \tag{1.12}$$

where B_t is the same Wiener process as in (1.7) . By Ito's formula and (1.9), we have

$$d|Y_t|^2 = 2\sum_i Y_t^i\, dY_t^i + \sum_i dY_t^i\, dY_t^i$$

$$= 2\sum_{i,k} Y_t^i\, \sigma^{ik}(t,Y_t)\, dB_t^k + 2\sum_i Y_t^i\, \gamma^i(t,Y_t)\, dt + \sum_i g^{ii}(t,Y_t)dt$$

$$= 2\sum_k Y_t^k\, dB_t^k + n\, dt.$$

If we set

$$M_t = \sum_k \int_0^t \frac{Y_s^k}{|Y_s|}\, dB_s^k \tag{1.13}$$

then M_t is a one-dimensional Wiener process and

$$d|Y_t|^2 = 2|Y_t|\, dM_t + n\, dt. \tag{1.14}$$

From this we can conclude that $|Y_t|$ is $BES^0(n)$ and $\underline{\underline{F}}_t(|Y|) = \underline{\underline{F}}_t(M)$ where $BES^0(n)$ denotes a Bessel diffusion process with index n starting at 0 and $\underline{\underline{F}}_t(Z)$, in general, denotes the filtration generated by the process $Z = (Z_t)$, [6]. By the Girsanov theorem,

$$(1.15)\qquad P(\|X\|_T < \varepsilon)$$
$$= E(\exp[\int_0^T \sigma_{ij}(t,Y_t)\delta^j(t,Y_t)dB_t^i - \frac{1}{2}\int_0^T g_{ij}(t,Y_t)\delta^i(t,Y_t)$$
$$\times\ \delta^j(t,Y_t)dt] : \|Y\|_T < \varepsilon)$$

where

$$(1.16)\qquad \delta^i(t,x) = b^i(t,x) - \gamma^i(t,x)$$

and $\sigma_{ij}(t,x)$ is the inverse of $\sigma^{ij}(t,x)$, i.e., the square-root of $g_{ij}(t,x)$. Denoting the exponential in the expectation in the RHS of (1.15) by e^{Φ_T}, we have

$$P(\|X\|_T < \varepsilon) = E(e^{\Phi_T} \mid \|Y\|_T < \varepsilon)\ P(\|Y\|_T < \varepsilon).$$

Since $|Y_t|$ is $BES^0(n)$, it is easy to see by the eigenfunction expansion that

$$P(\|Y\|_T < \varepsilon) = \sum_{m=1}^{\infty} \exp(-\lambda_m T/\varepsilon^2)\ \psi_m(0)\int_D \psi_m(y)dy$$

where $\{\lambda_m, \psi_m\}$ is the eigensystem for $-\frac{1}{2}\Delta_{R^n}$ in the unit ball $D = \{x; |x| < 1\}$ of R^n with Dirichlet's boundary condition. Thus

$$(1.17) \qquad P(\| Y\|_T < \varepsilon) \sim \exp(-\lambda_1 T/\varepsilon^2)\ \psi_1(0)\int_D \psi_1(y)dy$$

and the theorem is proved if we can show that

$$(1.18) \qquad E(e^{\Phi_T} \,|\, \| Y\|_T < \varepsilon) \sim \exp\left[-\int_0^T L(\phi_t,\dot\phi_t)\, dt\right] \quad \text{as} \quad \varepsilon\downarrow 0 .$$

The following simple lemma is easily proved ([6]).

<u>Lemma 1.1</u> <u>Let</u> $Z_1,Z_2,\ldots,Z_m$ <u>be random variables and</u> $a_1,a_2,\ldots,a_m \in R$. <u>If, for each</u> i, $E(e^{cZ_i} \,|\, \| Y\|_T < \varepsilon) \sim e^{ca_i}$ <u>as</u> $\varepsilon\downarrow 0$ <u>for every</u> $c\in R$, <u>then</u> $E(\exp[\sum_i Z_i] \,|\, \|Y\|_T < \varepsilon) \sim \exp(\sum_i a_i)$ <u>as</u> $\varepsilon\downarrow 0$.

Let

$$(1.19) \qquad \Phi_T^{(1)} = \int_0^T \sigma_{ij}(t,Y_t)\ \delta^j(t,Y_t)\ dB_t^i$$

and

$$(1.20) \qquad \Phi_T^{(2)} = -\frac{1}{2}\int_0^T g_{ij}(t,Y_t)\ \delta^i(t,Y_t)\ \delta^j(t,Y_t)\ dt.$$

Then $\Phi_T = \Phi^{(1)} + \Phi_T^{(2)}$. Since $g_{ij}(t,x) = \delta_{ij} + O(|x|^2)$, $\gamma^i(t,x) = O(|x|)$ and hence, by (1.6),

$$g_{ij}(t,x)\delta^i(t,x)\delta^j(t,x) = \sum_i (f^i(t,0)-\dot\phi^i(t))^2 + O(|x|).$$

Thus, on the set $\{\| Y\|_T < \varepsilon\}$,

$$\phi_T^{(2)} = -\frac{1}{2}\sum_i \int_0^T |f^i(t,0) - \dot{\phi}^i(t)|^2 \, dt + O(\varepsilon)$$

$$= -\frac{1}{2}\int_0^T \| f(\phi_t) - \dot{\phi}_t\|^2 \, dt + O(\varepsilon) \qquad \text{as} \quad \varepsilon \downarrow 0$$

and we can conclude from this that

$$\text{(1.21)} \qquad E(e^{c\phi_T^{(2)}} \,|\, \|Y\|_T < \varepsilon) = \exp\left\{- \frac{c}{2}\int_0^T \| f(\phi_t)-\dot{\phi}_t\|^2 \, dt + O(\varepsilon)\right\}$$

as $\varepsilon\downarrow 0$ for every $c \in \mathbf{R}$.

Noting the lemma, (1.18) will be proved if we can show

$$\text{(1.22)} \qquad E(e^{c\phi_T^{(1)}} \,|\, \|Y\|_T < \varepsilon) \sim \exp[- \frac{c}{2}\int_0^T \operatorname{div} f(\phi_t)dt + \frac{c}{12}\int_0^T R(\phi_t)dt]$$

as $\varepsilon \downarrow 0$ for every $c \in \mathbf{R}$.

Since $\phi_T^{(1)}$ is a stochastic integral, the estimate (1.22) is not so simple as (1.21): it can not be obtained pathwise. To obtain (1.22), we shall generally discuss on some asymptotic behaviors of stochastic integrals under the conditional probability $P(\; \|Y\|_T < \varepsilon)$ as $\varepsilon \downarrow 0$.

2. Asymptotic properties of stochastic integrals under conditional probabilities.

Let Y_t be, as in section 1, the solution of (1.12). The results of this section are based on the following theorem.

Theorem 2.1 *Let* $f(r,\theta,t)$ *be a continuous function on* $[0,\infty) \times S^{n-1} \times [0,T]$. *Then,*

$$\text{(2.1)} \qquad E(\exp[\int_0^T f(|Y_t|, \frac{Y_t}{|Y_t|}, t) \, dt] \,|\, \|Y\|_T < \varepsilon)$$

$$\sim \exp\left[\int_0^T dt \int_{S^{n-1}} f(0,\theta,t)\, d\theta \right] \quad \underline{\text{as}}\ \varepsilon \downarrow 0$$

where $d\theta$ is the normalized uniform measure on the unit sphere $S^{n-1} = \{ x \in R^n ;\ |x| = 1 \}$.

This theorem is intuitively clear: under the conditioning by $\|Y\|_T < \varepsilon$ on the radial process $|Y_t|$, the spherical motion $Y_t/|Y_t|$ moves so quickly that a kind of homogeneization theorem or ergodic theorem works. A rigorous proof, which is analytic in nature, will be given in the next section.

Let $F_i(t,x)$, $i = 1,2,\dots,n$, be continuous functions on $[0,T] \times R^n$ possessing the following asymptotic properties:

$$(2.2) \qquad F_i(t,x) = F_i(t,0) + F_{ij}\left(t, \frac{x}{|x|}\right)x^j + O(|x|^2) \qquad \text{as } |x| \downarrow 0 ,$$

where $F_i(t,0)$ is C^1 in t and $F_{ij}(t,\theta)$ are functions on $[0,T] \times S^{n-1}$ which are C^1 in t and C^3 in θ. Define the following differential operators (i.e. vector fields) L^i (denoted also by L_i to fit the summation convention) , $i = 1,2,\dots,n$, on S^{n-1} as the restriction to $S^{n-1} \subset R^n$ of the differential operators on R^n :

$$(2.3) \qquad L^i = (\delta^{ij} - x^i x^j) \frac{\partial}{\partial x^j} \Big|_{S^{n-1}} .$$

Clearly L^i is well-defined as a differential operator on the sphere S^{n-1}. Note that the spherical Laplacian $\tilde{\Delta}$ on S^{n-1} is given by $\tilde{\Delta} = L^i(L_i)$.

Theorem 2.2 Let Y_t be as above and B_t be the Wiener process *appearing in the SDE* (1.12). *Then*

$$(2.4)\qquad E[\exp(\int_0^T F_i(s,Y_s)\,dB_s^i)\mid \|Y\|_T<\varepsilon]$$

$$\sim \exp[-\frac{1}{2}\int_0^T dt\int_{S^{n-1}}\{\sum_i F_{ii}(t,\theta)\}\,d\theta$$

$$-\frac{1}{2}\int_0^T dt\int_{S^{n-1}}\{\theta^j L^i F_{ij}(t,\theta)\}\,d\theta\,] \quad \text{as}\quad \varepsilon\downarrow 0.$$

Here, $\theta=(\theta^1,\theta^2,\ldots,\theta^n)\in S^{n-1}$, *i.e.*, $|\theta|=1$.

Remark 2.1 (i) If $F_i(t,x)$ is smooth in x, then $F_{ij}(t,\theta)=\frac{\partial F_i}{\partial x^j}(t,0)$ is independent of θ and hence

$$\text{RHS of (2.4)} = \exp[-\frac{1}{2}\int_0^T\{\sum_i \frac{\partial F_i}{\partial x^i}(t,0)\}dt\,].$$

(ii) If $F_{ij}(t,\theta) = \delta_{ij}F(t,\theta)$, then, since $\theta^i L_i = 0$,

$$\text{RHS of (2.4)} = \exp[-\frac{n}{2}\int_0^T dt\int_{S^{n-1}}F(t,\theta)\,d\theta\,].$$

First we shall complete the proof of (1.22) admitting Th. 2.2 for a moment. We have

$$c\,\Phi_T^{(1)} = \int_0^T F_i(t,Y_t)\,dB_t^i$$

where

$$F_i(t,x) = c\,\sigma_{ij}(t,x)\,\delta^j(t,x).$$

Since

$$\sigma_{ij}(t,x) = \delta_{ij} + O(|x|^2)$$

and

$$\delta^j(t,x) = b^i(t,x) - \gamma^i(t,x)$$

$$= f^i(t,0) - \dot{\phi}^i(t,0) + \sum_j [\frac{\partial f^i}{\partial x^j}(t,0) - \frac{1}{3} R_{ij}(t,0)]x^j$$

$$+ \frac{x^i}{6|x|^2} \sum_j R_{jmlj}(t,0)x^m x^l + O(|x|^2)$$

by (1.6), (1.11) and (1.2), we see easily that F_i is of the form (2.2) with

$$F_{ij}(t,\theta) = c\left\{ [\frac{\partial f^i}{\partial x^j}(t,0) - \frac{1}{3} R_{ij}(t,0)] + \frac{1}{6} \delta_{ij} \theta^m \theta^l R_{ml}(t,0) \right\} .$$

Hence, we can apply the above Remark 2.1 to obtain

$$E(e^{c\Phi_T^{(1)}} \Big| \|Y\|_T < \varepsilon)$$

$$\exp[-\frac{c}{2} \int_0^T \sum_i \left\{ \frac{\partial f^i}{\partial x^i}(t,0) - \frac{1}{3} R_{ii}(t,0) \right\} dt - \frac{cn}{12} \int_0^T dt \int_{S^{n-1}} \theta^m \theta^l$$

$$\times R_{ml}(t,0)\, d\theta \,]$$

$$= \exp[-\frac{c}{2} \int_0^T \operatorname{div} f(\phi_t)\, dt + (\frac{c}{6} - \frac{c}{12}) \int_0^T R(\phi_t)\, dt \,] \quad \text{as } \varepsilon \downarrow 0.$$

Here we used the following: by definition, $\sum_{i=1}^n R_{ii}(t,0) = R(\phi_t)$ is the scalar curvature and

$$\int_{S^{n-1}} \theta^m \theta^l \, d\theta = \delta^{ml} \frac{1}{n} \quad .$$

Thus (1.22) is proved.

Now we proceed to the

Proof of Theorem 2.2.

Set

$$M^{ij}(t) = \frac{1}{2}\int_0^t [\, Y^i(s)\sigma_k^j(s,Y_s) - Y^j(s)\sigma_k^i(s,Y_s)\,]\, dB^k(s) \tag{2.5}$$

$$i,j = 1,2,\dots,n.$$

$M^{ij}(t)$ is the martingale part of the semimartingale $S^{ij}(t)$ defined by

$$S^{ij}(t) = \frac{1}{2}\int_0^t [\, Y^i(s)\circ dY^j(s) - Y^j(s)\circ dY^i(s)\,] \tag{2.6}$$

$$i,j = 1,2,\dots,n,$$

where $\circ$ is the Stratonovich stochastic differential, [6] . Let $(\underline{\underline{F}}_t)$ be the filtration generated by the Wiener process $B(t)$. Then M^{ij} is a system of $(\underline{\underline{F}}_t)$-martingales such that

$$\langle M^{ij}, M \rangle = 0, \qquad i,j = 1, 2,\dots,n \tag{2.7}$$

where the martingale M is defined by (1.13). (2.7) follows at once from (1.9). An important consequence of (2.7) is the following

Lemma 2.1 *If $\Phi_{ij}(t)$, $i,j = 1,2,\dots,n$, are bounded and $(\underline{\underline{F}}_t)$-adapted measurable processes, then*

$$E[\, \exp\Big(\int_0^t \Phi_{ij}(s)dM^{ij}(s) - \frac{1}{2}\int_0^t \Phi_{ij}(s)\Phi_{kl}(s)d\langle M^{ij},M^{kl}\rangle(s) \Big) \,\Big|\, \|Y\|_T < \varepsilon \,] = 1 \tag{2.8}$$

for every $t \in [0,T]$ *and* $\varepsilon > 0$.

Proof. Let $\Psi = I_{\{\|Y\|_T < \varepsilon\}}$. Since $\underline{\underline{F}}_t(|Y|) = \underline{\underline{F}}_t(M)$, Ψ can be expressed as

$$\Psi = E[\Psi] + \int_0^T \psi(s)dM_s$$

for some $(\underline{\underline{F}}_t(M))$-adapted measurable process $\psi(s)$, [8]. Let the exponential in the expectation of (2.8) be denoted by $N(t)$. It is clear that $N(t) - N(0)$ is a sum of stochastic integrals by M^{ij}. Then, by (2.7),

$$E(N(t)\Psi) = E[\ N(t)(\ E(\Psi) + \int_0^t \psi(s)dM_s\ )] = E(\Psi)E(N(t)) = E(\Psi).$$

Next we prepare a stochastic version of the Stokes theorem for $Y(t)$. For each $t > 0$, let $c(t)$ be a stochastic singular 2-chain defined by

$$c(t) = \{ uY(s)\ ;\ 0 \leq u \leq 1,\ 0 \leq s \leq t \} .$$

For a 2-form $\alpha = \sum_{i<j} \alpha_{ij}(x)\ dx^i \wedge dx^j$ on R^n, the surface integral $\int_{c(t)} \alpha$ is defined by

$$\text{(2.9)} \qquad \int_{c(t)} \alpha = \sum_{i<j} \int_0^t \overline{\alpha}_{ij}(Y(s)) \circ dS^{ij}(s)$$

where

$$\overline{\alpha}_{ij}(x) = 2\int_0^1 \alpha_{ij}(ux)udu, \qquad i,j = 1,2,\ldots,n.$$

For a 1-form $\beta = \beta_i(x)dx^i$, we set

$$\overline{\beta}_i(x) = \int_0^1 \beta_i(ux)du, \qquad i = 1,2,\ldots,n.$$

<u>Lemma 2.2</u> (<u>Stokes's theorem</u>) <u>Let</u> $\beta = \beta_i(x)dx^i$ <u>be a 1-form on</u> R^n. <u>Then</u>

$$\text{(2.10)} \qquad \int_{c(t)} d\beta = \int_0^t \beta_i(Y(s)) \circ dY^i(s) - Y^i(t)\ \overline{\beta}_i(Y(t)).$$

Proof. If $Y(t)$ is a smooth curve, (2.10) is nothing but the classical Stokes theorem. (2.10) can be obtained by a standard argument of the transfer from smooth curves to stochastic curves,([5]).

In the following, we need to consider the case when 1-form β is time dependent : $\beta = \beta_i(s,x)dx^i$. In this case, we regard β as a 1-form on $[0,\infty)\times R^n$ and consider a 2-chain $c(t)$ defined by

$$c(t) = \left\{ (s,\ uY(s));\quad 0 \leqslant u \leqslant 1,\quad 0 \leqslant s \leqslant t \right\} .$$

In a similar way, we obtain the following

Lemma 2.3 (Stokes's theorem; the time dependent case)

$$(2.11)\qquad 2\sum_{i<j}\int_0^t \left[\frac{\partial \bar{\beta}_j}{\partial x^i} - \frac{\partial \bar{\beta}_i}{\partial x^j} \right] (s,Y(s))\circ dS^{ij}(s) - \int_0^t \frac{\partial \bar{\beta}_i}{\partial s}(s,Y(s))Y^i(s)ds$$

$$= \int_0^t \beta_i(s,Y(s))\circ dY^i(s) - \bar{\beta}_i(t,Y(t))Y^i(t)$$

where $\bar{\beta}_i(s,x) = \int_0^1 \beta_i(s,ux)du$. More generally, if β is smooth in $(s,x)\in[0,\infty)\times(R^n\setminus\{0\})$

$$(2.12)\qquad 2\sum_{i<j}\int_{t_0}^t \left[\frac{\partial \tilde{\beta}_j}{\partial x^i} - \frac{\partial \tilde{\beta}_i}{\partial x^j} \right](s,Y(s))\circ dS^{ij}(s)$$

$$- \int_{t_0}^t \frac{\partial \tilde{\beta}_i}{\partial s}(s,Y(s))Y^i(s)ds$$

$$= \int_{t_0}^t \beta_i(s,Y(s))\circ dY^i(s) - u_0\int_{t_0}^t \beta_i(s,\ u_0Y(s))\circ dY^i(s)$$

$$+ \tilde{\beta}_i(t_0,Y(t_0))Y^i(t_0) - \tilde{\beta}_i(t,Y(t))Y^i(t)$$

for every $0 < t_0 < t$ *and* $0 < u_0 < 1$,

where $\tilde{\beta}_i(s,x) = \int_{u_0}^{1} \beta_i(s,ux)du$.

Lemma 2.4 *Let* L^i *be defined by* (2.3) *and set*

$$R^{ij} = \theta^i L^j - \theta^j L^i, \quad i,j=1,2,\ldots,n, \quad \theta = (\theta^1,\theta^2,\ldots,\theta^n) \in S^{n-1} \subset R^n.$$

Then, if f *and* g *are smooth functions on* S^{n-1},

$$(2.13) \qquad \int_{S^{n-1}} (R^{ij}f)(\theta)\, g(\theta)\, d\theta = -\int_{S^{n-1}} f(\theta)(R^{ij}g)(\theta)\, d\theta.$$

In particular,

$$(2.14) \qquad \int_{S^{n-1}} (R^{ij}f)(\theta)\, d\theta = 0.$$

Proof. Let $\omega = \tan^{-1}\frac{x^j}{x^i}$. Then $R^{ij} = \frac{\partial}{\partial\omega}$ and the assertion is obvious.

Lemma 2.5 *Let* $g(\theta)$ *be a smooth function on* S^{n-1}. *Then*

$$\frac{\partial}{\partial y^i} g\left(\frac{y}{|y|}\right) = \frac{1}{|y|}(L^i g)\left(\frac{y}{|y|}\right), \quad y \neq 0, \quad i = 1,2,\ldots,n.$$

Hence if $f(t,\theta)$ *is a smooth function on* $[0,\infty) \times S^{n-1}$, $f\left(t, \frac{Y_t}{|Y_t|}\right)$ *is a continuous semimartingale on* $[t_0,\infty)$ *for every* $t_0 > 0$ *and*

$$(2.15) \qquad df\left(t,\frac{Y_t}{|Y_t|}\right) = \frac{\partial f}{\partial t}\left(t, \frac{Y_t}{|Y_t|}\right) dt + \frac{1}{|Y_t|}(L_i f)\left(t,\frac{Y_t}{|Y_t|}\right)\circ dY_t^i .$$

Proof is obvious.

Now we shall prove (2.4). The following proof was suggested by N. Ikeda. In the following c is an arbitrary real constant and $O(\ )$ is always independent of the stochastic parameter ω. First, we note that if $g(s,\omega) = (g_i(s,\omega))$ is a system of measurable $(\underline{\underline{F}}_t)$-adapted processes such that $|g(s,\omega)| = O(\varepsilon^2)$ on the set $\{\|Y\|_T < \varepsilon\}$, then

$$E(\exp(\int_0^T g_i(s,\omega)dB_s^i) \,|\, \|Y\|_T < \varepsilon) \sim 1 \quad \text{as } \varepsilon \downarrow 0, \text{([6], p.451).}$$

It is obvious that

$$E(\exp(c \int_0^T F_i(s,0)dY_s^i) \,|\, \|Y\|_T < \varepsilon) \sim 1 \quad \text{as } \varepsilon \downarrow 0 .$$

Noting also that $|\gamma(t,Y_t)| = O(\varepsilon)$ on the set $\{\|Y\|_T < \varepsilon\}$, we have therefore that

$$(2.16) \qquad E(\exp[c \int_0^T F_i(s,Y_s)dB_s^i] \,|\, \|Y\|_T < \varepsilon)$$

$$\sim E(\exp[c \int_0^T F_{ij}(s,\frac{Y_s}{|Y_s|})\, Y_s^j dY_s^i] \,|\, \|Y\|_T < \varepsilon) \quad \text{as } \varepsilon \downarrow 0.$$

Let β be the time dependent 1-form on $R^n \backslash \{0\}$ defined by

$$\beta = F_{ij}(t, \frac{x}{|x|})x^j dx^i .$$

Applying the Stokes theorem (2.12) and then letting $t_0 \downarrow 0$ and $u_0 \downarrow 0$, we have that

$$I_1 := \int_0^T F_{ij}(s, \frac{Y_s}{|Y_s|})\, Y_s^j \circ dY_s^i$$

$$= \overline{\beta}_i(T,Y_T)\, Y_T^i + 2 \sum_{i<j} \int_0^T [\frac{\partial \overline{\beta}_j}{\partial x^i} - \frac{\partial \overline{\beta}_i}{\partial x^j}](s,Y_s) \circ dS^{ij}(s)$$

$$- \int_0^T \frac{\partial \overline{\beta}_i}{\partial s} (s,Y_s)\, Y_s^i \, ds$$

where $\bar{\beta}_i(s,x) = \frac{1}{2} F_{ij}(s, \frac{x}{|x|})x^j$.

Clearly

$$I_1 = 2 \sum_{i<j} \int_0^T [\frac{\partial \bar{\beta}_j}{\partial x^i} - \frac{\partial \bar{\beta}_i}{\partial x^j}](s,Y_s)\circ dS^{ij}(s) + O(\varepsilon^2)$$

on the set $\{\|Y\|_T < \varepsilon\}$. By Lemma 2.5, it is clear that

$$[\frac{\partial \bar{\beta}_j}{\partial x^i} - \frac{\partial \bar{\beta}_i}{\partial x^j}](s,x) = \gamma_{ij}(s, \frac{x}{|x|})$$

where $\gamma_{ij}(s,\theta)$ is a smooth function on $[0,T] \times S^{n-1}$. We have that (denoting $\frac{Y_s}{|Y_s|}$ by Θ_s)

$$I_2 := 2 \sum_{i<j} \int_0^T \gamma_{ij}(s,\Theta_s)\circ dS^{ij}(s)$$

$$= 2 \sum_{i<j} \int_0^T \gamma_{ij}(s,\Theta_s)dS^{ij}(s) + \sum_{i<j} \int_0^T d\gamma_{ij}(s,\Theta_s)\cdot dS^{ij}(s)$$

$$:= I_{21} + I_{22}$$

and

$$I_{21} = 2 \sum_{i<j} \int_0^T \gamma_{ij}(s,\Theta_s)dM^{ij}(s) + O(\varepsilon^2)$$

on the set $\{\|Y\|_T < \varepsilon\}$. Set $I_{211}(t) = 2 \sum_{i<j} \int_0^t \gamma_{ij}(s,\Theta_s)dM^{ij}(s)$.

Then by Lemma 2.1,

$$E[\exp(cI_{211}(T) - (c^2/2)\langle I_{211}\rangle(T)) \mid \|Y\|_T < \varepsilon] = 1$$

and $\langle I_{211}\rangle(T) = O(\varepsilon^2)$ on the set $\{\|Y\|_T < \varepsilon\}$. Hence

$$E[\ \exp(cI_{211}(T)) \mid \|Y\|_T < \varepsilon\ ] \sim 1 \quad \text{as} \quad \varepsilon \downarrow 0.$$

Consequently

$$E[\ \exp(cI_{21}) \mid \|Y\|_T < \varepsilon\ ] \sim 1 \quad \text{as} \quad \varepsilon \downarrow 0.$$

Next, we consider I_{22}. By (2.15) (denoting $\frac{Y_s^i}{|Y_s|}$ by Θ_s^i),

$$I_{22} = \frac{1}{2}\sum_{i<j}\int_0^T (L_k\gamma_{ij})(s,\Theta_s)(\ \Theta_s^i g^{kj}(s,Y_s) - \Theta_s^j g^{ki}(s,Y_s))ds$$

$$= \frac{1}{2}\sum_{i<j}\int_0^T \left\{\Theta_s^i(L^j\gamma_{ij})(s,\Theta_s) - \Theta_s^j(L^i\gamma_{ij})(s,\Theta_s)\right\}ds + O(\varepsilon^2)$$

on the set $\left\{\|Y\|_T < \varepsilon\right\}$. By Th.2.1 and (2.14),

$$E[\ \exp(cI_{22}) \mid \|Y\|_T < \varepsilon\ ] \sim \exp\left\{c/2\sum_{i<j}\int_0^T ds\int_{S^{n-1}}(R^{ij}\gamma_{ij})(s,\theta)d\theta\right\} = 1\ .$$

Hence we obtained

$$E[\ \exp(cI_1) \mid \|Y\|_T < \varepsilon\ ] \sim 1 \quad \text{as} \quad \varepsilon \downarrow 0.$$

Finally, if we set $I = \int_0^T F_{ij}(s,\Theta_s)Y_s^j dY_s^i$,

$$I = I_1 - \frac{1}{2}\int_0^T F_{ij}(s,\Theta_s)dY_s^j dY_s^i - \frac{1}{2}\int_0^T (L_kF_{ij})(s,\Theta_s)\Theta_s^j\ dY_s^k dY_s^i$$

$$= I_1 - \frac{1}{2}\int_0^T F_{ij}(s,\Theta_s)\ g^{ij}(s,Y_s)\ ds - \frac{1}{2}\int_0^T (L_kF_{ij}(s,\Theta_s)\Theta_s^j g^{ki}(s,Y_s)\ ds$$

$$= I_1 - \frac{1}{2}\sum_i\int_0^T F_{ii}(s,\Theta_s)ds - \frac{1}{2}\int_0^T (L^iF_{ij})(s,\Theta_s)\Theta_s^j\ ds + O(\varepsilon^2)$$

on the set $\left\{\|Y\|_T < \varepsilon\right\}$.

By Th.2.1,

$$E[\exp(cI) \mid \|Y\|_T < \varepsilon]$$

$$\sim \exp[-c/2 \int_0^T ds \int_{S^{n-1}} \sum_i F_{ii}(s,\theta)d\theta - c/2 \int_0^T ds \int_{S^{n-1}} \theta^j (L^i F_{ij})(s,\theta)d\theta]$$

and by (2.16) this completes the proof.

3. Proof of Theorem 2.1.

We shall prove (2.1) where the integral $\int_0^T$ is replaced by $\int_{t_0}^T$ with $t_0 > 0$. Theorem 2.1 then follows at once by letting $t_0 \downarrow 0$. Also we may assume $f(r,\theta,s)$ is smooth in θ.

First of all, we shall study the spherical motion $\Theta_t = Y_t/|Y_t|$, $t \in [t_0, T]$. Let $g(t,\theta)$ be a smooth function on $[t_0,T] \times S^{n-1}$. By Lemma 2.5,

$$(3.1) \qquad dg(t,\Theta_t) = \frac{\partial g}{\partial t}(t,\Theta_t)dt + \frac{1}{|Y_t|}(L_i g)(t,\Theta_t)\circ dY_t^i$$

$$= \frac{1}{|Y_t|} \sum_k (L_i g)(t,\Theta_t)\sigma^{ik}(t,Y_t)dB_t^k \quad (:= dN_t)$$

$$+ [\frac{\partial g}{\partial t}(t,\Theta_t)dt + \frac{1}{|Y_t|}(L_i g)(t,\Theta_t)\gamma^i(t,Y_t)dt$$

$$- \sum_k \frac{Y_t^k}{2|Y_t|^3}(L_i g)(t,\Theta_t)dY_t^k dY_t^i + \frac{1}{2|Y_t|^2} L_j(L_i g)(t,\Theta_t)dY_t^j dY_t^i]$$

$$= dN_t + [\frac{\partial g}{\partial t}(t,\Theta_t) + \frac{1}{|Y_t|}(L_i g)(t,\Theta_t)\gamma^i(t,Y_t)$$

$$- \frac{1}{2|Y_t|^3} \sum_k Y_t^k (L_i g)(t,\Theta_t)g^{ki}(t,Y_t) + \frac{1}{2|Y_t|^2} L_j(L_i g)(t,\Theta_t)$$

$$\times g^{ij}(t,Y_t)\ ]\ dt$$

$$= dN_t + [\ \frac{\partial g}{\partial t}(t,\Theta_t) + \frac{1}{|Y_t|}(L_i g)(t,\Theta_t)\gamma^i(t,Y_t)$$

$$+ \frac{1}{2|Y_t|^2} L_j(L_i g)(t,\Theta_t)\ g^{ij}(t,Y_t)\ ]\ dt$$

since $\sum_k y^k (L_i g) g^{ki} = y^i (L_i g) = 0$. The martingale part N_t in the above semimartingale decomposition is orthogonal to M_t, i.e. $\langle N,M\rangle_t = 0$ where M_t is given by (1.13). Hence, by the same proof as Lemma 2.1, the above semimartingale decomposition remains valid (i.e., N_t is again a martingale) under the probability $P(\ \|Y\|_T < \varepsilon\ )$.

In the following, $r = (r_t)$ is any continuous function on $[0,T]$ such that $r_0=0$ and $0<r_t<1$ for $0<t\leq T$. The totality of such functions r is denoted by $\mathcal{R}$. We denote by $P^{\varepsilon r}, \varepsilon>0$, $r\in\mathcal{R}$, the regular conditional probability $P(\ |Y_t| = \varepsilon r_t,\ t\in[0,T]\ )$. By the same reason as above, the semimartingale decomposition (3.1) remains valid under $P^{\varepsilon r}$ for almost all $r\in\mathcal{R}$ with respect to the law $P(\varepsilon^{-1}|Y_\cdot|\in dr \mid \|Y\|_T<\varepsilon)$. This implies that Θ_t, $t\in[t_0,T]$, is a time dependent diffusion on S^{n-1} under $P^{\varepsilon r}$ with the infinitesimal generator

$$(3.2)\quad (A_t^{\varepsilon r} g)(t,\theta) = (\varepsilon r_t)^{-1}(L_i g)(t,\theta)\gamma^i(t,\varepsilon r_t\theta)$$

$$+ (2\varepsilon^2 r_t^2)^{-1} g^{ij}(t,\varepsilon r_t\theta) L_j(L_i g)(t,\theta).$$

If $g^{ij} = \delta^{ij}$ and $\gamma^i = 0$, the corresponding operator is

$$(3.3)\quad (L_t^{\varepsilon r} g)(t,\theta) = (2\varepsilon^2 r_t^2)^{-1} L_i(L^i g)(t,\theta) = (2\varepsilon^2 r_t^2)^{-1}(\tilde{\Delta} g)(t,\theta)$$

where $\tilde{\Delta}$ is the spherical Laplacian on S^{n-1}. Noting (1.2), it is easy to see that

(3.4) $$A_t^{\varepsilon r} := (\Lambda_t^{\varepsilon r} - L_t^{\varepsilon r}) = A_t + O(\varepsilon) \quad \text{as} \quad \varepsilon \downarrow 0$$

where A_t , $t_o \leqslant t \leqslant T$, is a smooth differential operator on S^{n-1} of the second order with $A^t 1 = 0$: $O(\varepsilon)$ means that all the coefficients, together with their derivatives in θ, are $O(\varepsilon)$ as $\varepsilon \downarrow 0$ uniformly in $(t,\theta,r) \in [t_0,T] \times S^{n-1} \times \mathbb{R}$. Let

(3.5) $$u_\varepsilon(t,\theta) = E^{\varepsilon r}[\exp(\int_t^T f(\varepsilon r_s,\Theta_s,s)ds \,|\, \Theta_t = \theta], \quad t_0 \leqslant t \leqslant T, \ \theta \in S^{n-1}.$$

u_ε is the solution of

(3.6) $$\begin{cases} -\dfrac{\partial u}{\partial t} = \Lambda_t^{\varepsilon r} u + f(\varepsilon r_t,\theta,t)\, u \\ \lim\limits_{t \uparrow T} u(t,\theta) = 1 \end{cases}$$

The solution of

(3.7) $$\begin{cases} -\dfrac{\partial u}{\partial t} = L_t^{\varepsilon r} u \\ \lim\limits_{t \uparrow T} u(t,\theta) = 1 \end{cases}$$

is clearly $u \equiv 1$. Hence

(3.8) $$u_\varepsilon(t,\theta) = 1 + \int_t^T ds \int_{S^{n-1}} d\xi \; p(\int_t^s (\varepsilon r_\tau)^{-2} d\tau,\theta,\xi)$$

$$\times [(A_s^{\varepsilon r} u_\varepsilon)(s,\xi) + f(\varepsilon r_s,\xi,s) u_\varepsilon(s,\xi)], \quad t \in [t_0,T]$$

$$\theta \in S^{n-1} .$$

Here $p(t,\theta,\xi)$ is the fundamental solution of the heat equation

$$\frac{\partial}{\partial t} = \frac{1}{2}\tilde{\Delta} \quad \text{on} \quad S^{n-1}.$$

Note that $p(t,\theta,\xi) \to 1$ as $t \uparrow \infty$ uniformly in θ and ξ. Letting $\varepsilon \downarrow 0$ in (3.8), we have formally that $u_\varepsilon \to u$ where u satisfies

$$(3.9) \qquad u(t,\theta) = 1 + \int_t^T ds \int_{S^{n-1}} d\xi \; [\; (A_s u)(s,\xi) + f(0,\xi,s)u(s,\xi) \;].$$

Then $u(t,\theta) = u(t)$ is independent of $\theta \in S^{n-1}$ and

$$u(t) = 1 + \int_t^T u(s)\bar{f}(s)ds$$

where $\bar{f}(s) = \int_{S^{n-1}} f(0,\theta,s)d\theta$. Therefore $u(t) = \exp(\int_t^T \bar{f}(s)ds)$.

If we can justify the convergence $u_\varepsilon(t,\theta) \to u(t)$ as $\varepsilon \to 0$ and furthermore, if this convergence is uniform in $\theta \in S^{n-1}$, $t \in [t_0,T]$ and $r \in \mathcal{R}$, then

$$E[\; \exp(\int_{t_0}^T f(|Y_s|,\theta_s,s)ds \;) \;\Big|\; \|Y\|_T < \varepsilon \;]$$

$$= \int_{\mathcal{R}\times S^{n-1}} u_\varepsilon(t_0,\theta) \; P(\; \tfrac{1}{\varepsilon}|Y.| \in dr, \;\; \theta_{t_0} \in d\theta \;\Big|\; \|Y\|_T < \varepsilon \;)$$

$$\longrightarrow \; u(t_0) \qquad \text{as} \quad \varepsilon \downarrow 0$$

and the assertion will be proved.

So we proceed to prove the convergence of $u_\varepsilon(t,\theta)$ to $u(t)$ and its uniformity. First, introduce the following differential operators;

$$\Delta_t = A_t + f(0,\theta,t)\cdot \quad , \qquad \text{and}$$

$$\Delta_t^{\varepsilon} = A_t^{\varepsilon r} + f(\varepsilon r_t, \theta, t)\cdot .$$

Also introduce the following operators for smooth functions $g(t,\theta)$ on $[t_0,T] \times S^{n-1}$:

$$(3.10) \qquad (G_1^{\varepsilon} g)(t,\theta) = \int_{S^{n-1}} d\xi \int_{(t+\varepsilon^2)\wedge T}^{T} ds\ p(\varepsilon^{-2}\int_t^s r_\tau^{-2} d\tau, \theta, \xi)(\Delta_s^{\varepsilon} g)(s,\xi)$$

$$(3.11) \qquad (G_2 g)(t,\theta) = \int_{S^{n-1}} d\xi \int_{t}^{(t+\varepsilon^2)\wedge T} ds\ p(\varepsilon^{-2}\int_t^s r_\tau^{-2} d\tau, \theta, \xi)(\Delta_s^{\varepsilon} g)(s,\xi)$$

and

$$(3.12) \qquad (Gg)(t,\theta)\ (= (Gg)(t)\ ) = \int_{S^{n-1}} d\xi \int_t^T ds\ (\Delta_s g)(s,\xi)\ .$$

Then the equation (3.8) is

$$(3.13) \qquad u_\varepsilon = 1 + G_1^{\varepsilon} u_\varepsilon + G_2^{\varepsilon} u_\varepsilon$$

and the equation (3.9) is

$$(3.14) \qquad u = 1 + Gu.$$

We shall introduce the following norms

$$\| g \|_\infty = \max_{t \in [t_0,T],\, \theta \in S^{n-1}} |g(t,\theta)|$$

$$\| g \|_2^t = \left\{ \int_{S^{n-1}} |g(t,\theta)|^2\, d\theta \right\}^{\frac{1}{2}}, \quad t \in [t_0,T]$$

and

$$||| g ||| = \left\{ \int_{t_0}^T dt \int_{S^{n-1}} |g(t,\theta)|^2\, d\theta \right\}^{\frac{1}{2}} .$$

It is obvious that $||| g ||| \leqslant (T-t_0)^{\frac{1}{2}} \| g \|_\infty$. In the following, $K_1, K_2, \ldots,$

$K(j)$, $K_{j,1}, \ldots$ are some positive constants independent of ε, $r = (r_t)$ and functions g.

Lemma 3.1 *)

(3.15) $$\| G_1^\varepsilon g - Gg \|_\infty \leq K_1 \varepsilon ||| g ||| .$$

Consequently,

(3.16) $$||| G_1^\varepsilon g - Gg ||| \leq K_2 \varepsilon ||| g ||| .$$

Lemma 3.2

(3.17) $$\| Gg \|_\infty \leq K_3 ||| g ||| .$$

Lemma 3.3 There exists a positive integer j such that

(3.18) $$\| G_2^\varepsilon g \|_\infty \leq K_4 \varepsilon ||| (1-\tilde{\Delta})^j g ||| .$$

Lemma 3.4 For each $j=0,1,2,\ldots$,

(3.19) $$||| (1-\tilde{\Delta})^j G_2^\varepsilon g ||| \leq K(j) \varepsilon^2 ||| (1-\tilde{\Delta})^j g ||| .$$

Lemma 3.2 is almost obvious by the integration by parts. Other lemmas are more or less easily proved using the eigenfunction expansion. So, let $\{\mu_m, S_m(\theta)\}_{m=0,1,2,\ldots}$ be the (real) eigensystem for $-\tilde{\Delta}$ on S^{n-1}. Thus, $S_m(\theta)$ are spherical harmonics and, $\mu_0 = 1$ and $S_0 \equiv 1$. We have

$$p\left(\int_t^s (\varepsilon r_\tau)^{-2} d\tau, \theta, \xi\right) = \sum_m \exp\left[-\mu_m/2 \int_t^s (\varepsilon r_\tau)^{-2} d\tau\right] S_m(\theta) S_m(\xi).$$

It holds that $\sup_\theta \sum_{m \geq 1} \mu_m^{-2k} |S_m(\theta)|^2 < \infty$ if $k > (n-1)/4$.

---

*) We assume, as we may, that $f(\varepsilon,\theta,t) - f(0,\theta,t) = O(\varepsilon)$ together with its derivatives in θ uniformly in $\theta \in S^{n-1}$, $t \in [t_0, T]$.

Also, the following estimate is well-known: for each $k = 0,1,..,$ there exists a constant c_k independent of ε, r, g and $t \in [t_0, T]$ such that

$$\| (1-\tilde{\Delta})^{k-1} \Delta_t^\varepsilon g \|_2^t \leq c_k \| (1-\tilde{\Delta})^k g \|_2^t \quad , \quad t \in [t_0, T].$$

We omit the proof of Lemma 3.1 since it is rather easy. As for Lemma 3.3,

$$|G_2^\varepsilon g(t,\theta)| = \left| \int_{S^{n-1}} d\xi \int_t^{(t+\varepsilon^2)\wedge T} ds\, p\left(\int_t^s (\varepsilon r_\tau)^{-2} d\tau, \theta, \xi\right) (\Delta_s^\varepsilon g)(s,\xi) \right|$$

$$= \left| \int_t^{(t+\varepsilon^2)\wedge T} ds \sum_m \exp[-\mu_m/2 \int_t^s (\varepsilon r_\tau)^{-2} d\tau\,]\, S_m(\theta) \times \int_{S^{n-1}} S_m(\xi)(\Delta_s^\varepsilon g)(s,\xi) d\xi \right|$$

$$= \left| \int_t^{(t+\varepsilon^2)\wedge T} ds \sum_m \frac{S_m(\theta)}{(1+\mu_m)^k} \exp[-\mu_m/2 \int_t^s (\varepsilon r_\tau)^{-2} d\tau\,] \times \int_{S^{n-1}} (1-\tilde{\Delta})^k S_m(\xi)(\Delta_s^\varepsilon g)(s,\xi) d\xi \right|$$

$$\leq \int_t^{(t+\varepsilon^2)\wedge T} ds \sqrt{\sum_m \frac{S_m(\theta)^2}{(1+\mu_m)^{2k}}} \sqrt{\sum_m \left\{ \int_{S^{n-1}} (1-\tilde{\Delta})^k S_m(\xi)(\Delta_s^\varepsilon g)(s,\xi) d\xi \right\}^2}$$

$$\leq K_5 \varepsilon \sqrt{\int_t^{(t+\varepsilon^2)\wedge T} \left\{ \| (1-\tilde{\Delta})^k \Delta_s^\varepsilon g \|_2^s \right\}^2 ds}$$

$$\leq K_5\, \varepsilon \, ||| (1-\tilde{\Delta})^k \Delta_s^\varepsilon g ||| \leq K_6 \varepsilon ||| (1-\tilde{\Delta})^{k+1} g ||| .$$

As for Lemma 3.4, we have

$$(1-\tilde{\Delta})^j G_2^\varepsilon g(t,\theta)$$

$$= \int_t^{(t+\varepsilon^2)\wedge T} \Big\{ \sum_m S_m(\theta) \exp[-\mu_m/2 \int_t^s (\varepsilon r_\tau)^{-2} d\tau\,] \int_{S^{n-1}} S_m(\xi)$$

$$\times\ [(1-\tilde{\Delta})^j \Delta_s^\varepsilon g](s,\xi)d\xi \Big\}\ ds$$

Let

$$h_m(s) = \int_{S^{n-1}} S_m(\xi)[(1-\tilde{\Delta})^j \Delta_s^\varepsilon g](s,\xi)\ d\xi\ .$$

Then

$$|||(1-\tilde{\Delta})^j G_2^\varepsilon g|||^2$$

$$= \int_{t_0}^T \sum_m \Big\{ \int_t^{(t+\varepsilon^2)\wedge T} \exp[-\mu_m/2 \int_t^s (\varepsilon r_\tau)^{-2} d\tau]\ h_m(s)\ ds \Big\}^2\ dt$$

$$\leq \sum_m \int_{t_0}^T \Big\{ \int_t^T I_{[\ s-t \leqslant \varepsilon^2]}\ \exp[-\mu_m(s-t)/2\varepsilon^2] \left| h_m(s) \right| ds \Big\}^2\ dt$$

$$= \sum_m \int_{t_0}^T dt \Big\{ \int_0^T (1+\mu_m)\ I_{[\tau \leqslant \varepsilon^2]} \exp[-\mu_m \tau/2\varepsilon^2]\ \frac{|h_m(\tau+t)|}{1+\mu_m}\ I_{[\tau+t \leqslant T]}\ d\tau \Big\}^2 \quad *)$$

$$\leq \sum_m \Big\{ \int_0^T (1+\mu_m)\ I_{[\tau \leqslant \varepsilon^2]}\ \exp[-\mu_m \tau/2\varepsilon^2]\ [\int_{t_0}^T \Big\{ \frac{h_m(\tau+t)}{1+\mu_m} \Big\}^2\ I_{[\tau+t \leqslant T]} dt\ ]^{\frac{1}{2}}\ d\tau \Big\}^2$$

$$\leq \sum_m \Big\{ [(1+\mu_m) \int_0^{\varepsilon^2} \exp(-\mu_m \tau/2\varepsilon^2) d\tau\ ]^2 \int_{t_0}^T \frac{h_m(t)^2}{(1+\mu_m)^2}\ dt \Big\}$$

$$= \int_{t_0}^T \Big(\sum_m \Big\{ (1+\mu_m)\ \frac{2\varepsilon^2}{\mu_m}\ (1-\exp[-\mu_m/2]) \Big\}^2 \cdot \frac{h_m(t)^2}{(1+\mu_m)^2} \Big) dt$$

$$\leq K_7\ \varepsilon^4 \int_{t_0}^T \sum_m \Big[\frac{h_m(t)}{(1+\mu_m)} \Big]^2\ dt$$

$$= K_7\ \varepsilon^4 \int_{t_0}^T \Big\{ \| (1-\tilde{\Delta})^{-1} (1-\tilde{\Delta})^j (\Delta_t^\varepsilon g) \|_{\frac{t}{2}} \Big\}^2\ dt$$

$$\leq K_8\ \varepsilon^4\ |||(1-\tilde{\Delta})^j g|||^2\ .$$

---

*) Generally, the following inequality holds:

$$\int [\int F(a,x) da\ ]^2\ dx \ \leqslant \Big\{ \int [\int F(a,x)^2 dx\ ]^{\frac{1}{2}}\ da \Big\}^2\ .$$

It is easy to see that

$$(G_1^\varepsilon g)(t,\theta) = \int_{S^{n-1}} d\xi \int_t^T k_\varepsilon(t,s,\theta,\xi) g(s,\xi) ds$$

and

$$|k_\varepsilon(t,s,\theta,\xi)| \leqslant K_9 \quad \text{for all} \quad t_0 \leqslant t < s \leqslant T, \quad \theta,\ \xi \in S^{n-1} .$$

Hence

$$\int_{S^{n-1}} \left\{ G_1^\varepsilon g(t,\theta) \right\}^2 d\theta \leqslant K_{10} \int_t^T \int_{S^{n-1}} g(s,\theta)^2 \, d\theta$$

and by repeating this,

$$|||(G_1^\varepsilon)^j g||| \leqslant \left(\frac{K_{11}^j}{j!} \right)^{\frac{1}{2}} |||g|||, \quad j = 1,2,\ldots .$$

Also, by Lemma 3.1,

$$|||(G_1^\varepsilon - G) g||| \leqslant K_{12} \varepsilon |||g||| .$$

Hence it is easy to deduce that

$$|||(I-G_1^\varepsilon)^{-1} - (I-G)^{-1}||| \longrightarrow 0 \qquad \text{as } \varepsilon \to 0.$$

Lemma 3.4 implies $|||G_2^\varepsilon||| \leqslant K_{13}\varepsilon^2$. Since

$$u_\varepsilon = (I - G_1^\varepsilon - G_2^\varepsilon)^{-1} 1$$

$$= \left\{ I - (I-G_1^\varepsilon)^{-1} G_2^\varepsilon \right\}^{-1} (I - G_1^\varepsilon)^{-1} 1$$

we can easily conclude that

$$|||u_\varepsilon - u||| \longrightarrow 0 \qquad \text{as} \quad \varepsilon \to 0$$

where $u = (I-G)^{-1}1$.

Finally, we show that $\|u_\varepsilon - u\|_\infty \to 0$ as $\varepsilon \to 0$. For this, we show first that, for each $j = 1,2,\ldots$, there exists K_j such that

(3.20) $$|||(1-\tilde{\Delta})^j u_\varepsilon||| \leqslant K_j.$$

The kernel $k_\varepsilon(t,s,\theta,\xi)$ of G_1^ε clearly satisfies that

$$\left|(1-\tilde{\Delta})_\theta^j k_\varepsilon(t,s,\theta,\xi)\right| \leqslant K_{1,j} \quad \text{for all } t<s,\ \theta,\xi$$

and hence

$$\|(1-\tilde{\Delta})^j G_1^\varepsilon u_\varepsilon\|_\infty \leqslant K_{2,j} \int_{S^{n-1}} d\theta \int_{t_0}^T |u_\varepsilon(t,\theta)|\, dt.$$

In particular,

$$|||(1-\tilde{\Delta})^j G_1^\varepsilon u_\varepsilon||| \leqslant K_{3,j} |||u_\varepsilon|||.$$

Also, by Lemma 3.4,

$$|||(1-\tilde{\Delta})^j G_2^\varepsilon u_\varepsilon||| \leqslant K_{4,j}\, \varepsilon^2 |||(1-\tilde{\Delta})^j u_\varepsilon|||.$$

Hence, from the identity

$$(1-\tilde{\Delta})^j u_\varepsilon = 1 + (1-\tilde{\Delta})^j G_1^\varepsilon u + (1-\tilde{\Delta})^j G_2^\varepsilon u ,$$

we have

$$(1-K_{4,j}\varepsilon^2)\, |||(1-\tilde{\Delta})^j u_\varepsilon||| \leqslant \sqrt{T-t_0} + K_{3,j} |||u_\varepsilon|||$$

concluding (3.20)

By Lemma 3.3 and (3.20),

(3.21) $$\|G_2^{\varepsilon}u_{\varepsilon}\|_{\infty} \leqslant K_4\varepsilon \|\!|\!|(1-\tilde{\Delta})^j u_{\varepsilon}\|\!|\!| \leqslant K_{13}\varepsilon \to 0.$$

Also

(3.22) $$\|G_1^{\varepsilon}u_{\varepsilon} - Gu\|_{\infty} \leqslant \|G_1^{\varepsilon}u_{\varepsilon} - Gu_{\varepsilon}\|_{\infty} + \|G(u_{\varepsilon} - u)\|_{\infty}$$

$$\leqslant K_1\varepsilon \|\!|\!| u_{\varepsilon}\|\!|\!| + K_3 \|\!|\!| u_{\varepsilon} - u \|\!|\!| \longrightarrow 0$$

by Lemma 3.1 and Lemma 3.2. Since

$$u_{\varepsilon} = 1 + G_1^{\varepsilon}u + G_2^{\varepsilon}u$$

and

$$u = 1 + Gu$$

(3.21) and (3.22) imply that

$$\|u_{\varepsilon} - u\|_{\infty} \to 0 \quad \text{as} \quad \varepsilon \to 0.$$

This convergence is clearly uniform in $r \in \mathcal{R}$ since the constants K_1, ..., $K(j)$, ... are independent of r. This completes the proof.

Acknowledgment We would like to thank N. Ikeda and S. Kotani for their valuable suggestions and discussions. The above proof of Th.2.2 was suggested by N. Ikeda which simplified our original proof. L^2-convergence in section 3 was suggested by Fujita and Kotani [3] where the same problem is discussed by a purely analytical method.

References

[1] E.Cartan; Lecons sur la geometrie des espaces de Riemann, Gauthier-Villars, Paris, 1963

[2] D.Durr and A.Bach; The Onsager-Machlup function as Lagrangian for the most probable path of a diffusion process, Comm. Math. Phys. 60(1978) 153-170.

[3] T.Fujita and S.Kotani; The Onsager-Machlup functions for diffusion processes, to appear.

[4] R.Graham; Path integral formulation of general diffusion processes, Z. Physik B 26(1979), 281-290

[5] N.Ikeda and S.Manabe; Integral of differential forms along the path of diffusion processes, Publ. RIMS,Kyoto Univ. 15(1979), 827-852.

[6] N.Ikeda and S.Watanabe; Stochastic differential equations and diffusion processes, Kodansha-John Wiley, 1980.

[7] H.Ito; Probabilistic construction of Lagrangean of diffusion processes and its application, Prog.Theoretical Phys. 59(1978), 725-741.

[8] H.Kunita and S.Watanabe; On square integrable martingales, Nagoya Math.J. 30 (1967), 209-245.

[9] L.Onsager and S.Machlup; Fluctuations and irreversible processes, I, II, Phys. Rev. 91(1953), 1505-1512, 1512-1515.

[10] R.L.Stratonovich; On the probability functional of diffusion processes, Select. Transl. in Math. Stat. Prob. 10(1971), 273-286.

ITO AND GIRSANOV FORMULAE FOR

TWO PARAMETER PROCESSES.

ATA AL-HUSSAINI

UNIVERSITY OF ALBERTA, CANADA

AND

ROBERT J. ELLIOTT

UNIVERSITY OF HULL, ENGLAND

Consider a single event that occurs at a random two parameter 'time' $T = (T_1, T_2) \in [0,\infty]^2$. The underlying probability space can be taken to be $\Omega = [0,\infty]^2$. A probability measure μ, which describes when the even occurs, is supposed given on Ω. In addition, we assume that

$$\mu\{(0,s_2) \cup (s_1,0) : (s_1,s_2) \in [0,\infty]^2\} = 0$$

so that the event occurs on neither axis. The σ-field $\overset{\circ}{F}_t$, $t \in [0,\infty]^2$, is defined to be the product of the σ-fields $F_{t_1} = \sigma\{I_{T_1 \geq s_1} : s_1 \leq t_1\}$ and $F_{t_2} = \sigma\{I_{T_2 \geq s_2} : s_2 \leq t_2\}$ on the two factors of Ω. $\overset{\circ}{F} = \vee_t \overset{\circ}{F}_t$ and F is the completion of $\overset{\circ}{F}$. The filtration $\{F_t\}$ considered is the right continuous completion of $\{\overset{\circ}{F}_t\}$. We suppose that T_1 and T_2 are independent under μ.

Write

$$F^1_{t_1} = \mu(]t_1,\infty] \times [0,\infty]),$$

$$F^2_{t_2} = \mu([0,\infty] \times]t_2,\infty]),$$

$$p_{t_i} = I_{t_i \geq T_i}$$

$$\tilde{p}_{t_i} = - \int_{]0,t_i \wedge T_i]} (F^i_{u_i^-}) \, dF^i_{u_i}$$

and $$q_{t_i} = p_{t_i} - \tilde{p}_{t_i}, \qquad i = 1,2.$$

Also, p_{t_i} will be abreviated as p_i etc., and for suitable integrands $\theta(u_1,u_2) \in L^1(\Omega,\mu)$

$$\theta \cdot p_1\tilde{p}_2 = \int_0^{t_2} \int_0^{t_1} \theta(u_1,u_2) dp_{u_1} d\tilde{p}_{u_2} \quad \text{etc.}$$

DEFINITION

A two parameter process X_{t_1,t_2} is a SEMIMARTINGALE if it is of the form

$$X_{t_1,t_2} = X_{o,o} + \theta^1.p_1p_2 + \theta^2.\tilde{p}_1p_2 + \theta^3.p_1\tilde{p}_2 + \theta^4.\tilde{p}_1\tilde{p}_2$$

where $\theta^i \in L^1(\Omega,\mu)$, $i = 1,2,3,4$. X can then be written either as

$$X_{t_1,t_2} = X_{o,o} + u.p_1 + v.\tilde{p}_1, \text{ where } u = \theta^1.p_2 + \theta^3.\tilde{p}_2,\ v = \theta^2.p_2 + \theta^4.\tilde{p}_2 \text{ or}$$

$$X_{t_1,t_2} = X_{o,o} + \bar{u}.p_2 + \bar{v}.\tilde{p}_2, \text{ where } \bar{u} = \theta^1.p_1 + \theta^2.\tilde{p}_1,\ \bar{v} = \theta^3.p_1 + \theta^4.\tilde{p}_1 .$$

NOTATION

If Y_{t_1,t_2} is a process write $\Delta^1Y_{t_1,t_2}$ for the process $Y_{t_1,t_2} - Y_{t_1-,t_2}$

$\Delta^2Y_{t_1,t_2}$ for the process $Y_{t_1,t_2} - Y_{t_1,t_2-}$

and $\Delta Y_{t_1,t_2}$ for the process $Y_{t_1,t_2} + Y_{t_1-,t_2-} - Y_{t_1-,t_2} - Y_{t_1,t_2-}$.

Our form of Ito's formula is given by the following theorem.

THEOREM

Suppose X_{t_1,t_2} is a semimartingale, as above, and $F : \mathbb{R} \to \mathbb{R}$ is a twice differentiable function.

Then

$$\begin{aligned} F(X_{t_1,t_2}) &= F(X_{o,o}) + \Delta F(X).p_1\, p_2 \\ &+ (\Delta^1(\bar{v}F'(u_1,u_2-)).p_1\, \tilde{p}_2 + (\Delta^2(vF'(u_1-,u_2)).\tilde{p}_1\, p_2 \\ &+ (F'(X_{u_1-,u_2-})\theta^4(u_1,u_2) + F''(X_{u_1-,u_2-})v(u_1,u_2-)\bar{v}(u_1-,u_2)).\tilde{p}_1\, \tilde{p}_2. \end{aligned}$$

PROOF.

Recall the differentiation formula for semimartingales in one dimensional time. There, if f is a twice differentiable function and X a semimartingale

$$\begin{aligned} f(X_t) &= f(X_o) + \int_o^t f'(X_{s-})dX_s + \tfrac{1}{2} \int_o^t f''(X_{s-})d\langle X^c,X^c\rangle_s \\ &+ \sum_{0<s\leq t} (f(X_s) - f(X_{s-}) - f'(X_{s-})\Delta X_s). \end{aligned} \tag{1}$$

We also quote the vector form of this result for the special case of a product:

$$X_tY_t = X_oY_o + \int_o^t X_{s-}dY_s + \int_o^t Y_{s-}\, dX_s + [X,Y]_t$$

where $$[X,Y]_t = \langle X^c, Y^c\rangle_t + \sum_{0<s<t} \Delta X_s \Delta Y_s \,. \tag{2}$$

In both time parameters the continuous part of our two paramter semimartingale X_{t_1,t_2} is zero, so no predictable quadratic variation terms occur in our formulae. Holding t_2 fixed and applying (1) to the X and F of our theorem:

$$F(X_{t_1,t_2}) = F(X_{o,o}) + \int_o^{t_1} F'(X_{u_1-,t_2})dX_{u_1,t_2}$$

$$+ I_{t_1\geq T_1}[F(X_{T_1,t_2}) - F(X_{T_1-,t_2}) - F'(X_{T_1-,t_2})\Delta X_{T_1,t_2}]$$

$$= F(X_{o,o}) + \int_o^{t_1} \Delta^1 F(X_{u_1,t_2})dp_1 + \int_o^{t_1} F'(X_{u_1-,t_2})v(u_1,t_2)d\tilde{p}_1. \tag{3}$$

(Note $X_{o,o} = X_{t_1,o} = X_{o,t_2}$).

Similarly

$$F'(X_{u_1t_2}) = F'(X_{o,o}) + \int_o^{t_2} \Delta^2 F'(X_{u_1,u_2})dp_2 + \int_o^{t_2} F''(X_{u_1,u_2-})\bar{v}(u_1,u_2)d\tilde{p}_2. \tag{4}$$

A similar expression for $F'(X_{u_1-,t_2})$ is immediate.

Also

$$F(X_{u_1,t_2}) = F(X_{o,o}) + \int_o^{t_2} \Delta^2 F(X_{u_1,u_2})dp_2 + \int_o^{t_2} F'(X_{u_1,u_2-})\bar{v}(u_1,u_2)d\tilde{p}_2$$

with a similar formula for $F(X_{u_1-,t_2})$. Using the product rule (2):

$$F'(X_{u_1-,t_2})v(u_1,t_2) = \int_o^{t_2} F'(X_{u_1-,u_2})dv + \int_o^{t_2} v(u_1,u_2-)dF'$$

$$+ I_{t_2\geq T_2}[\theta^2(u_1,T_2)\Delta^2 F'(X_{u_1-,T_2})],$$

because $v = \theta^2.p_2 + \theta^4.\tilde{p}_2$.

Substituting for dv , and for dF' from (4), in this product formula, and finally substituting in (3) the desired identity is obtained.

GIRSANOV'S THEOREM

Suppose $\bar{\mu}$ is a new probability measure absolutely continuous with respect to μ and write $L_{t_1,t_2} = E[\frac{d\bar{\mu}}{d\mu} | F_{t_1,t_2}]$

for the martingale of conditional expectations. We also suppose that μ is absolutely continuous with respect to $\bar{\mu}$, so that $\frac{d\bar{\mu}}{d\mu} > 0$ *a.s.* Note that T_1 and T_2 are not necessarily independent under $\bar{\mu}$.

$L_{t_1,t_2} - 1$ is a centred martingale and so has a representation

$$L_{t_1,t_2} = 1 + g.q_1q_2,$$

where $g \in L^1(\Omega)$.

Therefore,

$$\begin{aligned} L_{t_1,t_2} &= 1 + \int_0^{t_1} L_{s_1-,t_2} H(s_1,t_2)dq_{s_1} \\ &= 1 + \int_0^{t_2} L_{t_1,s_2-} \widetilde{H}(t_1,s_2)dq_{s_2} \end{aligned}$$

where

$$H(t_1,t_2) = L^{-1}_{t_1-,t_2} \int_0^{t_2} g(t_1,s_2)dq_{s_2}$$

and

$$\widetilde{H}(t_1,t_2) = L^{-1}_{t_1,t_2-} \int_0^{t_1} g(s_1,t_2)dq_{s_2} \quad .$$

Write

$$\alpha(s_1,s_2) = L^{-1}_{s_1-,s_2-} g(s_1,s_2)$$

$$R(s_1,s_2) = \alpha(s_1,s_2) - H(s_1,s_2-)\widetilde{H}(s_1-,s_2) .$$

Analogues of Girsanov's formula are given by the following results:

THEOREM

Consider a centred martingale X under measure μ, so that

$$X_{t_1,t_2} = \int_0^{t_1} \int_0^{t_2} \theta(s_1,s_2)dq_{s_2}dq_{s_1} \quad \text{for some } \theta \in L^1(\Omega,F,\mu).$$

Then, under the measure $\bar{\mu}$, the process

$$\begin{aligned} N_{t_1,t_2} &= \int_0^{t_1} \int_0^{t_2} \theta(s_1,s_2)(dp_2 - (1 + \widetilde{H}(s_1-,s_2)d\widetilde{p}_2)(dp_1 - (1 + H(s_1,s_2-)d\widetilde{p}_1) \\ &\quad - \int_0^{t_1} \int_0^{t_2} \theta(s_1,s_2)R(s_1,s_2)d\widetilde{p}_2d\widetilde{p}_1 \end{aligned}$$

is a weak martingale.

PROOF

We must show that $L_{t_1,t_2}N_{t_1,t_2}$ is a weak martingale under μ.

Write $L = 1 + (LH).p_1 - (LH).\widetilde{p}_1$

$$N = (\theta.p_2 - \theta(1+\widetilde{H}).\widetilde{p}_2).p_1 - (\theta(1+H).p_2 + \theta(R - (1+H)(1+\widetilde{H})).\widetilde{p}_2).\widetilde{p}_1$$

$$= \phi.p_1 - \psi.\widetilde{p}_1$$

where $\phi = \theta.p_2 - \theta(1+\widetilde{H}).\widetilde{p}_2$

and $\psi = \theta(1+H).p_2 + \theta(R - (1+H)(1+\widetilde{H})).p_2$.

Here, and in the sequel, the integrands are, where approproate, to be interpreted as left limits, so that, for example

$$(LH).p_1 = \int_0^{t_1} L_{s_1-,t_2} H(s_1,t_2)dp_1.$$

By the one-parameter product formula (2)

$$LN = (L\phi).p_1 - (L\psi).\widetilde{p}_1 + (NLH).q_1$$
$$+ (\phi LH).p_1$$
$$= (NLH).q_1 + L\psi.q_1 + (LK).p_1$$

where

$$K = \phi - \psi + \phi H.$$

However, using one-parameter calculus $H = \frac{R}{1+\widetilde{H}}.p_2 - R.\widetilde{p}_2$, and by the product formula:

$$\phi H = \frac{\phi R}{1+\widetilde{H}}.p_2 - \phi R.\widetilde{p}_2 + \theta H.p_2 - \theta H(1+\widetilde{H}).\widetilde{p}_2 + \frac{\theta R}{1+\widetilde{H}}.p_2 .$$

Substituting, we have that

$$K = \theta.p_2 - \theta(1+\widetilde{H}).\widetilde{p}_2 - \theta(1+H).p_2$$
$$- \theta R.\widetilde{p}_2 + \theta(1+H)(1+\widetilde{H}).\widetilde{p}_2 + \frac{R(\theta+\phi)}{1+H}.p_2$$
$$- \phi R.\widetilde{p}_2 + \theta H.p_2 - \theta H(1+\widetilde{H}).\widetilde{p}_2$$
$$= \frac{R(\theta+\phi)}{1+\widetilde{H}}.p_2 - R(\theta+\phi).\widetilde{p}_2 .$$

However, we can also write

$$L = 1 + (L\widetilde{H}).p_2 - (L\widetilde{H}).\widetilde{p}_2 .$$

Therefore,

$$LK = \frac{LR(\theta + \phi)}{1 + \widetilde{H}}.p_2 - LR(\theta + \phi).\widetilde{p}_2$$

$$+ (KL\widetilde{H}).q_2 + \frac{L\widetilde{H}R(\theta + \phi)}{1 + \widetilde{H}}.p_2$$

$$= LR(\theta + \phi).p_2 - LR(\theta + \phi).\widetilde{p}_2 + (KL\widetilde{H}).q_2$$

$$= \Lambda.q_2 \text{ , where } \Lambda = LR(\theta + \phi) + KL\widetilde{H}.$$

By Fubini's Theorem the order of integration can be changed, so that

$$(LK).p_1 = (\Lambda.q_2).p_1 = (\Lambda.p_1).q_2$$

$$= \Lambda_2.q_2$$

where $\Lambda_2 = \Lambda.p_1$

Write $\Lambda_1 = NLH + L\psi$.

Then

$$LN = \Lambda_1.q_1 + \Lambda_2.q_2$$

Because T_1 and T_2 are independent, $\Lambda_i.q_i$ $(i = 1,2)$ is an i-martingale. Therefore, LN is a weak μ-martingale

THEOREM

For $\theta \in L^1(\Omega, ,\mu)$ and $X = \theta.q_1q_2$, as in the theorem above $\widetilde{N} = X - \alpha.\widetilde{p}_1\widetilde{p}_2$, is a $\bar{\mu}$ weak martingale.

PROOF

The above result states that the process N is a $\bar{\mu}$ weak martingale. Simple algebra shows that

$$\widetilde{N} = N + (\theta\widetilde{H}.\widetilde{p}_2).(q_1 - \widetilde{H}p_1) + (\theta H.\widetilde{p}_1).(q_2 - \widetilde{H}p_2).$$

The final two terms are, respectively, $\bar{\mu}$ 1 and 2 martingales. These are, a fortiori, $\bar{\mu}$ weak martingales and so the result is proved.

L^p-INEQUALITIES FOR TWO-PARAMETER MARTINGALES

by L. CHEVALIER

(Splinter group on two-parameter processes).

I. INTRODUCTION.

The purpose of this talk is to give some recent results concerning inequalities of Burkholder-Gundy type for two-parameter martingales, and related analysis theorems.

Let $T = \mathbb{N}^i$ or $\mathbb{R}^i_+$, where $i=1$ or 2, and let M be a (L^2-bounded, null at the boundary of T) martingale indexed by T. As usually, we define

$$M^* = \sup_{t \in T} |M_t| \quad \text{and} \quad S(M) = [M, M]^{\frac{1}{2}}$$

(for a construction of $[M, M]$ in the two-parameter setting (continuous case), see [5], [6]). We are interested in the extension to two parameter martingales of the following classical inequalities of the one-dimensional time theory (the symbol $\simeq$ means that both sides define equivalent "norms") :

(D) for $p > 1$, $E^{1/p}[(M^*)^p] \simeq \sup_{t \in T} E^{1/p}[|M_t|^p]$ (J. L. Doob)

(B) for $p > 1$, $E^{1/p}[S^p(M)] \simeq \sup_{t \in T} E^{1/p}[|M_t|^p]$ (D. L. Burkholder)

(BG) for $p > 0$, and <u>regular</u> M,
$E^{1/p}[S^p(M)] \simeq E^{1/p}[(M^*)^p]$ (R. F. Gundy).

As an important application of this last result to classical analysis, let us mention L^p estimates between the Lusin Calderon area function and the Hardy-Littlewood maximal function for an harmonic function on $\mathbb{R}^n \times \mathbb{R}_+$ (D. L. Burkholder, R. F. Gundy, M. L. Silverstein).

(BG) inequality follows from two repartition inequalities of the type :

(G) $$P[A>\lambda] \leq C(P[B>\lambda] + \frac{1}{\lambda^2} E[B^2 ; B\leq\lambda]) ,$$

where A (resp. B) = M^* and B (resp. A) = S(M) .

We assume that the usual conditional independance F4 property ([3]) holds.

II. - <u>RESULTS FOR</u> $p > 1$.

These results (inequalities (D) and (B)) have been known to the specialists for a long time ; briefly speaking, they can be obtained from one-dimensional time inequalities by iteration methods.

III. - <u>RESULTS FOR</u> $p \leq 1$.

These results are deeper and more difficult to prove : iterations methods do not work, and stopping times (the main tool used to prove (G) inequalities) are no more available. This explains why new ideas were necessary.

The first one, which led to the proof of the estimate

(F) $$E^{1/p}[(M^*)^p] \lesssim E^{1/p}[S^p(M)]$$

is due to C. Fefferman ; using some very special properties of the <u>process</u> [M, M] , a kind of stopping method is used to derive a (G)-type inequality. Of course, a regularity condition is needed. We shall assume that the filtration $(\mathcal{F}_t)$ has the following property : every (L^2-bounded) martingale with respect to $(\mathcal{F}_t)$ has a continuous version. It is well known and easily seen that this condition is fulfilled in the two following particular cases : $(\mathcal{F}_t)$ is generated by (1) the tensor product of two independant one-parameter brownian motions or (2) the two-parameter Wiener process (the so-called "brownian sheet"). The proof runs as follows :

Given $\lambda > 0$, put $E = \{S(M) \leq \lambda\}$, $v_t = E[\mathbb{1}_E / \mathcal{F}_t]$ $(t \in T)$, and $F = \{(1-v)^* > \frac{1}{2}\}$. Since we have

$$P[M^* > \lambda] \leq P[M^* > \lambda \; ; F] + P(F^c)$$

and

$$P[F^c] \leq 4E[((1-v)^*)^2] \leq 64E[(1-v)^2] = 64P[S(M) > \lambda] \; ,$$

it is sufficient to obtain a good estimate for $P[M^* > \lambda \; ; F]$. Introducing the "stopped" martingale $\bar{M}$ defined by the double stochastic integral

$$\bar{M}_t = \iint_{[0,t]} \mathbb{1}_{\{v_s \leq \frac{1}{2}\}} dM_s \; ,$$

we can write

$$\begin{aligned} P[M^* > \lambda \; ; F] &= P[\bar{M}^* > \lambda \; ; F] \\ &\leq \frac{1}{\lambda^2} E[(\bar{M}^*)^2] \\ &\leq \frac{16}{\lambda^2} E[S^2(\bar{M})] \\ &\leq \frac{64}{\lambda^2} E[\iint_{[0,t]} v_s d[M,M]_s] \\ &= \frac{64}{\lambda^2} E[S^2(M) \; ; E] \; , \end{aligned}$$

and we are done.

Unfortunately, the properties of $[M,M]$ used here are by no means owned by the process $(M^*)^2$, and so the missing inequality cannot be obtained in this way.

The second idea appears in a paper ([8]) by M. P. and P. Malliavin ; they prove that, for bi-harmonic functions, the area function is a.e. finite where the maximal function is finite. By quantifying these methods, R. F. Gundy and E. M. Stein obtained ([7]) the extension to bi-harmonic functions of the above mentioned Burkholder-Gundy-Silverstein theorem. Nevertheless, the problem for martingales remained open, since the proof of the preceeding result was partially based upon geometric facts.

The next step was taken by J. Brossard ([1]), who obtained a (G)-type inequality and then derived the missing L^p estimate (for regular discrete martingales). The same result for bi-brownian martingales was obtained by J. Brossard and the author ([2]), using a different method : a suitable "Itô formula" allowed us to deduce the second inequality from the first one (a probabilistic proof of Gundy-Stein theorem in [7] follows from this result). The same method may be adapted in order to deal with arbitrary continuous martingales ([5], [6]). We shall inclıde a proof for the sake of completeness. We first indicate the idea in the (much easier) case of one-parameter martingale (cf. [4]) ; starting from the well-known identity

(1) $$M_t^2 = 2N_t + [M,M]_t \quad , \quad \text{where} \quad N_t = \int_0^t M_s dM_s \; ,$$

we can write, given $p > 0$,

$$S^p(M) \leq C_p((M^*)^p + (N^*)^{p/2}) \; .$$

Taking the expectations and using (F) estimate with the exponent $p/2$ leads to the inequality

$$E[S^p(M)] \leq C_p(E[(M^*)^p] + E[(\int_0^t M_s^2 d[M,M]_s)^{p/4}]) \; .$$

Since we have

$$\int_0^t M_s^2 d[M,M]_s \leq (M_t^*)^2 S_t^2(M) \; ,$$

we obtain, using Schwarz inequality and putting $X = E^{\frac{1}{2}}[S^p(M)]$, and $Y = E^{\frac{1}{2}}[(M^*)^p]$,

(2) $$X^2 \leq C_p(XY + Y^2) \; .$$

The missing L^p-estimate obviously follows from (2).

It we try to carry out the same proof over the two-parameter setting, we have to write down the analogous of (1), namely

(3) $$M_t^2 = 2\iint_{[0,t]} M_s dM_s + 2\tilde{M}_t + [M,M]_t^1 + [M,M]_t^2 - [M,M]_t \quad ,$$

where $\tilde{M}$ is a martingale (naively, $\tilde{M}$ is the sum over T of the products of the horizontal and vertical increments of M at time t) and $[M,M]^1_t$, for instance, is the value at time t_1 of the one-parameter "horizontal" martingale at level t_2 $(t=(t_1,t_2))$. Identity (3) is a special case of the following "Itô formula" ([2], [5], [6]), where φ is any real-valued C^4 function defined on the real line :

$$(4)\qquad \varphi(M_t) = \iint_{[0,t]} \varphi'(M_s)dM_s + \iint_{[0,t]} \varphi''(M_s)d\tilde{M}_s + \frac{1}{2}\int_0^{t_1} \varphi''\left(M_{s_1,t_2}\right)d[M,M]^1_{s_1,t_2} + \frac{1}{2}\int_0^{t_2} \varphi''\left(M_{t_1,s_2}\right)d[M,M]^2_{t_1,s_2} - \frac{1}{2}\iint_{[0,t]} \varphi''(M_s)d[M,M]_s - \iint_{[0,t]} \varphi'''(M_s)d[M,\tilde{M}]_s - \frac{1}{4}\iint_{[0,t]} \varphi^{iv}(M_s)d[\tilde{M},\tilde{M}]_s .$$

Using identity (3), inequality (F) with the exponent $p/2$, and classical (one dimensional time) Burkholder-Gundy inequalities to majorize $E[([M,M]^i)^{p/2}]$ $(i=1,2)$, we obtain the inequality

$$(5)\qquad E[S^p(M)] \le C_p(E[(M^*)^p] + E[S^{p/2}(\tilde{M})]) .$$

Thus, all we need is a good estimate for $E[S^{p/2}(\tilde{M})]$. Applying equality (4) with $\varphi(x) = x^4$, using (F) inequality with the exponent $p/4$ and Burkholder-Gundy inequalities for one-dimensional time processes, the same methods lead to the estimate

$$(6)\qquad E[S^{p/2}(\tilde{M})] \le C_p\left(E[(M^*)^p] + E^{\frac{1}{2}}[(M^*)^p]E^{\frac{1}{2}}[S^p(M)]\right) ,$$

and combining (5) and (6) gives the desired result.

IV. - <u>SOME UNSOLVED PROBLEMS</u>.

The complete extension of above results to <u>local</u> (= locally L^2, for instance) martingales is an open problem. It is also unknown whether a martingale M such as $E(M^*)$ be finite is a local martingale or not. Another unsolved question is the extension to the two-parameter setting of B. Davis L^1 estimates (for arbitrary martingales).

REFERENCES.

[1] J. BROSSARD, Généralisation des inégalités de Burkholder et Gundy aux martingales régulières à deux indices, C.R. Acad. Sc. Paris, 289, série A (1979), pp. 233-236.

[2] J. BROSSARD et L. CHEVALIER, Calcul stochastique et inégalités de norme pour les martingales bi-browniennes. Application aux fonctions bi-harmoniques, Ann. Inst. Fourier, Grenoble, 30, 4 (1980) (to appear).

[3] R. CAIROLI and J.B. WALSH, Stochastic integrals in the plane, Acta Math. 134 (1975), pp. 121-183.

[4] L. CHEVALIER, Démonstration "atomique" des inégalités de Burkholder-Davis-Gundy, Ann. Scient. Univ. Clermont, 67 (1979), pp. 19-24.

[5] L. CHEVALIER, Variation quadratique, calcul stochastique et inégalités de norme pour les martingales continues à deux paramètres, C.R. Acad. Sc. Paris, 290, série A (1980), pp. 847-850.

[6] L. CHEVALIER, Martingales continues à deux paramètres, Bull. Sc. Math. (to appear).

[7] R.F. GUNDY and E.M. STEIN, H^p theory for the poly-disc, Proc. Natl. Acad. Sc. USA, vol. 76, n°3 (1979), pp. 1026-1029.

[8] M.P. et P. MALLIAVIN, Intégrales de Lusin-Calderon pour les fonctions bi-harmoniques, Bull. Sc. Math., 2ème série, 101 (1977), pp. 357-384.

DIRICHLET PROCESSES

by H.Föllmer

From a measure-theoretic point of view, the class of semimartingales is the natural framework for the "general theory" of stochastic processes; see [1] , and in particular the theorem of Dellacherie and Bichteler on p.401 which states that any L^0-integrator is a semimartingale. There are, however, some natural procedures which lead out of this class. One important example is the theory of Dirichlet spaces. As shown by Fukushima [3] , it leads to functions which, if observed along the paths of the underlying Markov process, do no longer yields semimartingales but processes of type (3) below. This motivates the following decomposition theorem.

Let $X = (X_t)_{t\geq 0}$ be a square-integrable adapted process over $(\Omega, \underline{\underline{F}}, \underline{\underline{F}}_t, P)$ whose paths are right-continuous with limits from the left. Let us say that X is a Dirichlet process if its conditional energy vanishes in the following sense: For any $t > 0$, and for partitions $\tau = (t_0,..,t_n)$ and $\sigma = (s_0,..,s_m)$ of $[0,t]$,

(1) $$\sup_{\sigma \succ \tau} \sum_{t_i} E[(\sum_{t_i \leq s_j < t_{i+1}} E[X_{s_{j+1}} - X_{s_j} | \underline{\underline{F}}_{s_j}])^2]$$

converges to 0 as the step $|\tau|$ of the partition τ goes to 0 (the supremum is taken over all refinements σ of τ). Up to localization, Dirichlet processes are an extension of the class of continuous semimartingales, as shown by the following variant of the Doob-Meyer decomposition.

(2) Theorem. X is a Dirichlet process if and only if it is of the form

(3) $$X = M + A$$

where $M = (M_t)_{t\geq 0}$ is a square-integrable martingale, and $A = (A_t)_{t\geq 0}$ is a continuous adapted process with $A_o = 0$ and

(4) $$\lim_{\tau} \sum_{t_i} E[(A_{t_{i+1}} - A_{t_i})^2] = 0 \qquad (t > 0)$$

as the step $|\tau|$ of the partition τ of $[0,t]$ goes to 0. This decomposition into a martingale and a process "of zero energy" is unique.

Proof. 1) For a partition $\tau = (t_0,..,t_n)$ of $[0,t]$ consider the discrete Doob decomposition

(5) $$X_{t_i} = M^{\tau}_{t_i} + A^{\tau}_{t_i} \qquad (i=0,..,n)$$

along τ , where

$$A^{\tau}_{t_i} = \sum_{j<i} E[X_{t_{j+1}} - X_{t_j} | \underline{\underline{F}}_{t_j}]$$

(= 0 for i=0). For $\sigma \succ \tau$ we have

(6) $$E[(A^{\sigma}_t - A^{\tau}_t)^2] = \sum_{t_i} E[(A^{\sigma}_{t_{i+1}} - A^{\sigma}_{t_i})^2] + \sum_{t_i} E[(A^{\tau}_{t_{i+1}} - A^{\tau}_{t_i})^2]$$
$$- 2 \sum_{t_i} E[(A^{\tau}_{t_{i+1}} - A^{\tau}_{t_i})(A^{\sigma}_{t_{i+1}} - A^{\sigma}_{t_i})]$$

since $A^{\sigma} - A^{\tau}$ is a martingale along τ . Thus, condition (1) implies that (A^{τ}_t) is a Cauchy sequence in L^2. Define $A_t = \lim_{|\tau|\downarrow 0} A^{\tau}_t$, $(M_s)_{0\leq s\leq t}$ as a right-continuous version of the square-integrable martingale $E[X_t - A_t | \underline{\underline{F}}_s]$, $A_s = M_s - X_s$ $(0\leq s\leq t)$. Then

$$\sum_{t_i} E[(A_{t_{i+1}} - A_{t_i})]^2 \leq 2(E[(M_t - M^{\tau}_t)^2] + \sum_{t_i} E[(A_{t_{i+1}} - A_{t_i})^2]) ,$$

and the right side converges to 0 as $|\tau| \downarrow 0$. This implies that almost all paths of A are continuous, and so we have the existence of the decomposition (3). Its uniqueness as well as the extension from $[0,t]$ to $[0,\infty)$ is clear since a martingale with property (4) must be constant.

2) Suppose that X is of the form (3). The argument in [4] p.91 shows that the random variables A^{τ}_t in (5) converge to A_t in L^2 as $|\tau|\downarrow 0$. By (4), the last two sums in (6) converge to 0 . These two facts imply via (6) that

$$\sup_{\sigma\succ\tau} \sum_{t_i} E[(A_{t_{i+1}} - A_{t_i})^2]$$

converges to 0 as $|\tau| \downarrow 0$, and this is condition (1).

Although a Dirichlet process is in general not a semimartingale, hence not an integrator for general predictable integrands by the theorem mentioned above, it does admit a pathwise Itô calculus in the following manner. For a suitable sequence (τ_n) of partitions of $[0,\infty)$, almost all trajectories of X have a <u>quadratic variation</u> along (τ_n), i.e., there is an increasing process of the form

$$[X,X]_t = [X,X]^c_t + \sum_{s\leq t} (\Delta X)^2_s , \tag{7}$$

such that, P-almost surely,

$$[X,X]_t = \lim_n \sum_{\tau_n \ni t_i < t} (X_{t_{i+1}} - X_{t_i})^2 \tag{8}$$

in all continuity points of $[X,X]$. But for any trajectory with properties (7) and (8) and for any $F \in C^2$, the stochastic integral

$$\int_0^t F'(X_{s-})dX_s = \lim_n \sum_{\tau_n \ni t_i < t} F'(X_{t_i})(X_{t_{i+1}} - X_{t_i})$$

exists and satisfies the Itô formula

$$F(X_t) = F(X_0) + \int_0^t F'(X_{s-})dX_s + \frac{1}{2} \int_{(0,t]} F''(X_{s-})d[X,X]_s$$

$$+ \sum_{s\leq t} [F(X_s) - F(X_{s-}) - F'(X_{s-})\Delta X_s - \tfrac{1}{2}F''(X_{s-})\Delta X_s^2];$$

see [2].

<u>References</u>

[1] DELLACHERIE,C., et MEYER,P.A.: Probabilités et Potentiel, Ch.V-VIII Hermann, Paris (1980)

[2] FÖLLMER,H.: Calcul d'Itô sans probabilités. Preprint (1980)

[3] FUKUSHIMA,M.: Dirichlet forms and Markov processes. North Holland (1980)

[4] MEYER,P.A.: Intégrales stochastiques. Sém.Prob.I, Lecture Notes in Mathematics 39, Springer (1967)

Brownian motion, negative curvature, and harmonic maps.

by

W.S.Kendall,

Department of Mathematical Statistics,

The University, Hull.

This article discusses some relationships between the three topics mentioned in the title. It is now well-known that Brownian motion and complex function theory are closely linked, and that their interaction has been very fruitful. A natural generalisation of complex function theory is to 'harmonic' functions taking their values in Riemannian manifolds. The report of Eells and Lemaire (1978) surveys work in this field. It is tempting to ask whether Brownian motion and probabilistic techniques have anything to offer; the article is an attempt to demonstrate that they do. Details of proofs are not given; they will appear in a later production. It is hoped that their absence will allow the basic ideas to stand out clearly.

1. Brownian motion and the little Picard theorem.

1.1 Theorem (Picard)

If $f : C \to C \setminus \pm 1$ is holomorphic then it is constant.

Proof (after Burgess Davis (1975))

The complex Brownian motion $Z = BM(C)$ is recurrent on open sets. By Lévy's theorem (see 2.1) the image $f(Z)$ is a time-changed $BM(C \setminus \pm 1)$. If f is nonconstant then the time-change is nontrivial. The idea of the proof then runs as follows.

Let U be an open contractible neighborhood of the starting point of Z. If U is sufficiently small, and the starting point is not at a degenerate point of f, then $f(U)$ will also be open and contractible. There will be a peculiar random time T such that whenever $t > T$ if Z_t is in U then the path of $f(Z)$ up to time t will not be contractible in $C \setminus \pm 1$ relative to $f(U)$.

The proof that such a time T exists is the crux of the matter. Burgess Davis showed that such a T could be found, with $f(Z)|[0,t]$ not null-homotopic in $C \setminus \pm 1$ when $t > T$ and $Z_t \in U$. More recently McKean and Lyons (1980) have shown that 'not null-homotopic' can be replaced by 'not null-homologous', a stronger result correcting a contrary assertion of McKean (1969). Either way the existence of a nonconstant holomorphic function recieves a topological-probabilistic contradiction, as can be seen from the diagram below.

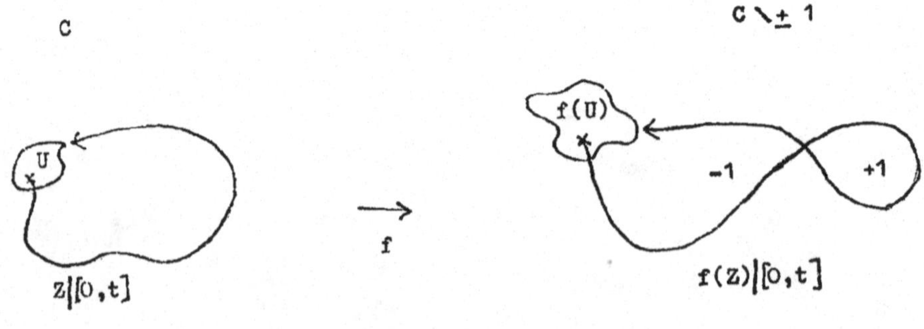

t > T

2. The key.

2.1 Theorem (Lévy)

Holomorphic functions preserve BM(C) up to time changes.

There is a straightforward proof using the Itô calculus (see McKean (1969)).

Naturally one asks oneself if such methods would work in higher dimensions, and for a first step one seeks a generalisation of 2.1.

2.2 Theorem (Fuglede (1978), Bismut (1980), Ducourtioux,...)

Let (M,g) and (N,h) be riemannian manifolds. Then

$$F : M \longrightarrow N$$

is a harmonic morphism, preserving harmonic functions and thus Brownian motions up to time changes, if

(i) dim M $\geqslant$ dim N ;

(ii) F is harmonic (in a generalised sense; see below);

(iii) F is horizontally conformal (see below).

Again one uses the Itô calculus.

The definitions involved appear at 2.3 and 2.4 ;

2.3 Definition (see Eells and Lemaire (1978))

F is harmonic if

$$g^{ij}\left[\frac{\partial^2 F^\gamma}{\partial x^i \partial x^j} - {}^M\Gamma^k_{ij}\frac{\partial F^\gamma}{\partial x^k} + {}^N\Gamma^\gamma_{\alpha\beta}\frac{\partial F^\alpha}{\partial x^i}\frac{\partial F^\beta}{\partial x^j}\right] = 0$$

where the Γ are the Christoffel symbols of the manifolds.

It is illuminating to compare this with the formula for the differential generator of BM(M,g) . This generator is the Laplacian operator Δ for M and is given by .

$$\Delta f = g^{ij}\left[\frac{\partial^2 f}{\partial x^i \partial x^j} - {}^M\Gamma^k_{ij}\frac{\partial f}{\partial x^k}\right]$$

where both here and everywhere in the paper we are using the summation convention for repeated indices. Note that the condition for F to be harmonic can be written

$$\Delta F^\gamma + g^{ij}\, {}^N\Gamma^\gamma_{\alpha\beta}\frac{\partial F^\alpha}{\partial x^i}\frac{\partial F^\beta}{\partial x^j} = 0 .$$

2.4 Definition

F is semiconformal (Fuglede terminology) or horizontally conformal (Eells-Lemaire terminology) if the following condition on the derivative dF holds;

when $dF \neq 0$ the restriction $dF\,|\,(\,\mathrm{Ker}\, dF\,)^\perp$ is conformal and surjective.

As an exercise, check Fuglede's theorem 2.2 using these definitions and Itô's lemma !

The fly in the ointment is best expressed by quoting the following result, now a classic;

2.5 Theorem

If $M = N = R^m$ and $m > 2$ then a map that is conformal in an open region is merely a restriction of a rigid motion in that region.

Since harmonic morphisms are essentially special conformal maps in such a context this result means THERE IS NO direct analogue of the beautiful interplay between complex analysis and Brownian motion as soon as the dimensions get at all high ! At least for Euclidean spaces of the same dimension the harmonic morphisms become very rare birds !

3. Geometrical generalisations.

The holomorphic universal covering of $C\setminus\pm1$ is U = ball C . Because C is simply connected we can complete the dotted arrow by a holomorphic map in the diagram below;

$$\begin{array}{ccc} & \dashrightarrow & U \\ & & \downarrow \\ C & \longrightarrow & C\setminus\pm1 \end{array}$$

(a commuting diagram of holomorphic maps)

Once we have shown the existence of the covering map from U to $C\setminus\pm1$ indicated in the diagram then the proof of the little Picard theorem is merely an application of Liouville's theorem. However the existence of the covering map (which must be holomorphic) is related to elliptic function theory. Thus the Burgess Davis proof is a genuine alternative.

This idea of a universal covering suggests a generalisation of the little Picard theorem. The surface U can be given a metric compatible with the analytic structure, and of constant negative curvature (think of a saddleback to visualise negative curvature). Skipping over the technicalities of curvature, we arrive at

3.1 Theorem (Goldberg, Ishihara, Petridis (1975))

Let (M,g), (N,h) be riemannian manifolds. Suppose

(a) the Ricci curvature of M is everywhere nonnegative-definite and also M is connected;

(b) the sectional curvatures of N are everywhere bounded above by $-H^2 < 0$.

Then there are no non-constant harmonic $F : M \longrightarrow N$ that are of K-bounded dilatation.

3.2 Definition

The map F has K-bounded dilatation if the quadratic form $(\xi^{\alpha\beta}) = \left(g^{ij} \frac{\partial F^\alpha}{\partial x^i} \frac{\partial F^\beta}{\partial x^j} \right)$ when diagonalised with respect to the riemannian inner product on N has eigenvalues $\lambda_1 \geq \lambda_2 \geq \dots \geq \lambda_{\dim N} \geq 0$ such that

$$\sqrt{(\lambda_1/\lambda_2)} \leq K \quad \text{everywhere} .$$

In coordinate-free terms this concerns the first two eigenvalues of $dF.dF^T$. The slogan is, infinitesimal spheres go to ellipsoids that are not too needle-shaped .

Note how the near harmonic morphisms of 3.1 are near-holomorphic functions in the two-dimensional case.

4. A probabilistic proof ?

Goldberg *et al* prove their theorems by geometrical arguments. It is possible to produce a proof by probabilistic methods, as is sketched below. The resemblance to Davis' proof is not obvious, but is there under the skin !

Recall the universal covering described in 3 . For the general case similar (but easier) arguments show that we need only consider the universal covering of N . This universal covering is described explicitly by the beautiful theorem of Cartan and Hadamard;

4.1 *Theorem.*

If N *is as described in* 3.1 *then the exponential map*

$$\exp : T_oN \longrightarrow N$$

is a universal covering map.

This theorem eliminates what topological problems one might have with N . The problems caused by the topology of M are over once one knows that BM(M) satisfies a 0-1 law at infinite time. This is certainly so if M is a Euclidean space. Results reported in Eells and Lemaire (1978) show that it suffices to know that the Ricci curvatures of M are nonnegative. There is one additional condition on N ; its sectional curvatures must be bounded not only above but also below. Then one has the theorem;

4.2 *Theorem*

With M *a Euclidean space and* N *a simply connected Riemannian manifold with sectional curvatures lying in the interval* $[-L^2,-H^2]$ *with* $H^2 > 0$, *there is no harmonic map from* M *to* N *that is nonconstant and of bounded dilatation.*

A full discussion of the probabilistic proof is in preparation (W.S.Kendall). The idea is as follows.

To see the method of proof just suppose that harmonic maps of bounded dilatation really did preserve Brownian motion up to time change. Then two theorems due to Prat and other French workers could be employed;

4.3 Theorem (Prat (1975))

BM(N,h) is not recurrent on compact sets.

4.4 Theorem (Prat (1975))

BM(N,h) has a limiting asymptotic direction.

4.5 Corollary

BM(N,h) has a nondegenerate random variable (the limiting direction) depending only on its remote future behaviour.

These results, which use the Rauch comparison theorem and the Cartan-Hadamard theorem of 4.1 , show that BM(N,h) does not have a 0-1 law. This is in contrast to $BM(R^{\dim M})$ for which any random variable depending only on its remote future behaviour must be degenerate . Consequently no timechanged version of BM(N,h) can be an image of $BM(R^{\dim M})$. So the theorem at 4.2 would be proved were our supposition correct.

In fact the true proof merely dissects the proofs of 4.3, 4.4, and 4.5 , and shows that they work for $F(BM(R^{\dim M}))$ as well.

The case of M a manifold of nonnegative Ricci curvature follows from proving a 0-1 law for BM(M) .

5. Results about distorted Brownian motion.

Suppose (M,g), (N,h) are riemannian manifolds.

5.1 Definition

Let B be $BM(M,g)$ and let $F : M \to N$ be harmonic and of K-bounded dilatation. Then we say $F(B)$ is a K-distorted Brownian motion on N.

An application of the Itô construction of $BM(M,g)$ and of Itô's lemma will show why 5.1 is a good definition. At least locally $BM(M,g)$ can be written as satisfying the Ito differential equation

$$dB^k = \sqrt{2}\,\sigma^{ki}\,dW^i - g^{ij}\,{}^M\Gamma^k_{ij}\,dt$$

where W is a Brownian motion on euclidean space of dimension equal to that of M, and (σ^{ki}) is the positive square root of (g^{ij}). This is essentially the Itô construction; an application of Itô's lemma shows

5.2 Remark

$$dF^\gamma(B) = \sqrt{2}\,\frac{\partial F^\gamma}{\partial x^i}\,\sigma^{ij}\,dW^j + \Delta F^\gamma\,dt$$

$$= \sqrt{2}\,\frac{\partial F^\gamma}{\partial x^i}\,\sigma^{ij}\,dW^j - g^{ij}\,\frac{\partial F^\alpha}{\partial x^i}\,\frac{\partial F^\beta}{\partial x^j}\,{}^N\Gamma^\gamma_{\alpha\beta}\,dt$$

where the last step uses the harmonic nature of F.

Note from 3.2 how the drift term involves the form $(\xi^{\alpha\beta})$ used to define the notion of K-bounded dilatation.

By analogy with 4.3 it is desirable to discuss the recurrence or otherwise of F(B) when N has sectional curvatures bounded above by $-H^2$ with $H^2 > 0$. Suppose that F(B) begins at o. Then by the Cartan-Hadamard theorem (4.1) we can consider the geodesic distance $\rho(F(B))$ of F(B) from o and in effect go into polar coordinates.

5.3 Theorem (compare 4.3)

Either lim inf $t^{-1}\rho(X_t) \geqslant$ constant for some time changed version $\hat{X}$ of F(B)

Or F(B) converges to a limit point in N.

Note that a truly random limit is not possible if BM(M,g) satisfies a 0-1 law.

The proof will not be described in detail here. The method is first to apply Itô's lemma to the process (F(B)) and then to note

(a) the martingale part of the radial process is a Brownian motion run at a random rate no faster than 2λ, where λ is the largest eigenvalue of the quadratic form $(\delta^{\alpha\beta})$ discussed at 5.2 and 3.2;

(b) the drift part of the radial process has differential

$$\delta^{\alpha\beta}\left\{\frac{\partial^2\rho}{\partial u^\alpha \partial u^\beta} - {}^N\Gamma^k_{\alpha\beta}\frac{\partial\rho}{\partial u^k}\right\} dt$$

which is related to the Hessian of the length function ρ.

The theory of variation of geodesics and the Rauch comparison theorem (introductions to these topics can be found in Cheeger and Ebin (1975) for example) enable the estimate that the coefficient of dt is greater than

$$\text{constant.}\ \lambda \cdot H \cdot \coth(H.\ \rho(o,\underline{n}))$$

at $\underline{n}$ in N , where the constant depends on K .

The conclusion of the theorem follows by comparison with Brownian motion on the line of constant drift.

5.4 Theorem (compare 4.4)

Because exp : $T_oN \to N$ is a diffeomorphism when N is simply connected it is possible to talk of the direction of F(B) from o . This direction tends to a limit if the sectional curvatures of N are bounded below by $-L^2$.

The proof first uses the lower bound on the curvatures and a similar argument to that of 5.3 to show that for some time changed version X of F(B)

$$\lim\ \inf\ t^{-1}\ \rho(\ X_t\ ) \leq \text{constant} \ .$$

A comparison is then made of the geometry of N with the geometry of a manifold of constant curvature $-H^2$. Basic arguments about Brownian motion then serve to establish the existence of a limiting direction. The arguments follow those of Prat (1975) very closely.

To use these results to show the nonexistence of nonconstant maps from suitable M it is necessary to make two more points. Clearly the distorted Brownian motion either runs to a stop or travels out to infinity in a limiting direction. Suitable M will have 0-1 laws, so it is necessary to show that either limit must have a nondegenerate distribution. This follows in the case of a limiting point by an application of the ideas of 5.3 using as new origin a supposed nonrandom limiting point. In the case of a limiting direction it is necessary to inspect the argument of 5.4 to show that the limiting direction will not be too far away from the direction of F(B(O)) if F(B(O)) is far enough away from the origin o . By such means a contradiction of the 0-1 law may be obtained.

6. Conclusion.

As mentioned above, it has been the purpose of this article to show that Brownian motion and probabilistic techniques can be applied to prove results in geometric function theory. The main result obtained, at 4.2 , is weaker than the corresponding result at 3.1 proved by geometric methods. However it is possible to extend 4.2 , for example by relaxing the curvature conditions to hold only off a compact set.

W.S.Kendall,
The University,
Cottingham Road,
Hull HU6 7RX .

7. References.

Bismut, J-M. Formulation geometrique du calcul de Ito, relevement de connexions et calcul des variations. C.R.A.S.(A) 290 427-429 (1980).

Cheeger,J. and Ebin,D. Comparison theorems in Riemannian Geometry. North-Holland, Amsterdam (1975) .

Davis,Burgess. Picard's Theorem and Brownian motion. Trans. Amer. Math. Soc. 213 353- 362 (1975) .

Ducourtioux unpublished.

Eells,J. and Lemaire,L. A report on harmonic maps. Bull. London Math. Soc. 10 1-68 (1978) .

Fuglede,B. Harmonic morphisms between Riemannian manifolds. Ann. l'inst. Fourier 28 107-144 (1978) .

Goldberg,S.I. ,Ishihara,T. and Petridis,N.C. Mappings of bounded dilatation of Riemannian manifolds. J. diff. Geom. 10 619-630 (1975) .

Kendall,W.S. Brownian motion and a generalised Picard's Theorem. in preparation.

McKean,H.P. Stochastic Integrals. Academic Press, New York (1969) .

McKean,H.P. and Lyons,T. On the winding of Brownian motion about two points in R^2 . in preparation.

Prat, J-J. Etude asymptotique et convergence angulaire du mouvement brownien sur une variete a courbure negative. C.R.A.S.(A) 280 1539-1542 (1975) .

LOCAL BEHAVIOUR OF HILBERT SPACE VALUED STOCHASTIC INTEGRALS AND THE CONTINUITY OF MILD SOLUTIONS OF STOCHASTIC EVOLUTION EQUATIONS *

P. Kotelenez

Forschungsschwerpunkt Dynamische Systeme
Universität Bremen
Bibliothekstraße
Postfach 330 440
D-2800 Bremen 33
F.R. Germany

0. The set-up

H and K are real separable Hilbert spaces with inner products $<.,.>_H$ and $<.,.>_K$ respectively, and $L(K,H)$ is the space of bounded linear operators from K to H. (Ω,P,F,F_t) is a probability space and $w(t)$ is a K-valued Wiener process with covariance operator W (cf. [5]) where the filtration F_t is assumed to be compatible with $w(t)$ in the usual way. E denotes the mathematical expectation and $||\cdot||_H$, $||\cdot||_{L(K,H)}$ etc. the norms in the corresponding spaces. Then, for an $L(K,H)$-valued, F_u adapted process $\Phi(u)$ with $u\in[a,b]$ and $\int_a^b E||\Phi(u)||^2_{L(K,H)}du < \infty$, the stochastic integral

$x(t) := \int_a^t \Phi(u)dw(u)$ exists and is a martingale which has a continuous version (cf. [5]).

1. The loglog law for $x(t)$

Theorem 1

A) If there is a function $N_t(\omega) > 0$ and a stopping time $\delta_t(\omega) > 0$, a.s., such that

$\int_t^b E||\chi_{[s \le \delta_t(\omega)]}(\Phi(s)-\Phi(t))||^4 ds < \infty$ and

$$\chi_{\{s \le \delta_t(\omega)\}}||\Phi(s)-\Phi(t)||_{L(K,H)} \le N_t(\omega), \text{ a.s.,}$$

* from a joint work with Ruth F. Curtain ([6])

then

$$\overline{\lim_{s\downarrow o}} \frac{||x(t+s)-x(t)||_H}{[b(s)]^r} = 0, \quad \text{a.s.,} \quad \text{for all} \quad r < 1,$$

where $b(t) = (2t|\log|\log t||)^{1/2}$.

B) Moreover, if there is an $\alpha > 0$ such that $\Phi(\cdot)$ is a.s. Hölder continuous with exponent α, then the set of limit points of

$$\left\{\frac{x(t+s)-x(t)}{b(s)}\right\}_{s\downarrow 0}$$

is equal, a.s., to the random closed set

$$E_t = \Phi(t)W^{\frac{1}{2}}S_1, \; S_1 \text{ being the unit sphere in } K .$$

<u>Proof</u> (sketch)

(i) $$x(t+u)-x(t) = \Phi(t)(w(t+u)-w(t)) + \int_t^{t+u} [\Phi(v)-\Phi(t)]dw(v)$$

$$=: I(u) \qquad + \qquad J(u) .$$

The set of limit points of $\left\{\frac{w(t+u)-w(t)}{b(u)}\right\}_{u\downarrow o}$ is equal to $W^{\frac{1}{2}}S_1$ (cf. [7], [6]).

(ii) $\eta(v) := \chi_{\{v \le \delta_t\}}[\Phi(v)-\Phi(t)]$, $\quad y(u) := \int_t^{t+u}\eta(v)dw(v)$.

Hence, locally, $y(u) = J(u)$.

(iii) $$||y(u)||_H^2 = \int_t^{t+u} \langle 2\eta^*(v)y(v),dw(v)\rangle_K + \int_t^{t+u} \operatorname{trace}(\eta(v)W\eta^*(v))dv$$

$$=: \quad m(u) \qquad + \qquad A_y(u)$$

by Itô's lemma in Hilbert space (cf. [4]).

(iv) $m(u)$ is a square integrable real valued martingale with quadratic variation

$$A_m(u) := \int_t^{t+u} 4||W^{\frac{1}{2}}\eta^*(v)y(v)||_K^2 \, dv,$$

hence,

$$\overline{\lim_{u\downarrow o}} \frac{|m(u)|^2}{(b(A_m(u)))^2} \le 1 \quad \text{(cf. [1]).}$$

(v) $$f(u) := u^{-1}A_m(u) \le cu^{-1} \int_t^{t+u} ||y(v)||_H^2 dv = c||y(\tilde{u})||_H^2, \text{ for } \tilde{u}\in(0,u),$$

by the mean value theorem, whence

$$\frac{(b(A_m(u))^2}{(b(u))^2} \le (f(u))^{1-\varepsilon}[(f(u))^{\varepsilon}(1+\frac{\log|\log f(u)|}{\log|\log u|}) \quad , 0 < \varepsilon < 1 .$$

Denoting the second factor on the r.h.s. by $F_\varepsilon(u)$ we have $F_\varepsilon(u) \to 0$, as $u \downarrow 0$, for all $\varepsilon \in (0,1)$. (cf. [6]).

(vi) From (iv) and (v) we obtain (cf. [1])

$$\overline{\lim_{u\downarrow 0}} \frac{(m(u))^2}{(b(u))^2} \le \overline{\lim_{u\downarrow 0}} \, c\|y(\tilde{u})\|_H^{2(1-\varepsilon)} F_\varepsilon(u) = 0.$$

Hence,

(vii) $$\overline{\lim_{u\downarrow 0}} \frac{\|y(u)\|_H^4}{(b(u))^2} \leqq 2c \, \overline{\lim_{u\downarrow 0}} \|y(\tilde{u})\|_H^{2(1-\varepsilon)} F_\varepsilon(u) + 2 \, \overline{\lim_{u\downarrow 0}} \frac{(A_y(u))^2}{(b(u))^2} = 0,$$

from which A) follows by induction (cf. Lemma 3.1 in [6]).
B) follows from A) and

$$f(u) = c\,\frac{1}{u} \int_t^{t+u} \|\eta^*(v)\|_{L(K,H)}^2 \|y(v)\|_H^2 dv = c\|\eta^*(\tilde{u})\|_{L(K,H)}^2 \|y(\tilde{u})\|_H^2$$

for $\tilde{u} \in (0,u)$ (see Lemma 3.1 B) in [6]). ∎

2. Levy's modulus of continuity for $w(t)$

Theorem 2

The set of limit points of

$$\left\{\frac{w(t_2)-w(t_1)}{(2(t_2-t_1)\,|\log\,(t_2-t_1)|)^{1/2}}\right\} \qquad (t_2-t_1) \downarrow 0, \; 0 \le t_1 < t_2 \le 1$$

is contained in $W^{\frac{1}{2}}S_1$, S_1 being the unit sphere in K .

The proof is based on Fernique's theorem which yields for our problem: There exists a $\gamma > 0$ such that

$$E \exp \gamma \|w(1)-w^m(1)\|_K^2 < \infty ,$$

γ independent of m, where $w^m := \sum_{i=1}^{m} \langle w(t),e_i\rangle_K e_i$, e_i being the eigenvectors or W . The proof is a combination of the proof of the loglog law for a Banach space valued Wiener process ([7]) and of Levy's modulus for a real-valued standard Wiener process (for details cf. [6]).

3. Continuity of mild solutions of stochastic evolution equations

Let us consider the follwing stochastic evolution equation

(1) $dz(t) = Az(t) + B(t,z(t))dw(t)$, $z(t_o) = z_o$, $0 \le t_o \le t \le b$,

where A is the generator of an analytic semigroup T_t on H, $B(t,x) \in L(K,H)$, measurable in (t,x) and uniformly Lipschitz in x. Under these assumptions (1) has a unique mild solution, which satisfies by definition (cf. [2])

$$z(t) = T_{t-t_o} z_o + \int_{to}^{t} T_{t-s} B(s,z(s))dw(s).$$

Theorem 3

If $\int_{to}^{t} E||B(s,z(s))||^4_{L(K,H)}$ is Hölder continuous, then the solution of (1) has a continuous version.

Proof

(i) Set $y(t) := \int_{to}^{t} T_{t-s} B(s,z(s))dw(s)$, $x(t) := \int_{to}^{t} B(s,z(s))dw(s)$.

From integration by parts ([3]) we obtain

$$y(t) = x(t) + A \int_{to}^{t} T_{t-s} x(s)ds \quad \text{a.s.}$$

(ii) Since A is closed and analytic

$A \int_{to}^{t} T_{t-s} x(s)ds$ is continuous, if $x(s)$ is Hölder continuous (cf. [5]), and the Hölder continuity of $x(s)$ follows from our assumption by Kolmogorov's law (cf. [6], Th. 2.1).

∎

References

[1] Arnold, L. The loglog law for multidimensional stochastic integrals and diffusion processes
Bull. of the Australian Math. Soc. 5 (1971) p. 351 - 356

[2] Arnold, L.; Curtain, R. F.; Kotelenez, P.
Nonlinear stochastic evolution equations in Hilbert space
Forschungsschwerpunkt Dynamische Systeme
Universität Bremen, Report Nr. 17 (1980)

[3] Chojnowska-Michalik, A.
Stochastic differential equations in Hilbert spaces and their applications
Ph. D. thesis, Institute of Mathematics, Polish Academy of Science, Warsaw 1976

[4] Curtain, R. F.; Falb, P. L.
Itô's lemma in infinite dimensions
J. Math. Analysis Appl. 31, No. 2, 1970, 434 - 448

[5] Curtain, R. F.; Pritchard, A. J.
Infinite dimensional linear system theory
Lecture notes in control and information sciences
Vol. 8, Springer-Verlag Berlin-Heidelberg-New York 1978

[6] Kotelenez, P.; Curtain, R. F.
Local behaviour of Hilbert space valued stochastic integrals and the continuity of mild solutions of stochastic evolution equations
Forschungsschwerpunkt Dynamische Systeme
Universität Bremen, Report Nr. 21 (1980)

[7] Kuelbs, J.; LePage, R.
The law of the iterated logarithm for Brownian motion in a Banach space
Trans. Amer. Math. Soc. 1973, 185, 253 - 264

SOME MARKOV PROCESSES AND MARKOV FIELDS IN QUANTUM THEORY, GROUP THEORY, HYDRODYNAMICS AND C^*-ALGEBRAS

by

Sergio Albeverio and Raphael Høegh-Krohn

1. Introduction

We would like to present here some examples of the recent interaction of stochastic analysis, in particular of stochastic integrals, with other fields of mathematics, with a personal orientation towards those inspired by physical problems. We will choose examples mainly from our joint work. Our plan is as follows (the titles refer to the main orientation of the applications):

I. Quantum Theory

I.1 Dirichlet forms

The theory of Dirichlet forms has roots in classical potential theory, see e.g. [I.1]. Its modern systematic development is mainly due to Beurling, Deny, Fukushima and Silverstein, see [I.1] - [I.3]. At this same Symposium Fukushima will give a report on the general theory [I.4].

I.1.1 Dirichlet forms and quantum mechanics

Let us mention quickly how Dirichlet forms enter quantum mechanics.
For any $f \in C_0^1(\mathbb{R}^d)$, let ∇f be the vector with component $\frac{\partial f}{\partial x^i}$ along the axis i.
Let $\nabla f \cdot \nabla g \equiv \sum_{i=1}^{d} \frac{\partial f}{\partial x^i} \frac{\partial g}{\partial x^i}$. Let ν be a positive Radon measure on $\mathbb{R}^d$, strictly positive on every non empty open set. The quadratic form

$$E(f,g) \equiv \frac{1}{2} \int \nabla f \cdot \nabla g d\nu \tag{1}$$

is called the energy form in $L^2(\mathbb{R}^d, \nu)$ given by ν. For useful working with such forms it is necessary to have them closable and general conditions are known for ν in order for this to be the case ([I.5], [I.6] , [I.2] ,[I.7]). Let consider from now on only forms E which are closable and let us denote the closure by the same symbol. There is a uniquely associated self adjoint positive operator H_ν such that $E(f,f) = (H_\nu^{1/2} f, H_\nu^{1/2} f)$, with $D(H_\nu^{1/2}) = D(E)$, where $D(\cdot)$ denotes domains and $(\ ,\ )_\nu$ denotes the scalar product in $L^2(\nu)$.
If ν is absolutely continuous on every open bounded subset of $\mathbb{R}^d$ with respect to Lebesgue measure dx, with density $\rho \equiv \varphi^2$, and besides $\varphi \in L^2_{loc}(\mathbb{R}^d)$ one has also $\nabla\varphi \in L^2_{loc}(\mathbb{R}^d)$, then $H_\nu = -\frac{1}{2}\Delta - \beta\cdot\nabla$ on $C_0^2(\mathbb{R}^d)$, with

$$\beta(x) \cdot \nabla \equiv \sum_{i=1}^{d} \beta_i(x) \frac{\partial}{\partial x_i} , \quad \beta_i(x) \equiv \frac{\partial}{\partial x_i} \ln \varphi(x).$$

Remark that this is already sufficient for closability of the energy form [I.5]. We also remark that if $\varphi \in L^2(\mathbb{R}^d)$ then we can write $\beta_i(x) = -\frac{1}{2} (\frac{\partial}{\partial x_i})^* 1$, where * is the adjoint in $L^2(\nu)$ and 1 is the function identically one in $L^2(\nu)$.
E is a particular Dirichlet form in the sense of [I.2], one which is local and has

$C_o^\infty(\mathbb{R}^d)$ as a core. By the general theory to E there is associated a (sub) Markov semigroup $P_t = e^{-tH_\nu}$, $t \geq 0$, strongly contractive on all $L^p(\mathbb{R}^d)$, $1 \leq p \leq \infty$, strongly continuous on all $L^p(\mathbb{R}^d)$, $1 \leq p < \infty$. There is a diffusion process ξ_t, with state space $\mathbb{R}^d$, and transition probabilities given by P_t, naturally associated with E [I.2]. We shall call it the diffusion process given by E. It satisfies in a natural sense (such that $\frac{d}{dt} E_s f(\xi_{t+s}-\xi_o-\int_o^{t+s}\beta(\xi_\tau)d\tau)$ $= E_s(\frac{1}{2}\Delta f(\xi_{t+s}-\xi_o-\int_o^{t+s}\beta(\xi_\tau)d\tau))$ for all $f \in \mathcal{S}(\mathbb{R})$, $t \geq s$, E_s being the conditional expectation with respect to the σ-algebra generated by $\xi_{s'}, s' \leq s$) the stochastic equation $d\xi_t = \beta(\xi_t)dt + dw_t$, where w_t is the standard Brownian motion on $\mathbb{R}^d$ ([I.5], [I.8]).
Consider now the map U_ρ: $f \in L^2(\mathbb{R}^d,\nu) \to \varphi f \in L^2(\mathbb{R}^d,dx)$. U_ρ is an isometry from $L^2(\mathbb{R}^d,\nu)$ onto the subspace $U_\rho L^2(\mathbb{R}^d,\nu)$ of $L^2(\mathbb{R}^d,dx)$. Let H be the self-adjoint positive operator in $U_\rho L^2(\mathbb{R},\nu)$ such that $U_\rho H_\nu \subset HU_\rho$ (i.e. HU_ρ is an extension of $U_\rho H_\nu$). If ν is such that $\varphi, \nabla\varphi, \nabla \ln\varphi$ and $\Delta\varphi/\varphi$ are all in $L^2_{loc}(\mathbb{R}^d)$ then one has $H = -\frac{1}{2}\Delta + V$ on $U_\rho U_\rho^* C_o^2(\mathbb{R}^d) \subset L^2(\mathbb{R}^d,dx)$, with

$$V(x) \equiv \frac{1}{2}\,\frac{\Delta\varphi(x)}{\varphi(x)} = \frac{1}{2}\left(\sum_{i=1}^d \beta_i(x)^2 + \frac{\partial}{\partial x_i}\beta_i(x)\right).$$

We have $U_\rho L^2(\mathbb{R}^d,\nu) = L^2(\mathbb{R}^d,dx)$, hence $U_\rho U_\rho^* C_o^2(\mathbb{R}^d) = C_o^2(\mathbb{R}^d)$ iff $\rho > 0$ ν-a.e., in which case H_ν and H are unitarily equivalent.
In this case H is the Schrödinger operator (in $L^2(\mathbb{R}^d,dx)$) for a particle of mass 1 (setting Planck's constant equal 2π) with potential V. Note that V is such that the infimum of the spectrum of H is 0, with corresponding eigenelement φ, not necessarily in $L^2(\mathbb{R}^d,dx)$. Conversely if $-\frac{1}{2}\Delta + \tilde{V}$ is a Schrödinger operator (defined in $L^2(\mathbb{R}^d,dx)$ as operator sum or as sum of quadratic forms) with lower bounded spectrum, then there exists a self-adjoint operator H_ν in $L^2(\mathbb{R}^d,\nu)$ such that $U_\rho H_\nu \subset HU_\rho$, where $H = -\frac{1}{2}\Delta + \tilde{V} - E$, E the infimum of the spectrum of $-\frac{1}{2}\Delta + \tilde{V}$, and $d\nu(x) \equiv \rho(x)dx$, $\rho \equiv \varphi^2$ and φ the solution of $\frac{1}{2}\frac{\Delta\varphi(x)}{\varphi(x)} = \tilde{V}(x) - E$. U_ρ is again defined by $(U_\rho f)(x) \equiv \varphi(x)f(x)$ for all $f \in L^2(\nu)$ of compact support and is thus an isometry from $L^2(\nu)$ onto $U_\rho L^2(\nu) \subset L^2(dx)$. If $\varphi = 0$ on a subset of positive Lebesgue measure then $U_\rho L^2(\nu)$ is a proper subset of $L^2(dx)$ and H_ν is only unitarily equivalent H on this subset. However under general conditions on $\tilde{V}$ one has $\varphi > 0$ Lebesgue a.e. (see e.g. [I.9],[I.43]), and in this case U_ρ is unitary, hence H_ν is unitarily equivalent H. Summarizing we see that the study of lower bounded Schrödinger operators is in a good sense equivalent with the study of (operators

associated with) energy forms in a suitable weighted L^2-space. This observation has essentially its roots in the canonical formalism for quantum mechanics and quantum field theory [I.10] and was exploited especially in the last few years (see e.g. [I.11], [I.5], [I.43], [I.45]).

Some basic uses of this relation in non relativistic quantum mechanics are:[12)]

1) To define the quantum Hamiltonian H (and hence the quantum dynamics) in cases where the potential is more general than a measurable function (see [I.11e)], [I.5], [I.11k),j),l)]).
2) Use the symmetric diffusion processes associated by the general theory [I.2], [I.3], [I.5] with energy forms to get probabilistic methods to discuss quantum mechanical quantities: e.g. ergodic behaviour [I.11 e)], [I.11 f)], behaviour at singularities [I.5], [I.11 k),j)], estimates on eigenvalues and eigenfunctions [I.11 i)], [I.12], [I.13], [I.17 c)], [I.43].
3) To discuss fundational questions in connection with stochastic mechanics (see e.g. [I.14], [I.13]).

Among the most recent results let us mention the detailed study of the relation between quantum mechanical tunneling and capacity [I.11 j)]. In particular in this work criteria in dimension $d > 1$ for attainability of boundaries are obtained, hence an analytic extension of Feller criteria to the case of dimensions higher than one.

The relation with the theory of stochastic equation with coefficients more singular than measurable functions has been made in [I.11 e)], [I.5], [I.4], [I.8]. One obtains, for stochastic equations with drift doefficient which are gradients and constant diffusion coefficients, existence of solutions. Uniqueness has also been discussed [I.11 f)], [I.11 g)], [I.12], [I.8], however only in the case of measurable (though singular) drifts. A study of the general case would be very valuable.

I.1.2 Dirichlet forms and quantum field theory

A (scalar) classical field $\varphi_x(t)$ is a function of time t and space $x \in \mathbb{R}^s$ satisfying a relativistic equation of motion, e.g.

$$\Box\, \varphi = -W'(\varphi)$$

with $\Box \equiv \frac{\partial^2}{\partial t^2} - \Delta$, Δ being the Laplacian in $\mathbb{R}^s$, W being a real-valued function on $\mathbb{R}$ with derivative W'. This can be looked upon as a Newton equation of motion for the position variable $\varphi_x(t) \equiv \varphi(t,x)$:

$$\frac{\partial^2}{\partial t^2} \varphi_x(t) = K(x,\varphi),$$

the "force" $K(x,\varphi)$ being given by

$$K(x,\varphi(t,x)) \equiv \Delta_x \varphi_x(t) - W'(\varphi(t,x)).$$

By analogy with (non relativistic) quantum mechanics heuristic canonical quantum field theory (as originated by [I.10 a)], see also e.g. [I.10]) seeks a space of functionals (formally $L^2(\mathbb{R}^{\mathbb{R}^s})$) in which the time zero field $\varphi_x(0)$ is quantized as multiplication by the "coordinate" $\varphi_x(0)$, time evolution and all Lorentz transformations being given by unitary groups in this space. In particular (as in quantum mechanics position x and moment π are quantized such that $[\pi,x] = -i$) one has to the "position operator" (time zero field) $\varphi_x(0)$ a canonical "momentum operator" $\pi_x(0)$ such that $[\pi_x(0), \varphi_y(0)] = \frac{1}{i} \delta(x-y)$. Formally then the Hamiltonian looks like

$$H = - \frac{1}{2} \int \frac{\delta^2}{\delta\varphi_x(0)^2} dx + \frac{1}{2} \int \varphi_x(0)(-\Delta_x)\varphi_x(0)dx + \int W(\varphi_x(0))dx. \tag{2}$$

If W is quadratic i.e. $W(\varphi_x(0)) = m^2\varphi_x(0)^2$ one has the so called free fields of mass m, if W is nonlinear one has so called self-interacting fields. Of course the problem consists in making sense of the formal expression for H (and the other quantities of interest). From our point of view at the moment we want just to remark that formally H is an infinite dimensional Schrödinger operator, hence we might hope to be able to do something like we did for the finite dimensional Schrödinger operators in Sect. I.1.1, i.e. look at the corresponding energy forms. These will be now energy forms on some space of functionals. It is therefore interesting to seek extensions of the formalism of Dirichlet forms from the case of $\mathbb{R}^d$ to the case of an infinite dimensional Hilbert space. This has been done by ourselves in [I.11 f)], see also [I.11 e)], [I.11 g)].

Let me describe shortly the formalism. The couple $(\mathbb{R}^d,\nu)$ in the definition (see I.1.1) of Dirichlet forms is replaced by a real separable Hilbert space $\mathcal{H}$, with a nuclear countable rigging $Q \subset \mathcal{H} \subset Q'$ (i.e. Q is a countable nuclear space densely contained in $\mathcal{H}$, Q' is its topological dual, the injection $Q \to \mathcal{H}$ being continuous and such that $\langle q,h\rangle = (q,h)$, $q \in Q$, $h \in \mathcal{H}$, with $\langle\ ,\ \rangle$ the dualization between Q and Q' and $(\ ,\ )$ the scalarproduct in $\mathcal{H}$) and a Radon probability measure ν on Q'. The space $C_0^1(\mathbb{R}^d)$ (which for Dirichlet forms on $\mathbb{R}^d$ with ν a probability measure can also be replaced by $C^1(\mathbb{R}^d)$) is replaced here by $FC^1(Q')$, i.e. the continuous bounded differentiable functions on Q' which are finitely based in the sense that to $f \in FC^1(Q')$ there exists an $\tilde{f} \in C^1(L_n)$, for some finite dimensional subspace L_n of $\mathcal{H}$, consisting of elements of Q, such that $f(\xi) = \tilde{f}(P\xi)$, where P is the projection from Q' onto L_n, $\xi \in Q'$. For $f \in FC^1(Q')$ and $q \in Q$ we denote the directional (Gâteaux) derivative of f in the direction q by $q \cdot \nabla f$ i.e.

$$q \cdot \nabla f(\xi) = \frac{d}{dt} f(\xi+tq)/_{t=0} \ .$$

Let e_i be a basis of $\mathcal{H}$ consisting of elements of Q. Let for $f \in FC^1(Q')$, $(\nabla f \cdot \nabla f)(\xi) \equiv \sum_i (e_i \cdot \nabla f(\xi))^2$ (note that the sum is finite, since f is finitely based). The quadratic form $f \to E(f,f) \equiv \frac{1}{2} \int \nabla f \cdot \nabla f d\nu$ in $L^2(d\nu)$ is called the energy form given by ν. As in finite dimensions we have to worry for which ν is such a form closable. A natural sufficient condition (a global version of the condition $d\nu = \varphi^2 dx, \varphi, \nabla\varphi \in L^2_{loc}(\mathbb{R}^d)$ in the finite dimensional case) is ν Q-quasi invariant (in the sense that $\nu(\cdot+q)$ equivalent $\nu(\cdot)$, for all $q \in Q$) and $\beta(q)(\xi) \equiv -\frac{1}{2}(q \cdot \nabla)^* 1(\xi) \in L^2(\nu)$. This is satisfied in all interesting cases of quantum fields; in the case where ν satisfies such a condition the closure of E, also denoted by E, has the properties of a Dirichlet form.[1)] Let H_ν the associated self-adjoint operator such that $(H_\nu^{1/2}f, H_\nu^{1/2}f) = E(f,f)$. One has $H_\nu = -\frac{1}{2}\Delta - \beta \cdot \nabla$ on $FC^2(Q')$, with $\Delta f = \sum_i (e_i \cdot \nabla)^2 f$, $(\beta \cdot \nabla f)(\xi) = \sum_i \beta(e_i)(\xi)\, e_i \cdot \nabla f(\xi)$, for all $\xi \in Q'$.

e^{-tH_ν}, $t \geq 0$ is a conservative Markov semigroup on $L^P(d\nu)$, for all $1 \leq p \leq \infty$, strongly continuous for all $1 \leq p < \infty$. There is a diffusion process ξ_t on a compactification of Q', satisfying the stochastic equation $d\langle q,\xi_t\rangle = \beta(q)(\xi_t)dt + d\langle q,w_t\rangle$, $q \in Q$, where $\langle\ ,\ \rangle$ and $\beta(q)(\cdot)$ are the natural extensions of the canonical pairing of Q, Q' resp. of $\beta(q)(\cdot)$, and w_t is the canonical Wiener process on $Q \subset \mathcal{H} \subset Q'$ i.e. such that its transition kernel has Fourier transform $e^{-\frac{t}{2}(q,q)}$, $q \in Q$. The stochastic equation is satisfied in the sense that for any $f \in \mathcal{S}(\mathbb{R})$ one has $\frac{d}{dt} E_o f(\langle q,w_t\rangle) = q^2 E_o(\Delta f\langle q,w_t\rangle)$ for $\langle q,w_t\rangle = \langle q,\xi_t\rangle - \langle q,\xi_o\rangle - \int_0^t \beta(q)(\xi_\tau)d\tau$, E_o being again the conditional expectation with respect to the σ-algebra generated by the $\langle q,\xi_o\rangle$ for all $q \in Q$. For more details see [I.11 f), g)]

Remark 1.

There are extensive results concerning the semigroup e^{-tH_ν}, the associated process and their relations with properties of the measure ν. Let us mention three types of results: 1) Ergodic decompositions of $L^2(d\nu)$ and ν according to the action of Q by translation resp. the action of e^{-tH_ν}. Ergodicity with respect to the semigroup e^{-tH_ν} ("T-ergodicity") is in general stronger than the one ("Q-ergodicity") with respect to Q-translation [I.11 e), f)];

2) There is a concept of strict positivity for a quasi-invariant measure ν. Shortly, let $Q' = R + R$, where R is a finite dimensional subspace of $\mathcal{H}$, consisting of elements of Q, and R its complement. ν is called strictly positive if the density of the conditional probability measure $\nu(\cdot|\eta)$ given (the σ-algebra generated by) R with respect to Lebesgue measure on R is strictly positive on compacts for ν - a.e. $\eta \in R$. A sufficient condition for ν to be strictly positive is e.g. that 1 is an analytic vector for the operator $(q \cdot \nabla)^*$ in $L^2(\nu)$. Again this assumption can be verified in the applications [I.11 f)]. Note that strict positivity of ν implies the equivalence of T-ergodicity and Q-ergodicity [I.11 f)].

3) There are general results concerning the conservation of ergodic, strict positivity and related properties under weak convergence. For details about such results, which apply to several interesting models of quantum fields, see [I.11 e), f)].

Remark 2.

Let us mention two recent investigations related to the above ones. Paclet [I.15] has given a formulation of the theory of Dirichlet forms and the relative notion of capacity for infinite dimensional spaces using Hilbert-Schmidt imbeddings. Kusuoka [I.16] has developed a theory of Dirichlet forms associated with a triple $Q \subset \mathcal{H} \subset Q'$ where Q' is a Banach space. The measure ν on Q' is supposed to be quasi-invariant and strictly positive. He defined infinite dimensional versions of Dirichlet forms described in the finite dimensional case by

$$\sum_{i,k} \frac{1}{2} \int a_{ik}(x) \frac{\partial f}{\partial x_i} \frac{\partial f}{\partial x_k} dx,$$

$((a))_{ik}$ some smooth strictly positive symmetric matrix. He exhibits a domain (a generalisation of Sobolev spaces) on which the form is closed. He realizes the sample paths in Q' and proves their continuity, thus obtaining a diffusion process on Q' with transition semigroup the one given by the Dirichlet form. It is not difficult to construct measures ν for which all the mentioned results hold. [I.11 f)] contains several examples. Perhaps the most interesting ones (but by no means the simplest ones) are those which are connected with relativistic local quantum fields. The construction of these will be shortly discussed in the next section.

I.2 Markov fields and symmetric Markov processes

I.2.1 Free fields

We saw in I.1.2 that there is interest in finding measures ν on some infinite dimensional space Q' yielding self-adjoint operators H_ν in $L^2(Q',\nu)$ s.t. $(H_\nu^{1/2}f, H_\nu^{1/2}f) = E(f,f)$ where E is the energy form given by ν and f is in its domain D(E). ν should be such that, as described in I.1.1 and I.1.2, the associated operator H_ν realizes the heuristic Hamiltonian (2). The difficulty in carrying through this program is of course that, even in the finite dimensional case, ν is only determined by solving an equation for its density φ, namely $\frac{1}{2}\Delta\varphi = V\varphi$(with V the potential). The simplest case is obviously the one in which $W'(\cdot)$ is a linear function (which corresponds in the finite dimensional case to the harmonic oscillator). For physical reasons (spectra of the masses for the free particles) one sets $W'(r) = m^2 r$, where m is a positive number. Then H is formally an infinite dimensional analog of the operator in $\mathbb{R}^n$

$$-\frac{1}{2}\Delta + \frac{1}{2}(x,A^2x) - \frac{1}{2}\operatorname{tr}A ,$$

where A^2 is a symmetric positive matrix, which plays the role of the operator $-\Delta_x + m^2$ in the infinite dimensional case.

The corresponding measure ν in the finite dimensional case is $d\nu(x) \equiv \left(\frac{A}{\pi}\right)^{1/2} e^{-(x,Ax)}dx$ (3)
and the corresponding operator H_ν is $-\frac{1}{2}\Delta + Ax\cdot\nabla$. This is the invariant measure for the Ornstein-Uhlenbeck velocity process ξ_t in $\mathbb{R}^n$.[2)] We have $e^{-tH_\nu}f(x) = E_x(f(\xi_t))$, where E_x means expectation for the process started at $x\in\mathbb{R}^n$, f being say a smooth bounded function (we assume, for transiency, $m\geq 0$, for $n\geq 3$, $m>0$ for n=1,2). The Gaussian measure on $C(\mathbb{R};\mathbb{R}^n)$ with mean zero and covariance $(-\frac{d^2}{dt^2}+A^2)^{-1}$ i.e. the fundamental solution of $(-\frac{d^2}{dt^2}+A^2)g(t) = 0$ is the measure μ_0 on the space of sample paths of the Ornstein-Uhlenbeck velocity process giving its distribution. In the finite dimensional case we can generate other pairs (ν,H_ν) with the desired properties by looking at the weak limit μ^V of

$$\frac{e^{-\int_{-t}^{t} V(\xi_t)dt}\,d\mu_0}{\int e^{-\int_{-t}^{t} V(\xi_t)dt}\,d\mu_0} \quad \text{as } t\to+\infty . \tag{4}$$

Note that $\int_{-t}^{t} V(\xi_t)dt$ is an additive functional of ξ_t.[8] The new measure ν, call it ν^V, is then the one given by $d\nu^V = \varphi^2 dx$, $\frac{1}{2}\Delta\varphi = V\varphi$. ν^V can be obtained from μ^V by restricting it to the σ-algebra generated by ξ_o^V, where ξ_t^V is the new process.[3] To attempt to do something similar in the infinite dimensional case the first thing to do is to look at an analogue of the Gaussian measure given by (3). As we saw above this should be the Gaussian measure with mean zero and covariance $A^{-1} = (-\Delta_x + m^2)^{-1/2}$, $x \in \mathbb{R}^s$.

It is well known that this can be realized as the canonical Gauss measure[4] ν_o (standard normal distribution) associated with the Sobolev Hilbert space obtained by closing $\mathcal{S}(\mathbb{R}^s)$ in the norm

$$\|f\|_{1/2} = \left(\int_{\mathbb{R}^s} f(-\Delta_x+m^2)^{1/2} f dx\right)^{1/2}.$$

ν_o is called the measure of the free time zero fields. What is the quantity which corresponds to the μ of the above Ornstein-Uhlenbeck process? By the finite dimensional analogy we are led to take for μ the Gaussian measure with mean zero and covariance $\left(-\frac{d^2}{dt^2} + A^2\right)^{-1}$ with $A^2 = -\Delta_x + m^2$, $t \in \mathbb{R}$, $x \in \mathbb{R}^s$.
μ is then the canonical Gauss measure with mean zero and covariance $(-\Delta_d + m^2)^{-1}$, with Δ_d the Laplacian on $\mathbb{R}^d$, $d \equiv s+1$, i.e. the canonical Gauss measure μ_o associated with the Sobolev Hilbert space obtained by closing $\mathcal{S}(\mathbb{R}^d)$ in the norm $\|f\|_1 \equiv \left(\int_{\mathbb{R}^d} f(-\Delta_x+m^2)\, f dx\right)^{1/2}$.

μ_o is the measure of the so called free (Euclidean Markov) fields [I.20]. Let us recall some of its properties, which also explain its name. μ_o can be realized as the measure on $\mathcal{S}'(\mathbb{R}^d)$ with Fourier transform (with respect to the $\mathcal{S}$-$\mathcal{S}'$-dualization):

$$\exp\left(\tfrac{1}{2}\int_{\mathbb{R}^d} \varphi(x)(-\Delta+m^2)^{-1}(x-y)\varphi(y)dxdy\right), \text{ where}$$

$(-\Delta+m^2)^{-1}(x-y)$ is the kernel of $(-\Delta+m^2)^{-1}$ i.e. the Green's function of $-\Delta + m^2$. μ_o is invariant under the transformations in $\mathcal{S}'(\mathbb{R}^d)$, induced by the Euclidean group acting in $\mathbb{R}^d$. The associated generalized random field $\xi(x)$, $x \in \mathbb{R}^d$ has mean 0 and covariance $(-\Delta + m^2)^{-1}(x-y)$, and $\rho \to \langle\rho,\xi\rangle$ can be looked as a linear process ([I.21]), where for any measure ρ of finite energy ($\int d\rho(x)(-\Delta+m^2)^{-1}(x-y)d\rho(y) < \infty$) one has $\langle\rho,\xi\rangle \in L^2(\mu)$.

It has been shown by Symanzik, Wong, Nelson, Molchan [I.20] that μ_o has the global Markov property in the following sense. Let for any Borel subset Λ of $\mathbb{R}^d$ $B(\Lambda)$ be the σ-algebra generated by all the linear functions $<\rho,\xi>$ with ρ of finite energy and supp $\rho \subset \Lambda$. Let C be a Lipschitz hypersurface in $\mathbb{R}^d$ dividing $\mathbb{R}^d$ into two disjoint components Ω_+^C and Ω_-^C. Let $f_\pm \in B(\Omega_\pm^C)$ (meaning $f_\pm$ measurable with respect to $B(\Omega_\pm^C)$) (and positive or integrable). Then

$$E(f_+f_-|C) = E(f_+|C)E(f_-|C),$$

where $E(\cdot|C)$ means conditional expectation with respect to B(C). This global Markov property has important consequences.

In general if μ is the measure of a global Markov random field (in the above sense) $\xi(x)$, $x = (x^o,\dots,x^{d-1}) \equiv (t,y) \in \mathbb{R}^d$, $t = x^o \in \mathbb{R}$, $y = (x^1,\dots,x^{d-1}) \in \mathbb{R}^{d-1}$ taking for C the hyperplane $x^o = 0$ we have that for $\varphi \in \mathcal{S}(\mathbb{R}^s)$, $s \equiv d-1$, $t \to E(f(<\delta_t \otimes \varphi, \xi>)|C)$ is a Markov semigroup on $L^2(S'(\mathbb{R}^s),\nu)$, where ν is identified with the restriction of μ to the σ-algebra B(C). Hence there is a Dirichlet form $(H_\nu^{1/2}f, H_\nu^{1/2}f)$ on $L^2(S'(\mathbb{R}^s),\nu)$ generating it. In the case where μ is the free field then, as shown in[I.11e)]the Markov semigroup is e^{-tH_ν}, with H_ν the energy form $E(f,f) = \frac{1}{2}\int \nabla f \cdot \nabla f d\nu$ given by ν.

It has been proven [I.11 e),f)] that the triple $Q = \mathcal{S}(\mathbb{R}^s)$, $\mathcal{H} = L^2(\mathbb{R}^s)$ (real), $Q' = \mathcal{S}'(\mathbb{R}^s)$ and this measure ν satisfy all properties of the general theory of Dirichlet forms in infinite dimensional spaces (I.1.2). Hence the results of the general theory apply and yield e.g. that the measure ν_o of the time zero free fields is $\mathcal{S}(\mathbb{R})$-quasi invariant, strictly positive, $\mathcal{S}(\mathbb{R})$-ergodic, has 1 as an analytic vector for $(\varphi \cdot \nabla)^*$, $\varphi \in \mathcal{S}(\mathbb{R}^s)$.

Let us call $\eta_t(\cdot)$ the process with values in $\mathcal{S}'(\mathbb{R}^s)$ given by the energy form E determined by ν. Then one can identify (in the sense of versions) $\eta_t(\cdot)$ with $\xi(t,\cdot)$. In particular the free Markov random field $\xi(t,\cdot)$ appears as a diffusion process in the direction of the x^o-axis with linear drift $\beta(\varphi)(\eta_o) = - <(-\Delta + m^2)^{1/2}\varphi, \eta_o>$, Δ being the Laplacian in $\mathbb{R}^s$. Note that there is nothing special about the x^o-axis, the whole discussion holds for any hyperplane C. H_ν is the so called energy operator for the free quantum fields.

I.2.2 Interacting fields

According to the discussion at the beginning of I.2.1 the finite dimensional situation suggests [3)] constructing new measures ν from the Gaussian one by a limit inspired by the one in (3). The analogue of the additive functional in (3) should be $\int_\Lambda V(\xi(x))dx$ where $\xi(x)$ is the free field on $\mathbb{R}^d$ and Λ is a cube in $\mathbb{R}^d$. The obvious difficulty here is that ξ is almost surely more singular than a measure, so it is not clear how one can define functions of it. As well known this problem has only been solved up to now in the cases d = 2 (in the case d = 3 a construction of a limit measure μ is possible [I.25], however requires modifications of the procedure we are describing and the probabilitstic interpretation of the construction is much less developed, as of yet). A particular simple example is the following one [I.26].

Let $U_{\Lambda,\kappa}(\xi) \equiv \int_\Lambda e^{\alpha\xi_\kappa(x)} e^{-\alpha^2/2 G_\kappa(0)} dx$,
$\xi \in \mathcal{S}'(\mathbb{R}^2)$, Λ a bounded Borel subset of $\mathbb{R}^2, \alpha \in \mathbb{R}$, ξ_κ a regularization of ξ by convolution with a smooth function χ_κ whose Fourier transform has support in a ball of radius κ . The following Theorem holds

Theorem: $U_{\Lambda,\kappa}$ is a positive L^2-martingale with respect to the σ-algebra generated by the $\xi_{\kappa'}$, $\kappa' \leq \kappa$. $e^{-U_{\Lambda,\kappa}}$ is a submartingale, converges a.s. and in L^2 as $\kappa \to \infty$. For d = 1,2 $|\alpha| < \sqrt{4\pi}$ the limit is e^{-U_Λ} with U_Λ a nonnegative, L^2-function not identically constant, in fact

$$U_\Lambda(\cdot) = \frac{d \int_\Lambda \mu_0(\cdot - \alpha G_x) dx}{d\mu_0(\cdot)} ,$$

with $G_x(\cdot) = G(x-\cdot)$ and $G(x-y)$ the kernel of $(-\Delta+m^2)^{-1}$. For $d \geq 3$ or (d = 2, $|\alpha| > \sqrt{8\pi}$) one has $U_\Lambda = 0$ a.s.

Remark: The proof for d = 2, $|\alpha| < \sqrt{4\pi}$ has been given in [I.26 b)] ($\sqrt{4\pi}$ is determined by the fact that $G(x) \approx -\frac{1}{2\pi} \ln x$ as $x \approx 0$). The situation for $\sqrt{4\pi} < \alpha \leq \sqrt{8\pi}$ is open. The "critical" value $\sqrt{4\pi}$ is of course the same as the one for the trigonometric interactions (Sine-Gordon model) [I.27].
The proofs for $d \geq 4$ resp. $d \geq 3$ and d = 2, $|\alpha| > \alpha_0$ has been given in [I.26 c)] resp. [I.26 d)]. The case d = 2, $|\alpha| > \sqrt{8\pi}$ has been covered in [I.28].

Let now u(r) be any real-valued function on $\mathbb{R}$ of one of the forms

1) $u(r) = \sum^{2N} a_s r^s$, $a_{2N} > 0$

2) $u(r) = \int e^{\alpha r} d\rho(\alpha)$

3) $u(r) = \lambda \int \cos(\alpha r + \theta) d\rho(\alpha)$

where ρ is any measure with support in $(-2\sqrt{\pi}, 2\sqrt{\pi})$. In all cases we have $u(r) = \sum_n c_n r^n$ for suitable c_n. Define $U_\Lambda^u(\xi) =: u(\xi): (\chi_\Lambda)$, with $:u(\xi): (\chi_\Lambda) = \sum c_n : \xi^n : (\chi_\Lambda)$, where $:\xi^n: (\chi_\Lambda)$ is such that $\sum \frac{\alpha^n}{n!} :\xi^n: (\chi_\Lambda) = U_\Lambda(\xi)$, U_Λ being the function in above theorem.[5] We shall call (Λ, U_Λ^u) an additive functional of the free field μ_o, or shortly an interaction (given by the function u).

It is shown in all cases 1) - 3) (a_n small in 1), λ small in 3)) that $\mu_\Lambda \equiv \frac{e^{-U_\Lambda}\mu_o}{e^{-U_\Lambda}\mu_o}$ converges weakly as $\Lambda \to \mathbb{R}^d$ to a measure μ on $S'(\mathbb{R}^2)$. This measure is Euclidean invariant (and yields relativistic quantum fields giving in particular the quantized solution of

$$\Box\varphi(t,y) + m^2\varphi(t,y) + u'(\varphi(t,y)) = 0).$$

For proofs of these statements see [I.29], [I.26], [I.27], (and references therein). The question now arises: can one put the constructed μ in connection with Dirichlet forms? This question was first discussed in [I.11 e)], [I.11 f)], [I.30] and following answer found. Let ν be the restriction of μ to the σ-algebra generated by the fields $<\delta_o \otimes \varphi, \xi>$, where ξ is the random fields with distribution μ and $\varphi \in \mathcal{S}(\mathbb{R}^s)$, where we use again the splitting $\mathbb{R}^d = \mathbb{R} \times \mathbb{R}^s$, $\mathbb{R}$ being the x^o-axis. Then it can be proven [I.11 f)], [I.30] that for above models ν has the property of the general theory of I.2., so that again these ν provide examples for the general theory. The relation between the diffusion process η_t associated with ν and the random field $\xi(t,y)$ is such that the Osterwalder-Schrader energy operator[6] H coincides as a form on twice-differentiable cylinder functions in $L^2(d\nu)$ with H_ν.

Next question: is μ global Markov, so that one can, as in the free field case, get directly Markov processes out of the associated random field?

This is a difficult question [7] and it remained open for about 10 years. By now one knows [I.30] that at least in the case where μ is obtained starting from a u of the trigonometric form 3), with λ sufficiently small, then μ is globally Markov. This settles the questions of existence of a non Gaussian random field on $\mathbb{R}^2$, homogeneous with respect to the full Euclidean group on $\mathbb{R}^2$ and having the global Markov property. [9]
The precise theorem is the following

<u>Theorem</u> [I.30 a)]. Let d = 2 and let μ be the weak limit of

$$\mu_\Lambda \equiv \frac{e^{-U_\Lambda}\mu_o}{e^{-U_\Lambda}\mu_o}, \quad U_\Lambda(\cdot) = :u(\cdot):(\chi_\Lambda), \quad u(r) = \lambda \cos(\alpha r+\theta),$$

λ sufficiently small, $0 \leq \theta < 2\pi$, $|\alpha| < \sqrt{2\pi}$.
Then μ is globally Markov.
For the proof we have to refer to the original work [I.30 a)] (or the expositions [I.30 c), d)]). Here we give just a summary. One first remarks that it is enough to show $E_{\mu_\Lambda}(f|\partial\Lambda \vee C) \to E(f|C)$, E the expectation with respect to μ, Λ bounded Borel $\uparrow \mathbb{R}^d$, $\Lambda_o \subset \Lambda$ Borel, $f \in B(\Lambda_o)$ where E_{μ_Λ} is the expectation with respect to μ_Λ (this is seen using the fact that μ is the weak limit [I.27 b)] of μ_Λ and μ_Λ has the global Markov property, since μ_o has it and U_Λ is an additive functional of μ_o; this then implies $E(f|(\mathbb{R}^d-\Lambda)\vee C) = E_{\mu_\Lambda}(f|\partial\Lambda \vee C))$.
Express now E_{μ_Λ} through E_{μ_o}. Furthermore one makes the basic observation that

$E_{\mu_o}(g|C_o)(\eta) = E_{\mu_o^{C_o}}(g_\eta^{C_o})$ with $g_\eta^{C_o}(\xi) \equiv g(\xi + \psi_\eta^{C_o})$, g positive μ_o-measurable,

C_o an arbitrary piecewise smooth curve and ψ^{C_o} the solution of the Dirichlet problem $(-\Delta + m^2)\,\psi_\eta^{C_o}(x) = 0$, $x \notin C_o$; $\psi_\eta^{C_o}(x) = \eta(x)$, $x \in C_o$, $\eta \in S'(\mathbb{R}^2)$, $\mu_o^{C_o}$ the Gaussian measure with mean 0 and covariance $(-\Delta_{C_o} + m^2)^{-1}$, where Δ_{C_o} is the Laplacian with Dirichlet boundary conditions on C_o.
Then it is seen that it suffices to show
$\psi_\eta^{\partial\Lambda \cup C}(x) \to \psi_\eta^C(x)$ for μ-a.e. $\eta \in S'(\mathbb{R}^2)$, locally uniformly in $x \notin C$ (and this is done by two-dimensional potential theory, using only that $E(\langle\varphi,\xi\rangle^2) \leq \text{const}\ E_{\mu_o^C}(\langle\varphi,\xi\rangle^2)$) and

$$e^{-U_{\Lambda,\eta}}\mu_o^{\partial\Lambda \cup C} \Big/ \int e^{-U^C_{\Lambda,\eta}}\, d\mu_o^{\partial\Lambda \cup C} \to \mu_\eta^C$$

where $\mu_\eta^C(\cdot)$ is μ conditioned with respect to B(C) (the latter convergence is done by using the essential locality of the additive functional U_Λ i.e. a "cluster expansion" similar to those of [I.29], [I.27 b)].

It is expected that the global Markov property holds for all known two dimensional models and there is work in progress ([I.33]).
Having the global Markov property essentially the same results concerning the construction of associated Markov processes as in the free fields case hold. With notations as in I.2.1, calling $\xi(t,y)$, $t\in\mathbb{R}$, $y\in\mathbb{R}$ the random field corresponding to μ we have that $t \to \xi_t(y) \equiv \xi(t,y)$ is a $\mathcal{S}'(\mathbb{R})$-valued symmetric Markov process with invariant measure $\mu \upharpoonright B(C) \equiv \nu_C$, $C \equiv \{(t,y) \in \mathbb{R}^2 | t = 0\}$. The structure $(Q, \mathcal{H}, Q', \nu_C)$ with $Q = \mathcal{S}'(\mathbb{R})$, $\mathcal{H} = L^2(\mathbb{R})$ has all the properties of those of the general theory discussed in Sect. I.2. In particular $e^{-tH_{\nu_C}}$ is a Markov semigroup in $L^2(\nu_C)$ with a properly associated ([I.11 f), g)], [I.2]) diffusion process $\eta_t(\cdot)$. H_{ν_C} coincides on a dense domain of $L^2(\nu_C)$ with the infinitesimal generator of the process $t \to \xi_t(\cdot)$. Both $\xi_t(\cdot)$ and $\eta_t(\cdot)$ solve in the weak sense the stochastic equation

$$d\xi_t = \beta(\varphi)(\xi_t)dt + dw_t, \text{ with } 2\beta(\varphi)(\cdot) \equiv -(\varphi \cdot \nabla)^* 1(\cdot).$$

Remark. The question of the complete identification of $\eta_t(y)$ and $\xi(t,y)$ hinges on uniqueness problems which are open also in the finite dimensional case (see [I.5], [I.12], [I.8], [I.3] for some results). H_{ν_C} has on a dense domain the form

$$H_{\nu_C} = -\frac{1}{2}\int \frac{\delta^2}{\delta\xi_o(y)^2}\, dy - \int \beta(\xi_o(y)) \frac{\delta}{\delta\xi_o(y)}\, dy,$$

hence formally the form of the infinite dimensional Schrödinger operator [11]) mentioned in I.1.2. A similar representation holds for the other generator Λ_{ν_C} of the Lorentz group in the 2-dimensional space time.
Let $\pi(\varphi)$ the infinitesimal generator of the strongly continuous unitary group of translations by elements in $\mathcal{S}(\mathbb{R})$, i.e.

$$\pi(\varphi)f(\xi) = \lim_{t\to 0} \frac{d}{dt} \left(\frac{d\mu_o(\xi+t\varphi)}{d\mu_o(\xi)}\right)^{1/2} f(\xi + t\varphi).$$

Then the structure $(\nu_C, H_{\nu_C}, \Lambda_{\nu_C}, \xi_o(\cdot), \pi(\cdot))$ is a canonical structure, which realizes, for these two-dimensional models, the desiderata of the

canonical formalism (in a sense the most natural and ambitious program of field quantization [I.10], [I.34]).
The extension of at least part of these results to the physical case $d = 4$ is of course a big problem. There are no doubts that more refined techniques are needed, but certainly it can be anticipated that probabilistic ideas would still provide much valuable inspiration.

Remark. Essentially the same proof as for the global Markov property of μ yields the uniqueness for Gibbs states associated with the interaction μ_Λ, namely the fact that μ is already uniquely determined by the specification [I.30a),c),d)]$E_{\mu_\Lambda}(f|\mathbb{R}^2-\Lambda)$, $f \in B(\Lambda_o)$, Λ_o Borel, $\Lambda_o \subset \Lambda$, Λ bounded Borel, within the class of measures $\tilde{\mu}$ with covariance bounded by the one of some free field. Another way of stating the uniqueness result is that there exists a unique weak limit for the conditional measures $\mu_\Lambda^{\partial\Lambda}(\cdot|\eta)$ of μ_Λ conditioned with respect to $B(\partial\Lambda)$, for any η which is the restriction to $\partial\Lambda$ of an element of $S'(\mathbb{R}^2)$ and such that $\psi_\eta^{\partial\Lambda}(x) \to 0$ locally uniformly in x as $\Lambda \uparrow \mathbb{R}^2$, where $\psi_\eta^{\partial\Lambda}$ is the solution of the Dirichlet problem introduced above.

Remark. There has been discussions in the literature [I.11 m)], [I.35 b)] on the relation of extremality of Gibbs states and global Markov property. For a recent counterexample to the conjecture that extremal states are automatically global Markov see [I.41].

Remark. The same ideas in the proof of the global Markov property above have also been used to give the proof of the global Markov property for discrete models of equilibrium classical statistical mechanics (lattice systems) [I.37], using previous uniqueness results of Dobrushin [I.42].

Footnotes to Ch.1

1) I.e. $E(f^{\#},f^{\#}) \leq E(f,f)$ for $f^{\#} = (f \vee 0) \wedge 1$, for all f in the domain of E. For other equivalent conditions see [I.2], [I.11 f)].

2) This process is, up to changes of scales the only Euclidean invariant Gaussian Markov process, see e.g. [I.17]. The Ornstein-Uhlenbeck velocity process is called oscillator process in [I.17 c)]. For a detailed presentation of non relativistic quantum mechanics as zero space dimensional field theory, and a detailed description in this simple case of the probabilistic structures involved in the passage from non relativistic quantum mechanics to quantum field theory see [I.18] and [I.45].

3) This idea has been formulated at an early stage by K. Symanzik, see e.g. [I.22]. It was taken up later by E. Nelson [I.24] and F. Guerra [I.23].

4) For the concept of canonical Gauss measures associated with a Hilbert space see e.g. [I.19] and references therein. In [I.19 c)] and [I.46] a new natural interpretation of measures on Hilbert spaces by non standard analysis is given.

5) $:\xi^n:$ is the so called n-th Wick power of the random field ξ. For discussion of this concept see e.g. [I.11 a)], [I.20 d)], [I.19 b)], [I.36].

6) For this concept see [I.40 a)] and more generally for information about methods and results of constructive field theory see any of a series of existing surveys like e.g. [I.20 d)], [I.31].
For the probabilistic meaning of the Osterwalder-Schrader property see also e.g. [I.40 b)] and references therein.

7) It was raised in [I.24] . See also e.g. [I.32], [I.35], [I.44].

8) See e.g. [I.39].

9) Non homogeneous Markovian random fields (i.e. with kinds of Markov property weaker than the global Markov one) have been discussed often in the literature, starting from Lévy. Two recent references are [I.38] (see also the references therein). It is however the homogeneity with respect to the

full Euclidean group, including reflections, that is required to get relativistic fields; this automatically implies that only fields that are truly generalized ones, with no measurable realizations (in all variables) come into play.

10) In quantum mechanics quantization from the Newton equation for a particle in $\mathbb{R}$, $\frac{\partial^2}{\partial t^2}\varphi(t) = K(\varphi(t))$, $\varphi(t) \equiv$ position at time t, $K = -V'$ is done by realizing $\varphi(0)$ as the multiplication operator by x in $L^2(\mathbb{R},dx)$ and realizing $\varphi(t)$ as the self-adjoint operator in $L^2(\mathbb{R},dx)$ given by $\varphi(t) = e^{itH}xe^{-itH}$, with H the Schrödinger operator in $L^2(\mathbb{R},dx)$ i.e. $H = -\frac{1}{2}\frac{d^2}{dx^2} + V$. The canonical momentum is then represented by the operator $\frac{1}{i}\frac{d}{dx}$ in $L^2(\mathbb{R},dx)$. This is the canonical Schrödinger representation of quantum mechanics we have in mind in discussing the extension to the case of quantum fields.

11) H_{ν_c} in $L^2(\nu_c)$ is formally unitarily equivalent to the operator (2) in $L^2(\prod_{x\in\mathbb{R}} d\xi_o(x))$, with $\varphi_x(0) \equiv \xi_o(x)$ and

$$\frac{1}{2}\int \xi_o(x)(-\Delta_x + m^2)\xi_o(x)dx + \int W(\xi_o(x))dx =$$

$$= \frac{1}{2}\int \beta(\xi_o(x))^2 dx + \frac{1}{2}\int \frac{\delta}{\delta\xi_o(x)}\beta(\xi_o(x))dx.$$

12) For a recent lucid survey of relations between quantum mechanics and energy forms see [I.11 1)].

References to Ch. I

[I.1] J. Deny, Méthodes hilbertiennes en théorie du potentiel, pp. 122-201 in Potential Theory, Ed. M. Brelot, Stresa 1969, C.I.M.E., Edizioni Cremonese, Roma 1970.

[I.2] M. Fukushima, Dirichlet forms and Markov processes, North-Holland/ Kodansha, Amsterdam / Tokyo (1980)

[I.3] a) M.L. Silverstein, Symmetric Markov Processes, Lecture Notes in Maths. 426, Springer, Berlin (1974)

b) M.L. Silverstein, Boundary Theory for Symmetric Markov Processes, Lecture Notes in Maths. 516, Springer, Berlin (1976)

[I.4] M. Fukushima,these Proceedings.

[I.5] S. Albeverio, R. Høegh-Krohn, L. Streit, Energy forms, Hamiltonians and distorted Brownian paths, J. Math. Phys. 18, 907-917 (1977)

[I.6] K. Sato, see [I.2]

[I.7] M.L. Silverstein, On the closability of Dirichlet forms, Z. Wahrscheinlichkeitsth. verw. Geb. 51, 185-200 (1980)

[I.8] M. Fukushima, On distorted Brownian motions, to appear in Proc. Les Houches Workshop "Stochastic Equations in Physics", Phys. Repts.

[I.9] a) M. Reed, B. Simon, Methods of Modern Mathematical Physics, Vol. IV: Analyis of Operators, Ch. XIII, 12, Academic Press, New York (1978); and references of Simon.

b) B. Simon, Brownian motions, L^p-properties of Schrödinger operators and the localization of binding, J. Funct. Anal. 35, 215-229 (1980)

[I.10] a) W. Heisenberg, W. Pauli, Zur Quantendynamik der Wellenfelder, Z. Phys. 56, 1-61 (1929)

b) W. Heisenberg, W. Pauli, Zur Quantentehorie der Wellenfelder II, Z. Phys. 59, 168-190 (1930)

c) K.O. Friedrichs, Mathematical aspects of the quantum theory of fields, Interscience, New York (1953)

d) I. Segal, Mathematical Problems of Relativistic Physics Am. Math. Soc., Providence (1963)

e) F. Coester, R. Haag, Representation of states in a field theory with canonical variables, Phys. Rev. 117, 1137-1145 (1960)

f) H. Araki, Hamiltonian formalism and the canonical commutation relations in quantum field theory, J. Math. Phys. 1, 492-504 (1960)

g) I.M. Gelfand,A.M. Yaglom, Integration in function spaces and its applications in quantum physics, J. Math. Phys. 1, 48-69 (1960)

[I.11] a) J. Glimm, A. Jaffe, Boson Quantum Field Models, pp. 77-143 in R.F. Streater ed., Mathematics of Contemporary Physics, Academic Press, New York (1972)

b) L. Gross, Logarithmic Sobolev inequalities, Amer. J. Math. 97, 1061-1083 (1975)

c) W. Faris, Self-adjoint Operators, Lecture Notes in Maths. 433, Springer Berlin (1975) (and references therein)

d) J.P. Eckmann, Hypercontractivity for anharmonic osciallators, Helv. Phys. Acta 45, 1074-1088 (1974)

e) S. Albeverio, R. Høegh-Krohn, Quasi invariant measures, symmetric diffusion processes and quantum fields, pp. 11-59 in Proc. Intern. Coll. Math. Methods of Quantum Field Theory, CNRS (1976)

f) S. Albeverio, R. Høegh-Krohn, Dirichlet forms and diffusion processes on rigged Hilbert spaces, Z. f. Wahrscheinlichkeitstheorie verw. Geb. 40, 1-57 (1977)

g) S. Albeverio, R. Høegh-Krohn, Hunt processes and analytic potential theory on rigged Hilbert spaces, Ann. Inst. H. Poincare B13, 269-291 (1977)

h) P.A. Deift, Applications of a commutation formula, Duke math. J. 45, 267-310 (1978)

k) S. Albeverio, R. Høegh-Krohn, L. Streit, Regularization of Hamiltonians and processes, J. Math. Phys. 21, 1636-1642 (1980)

i)1) R. Carmona, Regularity properties of Schrödinger and Dirichlet semigroups, J. Funct. Anal. 33, 259-296 (1979)

2) R. Carmona, Pointwise bounds for Schrödinger eigenstates, Comm. math., Phys. 62, 97-106 (1978)

j) S. Albeverio, M. Fukushima, W. Karwowski, L. Streit, Capacity and quantum mechanical tunneling, Bochum Preprint (1980)

l) L. Streit, Energy forms: Schrödinger theory, processes, to appear in Proc. Les Houches Workshop "Stochastic Equations in Physics"; Phys. Repts.

m) S. Albeverio, R. Høegh-Krohn, Topics in infinite dimensional Analysis pp. 279-303 in Edts. G. Dell'Antonio, S. Doplicher, G. Jona-Lasinio, Mathematical Problems in Theoretical Physics, Lect. Notes Phys. 80, Springer, Berlin (1978).

[I.12] J.G. Hutton, Dirichlet forms associated with hypercontractive semigroups, Trans. Am. Math. Soc. 253, 237-256 (1979)

[I.13] G. Jona-Lasinio, F. Martinelli, E. Scoppola, New approach to the semi classical limit of quantum mechanics I Multiple tunneling in one dimension, Inst. Physics Rome Preprint (1980)

[I.14] a) E. Nelson, Dynamical Theories of Brownian Motion, Princeton, (1967)

b) F. Guerra, P. Ruggiero, New interpretation of the Euclidean Markov field in the framework of physical Minkowski space-time Phys. Rev. Letts 31, 1022, (1973)

c) S. Albeverio, R. Høegh-Krohn, A remark on the connection between stochastic mechanics and the heat equation, J. Math. Phys. 15, 1745-1747 (1974).

d) F. Guerra, to appear in Proc. in Ref. [I.8]

[I.15] Ph. Paclet, Espaces de Dirichlet et capacités fonctionnelles sur triplets de Hilbert-Schmidt, Sém. Krée (1978)

[I.16] S. Kusuoka, Dirichlet forms and diffusion processes on Banach spaces, University of Tokyo Preprint (1980)

[I.17] a) G.E. Uhlenbeck, L.S. Ornstein, On the theory of Brownian motion I, Phys. Rev. 36, 823-841 (1930)

b) J.L. Doob, The Brownian movement and stochastic equations, Ann. of Math. 43, 351-369 (1942)

c) B. Simon, Functional integration and quantum physics, Academic Press, New York (1979)

[I.18] a) P. Courrège, P. Renouard, Oscillateur anharmonique, mesures quasi-invariantes sur C(R,R) et théorie quantique des champs en dimension d=1, Astérisque 22-23, 1-245 (1976)

b) P. Priouret, M. Yor, Processus de diffusion à valeurs dans $\mathbb{R}$ et mesures quasi-invariantes sur $C(\mathbb{R};\mathbb{R})$, Astérisque 22-23, 247-290 (1976).

For related work concerning quasi-invariant measures see also e.g.

c) G. Royer, Unicité de certaines mesures quasi-invariantes sur $C(\mathbb{R})$, Ann. Scient. Ec. Norm. Symp. 8, 319-338 (1975)

d) M. Yor, Etudes de mesures de probabilité sur $C(\mathbb{R}_+^*;\mathbb{R})$ quasi-invariantes sous les translations de $\mathcal{D}(\mathbb{R}_+^*;\mathbb{R})$, Ann. Inst. H. Poincaré B, 11, 127-172 (1975).

[I.19] a) H.H. Kuo, Gaussian measures in Banach spaces, Lecture Notes in Mathematics 463, Springer, Berlin (1975).

b) E. Nelson, Probability theory and Euclidean field theory, pp. 94-124 in Edts. C. Velo, A. Wightman, Constructive quantum field theory, Proc. Erice School, Lecture Notes in Phys. 25, Springer, Berlin (1973)

c) S. Albeverio, J.E. Fenstad, R. Høegh-Krohn, T. Lindstrøm, Nonstandard methods in stochastic analysis and mathematical physics, forthcoming book.

[I.20] E. Nelson, The free Markov field, J. Funct. Anal. 12, 211-227 (1973)

See also e.g.

a) E. Wong, Homogeneous Gauss-Markov random fields, Ann. Math. Stat. 40, 1625-1634 (1969)

b) E. Wong, Stochastic processes in information and dynamical systems, Ch. VII, McGraw Hill, New York (1971)

c) G.M. Molchan, Characterization of Gaussian fields with Markovian property, Dokl. Ak. Nauk SSR 197 (1971) (transl. Sov. Math. Dokl. 12, 563-567 (1971)).

d) B. Simon, The $P(\varphi)_2$ Euclidean (quantum) field theory, Princeton University Press (1974)

[I.21] I.M. Gelfand, N.Ya. Vilenkin, Generalized functions, Vol. 4, Academic Press, New York (1964).

[I.22] K. Symanzik, Euclidean quantum field theory, pp. 152-226 in Ed. R. Jost, Local Quantum Theory, Proc. Int. School E. Fermi, Academic Press, New York (1969)

[I.23] F. Guerra, Unpublished. See also F. Guerra, Uniqueness of the vacuum energy density and van Hove phenomena in the infinite volume limit for two dimensional self-coupled Bose fields, Phys. Rev. Lett. 28, 1213 (1972)

[I.24] a) E. Nelson, Quantum fields and Markoff fields, pp. 413-420 in Ed. D. Spencer, Partial differential equations, Symp. Pure Math. 23, AMS Providence Rhode-Island (1973).

b) E. Nelson, Construction of quantum fields from Markoff fields, J. Funct. Anal. 12, 97-112 (1973).

[I.25] See e.g. J.P. Eckmann, H. Epstein, Time-ordered products and Schwinger functions, Comm. math. Phys. 64, 95-130 (1979); and references therein.

[I.26] a) R. Høegh-Krohn, A general class of quantum fields without cut-offs in two space-time dimensions, Commun. math. Phys. 21, 244-251 (1971).

b) S. Albeverio, R. Høegh-Krohn, The Wightman axioms and the mass gap for strong interactions of exponential type in two dimensional space-time, J. Funct. Anal. 16, 39-82 (1974)

c) S. Albeverio, R. Høegh-Krohn, Martingale convergence and the exponential interaction in $\mathbb{R}^n$, pp. 331-353 in Ed. L. Streit, Quantum fields - Algebras, Processes, Springer, Wien (1980)

d) S. Albeverio, G. Gallavotti, R. Høegh-Krohn, Some results for the exponential interaction in two or more dimensions, Commun. math. Phys. 70, 187-192 (1979)

e) J. Fröhlich, C.M. Park, Remarks on the exponential interactions and the quantum Sine-Cordon equation in two space-time dimensions, Helv. Phys. Acta 50, 315-329 (1977).

[I.27] a) S. Albeverio, R. Høegh-Krohn, Uniqueness of the physical vacuum and the Wightman functions in the infinite volume limit for some non-polynomial interactions, Commun. math. Phys. 30, 171-Osipov 200 (1973)

b) J. Fröhlich, E. Seiler, The massive Thirring-Schwinger model (QED_2): convergence of perturbation theory and particle structure, Helv. Phys. Acta 49, 889-924 (1976).

[I.28] E.P. Osipov, On the triviality of the :exp $\varphi:_4$ quantum field theory in a finite volume, Novosibirsk Theor. Phys. Preprint (1979)

[I.29] J. Glimm, A. Jaffe, T. Spencer, The particle structure of the weakly coupled $P(\varphi)_2$ model and other applications of high temperature expansions, im Buch Ref. I.22 a).

[I.30] a) S. Albeverio, R. Høegh-Krohn, Uniqueness and the global Markov property for Euclidean fields. The case of trigonometric interactions, Commun. math. Phys. 68, 95-128 (1979).

b) S. Albeverio, R. Høegh-Krohn, Canonical relativistic quantum fields, to appear in Ann. Inst. H. Poincaré A

c) S. Albeverio, R. Høegh-Krohn, Uniqueness of Gibbs states and global Markov property for Euclidean fields, to appear in Bolyai Math. Soc., Proc. Coll. "Random fields", Esztergom (1979)

d) S. Albeverio, R. Høegh-Krohn, Uniqueness and global Markov property for Euclidean fields and lattice systems, pp. 303-329 in Ref [I.26 c)]

[I.31] J.P. Eckmann, Relativistic quantum fields in theories in two space-time dimensions, Quaderni CNR, Bologna (1977).

[I.32] F. Guerra, L. Rosen, B. Simon, The $P(\varphi)_2$ Euclidean quantum field theory as classical statistical mechanics, Ann. of Math. 101, 111-259 (1975).

[I.33] J. Bellissard, R. Høegh-Krohn, in preparation.

R. Gielerak, in preparation

[I.34] G. Wentzel, Quantum theory of fields, Interscience, New York (1949).

[I.35] a) C. Newman, The construction of stationary two-dimensional Markoff fields with an application to quantum field theory, J. Funct. Anal. 14, 44-61 (1973)

b) N. Dang Ngoc, G. Royer, Markov property of extremal local fields, Proc. Am. Math. Soc. 70, 185-188 (1978)

c) G.C. Hegerfeld, From Euclidean to relativistic fields and the motion of Markoff fields, Comm. Math. Phys. 35, 155-171 (1974)

d) J. Fröhlich, Schwinger functions and their generating functionals, I. Helv. Phys. Acta 74, 265-306 (1976); Adv. Math. 23, 119-180 (1977)

[I.36] I. Segal, Non linear functions of weak processes I, J. Funct. Anal. 4, 404-451 (1969).

R.L. Dobrushin, R.A. Minlos, Polynomials in linear random functions, Russ. Math. Surv. 32, 71-127 (1971).

R.L. Dobrushin, R.A. Minlos, The moments and polynomials of generalized random field, Theory of Prob. & its Appl. 23, 686-699 (1978)

[I.37] a) S. Albeverio, R. Høegh-Krohn, G. Olsen, The global Markov property for lattice systems, J. Multiv. Anal.

b) J. Bellissard, P. Pico Lattice quantum fields: Uniqueness and Markov property, CNRS-CPT Marseille Preprint (Dec. 1978)

c) H. Föllmer, On the global Markov property pp.293-302 in Ref. [I.26 c)]

[I.38] G.O.S. Ekhaguere, A characterization of Markovian homogeneous multicomponent Gaussian fields, Commun. math. Phys. 73, 63-77, (1980)
D. Surgailis, On covariant stochastic differential equations and Markov property of their solutions, Vilnius-Rome Preprint (1979).

[I.39] D. Williams, Diffusions, Markov processes and martingales, Vol. 1, J. Wiley, New York (1980)

[I.40] a) K. Osterwalder, R. Schrader, Axioms for Euclidean Green's function II, CMP 42, 281-305 (1975)
b) A. Klein, The semigroup characterization of Osterwalder-Schrader path spaces and the construction of Euclidean fields, J. Funct. Anal. 27, 277-291 (1978)

[I.41] H.v. Weizsäcker, A simple example concerning the global Markov property of lattice random fields, Kaiserslautern Preprint, to appear in Proc. Eigth Winter School in Abstract Analysis (1980).

[I.42] R. Dobrushin, Prescribing a system of random variables by conditional distribution, Th. Prob. Appl. 15, 3, 485-486 (1970) (and references therein).

[I.43] a) H.W. Goelden, On non-degeneracy of the ground state of Schrödinger operators, Math. Z. 155, 239-247 (1977).
b) F. Gesztesy, On non-degenerate ground states for Schrödinger operators, to appear in Rep. Math. Phys.

[I.44] S. Albeverio, R. Høegh-Krohn, Local and global Markoff fields, to appear in Rep. Math. Phys.

[I.45] R.F. Streater, Euclidean quantum mechanics and stochastic integrals, these Proceedings.

[I.46] T. Lindstrøm, A Loeb measure approach to theorems by Prokoroy Sazonov and Gross, Oslo Univ. Prepr. (1980), to appear in Trans. Am. Math. Soc.

II. Representation theory of groups of mappings

The homogeneous Markov random fields discussed in I.2 have extensions to the cases where the underlying space $\mathbb{R}^d$ is replaced by an orientable Riemannian manifold X and the space of values $\mathbb{R}$ is replaced by a compact semisimple Lie group G. We shall now discuss how this is done , according to [II.1-3]. Let g be the Lie algebra of G, equipped with the Euclidean structure given by the negative of the Killing form. Let G^X be the group of C^1 mappings from X into G, equal 1 outside compacts with pointwise multiplication. Let $\Omega(X,g)$ be the linear space of smooth maps from the tangent bundle TX into g, linear on each fiber, with compact support (the elements of Ω will be called simply 1-forms). Ω is a pre Hilbert space with respect to the scalar product

$$\omega_i \in \Omega \to (\omega_1,\omega_2) \equiv \int_X \mathrm{Tr}(\omega_1(x)\omega_2(x)^*)\ \rho(x)dx,$$

where the Tr is in the Lie algebra, * is the adjoint with respect to the Euclidean structure on the tangent plane T_xX at x and the one on g, $\rho(x)$ is a strictly positive smooth density on X and dx is the volume measure on X. Let $\mathcal{H}$ be the complex Hilbert space generated by Ω. We let G^X act in $\mathcal{H}$ by pointwise adjoint representation V i.e. V is defined for any $\omega \in \mathcal{H}$ and any $\psi \in G^X$ by

$$(V(\psi)\omega)(x) = \mathrm{Ad}\psi(x)\omega(x)$$

(we recall that $\mathrm{Ad}\psi(x)$ is the element of Aut g, the group of linear invertible transformations of g in itself, defined by $\exp \mathrm{Ad}\psi(x)\alpha = \psi(x) \exp\alpha\, \psi(x)^{-1}$, $\alpha \in g$). V is a unitary representation of G^X in $\mathcal{H}$, since the Killing form is invariant under the adjoint representation.

Let now $\gamma(\psi)(x)$ be the element in g obtained by first differentiating the map $\psi(x)$ and so obtaining an element in $T_{\psi(x)}g$ which is then translated back to the Lie algebra by using the connection given by right translations. As usual we write $\gamma(\psi)(x) = d\psi(x)\psi(x)^{-1}$. $\psi(x) \to \gamma(\psi)(x)$ is a one-cocycle (the Maurer-Cartan cocycle) for the group G^X and the representation V i.e. it is a continuous map from G^X, into $\mathcal{H}$ such that for any $\psi_i \in G^X$, i=1,2:

$$\gamma(\psi_1\psi_2) = \gamma(\psi_1) + V(\psi_1)\gamma(\psi_2).$$

Now it is well known that there exists a general procedure, due to Streater [II.11], Araki [II.12] and Parthasarathy-Schmidt [II.13], given a unitary representation V and a one cocycle, to construct another unitary representation, the "exponential representation" (see e.g. [II.6]). The representation space of this representation is the Fock space $e^{\mathcal{H}}$ over $\mathcal{H}$ i.e. $e^{\mathcal{H}}$ is the direct sum over n of the n-fold symmetric tensorproduct of $\mathcal{H}$ with itself. In $e^{\mathcal{H}}$ the new representation U is given by

$$U(\psi)e^{\omega} = e^{-\frac{1}{2}\|\gamma(\psi)\|^2} e^{-(V(\psi)\omega,\gamma(\psi))} e^{V(\psi)\omega+\gamma(\psi)} .$$

where $e^{\omega} \equiv \omega\otimes\ldots\otimes\omega$, $\omega\in\Omega(X,g)$, $\|\ \ \|$, $(\ ,\ )$ are the norm and scalar product in $\mathcal{H}$, respectively.

One has a well known unitary equivalence of $e^{\mathcal{H}}$ with $L^2(\mu)$, where μ is the standard Gaussian measure associated with the real part of $\mathcal{H}$. Using this we can also represent equivalently the representation U in $L^2(\mu)$, write it U', by $U'(\psi)f(\omega') = e^{i(\gamma(\psi),\omega')_g} f(V^{-1}(\psi)\omega')$, $\omega' \in \Omega\ (X;g)$, $(\ ,\ )_g$ being the scalar product in the Lie algebra g.

There is a third way ([II.3]) of giving the representation $U(\psi)$. Let $\mathcal{F}$ be the free module over $\mathbb{C}$, with generators written e^{ψ}, $\psi\in G^X$, with scalar product $(e^{\psi},e^{\psi'}) = e^{(\psi,\psi')}$ with $(\psi,\psi') \equiv (\gamma(\psi),\gamma(\psi'))$.

Define $U''(\psi)$ by $U''(\psi)e^{\psi'} = e^{-1/2(\psi,\psi)}e^{(\psi',\psi^{-1})}e^{\psi\psi'}$.

Then we see that U" is unitarily equivalent U hence is the expression on $\mathcal{F}$ of the representation U of G^X.

Remark. In [II.3] the representation U" is given also in the more general case where G is a Lie group with compact Lie algebra.

The representation U (or equivalently U' or U") is called the energy representation of G^X. It has been studied originally in [II.1] - [II.3].

From our point of view here the important property of this representation is that the positive definite normalized function $\psi \to (1,U'(\psi)1)$, $\psi\in G^X$, which is equal to $e^{-1/2\|\gamma(\psi)\|^2}$, and characterizes the representation is the non commutative analogue of the characteristic function $e^{-\frac{1}{2}\int_{\mathbb{R}^d}|\nabla\psi(x)|^2dx}$ of the standard normal measure on the Sobolev space $H_1(\mathbb{R}^d)$

(obtained as closure of $\mathcal{S}(\mathbb{R}^2)$ in the norm given by

$\psi \to \|\psi\|_1 \equiv (\int|\psi(x)|^2 dx + \int|\nabla\psi(x)|^2 dx)^{1/2})$.

Since this normal measure is the one which we called the free Euclidean field, we see that above representation of G^X is a non commutative version of the one (of $\mathbb{R}^{\mathbb{R}^d}$) given by the free field, so that $U'(\psi)$ corresponds to $e^{i<\psi,\xi>}$, where ξ is the free Euclidean field.

In the case where $G = \mathbb{R}$ and X is a Riemann manifold such that its standard Brownian motion is transient then the energy representation is the one given by $U'(\psi) = e^{i<\psi,\xi>}$, $\psi \in C_0^\infty(X)$, $\xi(x)$ the element of $\mathcal{S}'(T_x X)$ whose distribution is given by the standard normal measure μ_0 on the closure $H_1(X;\mathbb{R})$ of $C_0^\infty(X)$-function on X, in the energy norm $E(\psi,\psi) \equiv \int(d\psi(x), d\psi(x))dx$, where $\psi \to (d\psi(x), d\psi(x))$ is the form in the cotangent space at x given by the Riemann structure. We call $(\mu_0,\xi(x))$ the Brownian field on X. In the case where $X = \mathbb{R}^d$ we have that the Brownian field on X is the free Euclidean field of mass 0. Also in the general case μ_0 has the global Markov property (defined in a similar way as we did for the free Euclidean field).

Remark. The above notion of Brownian field appeared in [II.7 a)]. In that manuscript we associated more generally a Gaussian random field to any Dirichlet form. This idea has recently been promoted independently in a beautiful setting by Dynkin [II.7 b),c)].

What are the properties of the energy representation?

Let d = dim X. It was shown by Ismagilov (in '76) [II.1] that for $d \geq 5$ the energy representation is irreducible. This was improven in '79 by Vershik, Gelfand and Graev [II.2] to the case d = 4. We now have the following

Theorem [II.4] For $d \geq 3$ the energy representation is irreducible. For d = 2 it is irreducible if the length $|\lambda|$ of the root satisfies $|\lambda| > \sqrt{32\pi\rho(x)}$, for all roots λ. For d = 1 it is reducible.

Remark. The case d = 2, $|\lambda| \leq \sqrt{32\pi\rho(x)}$ is open. It is conjectured that for $|\lambda| < \sqrt{4\pi\rho(x)}$ one has reducibility.

The methods for proving these results is a combination of ideas by

Ismagilov [II.1] and Vershik, Gelfand, Graev [II.2] together with a strong result on singularity of translates of Gaussian measures proven for fields in the exponential interaction model by Gallavotti and ourselves [I.24 d)]. Let us give a short sketch, especially to show the connection with the theory of global Markov fields. Let T be a maximal torus in G and let t be the corresponding Cartan subalgebra in g. For $A \in t^X$ one has
$U(\exp A) = W(dA) \times \exp(V(\exp A)$ where W is a representation in $\exp(H(t))$ and $\exp V(\exp A)$ is a representation in $\exp(H(t^{\perp}))$, where $H(t)$ resp. $H(t^{\perp})$ are the Hilbert spaces for the energy representation of T^X resp. $T^{\perp}$.
W is defined by $W(dA)e^{\omega} \equiv e^{-1/2\|dA\|^2} e^{-(\omega,dA)} e^{\omega+dA}$, $\omega \in H(t)$. The spectral measure for $U \upharpoonright \exp t^X$ is $\mu_o * \nu$, where μ_o is the Gaussian measure on the dual of t^X with Fourier transform $e^{-1/2\|dA\|^2}$ and ν is the canonical Poisson measure with support on the space Φ of all $\chi \in (t^X)'$ which are such that for some $x_j \in X$ and some n, $\chi(A) = \sum \lambda_j A(x_j)$, where λ_j are roots of g and $A \in t^X$.
Results on convolutions of above form are given in [II.1] - [II.4], [II.8]. We shall now give a result which controls above convolutions and is at the basis of the results on the exponential interaction (triviality for $d \geq 3$ or $d = 2$ and $|\alpha|$ large) as well as of the singularity of $\mu_o * \nu$ with respect to μ_o, which then implies that the commutant of $U(\exp t^X)$ is in the set of decomposable operators of the integral representation for $U(\exp A)$ obtained from above decomposition. This in turns implies that $W(dA) \in U(G^X)''$ (the von Neumann algebra generated by $U(G^X)$) for all $A \in g^X$, which implies, by the simplicity of G, the irreducibility of U' in the cyclic component of 1. Since however 1 can be shown to be a cyclic vector for U' this then yields the irreducibility of U'. Let us now state the mentioned basic result:

<u>Theorem</u> [I.24d)], [II.4]):Let B be a bounded cube in $\mathbb{R}^d$. Let p_{ij}, i,j=1,...,d be real continuous function on X with $p_{ij} = p_{ji}$. Let P be the matrix of the p_{ij} and suppose for some m, M and all $x \in B$

$$m\mathbb{1} \leq P(x) \leq M\mathbb{1}.$$

Let μ_p be the Gaussian measure on $\mathcal{D}'(B)$ with Fourier transform

$$\exp(-\frac{1}{2} \int_B \sum_{i,j} p_{ij}(x) (\frac{\partial}{\partial x_i} f)^2 dx).$$

Then for all $d \geq 3$ and all α (resp. for $d = 2$ and $|\alpha| > 4\sqrt{2\pi}\,\frac{M^2}{m^{3/2}}$) there exists a measurable subset $Q \subset \mathcal{D}'(B)$ such that $\mu_p(Q) = 1$ and $\mu_p(Q + \alpha\delta_y(\cdot)) = 0$ for all $\alpha \neq 0$ (resp. all $|\alpha| > 4\sqrt{2\pi}\,\frac{M^2}{m^{3/2}}$) and <u>all</u> $y \in B$.

<u>Remark.</u> For a given $y \in B$ this is a well known result if Q is allowed to depend on y (one needs only to remark that $\delta_y(\cdot)$ is not an admissible translation). However it is harder to prove the existence of a universal Q for all $y \in B$ having above properties. Once this result is available the results on the triviality of the exponential interaction as well as on the irreducibility of the energy representation become simple. The proof is based on an exploitation of the singularity and scaling properties of the covariance of μ_p, as well as on tail estimates on μ_p.

Let us now spend a few words on the very interesting situation which arises for $d = 1$. For simplicity we take $X = \mathbb{R}$ (but we could as well take $X = S^1$). In this case the Markovian character of the energy representation is exhibited explicitely, through an identification [I.24 d)] of the representation with the one given by left translations on the set of sample paths of the standard Brownian motion $\eta(t)$ on the Lie group G (studied in [II.3]

$\eta(t)$ is defined [II.9] as the nonanticipating solution of the stochastic integral equation

$$\begin{cases} d\eta(t)\,\eta^{-1}(t) = \xi(t) \\ \eta(0) = e \in G \end{cases}$$

where $\xi(t)$ is the white noise on the Lie algebra g, so that $w(t) = \int_0^t \xi(\tau)d\tau$ is the standard Brownian motion on the Lie algebra g (with invariant inner product given by the negative of the killing form). $\eta(t)$ is continuous and starts afresh at stopping times and has right independent increments.

$\eta(t)$ is a left and right invariant process on G, in as much as $\sigma\eta$ and $\eta\sigma$ for all $\sigma \in G$ are identical in law). Let μ_W be the measure on $C(\mathbb{R};G)$ associated with the standard Brownian motion on G (made into a homogeneous process).

μ_W is quasi invariant under left and right translations by elements in $C_0^1(\mathbb{R};G)$. Hence we can in the usual way define unitary representations of $C_0^1(\mathbb{R};G)$ given by left resp. right translations:

$$(U^R_\varphi f)(\eta) = \left(\frac{d\mu(\eta\varphi)}{d\mu(\eta)}\right)^{1/2} f(\eta\varphi)$$

$$(U^L_\varphi f)(\eta) = \left(\frac{d\mu(\varphi\eta)}{d\mu(\eta)}\right)^{1/2} f(\varphi\eta) \ , \ \eta \in C(\mathbb{R};G),\ \varphi \in C^1_o(\mathbb{R};G).$$

We have the

Theorem[II.4]. The representation on U^R (and U^L) are unitary equivalent to the energy representation of $C^1_o(R;G) = G^X$. The unitary equivalence is induced by $\eta \in G^X \to d\eta(t)\cdot\eta^{-1}(t)$ (resp. $\eta^{-1}(t)d\eta(t)) \in \Omega^1(X;g)$.

U^L and U^R commute. In particular the energy representation is reducible, its commutant containing a representation unitary equivalent with the energy representation itself.

Remark. A more detailed study of the d = 1 case is in progress ([II.10]).

II

[II.1] R. Ismagilov, On unitary representations of the group $C_0^\infty(X,G)$, $G = SU_2$, Math. Sb. 100 (2), 117-131 (1976) (transl. Math. USSR Sb 29, 105-117 (1976)).

[II.2] A. Vershik, I. Gelfand, M. Graev, Remarks on the representation of the groups of functions with values in a compact Lie group, to appear in Compos. Math.

[II.3] S. Albeverio, R. Høegh-Krohn, The energy representation of Sobolev-Lie groups, Compositio Math. 36, 37-52 (1978)

[II.4] S. Albeverio, R. Høegh-Krohn, D. Testard, Irreducibility and reducibility for the energy representation of the group of mappings of a Riemannian manifold into a compact semisimple Lie group, Bochum Preprint (1979) (to appear in J. Funct. Anal.)

[II.5] I.B. Frenkel, Orbital theory for affine Lie algebras, Yale Preprint (1980).

[II.6] A. Guichardet, Symmetric Hilbert spaces and related topics, Lecture Notes in Maths. 261, Springer, Berlin (1972).

[II.7] a) S. Albeverio, R. Høegh-Krohn, preprint 77/962 Marseille (November 1977).

b) E.B. Dynkin, Markov processes and random fields, Cornell Univ. Preprint (1979);

c) E.B. Dynkin, to appear in same issue of Phys. Repts. as [I.11 1)].

[II.8] W.D. Wick, On the absolute continuity of a convolution with an infinite dimensional measure, Univ. of Washington, Seattle, Preprint (1979).

[II.9] H.P. McKean, Jr.; Stochastic integrals, Academic Press, New York (1969).

[II.10] S. Albeverio, A. Vershik, R. Høegh-Krohn, D. Testard, in preparation.

[II.11] R.F. Streater, Current commutation relations, continuous tensor products, and infinitely divisible group representations, pp. 247-263 in R. Jost, Edt., Local Quantum Theory, Academic Press, New York (1969)

[II.12] H. Araki, Factorisable representations of current algebra, Publ. RIMS, Kyoto, Ser. A, 5, 361-422 (1970).

[II.13] K.R. Parthasarathy, K. Schmidt, Positive definite kernels, continuous tensor products, and central limit theorems of probability theory, Lect. Notes in Maths. 272, Springer, Berlin (1972).

III. Stochastic solutions for hydrodynamics

In this chapter we shall mention some results about the construction of Gibbs measures for the random fields of hydrodynamics. The reason why we mention this here is that on one hand these measures are similar to those discussed above for Euclidean fields, in particular they are again of Markovian type, and moreover they provide stochastic solutions of partial nonlinear differential equations of a different class than the ones describing quantum fields. In the description of phenomena involving a large number of particles classical analytic mechanics is of little direct help and as far as equilibrium properties are concerned the classical approach of Gibbs and Einstein for statistical mechanics is the most successful one. In this approach time averages are replaced by so called ensemble averages, i.e. by expectation with respect to equilibrium measures. In the case where the temperature is chosen as a basic macroscopic variable the relevant Gibbs measure is $d\mu = \frac{d^{-\beta H(p,q)}dpdq}{\int e^{-\beta H(p,q)}dpdq}$, where $\beta = \frac{1}{kT}$, T being the temperature and k Boltzmann constant, H is the Hamiltonian function, p,q are canonical variables in phase space and dpdq is the volume element in phase space. The invariance of μ comes from the well known invariance of H and dpdq under the flow induced by the solutions of the classical Newton equations of motion.

Is it possible to do something similar for hydrodynamics, i.e. replace the solution of the partial differential equation of motion (corresponding to an infinite number of degrees of freedom, in the terminology of classical mechanics) by expectations with respect to suitable invariant measures? This hope arises in several approaches to hydrodynamics and the problem of turbulence. We shall here mention the construction of some Gibbsian measures, for hydrodynamics, following recent work by M. De Faria and ourselves [III.1], [III.2]. We consider Euler's equation for hydrodynamics (no viscosity term), slightly extended to take care of boundary conditions:

$$\frac{\partial}{\partial t} u = -(u \cdot \nabla) u - f$$
$$\text{rot } f = 0 \qquad (1)$$
$$\text{div } u = 0 \quad ,$$

where u(t,x) is the velocity field ($t \geq 0$, $x \in \Lambda$ a bounded domain of $\mathbb{R}^s$) and f is a vector valued function of (t,x) (which can actually be eliminated from the equations), with boundary conditions on $\partial\Lambda$ ($\partial\Lambda$ being the piecewise C^1 boundary of Λ) such that u is tangential on the boundary . An integral of motion is the functional $H(u) \equiv \frac{1}{2} \int_\Lambda u^2 dx$ (energy). For $s = 2$ we have also that $S(u) \equiv \frac{1}{2} \int_\Lambda (\text{rot } u)^2 dx$ (enstrophy) is an integral of motion and more generally $S_f(u) = \int_\Lambda f(\text{rot } u)dx$, for any f such that the integral exists, are integrals of the motion.

Obviously one can define (for all d) the Gauss measure on $\mathcal{D}'(\Lambda)$ with Fourier transform $e^{-1/2\, \beta \int g^2 dx}$, $g \in \mathcal{D}(\Lambda)$ i.e. with formal density $e^{-\beta H(u)}$. However it is a problem to give a meaning to the Euler equations for initial data in the support of this measure. The Euler equations are in fact of the form $\frac{d}{dt} y_n(t) = B_n(\underline{y}(t))$, where $\underline{y} = \{y_n\}$, y_n being the component of rot u with respect to the n-th eigenfunctions of the Laplacian with corresponding eigenvalue λ_n and B_n being a certain non linear operator. Now $B_n(\underline{y})$ has infinite square integral with respect to the above Gauss measure. So the above Gauss measures cannot be used as invariant measures. However for $d = 2$ there are other measures which are formally invariant and these can be indeed shown to be invariant. We shall here only mention shortly the Gaussian measures μ_γ given formally by const. $e^{-\gamma S(\text{rot } u)} d(\text{rot } u)$, $\gamma > 0$, i.e. the Gaussian measure on the $\underline{y}$ such that the y_n are independent random variables with mean zero and variance γ^{-1}. In [III.1], [III.2] we showed that the Euler equations make sense a.s. with respect to μ_γ, in fact $B_n(\underline{y}) \in L^2(d\mu_\gamma)$. Although $H(\underline{y}) = \frac{1}{2} \sum \frac{y_n^2}{\lambda_n}$ is a.s. $+\infty$ with respect to μ_γ the renormalized energy $:H:(\underline{y}) \equiv \frac{1}{2} \sum \lambda_n^{-1} y_n^2 - \frac{1}{2} \sum \lambda_n^{-1} \frac{1}{\gamma}$ is in $L^2(d\mu_\gamma)$, hence in particular finite μ_γ-a.s. Thus the measures

$$d\mu_{\beta,\gamma} = \frac{e^{-\beta :H:(\underline{y})} d\mu_\gamma(\underline{y})}{\int e^{-\beta :H:(\underline{y})} d\mu_\gamma(\underline{y})}$$

(with formal density $e^{-\beta :H:(\underline{y})} e^{-\gamma S(\underline{y})}$), are for all $\beta \geq 0$, $\gamma > 0$, well defined. They are all equivalent μ_γ, and one has $\mu_{\beta,\gamma} \perp \mu_{\beta',\gamma'}$ whenever $\gamma \neq \gamma'$.

$\mu_{\beta,\gamma}$ are shown in [III.1], [III.2] to be infinitesimal invariant under the flow $\underline{y}(0) \to \underline{y}(t)$ induced by the Euler equations in the sense that $B_n(\underline{y}) \in L^2(d\mu_{\beta,\gamma})$ and $\int Bf d\mu_{\beta,\gamma} = 0$ for all f in a dense domain of $L^2(d\mu_{\beta,\gamma})$, where $B(\underline{y}) \equiv \sum B_n(\underline{y}) \frac{\partial}{\partial y_n}$ is the densely defined Liouville operator for the flow.

Moreover it was also proven in [III.2] that, at least in the case of a simple connected domain, iB is symmetric and real in $L^2(d\mu_{\beta,\gamma})$, hence has self-adjoint extensions, which then give unitary strongly continuous groups in $L^2(d\mu_{\beta,\gamma})$, representing the Euler flow.

It is possible to show that among these groups there is one of the form

$$(U_t f)(\underline{y}(0)) = f(\alpha_t(\underline{y}(0))),$$

where α_t: $\underline{y}(0) \to \underline{y}(t)$ is an invertible continuous transformation of the space of the $\underline{y}$'s into itself, which implies in particular that the measures are invariant under the flow given by α_t i.e. associated to the Euler equations (since $\frac{d}{dt} U_t f(\underline{y}(0))/_{t=0} = Bf(\underline{y}(0))$ for all f in a dense domain in $L^2(d\mu_{\beta,\gamma})$).

The proof goes as follows (for details see [III.9]). Consider the truncated system of Euler equations $\frac{\partial}{\partial t}\ y_n^N(t) = B_n^N(\underline{y}^N)$, $n=1,\ldots,N$, where $B_n^N(\underline{y})$ is obtained from the expression for $B_n(\underline{y})$ by only retaining the contributions from the variables y_k, $k \leq N$ (where we choose the eigenvector φ_j to $-\Delta$ to correspond to the eigenvalue λ_j, with $\lambda_1 \leq \lambda_2 \leq \ldots$). Let $y_n^N(t)$ be the solution of these equations with initial data $y_n^N(0)$ (which exists for all t since $B_n^N(\underline{y}^N)$ is C^1 in the y_j and the energy of the N modes is conserved).

Consider now the mapping induced on continuous cylinder functions in $L^2(d\mu_{\beta,\gamma})$ by the mapping α_t^N: $\underline{y}^N \to \underline{y}^N(t)$, where $\underline{y}^N(t)$ is the solution of the cut-off Euler equation with initial condition $\underline{y}$. Then for $f \in FC^1$ (cylinder functions C^1 on the base, with notations as in Ch. I) we have

$$\int f \circ \alpha_t^N d\mu_\gamma = \int f d\mu_\gamma \tag{2}$$

and

$$\frac{\partial}{\partial t} \int f \circ \alpha_t^N d\mu_\gamma = \int B^N f \circ \alpha_t^N d\mu_\gamma$$

with $B^N \equiv \sum_{k=1}^{N} B_k^N(\underline{y}) \frac{\partial}{\partial y_k}$.

Since $B_k^N \to B_k$ as $N \to \infty$, strongly in $L^2(d\mu_{\beta,\gamma})$, we find, by passing if necessary to subsequences, $B_k^N \to B_k$ a.s. and for $\sup |f|<\infty, |\frac{\partial}{\partial t} f\circ \alpha_t^N(\underline{y})| \leq C(\underline{y})$, with $C(\underline{y}) < \infty$ μ_γ-a.s. and $C(\underline{y})$ is independent of N. Then the family $\{f \circ \alpha_t^N, N\in\mathbb{N}\}$ is equicontinuous uniformly bounded in t and N and with uniformly (in N and t) continuous bounded derivatives. Hence by Ascoli-Arzelà theorem, for all $t\in[0,T]$, $T < \infty$ there exists a subsequence (which we call again N) such that $f\circ\alpha_t^N(y)$ converges as $N \to \infty$ pointwise μ_γ a.s. to a function $\hat{f}_t(\underline{y})$, and the limit is continuous in t. Moreover the derivatives $\frac{\partial}{\partial t} f\circ\alpha_t^N$ converge as $N \to \infty$ μ_γ-a.s.

and one has $\frac{\partial}{\partial t} f\circ\alpha_t^N|_{t=0} \to Bf = \frac{\partial\hat{f}}{\partial t}|_{t=0}$.

If we now choose the f in a countable subalgebra of FC^1, dense in $L^2(d\mu_{\beta,\gamma})$, then after discarding if necessary sets of μ_γ-measure zero and going over to subsequences, we have the above convergence for all f in such a dense set simultaneously for all $\underline{y}$, except maybe for a μ_γ-zero measure set (independent of f).

By taking e.g. $f(\underline{y}) = \exp(i\beta_k \mathrm{Re}\, y_k)$ for $\beta_k\in\mathbb{R}$, we then see from the convergence of $f(\alpha_t^N(\underline{y}))$ as $N \to \infty$ that $\mathrm{Re}\,\alpha_t^N(\underline{y}_k)$ converges μ_γ a.s., with $\underline{y}_k = \{y_j, y_j=0 \text{ for } j\neq k\}$ and similarly using functions of the form $g(\underline{y}) \equiv \exp(i\beta_k' \,\mathrm{Im}\, y_k)$ we see that $\mathrm{Im}\,\alpha_t^N(\underline{y}_k)$ converges μ_γ-a.s., hence $\alpha_t^N(\underline{y}_k)$ converges μ_γ-a.s. We call $\alpha_t(\underline{y}_k)$ the limit, then for any function f in the algebra generated by above functions we have $f(\alpha_t^N(\underline{y})) \to f(\alpha_t(\underline{y}))$. Since this is a subalgebra of $C(\overline{\mathbb{C}^N})$ which separates the points it is dense in $L^2(d\mu_\gamma)$, hence by continuity we can extend the definition of $f\circ\alpha_t$ to all $f\in L^2(d\mu_\gamma)$.

For $f\in FC^1$ we have by the above $f\circ\alpha_t = \hat{f}_t$, hence

$$\frac{\partial}{\partial t} f\circ\alpha_t^N|_{t=0} \to \frac{\partial\hat{f}}{\partial t}|_{t=0} = Bf.$$

From (2) and $f \circ \alpha_t^N \to f\circ\alpha_t$ μ_γ-a.s. we get, using dominated convergence,

$$\int f\circ\alpha_t d\mu_\gamma = \int f d\mu_\gamma .$$

Hence the measure μ_γ is invariant (for all t) under the automorphism group $f\to f\circ\alpha_t$ on FC^1.

This defines then a one-parameter unitary group of mappings of $L^2(d\mu_\gamma)$, strongly continuous (since the α_t were continuous). By construction its generator coincides on FC^1 with iB, hence it is a self-adjoint extension of iB.

Moreover one shows that the Fourier transform of $\mu_{\beta,\gamma}$ are stationary solutions of the Hopf functional equation (this equation was introduced by Hopf [III.6] for a theory of turbulence).
Thus at least for inviscid 2-dimensional fluids there exist invariant measures which could play a role similar to that of the Gibbs measures of classical statistical mechanics.
Remark. The question of uniqueness of the flow is open. It is a question corresponding to one that for classical mechanics is only solved in a certain class of cases for $s = 1$ [III.8]. It would be settled if one could show that iB is essentially self-adjoint on smooth cylinder functions.

Footnote to Ch. III

1) These Gibbsian measures were discussed on several occasions before in the literature. E.g. [III.3].

III

[III.1] S. Albeverio, M. De Faria, R. Høegh-Krohn, Stationary measures for the periodic Euler flow in two dimensions, J. Stat. Phys. 20, 585-595 (1979).

[III.2] S. Albeverio, R. Høegh-Krohn, Stochastic flows with stationary distribution, Bochum Preprint (1979)

[III.3]a)T.D. Lee, On some statistical properties of hydrodynamical and magneto hydrodynamical fields, Quart. J. Appl. Math. 10, 69-74 (1952)

b)R.H. Kraichnan, Remarks on turbulence theory, Adv. Math. 16, 305-331 (1975)

c)H.M. Glaz, Two attempts at modeling two-dimensional turbulence, pp. 135-155 in Turbulence Seminar, Berkeley 1976/77, Lect. Notes in Maths. 615, Springer, Berlin (1977).

d)G. Gallavotti, Problèmes ergodiques de la mècanique classique, Ens. 3e cycle Phys. Suisse Romande, Ec. Polyt. Féd., Lausanne (1976).

d)H.A. Rose, P.L. Sulem, Fully developed turbulence and statistical mechanics, J. de Phys. (Paris) 39, 441-484 (1978).

f)R.H. Kraichnan, D. Montgomery, Two-dimensional turbulence, Rep. Progr. Phys. 43, 547-619 (1979)

g)C. Boldrighini, Introduzione alla fluidodinamica Quaderni CNR, Roma (1979)

See also for recent work [III.4], [III.5]

[III.4] C. Boldrighini, S. Frigio, Equilibrium states for the two dimensional in compressible Euler fluid, Atti Sem. Mat. Fis. Univ. Modena 27,106-125 (1978)

[III.5] C. Boldrighini, S. Frigio, Equilibrium states for a plane incompressible perfect fluid, Commun. math. Phys. 72, 55-76 (1980)

[III.6] S. Albeverio, R. Høegh-Krohn, Let us talk about the weather - A model for large scale atmospheric features, to appear in Proc. Les Embiez Conference Disordered Systems (1980); and paper in preparation.

[III.7] E. Hopf, Statistical hydrodynamics and functional calculus, J. Rat. Mech. Anal. 1, 87 - 123 (1952).

[III.8] C. Marchioro, A. Pellegrinotti, M. Pulvirenti, Self-adjointness of the Liouville operator for infinite classical systems, Commun. math. Phys. 58, 113-129 (1978)

[III.9] S. Albeverio, M. De Faria, R. Høegh-Krohn, in preparation.

IV. Markov processes and C^*-algebras

In recent years, stimulated particularly by studies in non equilibrium statistical mechanics, a rather extensive literature has appeared concerning extensions to non commutative situations of concepts (like Markov semigroups) of the theory of stochastic processes. At this conference L. Accardi will report on some results in this circle of problems. Here we would like to mention some work of ourselves concerned with the construction of completely positive Markov semigroups on C^*-algebras and the extension of the theory of Dirichlet forms to C^*-algebras, moreover I would like to report on work of ourselves with G. Olsen on relations between Markov semigroups on certain C^*-algebras and diffusion processes on Lie groups. Both topics provide new range of applicability and sources of new problems for the theory of stochastic processes (in particular of diffusion type).
Let me first explain shortly the theory of noncommutative Dirichlet forms [IV.1].

IV.1 Non commutative Dirichlet forms

To find a suitable extension of the commutative theory of Dirichlet forms mentioned in Ch. I one has to recall that the basic objects of the commutative theory were a space $\mathcal{A}$ ($\mathbb{R}^d$ resp. in I.1.2, a real separable Hilbert space $\mathcal{H}$), a measure μ associated with it, a dense subspace $\mathcal{A}_0$ ($C_0^1(\mathbb{R}^d)$ resp. FC^1) and a Dirichlet form E first defined on $\mathcal{A}_0$ (in I.1.2,by $E(f,f)\equiv\frac{1}{2}\int(\nabla f)^2 d\mu$, $f\in\mathcal{A}_0$).
In the non commutative case this structure $(\mathcal{A},\mu,\mathcal{A}_0,E)$ is replaced [IV.1] by $(\mathcal{A},\tau,\mathcal{A}_\tau,E)$,where $\mathcal{A}$ here is a (in general non commutative) C^*-algebra, a subalgebra of the algebra of all bounded operators on a Hilbert space $\mathcal{K}$, τ is a faithful lower semicontinuous trace on $\mathcal{A}$ s.t. the set $\mathcal{A}_\tau$ of all x such that $\tau(x^*x)<\infty$ is dense in $\mathcal{A}$, and E is realized by

$$E(x,x)\equiv tr(x^2M) + \sum_{i=1} tr([x,m_i]^* [x,m_i]) \quad (1)$$

for all $x\in L^2(\mathcal{A},\tau)\equiv$(closure of $\mathcal{A}_\tau$ in the norm given by $\tau(x^* x)$).Here M is any self-adjoint non negative operator on $\mathcal{K}$, the m_i are bounded operators with finite Hilbert-Schmidt norm i.e. $tr(m_i^* m_i)<\infty$, and [,] is the commutator.
More generally in the commutative case Dirichlet forms are characterized by being symmetric positive closed bilinear forms E on $L^2(d\mu)$ with a suitable

contraction property, e.g. $E(T(f),T(f)) \le E(f,f)$, for all f in the domain of the form and any map T from $\mathbb{R}$ into $\mathbb{R}$, Lipschitz continuous with Lipschitz norm 1 and mapping 0 into 0 (see e.g. [I.2]). In a similar way non commutative Dirichlet forms can be defined as symmetric positive closed bilinear forms E on $L^2(\mathcal{A},\tau)$, $\mathcal{A}$ being an arbitrary C^*algebra and τ as above, with $E(T(x), T(x)) \le E(x,x)$ for all hermitian elements in a dense domain of the hermitian part of $L^2(\mathcal{A},\tau)$. Again in analogy with the commutative case one defines Markov semigroups in $L^2(\mathcal{A},\tau)$ equipped with the order provided by the identification of elements of $L^2(\mathcal{A},\tau)$ with left multiplications (by the same elements) and using the order structure of closed operators. A (sub) Markov semigroup Φ_t on $L^2(\mathcal{A},\tau)$ is by definition a semigroup of maps of $L^2(\mathcal{A},\tau)$ into itself such that $0 \le x \le 1$, $x \in L^2(\mathcal{A},\tau)$ implies of $0 \le \Phi_t(x) \le 1$.
One has then the following extension of the corresponding Beurling-Deny-Fukushima theorem in the commutative case.

Theorem IV.1 [IV.1] There is a one-to-one correspondence between Dirichlet forms and symmetric Markov semigroups on $L^2(\mathcal{A},\tau)$.
It is well known that from the point of view of the construction of non commutative processes particularly in relation with applications to quantum non equilibrium statistical mechanics (see e.g. [IV.2], and references therein), the object which corresponds properly in the non commutative case to the commutative positive maps are the completely positive maps [IV.3]. For this reason it is useful to consider "completely Markov semigroups" which are Markov semigroups Φ_t, $t \ge 0$ in $L^2(\mathcal{A},\tau)$ which have the additional property of being such that $\Phi_t \otimes \mathbb{1}_n$ is also a Markov semigroup for all n, where $\mathbb{1}_n$ is the identity in the algebra M_n of $n \times n$ complex matrices.

Correspondingly it is useful to introduce the concept of a completely Dirichlet form, namely of a Dirichlet form E on $L^2(\mathcal{A},\tau)$ which has the additional property that $\sum_{i,j} E(x_{ij},x_{ij})$ is a Dirichlet form on $L^2(\mathcal{A},\tau) \otimes M_n$, where $x_{ij} \in L^2(\mathcal{A},\tau)$ are the components of $x \in L^2(\mathcal{A},\tau) \otimes M_n$. Then one has the

Theorem IV.2 [IV.1] A symmetric semigroup Φ_t on $L^2(\mathcal{A},\tau)$ is completely Markov iff the corresponding Dirichlet form E is completely Dirichlet. This is the case iff there exists a net of weights w_α resp. ρ_α on $\mathcal{A}$ resp. $\mathcal{A} \otimes \bar{\mathcal{A}}$ (where $\bar{\mathcal{A}}$

is $\mathcal{A}$ with scalar multiplication $(\lambda, a) \to \bar{\lambda} a$, $\lambda \in \mathbb{C}$, $a \in \mathcal{A}$, $\bar{\lambda}$ being complex conjugate to λ in $\mathbb{C}$) such that

$$E(x,x) = \lim_{\alpha\to\infty} [w_\alpha(x^2) + \rho_\alpha((x \times 1 - 1 \times x)^2).$$

An example of completely Dirichlet forms is given by (1) above.
The spectrum of completely Markov semigroups on von Neumann algebras has been studied (non commutative extension of the Perron-Frobenius theory) in [IV.4].

IV.2 Completely positive semigroups and diffusion processes on groups

Let me now mention shortly the relation between completely positive semigroups and diffusion processes on groups. To do so it is useful to look at the usual theory of commutative stochastic processes in the following way. Let X be a (say locally compact) space with some (regular finite) measure. One has the concept of sub-Markov semigroup P_t (s.t. $P_t\, 1 \leq 1$) and of associated Markov process $\xi_t^x(\omega)$ with state space X ($x \in X$: starting point, t time, ω sample point). $\xi_t^x(\omega)$ induces a 1-parameter family ($\alpha_t(\omega)$, $t \in \mathbb{R}$) of random transformations of C(X) into itself by $f \in C(X) \to \alpha_t(\omega) f$, with $(\alpha_t(\omega)f)(x) \equiv f(\xi_t^x(\omega))$. One can look upon ($\alpha_t(\omega)$, $t \geq 0$) as a Markov semigroup with state space the space of endomorphisms of C(X). Hence to the original Markov process ξ_t^x on the state space X there is associated a Markov semigroup with state space End C(X). Remarking that $\mathcal{A} \equiv C(X)$ is a C^*-algebra with unit we can now attempt to extend the commutative structure to the case of an arbitrary, non necessarily commutative, C^*-algebra $\mathcal{A}$ with unit 1.

The corresponding object to the Markov semigroup P_t is a semigroup (called dynamical semigroup) of completely positive maps ϕ_t of $\mathcal{A}$ into itself (where ϕ_t completely positive means that $\phi_t \times 1$ is a positive map of $\mathcal{A} \otimes M_n$, where M_n is the algebra of all $n \times n$ matrices and n is an arbitrary integer).

Let now $\mathcal{A} = M_n$. P_t symmetric corresponds to ϕ_t symmetric with respect to the inner product $a,b \in M_n \to Tr(a^*b)$. P_t conservative (i.e. $P_t 1 = 1$) corresponds to $\phi_t 1 = 1$. About the structure of semigroups of completely positive maps we have the following

Theorem 1 [IV.5]. If ϕ_t is a symmetric, conservative dynamical semigroup on the C^*-algebra $\mathcal{A} = M_n$ then $\phi_t = e^{tL}$ with L the operator on $\mathcal{A}$ defined by

$$L(a) \equiv \sum_{i=1}^{n} (ad\beta_i)^2(a) \ , \ a \in \mathcal{A} \ ,$$

where the β_i are elements of the Lie algebra su(n) of SU(n).

Note that the space Aut M_n of automorphisms of M_n is isomorphic SU(n) hence su(n) is isomorphic with aut M_n, by the mapping $\beta \to$ ad β. It is now possible to bring in connection the dynamical semigroups on M_n with the left invariant Markov processes on Aut M_n:

Theorem 2 [IV.5]. L is the infinitesimal generator of a dynamical semigroup on $\mathcal{A}$ iff $\tilde{L}$ is the infinitesimal generator of a symmetric left invariant Markov process on Aut $M_n \simeq SU(n)$. The bijection $L \leftrightarrow \tilde{L}$ is given by $L = \sum_i (ad\beta_i)^2$, $\beta_i \in su(n) \leftrightarrow \tilde{L} = \sum_i X_i^2$, where X_i is the left invariant vector field on Aut M_n corresponding to the operator ad β_i on $\mathcal{A}$. If $\xi_t^e(\cdot)(a)$ is the left invariant Markov process on Aut$(M_n) \simeq SU(n)$ started at $e \in SU(n)$ and evaluated at a $\in \mathcal{A}$, with expectation E, then $e^{tL}(a) = E(\xi_t^e(\cdot)(a))$. Moreover for any state σ on M_n one has

$$\sigma(La) = \tilde{L}(\sigma \circ \xi_t^e(\cdot)(a))\big|_{t=0}.$$

Conversely if $\xi_t(\cdot)$ is a left invariant Markov process on Aut $M_n \simeq SU(n)$ then $a \to E(\xi_t^e(\cdot)(a))$ is a dynamical semigroup on M_n.

In the classical theory of processes there are of course not only semigroups, the central object being the process itself running on points in state space i.e. the concept of sample path is essential. Since in non abelian C^*-algebras there is no natural concept of points all attempts of giving a proper extension of the whole theory of stochastic processes to the non commutative case have failed precisely at this point. We shall now see that in a sense above observation provides a remedy to this situation. Let $S(\mathcal{A})$ be the state space of the C^*-algebra $\mathcal{A}$ i.e. the space of all linear functionals on $\mathcal{A}$. Let $\partial_e S(\mathcal{A})$ be the extreme boundary of $S(\mathcal{A})$. We have then the

Theorem 3 [IV.5]. To any dynamical semigroup ϕ_t on M_n there exists a symmetric Markov process η_t on $S(M_n)$ with continuous paths such that if η_t^σ is the process on $S(M_n)$ started at $\sigma \in S(M_n)$ at time 0 then $\sigma \to \eta_t^\sigma(\cdot)$ is an affine function on S (i.e. for $\sigma_1, \sigma_2 \in S$, $0 < t < 1$: $\eta_t^{t\sigma_1+(1-t)\sigma_2} = t\eta_t^{\sigma_1} + (1-t)\,\eta_t^{\sigma_2}$), a.s.)

One has $\eta_t^\sigma = \sigma_\circ \xi_t^e$, where ξ_t^e is the Markov process on Aut M_n generated by $\tilde{L}$, where $\tilde{L}$ is the operator corresponding to $L = \frac{1}{t} \ln \phi_t$ in Theor. 1, 2.
One has $\sigma_\circ\phi_t = E(\eta_t^\sigma)$ and if $\sigma \in \partial_e S(M_n)$ one has that $t \to \eta_t^\sigma$ is a Markov process on $\partial_e S$.

Rem.: Since $\partial_e S$ is the extreme boundary, hence corresponds to points, we see that η_t "runs on points". The above result extends also to general C^*-algebras.

IV

[IV.1] S. Albeverio, R. Høegh-Krohn, Dirichlet forms and Markov semigroups on C^*-algebras, Commun. math. Phys. 56, 173-187 (1977).

[IV.2] a) G.G. Emch, Generalized K-flows, Commun. math. Phys. 49, 191-215 (1976)

b) G.G. Emch, S. Albeverio, J.P. Eckmann, Quasi-free generalized K-flows, Rept. Math. Phys. 13, 73-85 (1978)

c) E.B. Davies, Quantum theory of open systems, Academic Press, London (1976)

[IV.3] W. Arveson, Subalgebras of C^*-algebras I, Acta Math. 123, 141-224 (1969); II, Acta Math. 12f, 271-30f (1972).

[IV.4] S. Albeverio, R. Høegh-Krohn, Frobenius theory for positive maps of von Neumann algebras, Commun. math. Phys. 64, 83-94 (1978).

M. Enomoto, Y. Watatani, A Perron-Frobenius type theorem for positive linear maps on C^*-algebras, Osaka Univ. Preprint (1978)

[IV.5] S. Albeverio, R. Høegh-Krohn, G. Olsen, Dynamical semigroups and Markov processes on C^*-algebras, J. Reine und angew. Math. (Crelle's J.) 319, 25-37 (1980).

ACKNOWLEDGEMENTS

The first author would like to thank Professor David Williams for his kind invitation and for his personal engagement in making the Symposium a complete success. He also thanks the Mathematics Department of Oslo University for a long standing hospitality. Both authors also gratefully acknowledge the warm hospitality of Professor Mohammed Mebkhout at the Université d'Aix-Marseille as well as of the Centre de Physique, CNRS Luminy. They also express their gratitude to Mrs. Richter for her patience and her skilful typing.

Vol. 759: R. L. Epstein, Degrees of Unsolvability: Structure and Theory. XIV, 216 pages. 1979.

Vol. 760: H.-O. Georgii, Canonical Gibbs Measures. VIII, 190 pages. 1979.

Vol. 761: K. Johannson, Homotopy Equivalences of 3-Manifolds with Boundaries. 2, 303 pages. 1979.

Vol. 762: D. H. Sattinger, Group Theoretic Methods in Bifurcation Theory. V, 241 pages. 1979.

Vol. 763: Algebraic Topology, Aarhus 1978. Proceedings, 1978. Edited by J. L. Dupont and H. Madsen. VI, 695 pages. 1979.

Vol. 764: B. Srinivasan, Representations of Finite Chevalley Groups. XI, 177 pages. 1979.

Vol. 765: Padé Approximation and its Applications. Proceedings, 1979. Edited by L. Wuytack. VI, 392 pages. 1979.

Vol. 766: T. tom Dieck, Transformation Groups and Representation Theory. VIII, 309 pages. 1979.

Vol. 767: M. Namba, Families of Meromorphic Functions on Compact Riemann Surfaces. XII, 284 pages. 1979.

Vol. 768: R. S. Doran and J. Wichmann, Approximate Identities and Factorization in Banach Modules. X, 305 pages. 1979.

Vol. 769: J. Flum, M. Ziegler, Topological Model Theory. X, 151 pages. 1980.

Vol. 770: Séminaire Bourbaki vol. 1978/79 Exposés 525-542. IV, 341 pages. 1980.

Vol. 771: Approximation Methods for Navier-Stokes Problems. Proceedings, 1979. Edited by R. Rautmann. XVI, 581 pages. 1980.

Vol. 772: J. P. Levine, Algebraic Structure of Knot Modules. XI, 104 pages. 1980.

Vol. 773: Numerical Analysis. Proceedings, 1979. Edited by G. A. Watson. X, 184 pages. 1980.

Vol. 774: R. Azencott, Y. Guivarc'h, R. F. Gundy, Ecole d'Eté de Probabilités de Saint-Flour VIII-1978. Edited by P. L. Hennequin. XIII, 334 pages. 1980.

Vol. 775: Geometric Methods in Mathematical Physics. Proceedings, 1979. Edited by G. Kaiser and J. E. Marsden. VII, 257 pages. 1980.

Vol. 776: B. Gross, Arithmetic on Elliptic Curves with Complex Multiplication. V, 95 pages. 1980.

Vol. 777: Séminaire sur les Singularités des Surfaces. Proceedings, 1976-1977. Edited by M. Demazure, H. Pinkham and B. Teissier. IX, 339 pages. 1980.

Vol. 778: SK1 von Schiefkörpern. Proceedings, 1976. Edited by P. Draxl and M. Kneser. II, 124 pages. 1980.

Vol. 779: Euclidean Harmonic Analysis. Proceedings, 1979. Edited by J. J. Benedetto. III, 177 pages. 1980.

Vol. 780: L. Schwartz, Semi-Martingales sur des Variétés, et Martingales Conformes sur des Variétés Analytiques Complexes. XV, 132 pages. 1980.

Vol. 781: Harmonic Analysis Iraklion 1978. Proceedings 1978. Edited by N. Petridis, S. K. Pichorides and N. Varopoulos. V, 213 pages. 1980.

Vol. 782: Bifurcation and Nonlinear Eigenvalue Problems. Proceedings, 1978. Edited by C. Bardos, J. M. Lasry and M. Schatzman. VIII, 296 pages. 1980.

Vol. 783: A. Dinghas, Wertverteilung meromorpher Funktionen in ein- und mehrfach zusammenhängenden Gebieten. Edited by R. Nevanlinna and C. Andreian Cazacu. XIII, 145 pages. 1980.

Vol. 784: Séminaire de Probabilités XIV. Proceedings, 1978/79. Edited by J. Azéma and M. Yor. VIII, 546 pages. 1980.

Vol. 785: W. M. Schmidt, Diophantine Approximation. X, 299 pages. 1980.

Vol. 786: I. J. Maddox, Infinite Matrices of Operators. V, 122 pages. 1980.

Vol. 787: Potential Theory, Copenhagen 1979. Proceedings, 1979. Edited by C. Berg, G. Forst and B. Fuglede. VIII, 319 pages. 1980.

Vol. 788: Topology Symposium, Siegen 1979. Proceedings, 1979. Edited by U. Koschorke and W. D. Neumann. VIII, 495 pages. 1980.

Vol. 789: J. E. Humphreys, Arithmetic Groups. VII, 158 pages. 1980.

Vol. 790: W. Dicks, Groups, Trees and Projective Modules. IX, 127 pages. 1980.

Vol. 791: K. W. Bauer and S. Ruscheweyh, Differential Operators for Partial Differential Equations and Function Theoretic Applications. V, 258 pages. 1980.

Vol. 792: Geometry and Differential Geometry. Proceedings, 1979. Edited by R. Artzy and I. Vaisman. VI, 443 pages. 1980.

Vol. 793: J. Renault, A Groupoid Approach to C*-Algebras. III, 160 pages. 1980.

Vol. 794: Measure Theory, Oberwolfach 1979. Proceedings 1979. Edited by D. Kölzow. XV, 573 pages. 1980.

Vol. 795: Séminaire d'Algèbre Paul Dubreil et Marie-Paule Malliavin. Proceedings 1979. Edited by M. P. Malliavin. V, 433 pages. 1980.

Vol. 796: C. Constantinescu, Duality in Measure Theory. IV, 197 pages. 1980.

Vol. 797: S. Mäki, The Determination of Units in Real Cyclic Sextic Fields. III, 198 pages. 1980.

Vol. 798: Analytic Functions, Kozubnik 1979. Proceedings. Edited by J. Ławrynowicz. X, 476 pages. 1980.

Vol. 799: Functional Differential Equations and Bifurcation. Proceedings 1979. Edited by A. F. Izé. XXII, 409 pages. 1980.

Vol. 800: M.-F. Vignéras, Arithmétique des Algèbres de Quaternions. VII, 169 pages. 1980.

Vol. 801: K. Floret, Weakly Compact Sets. VII, 123 pages. 1980.

Vol. 802: J. Bair, R. Fourneau, Etude Géometrique des Espaces Vectoriels II. VII, 283 pages. 1980.

Vol. 803: F.-Y. Maeda, Dirichlet Integrals on Harmonic Spaces. X, 180 pages. 1980.

Vol. 804: M. Matsuda, First Order Algebraic Differential Equations. VII, 111 pages. 1980.

Vol. 805: O. Kowalski, Generalized Symmetric Spaces. XII, 187 pages. 1980.

Vol. 806: Burnside Groups. Proceedings, 1977. Edited by J. L. Mennicke. V, 274 pages. 1980.

Vol. 807: Fonctions de Plusieurs Variables Complexes IV. Proceedings, 1979. Edited by F. Norguet. IX, 198 pages. 1980.

Vol. 808: G. Maury et J. Raynaud, Ordres Maximaux au Sens de K. Asano. VIII, 192 pages. 1980.

Vol. 809: I. Gumowski and Ch. Mira, Recurences and Discrete Dynamic Systems. VI, 272 pages. 1980.

Vol. 810: Geometrical Approaches to Differential Equations. Proceedings 1979. Edited by R. Martini. VII, 339 pages. 1980.

Vol. 811: D. Normann, Recursion on the Countable Functionals. VIII, 191 pages. 1980.

Vol. 812: Y. Namikawa, Toroidal Compactification of Siegel Spaces. VIII, 162 pages. 1980.

Vol. 813: A. Campillo, Algebroid Curves in Positive Characteristic. V, 168 pages. 1980.

Vol. 814: Séminaire de Théorie du Potentiel, Paris, No. 5. Proceedings. Edited by F. Hirsch et G. Mokobodzki. IV, 239 pages. 1980.

Vol. 815: P. J. Slodowy, Simple Singularities and Simple Algebraic Groups. XI, 175 pages. 1980.

Vol. 816: L. Stoica, Local Operators and Markov Processes. VIII, 104 pages. 1980.